Plasmas and Laser Light

To my parents

Plasmas and Laser Light

T. P. Hughes
Senior Lecturer in Physics
University of Essex

A HALSTED PRESS BOOK

JOHN WILEY & SONS

New York

First published October 1975

Published in the U.S.A. by
Halsted Press, a Division of
John Wiley & Sons, Inc., New York

PHYSICS

Library of Congress Cataloging in Publication Data

Hughes, Thomas Peter, 1929—
 Plasmas and laser light.

 "A Halsted Press book."
 1. Plasma (Ionized gases)   2. Laser beams.
I. Title.
QC718.H83   530.4′4   75-8163
ISBN 0-470-42035-9

Published in Great Britain by Adam Hilger Ltd
Rank Precision Industries
29 King Street, London WC2E 8JH
Printed in Great Britain by J. W. Arrowsmith Ltd
Winterstoke Road, Bristol BS3 2NT

Contents

3   Incoherent Scattering of Light by Plasmas

Preface

The subject of this book, the interaction of intense laser radiation with plasma, is receiving much attention at present in many countries. The literature is growing bewilderingly rapidly. Contributions to the subject come from many disciplines, including not only plasma physics and classical optics but also fluid dynamics, astrophysics, geophysics, non-linear optics and microwave physics. It seems worthwhile to bring together the more important ideas in one volume for the benefit of research workers entering the field. At the same time it is hoped that the collections of experimental results, though still requiring further analysis, will make the book a useful source for those who are already familiar with the subject.

The reader will appreciate that laser-generated plasmas have many applications. The possibility of extracting useful quantities of nuclear energy from plasmas heated by laser light to thermonuclear temperatures is discussed in some detail because it is of special importance and has motivated much of the work described in this book. It involves the compression of matter to extremely high densities, a possibility which has great scientific significance. Other applications, for example in spectroscopy and electrical engineering and as sources for X-rays, neutrons, and highly-ionized atoms, have been omitted. Nor are lasers themselves described.

Equations have been written in SI units except where otherwise stated. However, several other familiar units have been used for numerical values and the results of experimental measurements: at the present time it would perhaps be more confusing than enlightening to convert the ångstrom, electron volt and Torr into metres, joules and pascals. Conversion factors will be found in Appendix A at the end of the book.

Notation has been a problem where so many disciplines merge. A list of the symbols used is given, beginning on the following page.

Wherever possible, reference has been made to the open literature and not to laboratory reports. Priorities are often uncertain because work has followed similar lines simultaneously in several laboratories in different countries. The reference list, though extensive, is not complete and I apologise to workers whose contributions to the subject have been omitted.

I acknowledge with thanks the permission given by many individuals and organizations to make use of published and unpublished material: the sources of figures and tables are given in the captions.

I am indebted to the University of Essex for sabbatical leave during the academic year 1973/4, without which the book could not have been completed, and for excellent library facilities. I am grateful to many people for their help in preparing the book. Dr. M. G. Haines read the entire manuscript and made many valuable suggestions for improving the text. Dr. T. A. Hall read and made helpful comments on Chapter 3 and part of Chapter 2. Mrs. Christine Lee and Mrs. Stella Sommerlad gave much assistance with the preparation of the manuscript, and checked and listed references. The book was very efficiently typed by Mrs. Mary Low and others. I am greatly indebted to my wife and daughters for all their help and for their patience and understanding.

Symbols and Constants

a constant (see (6.45))

a coefficient defined in (7.81)

a_0 Bohr radius $= 4\pi\hbar^2\varepsilon_0/m_e e^2$
$= 5{\cdot}292 \times 10^{-11}$ m

a_ω linear absorption coefficient

$a_{\omega T}$ Bremsstrahlung absorption coefficient in weak radiation field

$a_{\omega E}$ Bremsstrahlung absorption coefficient in strong radiation field

$\boldsymbol{a}$ acceleration

A area

A numerical factor (see (5.16))

A $= (\gamma + 1)/(\gamma - 1)$

A' $= (\gamma' + 1)/(\gamma' - 1)$

A_0 incident wave amplitude

A_1, A_2 normal mode amplitudes

A_{21} spontaneous emission coefficient

$\mathscr{A}$ function defined in (1.57)

$\mathscr{A}$ atomic mass number

$\mathscr{A}_S$ differential spontaneous emission coefficient

$\mathscr{A}_{SK}$ classical Bremsstrahlung value of $\mathscr{A}_S$

$\mathscr{A}'_{SK}$ $= \mathscr{A}_{SK} g_c$

b numerical factor ~ 1 (see (5.19))

b constant defined in (6.3)

$b_{\min}$ the greater of $Ze^2/4\pi\varepsilon_0\kappa_B T_e$ and $\hbar/(m_e\kappa_B T_e)^{\frac{1}{2}}$

b_j $= Z_j^2 n_j/\sum n_j Z_j$

$\mathbf{B}$ magnetic flux density

$\mathbf{B}_0$ static magnetic field

$\mathbf{B}_1$ magnetic field due to wave

B $= 2Q\rho_1/p_1$

B constant defined in (5.32)

B_{12} induced absorption coefficient

B_{21} induced emission coefficient

$\mathscr{B}$ function defined in (1.58)

$\mathscr{B}$ constant defined in (7.18)

$\mathscr{B}_A$ differential induced absorption coefficient

$\mathcal{B}_I$	differential induced emission coefficient	f_j	distribution function of species j
c	velocity of light in vacuum $= 2 \cdot 998 \times 10^8 \text{ m s}^{-1}$	f_{j0}	unperturbed distribution function of species j
c_{120}, c_{201}	coupling coefficients	f_{j1}	perturbation on f_{j0}
c_i	coefficient (see (5.45))	F	flux of photons per unit area and time
c_0, c_1, c_2	constants (see (7.19–21))		
C	fraction of nuclear reaction energy retained in plasma	F_0	flux of photons per unit area and time on beam axis in focal plane
C	constant (see (5.23))		
C	$= v_1^2 \rho_1 / p_1$	$\mathbf{F}_j$	force on particle j
C_p	specific heat of target material per unit mass at constant pressure	$\mathbf{F}_m$	ponderomotive force
		$F(\alpha, \beta, \gamma; x)$	hypergeometric function
		$\mathcal{F}_j(\mathbf{k}, \omega)$	unperturbed distribution function of velocity, parallel to $\mathbf{k}$, of species j
C_V	specific heat of target material per unit mass at constant volume		
		$g(a)$	$= \exp(-a) \mathrm{I}_0(a)$
C_V'	specific heat of plasma per unit mass at constant volume	g_B, g_c, g_S	Born, classical (Kramers) and Sommerfeld Gaunt factors
d	distance	$\bar{g}, \bar{g}_B, \bar{g}_c$	average Gaunt factors
d	penetration depth	$\bar{g}_E$	average Gaunt factor in strong radiation field
D	$= [(\gamma+1)/\gamma](\rho_1/p_1)(\gamma-1)Q$		
D	diffusion coefficient	$\bar{\bar{g}}$	Gaunt factor averaged over both electron velocity and radiation frequency
D	$= V f q \tau_L \lambda_0 / hc$		
$|\mathrm{D}|$	determinant of coefficients of wave equation	$\mathcal{g}_1, \mathcal{g}_2$	statistical weights of states 1 and 2
$\mathcal{D}_\omega$	optical depth for radiation of frequency ω	$(\mathcal{g}_0)_Z$	statistical weight of ground state of ion or atom of charge $Z (Z = 0, 1, \dots)$
$\mathcal{D}_{\omega \text{tot}}$	total optical depth of finite plasma for radiation of frequency ω		
		G_s	Raman scattering gain
		G_C'	Compton scattering parameter
e	proton charge $= 1 \cdot 602 \times 10^{-19} \text{C}$	$\mathcal{G}_e, \mathcal{G}_i$	screening integrals
$\mathbf{E}$	electric field	h	Planck's constant $= 6 \cdot 626 \times 10^{-34} \text{ J s}$
E_x, E_y	components of $\mathbf{E}$		
$\mathbf{E}_1$	electric field due to wave	$\hbar$	$= h/2\pi = 1 \cdot 055 \times 10^{-34} \text{ J s}$
E_0	amplitude of oscillating electric field	h_a	constant (see (5.49))
		$\mathbf{H}$	magnetic field
E_e	effective electric field, defined in (5.31)	H	heat flux
		$\mathcal{H}$	function defined in (1.119)
E_{ec}	value of E_e in a magnetic field	i	$= \sqrt{-1}$
E_H	electric field in first Bohr orbit	i_b	degree of ionization at breakdown threshold
E	a plasmon		
$\mathscr{E}$	ellipticity	I	irradiance
f	focal length of lens	I_0	irradiance on axis at focus
f, f'	fractions	I_{opt}	optimum irradiance for power absorption

I_r	reflected irradiance
I_{rel}	irradiance above which relativistic effects become important $\simeq 10^{18}/\lambda^2 \ \text{W cm}^{-2}$
I_z	irradiance on axis at distance z from focus
I_ω	spectral irradiance
I_1	defined in (3.53) and (3.54)
I_{10}	initial incident irradiance
$I_1(z), I_2(z)$	incident and reflected irradiances at position z
$\hat{I}_{th}$	peak irradiance at breakdown threshold
I	an ion-acoustic phonon
$I_0(a)$	modified Bessel function of first kind
j	emission coefficient
j_ω	spectral emission coefficient
$\dot{j}$	parameter defined in (6.41)
$\mathbf{J}$	current density
J_n	Bessel function of order n
J_n'	derivative of J_n w.r.t. argument
$\mathscr{J}$	intensity
$\mathscr{J}_\lambda, \mathscr{J}_\omega$	spectral intensity
$\mathscr{J}_{\omega B}$	spectral intensity of black body radiation
$\mathbf{k}$	propagation vector
$\mathbf{k}$	(Chapter 3) $= \mathbf{k}_0 - \mathbf{k}_s$
k	wave number
$k_\parallel$	component of $\mathbf{k}$ parallel to $\mathbf{B}_0$ (1.53) or $\mathbf{E}$ (2.144)
$k_\perp$	component of $\mathbf{k}$ perpendicular to $\mathbf{B}_0$
k_D	$= 1/\lambda_D$
$\mathbf{k}_0$	propagation vector of incident wave
$\mathbf{k}_s$	propagation vector of scattered wave
$\mathbf{k}_1, \mathbf{k}_2$	coupled wave vectors
K	$= \dfrac{H + C(1-H)}{1-H}$
$K_0(x)$	modified Bessel function of second kind
$\mathscr{K}$	function defined in (5.14)
l	distance

l	thickness of plasma
l	orbital angular momentum (5.3)
l_f	axial length of focal region
l_j	$= \dfrac{k_\perp v_\perp}{\omega_{cj}}$
l_m	thickness of hot layer at maximum $\overline{T}_i$
l_2	thickness of region 2
$l_{\alpha e}$	stopping distance for α-particle by electrons
ℓ_1, ℓ_2	major and minor semi-diameters of ellipse
L	density gradient scale length
L_e	rate of energy loss by an electron
$\mathscr{L}$	a constant
m	component of angular momentum l in direction of field
m_e	mass of electron $= 9 \cdot 110 \times 10^{-31} \ \text{kg}$
m_i	mass of an ion
m_j	mass of particle of species j
m_p	mass of proton $= 1 \cdot 673 \times 10^{-27} \ \text{kg} = 1836 m_e$
M	mass of plasma
$\overline{M}$	modified value of M
M'	mass of plasma per unit area
$\mathbf{M}_j$	matrix defined in (1.56)
$\mathscr{M}$	radiant exitance
$\mathscr{M}_\lambda, \mathscr{M}_\omega$	spectral exitance
$\mathscr{M}_{B_0}$	radiant exitance from black body in vacuum
$\mathscr{M}_{\lambda B_0}, \mathscr{M}_{\omega B_0}$	spectral exitance from black body in vacuum
n	principal quantum number
n	index
n	number density of nuclei
n_a	neutral atom density
n_D	deuteron density
n_e	electron density
n_{eb}	electron density at breakdown
n_{ebl}	electron density at which electron–ion and electron–atom inverse Bremsstrahlung absorption rates are equal
n_{ec}	critical electron density

n_{e0}	initial value of n_e	$P'_{\beta}(\lambda_0)$	Bremsstrahlung radiation power per unit volume of plasma in solid angle $\delta\Omega$ and wavelength interval $\delta\lambda$ at λ_0
n_{H_2}	density of hydrogen molecules		
n_i	ion density		
n_1, n_2	population densities of states 1, 2		
n_j	density of particles of species j	$P'_{\beta_{ee}}$	electron–electron Bremsstrahlung radiation power per unit volume and energy interval
n_m	density of gas molecules		
n_0	initial density of (D +T) nuclei		
n_s	density of nuclei in solid DT mixture $\simeq 4\cdot5 \times 10^{22}$ cm^{-3} ($4\cdot5 \times 10^{28}$ m^{-3})	$P_{\beta_{rel}}$	total relativistic Bremsstrahlung power per unit volume
n_Z	density of ground state ions or atoms with charge Z	P_{ce}, P_{ci}	electron or ion cyclotron radiation power per unit volume
n_1, n_2	population densities of states 1, 2	P_L	laser power
n'_e, n'_i	amplitudes of electron or ion density fluctuations	P_N	rate of nuclear energy release per unit volume
		P_s	total scattered light power
n'_{ek}, n'_{ik}	amplitudes of electron or ion density fluctuations with wave number k	$P'_s(\lambda, \alpha)$	scattered light power per unit volume of plasma in solid angle $\delta\Omega$ and bandwidth $\delta\lambda$
N	number of particles	P_{sf}	self-focusing threshold beam power
N_e, N_i	numbers of electrons or ions		
N_p	number of photons	P_T	light power scattered into element of solid angle $\delta\Omega$
N_R	number of nuclear reactions		
N_s	number of scattered photons in mode s	P_0	incident beam power
N	effective number of photons involved in ionization	P	mean power absorbed per particle
N_0	smallest (integer) number of photons whose energy exceeds $\bar{\chi}_0$	P_e	rate of energy gain by electron from electric field
		P	principal value
N_s	number of photons producing resonant excitation of state s	$\mathscr{P}$	radiation power per unit volume per steradian per unit band-width
$\mathscr{N}_A$	$= n_1 \mathscr{I}_\omega B_{12}$	$\mathscr{P}_A$	total absorbed $\mathscr{P}$
$\mathscr{N}_I$	$= n_2 \mathscr{I}_\omega B_{21}$	$\mathscr{P}_I$	induced emission $\mathscr{P}$
$\mathscr{N}_S$	$= n_2 A_{21}$	$\mathscr{P}_S$	spontaneously emitted $\mathscr{P}$
p	pressure	$\mathscr{P}$	$= p_2/p_1$
p_F	pressure of degenerate electron gas	$\mathscr{P}_C$	value of p_2/p_1 at Chapman–Jouguet point
p_0	initial pressure	q	quantum efficiency of photocathode
p_1, p_2	pressures in regions 1, 2		
P	power	q_j	charge on a particle of species j
$P(\mathbf{k}, \omega)$	spectral power density	Q	energy supplied to unit mass of material passing through a detonation or deflagration front
P_A	power absorbed per unit volume		
P_β	total Bremsstrahlung radiation power per unit volume		

Q	average nuclear energy released per reaction	t_b	time of breakdown
$\mathbf{r}$	position vector	t_{bm}	time interval from breakdown to maximum of laser pulse
r	radius	t_{bz}	time of breakdown at axial distance z from focus
r_A	radius of Airy disc		
r_b	radius of plasma boundary	t_B	time when thickness of dense layer is reduced to zero
r_e	classical electron radius $= e^2/4\pi\varepsilon_0 m_e c^2$ $= 2\cdot818 \times 10^{-15}$ m	t_c	containment time
		t_h	time when heating ceases
r_f	radius of focal region	t_m	duration of maximum $\overline{T_i}$
r_{ge}	electron gyroradius	t_r	time when hydrodynamic separation occurs
r_h	radius of plasma when heating ceases	t_s	time when shock wave reaches rear surface of target
r_l	radius of lens		
r_L	laser beam radius at focusing lens	t_F, t_L, t_P	see Figures 8.112, 8.113
r_{min}	minimum radius for triggering reaction	$t(T_i)$	time required to heat ion to temperature T_i
r_s	radius of blast wave	$t(Z)$	time required for atom to reach ionization state with charge Z
r_t	radius of target		
r'	radius of shell	t'	time when plasma boundary reaches r'
r_0	initial radius		
R	nuclear reaction rate per unit volume	T	temperature
		T_c	electron temperature in conduction region
R	recombination coefficient		
R	gas constant per mole $= 8\cdot314$ J mole^{-1} K^{-1}	T_D	deuteron temperature
		T_e	electron temperature
$\mathscr{R}$	(intensity) reflectivity	T_{e0}	initial electron temperature
$\mathscr{R}_0$	initial value of $\mathscr{R}$	$T_{es}(t)$	surface value of T_e at time t
$\mathscr{R}_\omega$	reflectivity for radiation of frequency ω	T_g	gas temperature
		$T_{g\text{crit}}$	critical gas temperature at which plasma is assumed to form rapidly
$\mathscr{R}_{CO_2}$	reflectivity for CO_2 laser light (λ 10·6 μm)		
$\mathscr{R}_{Nd}$	reflectivity for Nd laser light (λ 1·06 μm)	T_h	temperature at which heating ceases
s	state of an atom	T_H	electron temperature in hot corona
S	constant		
$S(\mathbf{k}, \omega)$	dynamic form factor	T_i	ion temperature
$S(\omega, \psi)$	dynamic form factor for magnetized plasma	T_j	temperature of particle of species j
$S(\mathbf{k})_e, S(\mathbf{k})_i$	integral of electron or ion component of $S(\mathbf{k}, \omega)$ over frequency	T_p	plasma temperature
		T_0	initial temperature
		T_1, T_2	temperatures in regions 1, 2
$S(\mathbf{k})$	$= S(\mathbf{k})_e + S(\mathbf{k})_i$	$\overline{T_e}, \overline{T_i}$	average values of T_e, T_i
$\mathscr{S}_\omega$	source function for radiation of frequency ω	$(\overline{T_i})_{max}$	maximum value of $\overline{T_i}$
		T, T$'$	photons
t	time		

$\mathscr{T}$ — fractional (intensity) transmittance

$\mathbf{u}$ — electron drift velocity

u — optical coordinate defined in (5.49)

u_1, u_2 — laboratory frame velocities of material in regions 1, 2

$\bar{u}_1$ — average value of u_1

u_{pt} — laboratory frame velocity of plasma emitted at time t

U — detonation wave velocity

U_b — breakdown boundary velocity

U_D — deflagration wave velocity

U_S — shock wave velocity

$\mathscr{U}_a, \mathscr{U}_i$ — partition functions for neutral atom, or ion

v — optical coordinate defined in (5.49)

$\mathbf{v}, \mathbf{v}'$ — velocities

$\mathbf{v}_e$ — electron velocity

v_{e1}, v_{e2} — initial and final electron velocities

v_{exp} — limiting velocity of expansion

v_E — maximum velocity of electron oscillating in radiation field

$\mathbf{v}_j$ — velocity of particle of species j

$v_{j\parallel}$ — component of $\mathbf{v}_j$ parallel to $\mathbf{B}$

v_k — component of particle velocity parallel to $\mathbf{k}$

v_p — plasma velocity

v_T — electron thermal velocity

v_1 — velocity of undisturbed gas relative to detonation wave front

v_2 — velocity of material behind detonation or deflagration wave, relative to wave front

$v_{2_{CJ}}$ — Chapman–Jouguet point value of v_2

v_α — velocity of α-particle

$v_{\alpha 0}$ — initial velocity of α-particle

$v_\parallel$ — velocity parallel to $\mathbf{B}$

$v_\perp$ — velocity perpendicular to $\mathbf{B}$

V — volume

V_a — ablation velocity

V_A — Alfvén velocity

V_b — plasma boundary velocity

V_{bl} — lateral plasma boundary velocity

$V_{b\infty}$ — asymptotic plasma boundary velocity

V_f — volume of focal region

$\mathbf{V}_g$ — group velocity

V_{ia} — ion acoustic wave velocity

V_N — effective volume for N-photon ionization

$\mathbf{V}_p$ — phase velocity

V_s — velocity of sound

v — potential difference

$\mathscr{V}$ — $= \rho_1/\rho_2$

$\mathscr{V}_C$ — value of $\mathscr{V}$ at Chapman–Jouguet point

w — radiant energy density

w_ω — spectral energy density

$w_{\omega B}$ — spectral energy density of black body radiation in a single polarization

$w_{\omega B_0}$ — $w_{\omega B}$ in vacuum

W — energy

W_I — energy required to ionize unit mass of target

W_L — laser pulse energy

W_0 — energy required to trigger a thermonuclear reaction wave

W_s — specific internal energy

W_v — energy of vaporization and acceleration

W' — energy per unit area

w, w' — energies of states

$\mathrm{w}_1, \mathrm{w}_2$ — energies of states 1, 2

w_F — Fermi energy

w_i — expansion energy of ion

w_α — energy of α-particle

$\mathrm{w}_{\alpha 0}$ — initial energy of α-particle

$\mathscr{W}(y)$ — function defined in (3.25)

$\mathscr{W}_1, \mathscr{W}_2$ — functions defined in (2.106), (2.107)

x — $= -4\eta_1\eta_2/(\eta_2-\eta_1)^2$

x — $= \dfrac{\omega}{k}\left(\dfrac{2\kappa_B T_i}{m_p}\right)^{\frac{1}{2}}$ (Figures 3.17, 3.18)

x	index	γ'	adiabatic exponent for plasma
x_0	amplitude of electron oscillation in radiation field	γ^*	effective adiabatic exponent = 1·781
x_0'	scale length for electron oscillations	γ_g	
		γ_k	imaginary part of complex ω: growth rate of instability of wave number k
X	function defined in (6.64)		
X'	function defined after (6.74)	γ_{max}	maximum growth rate
y	index	γ_t	adiabaticity parameter
y_e	$= \omega/Y_e$	γ_λ	growth rate of instability of wavelength λ
y_j	$= \omega/Y_j$		
y_u	drift velocity parameter	Γ_s	Stark breadth of level s
y_0	value of y_e at resonance	Γ_1, Γ_2	damping coefficients for waves of frequency ω_1, ω_2
$Y(\gamma)$	numerical factor (see (6.79))		
Y_e	$= \left(\dfrac{2k^2\kappa_B T_e}{m_e}\right)^{\frac{1}{2}}$	$\Gamma_\alpha(y)$	function defined in (3.29)
		δ	Dirac delta function
		δ	$= \omega_0 - \omega_2$: frequency mismatch
Y_j	$= \left(\dfrac{2k^2\kappa_B T_j}{m_j}\right)^{\frac{1}{2}}$	δw_i	change in ion energy
		$(\delta\lambda)_{rel}$	relativistic wavelength shift
$\mathscr{Y}$	parameter defined in (7.115)	$(\delta\omega)_e$	displacement of electron component of scattered light spectrum
z	coordinate		
z	distance along beam axis from Gaussian focus		
		Δ	$= \omega_0 - \omega_1 - \omega_2$: frequency mismatch
z_b	boundary of plane thermal wave		
z_c	position of critical density layer	Δ_{N_s}	energy mismatch for N-photon resonance with Stark-displaced level
z_d	distance in which beam radius doubles by diffraction		
		$(\Delta w)_{max}$	largest interval between energy levels
z_f	self-focusing distance		
Z	ion charge (in units of proton charge)	Δz	thickness of detonation or deflagration wave front
$\bar{Z}$	average ion charge	$\Delta\alpha$	divergence angle of laser beam
Z_j	charge on jth species of ion	$\Delta\lambda$	line breadth
Z_n	nuclear charge = atomic number	$\Delta\omega$	line breadth
Z'	effective charge for electron–atom Bremsstrahlung	$(\Delta\omega)_e$	breadth of electron component of scattered light spectrum
α	fine structure constant $= e^2/2hc\varepsilon_0 = 137\cdot0$	$\Delta\omega_0$	bandwidth of radiation centred at ω_0
α	scattering parameter $= 1/k\lambda_D$	$\Delta\Omega$	solid angle from which detector receives scattered light
α	parameter (see (5.49))		
α	semi-angle of beam convergence	ε	dielectric tensor
β	scattering parameter defined in (3.29)	ε	dielectric constant
		ε_0	permittivity of free space $= 8\cdot854 \times 10^{-12}$ F m^{-1}
β	parameter (see (5.49))		
β	parameter (see (6.41))	ε_I	imaginary part of complex ε
γ	adiabatic exponent $= C_p/C_V$	ε_{IL}	imaginary part of ε_L

ε_L	longitudinal dielectric constant
ε_R	real part of complex ε
ε_{RL}	real part of ε_L
ε_T	transverse dielectric constant
ε_{xy}	etc, components of $\boldsymbol{\varepsilon}$
ζ	parameter defined in (1.136)
ζ	angle of misalignment in latitude with respect to $\mathbf{B}$
ζ	$= Z(\chi_H/\chi_s)^{\frac{1}{2}}$
ζ	parameter (see (7.36))
η	$= v_E/v_T$: field strength parameter
η_1, η_2	$= \dfrac{Ze^2}{4\pi\varepsilon_0\hbar v_{e1,2}}$
H	efficiency of energy conversion
θ	scattering angle
Θ	dimensionless parameter
κ	absorption index
κ_B	Boltzmann's constant $= 1{\cdot}381 \times 10^{-23}$ J K^{-1}
λ	wavelength
λ_a	absorption length for radiation
λ_D	Debye length $= \left(\dfrac{\kappa_B T_e \varepsilon_0}{n_e e^2}\right)^{\frac{1}{2}}$
λ_{ee}	mean free path for electron–electron collisions
λ_{max}	wavelength of maximum $\mathcal{M}_{\lambda B}$
λ_0	wavelength of incident light
Λ	parameter defined after (2.99)
Λ_d	electron diffusion length
Λ_e	parameter defined after (3.46)
Λ_T	thermal diffusion length
μ	real refractive index
$\tilde{\mu}$	complex refractive index
μ_0	permeability of free space $= 4\pi \times 10^{-7}$ H m^{-1}
μ_2	non-linear refractive index coefficient
μ'	relative permeability
v_a	rate of attachment
v_d	rate of diffusion
v_{ea}	electron–atom collision frequency for momentum transfer
v_{ee}	electron–electron collision frequency

v_{ei}	electron–ion collision frequency for momentum transfer
v_{ei}^*	effective electron–ion collision frequency
$\langle v_{ei} \rangle$	average value of v_{ei}
v_{ek}, v_{ik}	effective electron–ion or ion–ion collision frequencies allowing for Landau damping
v_i	ionization rate
v_I	ionization probability per atom
$v_{I_{d.c.}}$	ionization probability per atom by d.c. field
$v_I^{(N_0)}$	probability of ionization by N_0 photons
$v_{0s}^{(N_s)}$	excitation probability to state s by N photons
$v_{sI}^{(N_0-N_s)}$	ionization probability from state s by $(N_0 - N_s)$ photons
v_r	rate of recombination
v_ϕ	phase jump frequency
ξ	small positive frequency
ξ	parameter defined by (6.70) and (2.72)
ρ	mass density
ρ_A	mass density of air at 760 Torr, 273 K $= 1{\cdot}29 \times 10^{-3}$ g cm^{-3} ($1{\cdot}29$ kg m^{-3})
ρ_0	initial mass density
ρ_q	charge density
ρ_s	mass density of solid DT $\sim 0{\cdot}2$ g cm^{-3} (200 kg m^{-3})
ρ_1, ρ_2	mass densities in regions 1, 2
$\boldsymbol{\sigma}$	conductivity tensor
σ	cross-section
$\langle \sigma v \rangle$	average of (cross section $\times$ relative velocity)
σ_{IC}, σ_{IL}	ionization cross-sections for circularly or linearly polarized light
σ_S	Stefan–Boltzmann constant $= 5{\cdot}670 \times 10^{-8}$ J m^{-2} s^{-1} K^{-4}
σ_T	Thomson scattering cross-section $= 6{\cdot}65 \times 10^{-29}$ m^2
$\sigma_T'(\phi), \sigma_T'(\theta),$ $\sigma'(\mathbf{k}, \omega)$	differential scattering cross-sections

σ_1	$= \int \psi(x, y, 0) \, dx \, dy$	χ_0	normal ionization energy from ground state
Σ	signal-to-noise ratio	$\tilde{\chi}_0$	effective ionization energy from ground state
Σ_p	Σ using polarizer	ψ	angle between $\mathbf{B}$ and normal to $\mathbf{k}$
τ_b	bounce period	ψ	spatial part of photon flux distribution function
τ_c	inertial containment time constant	ψ	$= 2 \cdot 5 \times 10^{-55} \, (\omega_{ruby}/\omega_0)^2$
τ_{ee}	electron thermalization time	Ψ	intensity parameter
τ_{ei}	electron–ion momentum transfer time	Ψ_{th}	threshold value of Ψ
τ_L	laser pulse duration	ω	angular frequency
τ_N	effective pulse duration for N-photon ionization	ω_{BG}	Bohm–Gross frequency
τ_p	lifetime of plasma	ω_{cj}	cyclotron frequency for particles of species j
τ_t	tunnelling time	ω_d	Doppler shifted frequency
$\tau_{\alpha e}$	slowing down time for α-particle, by electrons	ω_i	ion wave frequency
τ_0	time constant for isentropic compression	ω_{ia}	ion acoustic wave frequency
τ_1	$\int \phi(t) \, dt$	ω_l, ω_r	cut-off frequencies for left- or right-handed circularly polarized waves
$\phi(t)$	temporal part of photon flux distribution function	$\omega_{l'}$	ω_l allowing for ion motion
ϕ	angle	ω_{LH}, ω_{UH}	lower and upper hybrid resonance frequencies
ϕ_B	angle to magnetic field direction	ω_{max}	frequency of maximum $\mathcal{M}_{\omega B_0}$
ϕ_C	angle between direction of propagation of Čerenkov radiation and particle velocity	ω_p	$= (\omega_{pe}^2 + \omega_{pi}^2)^{\frac{1}{2}}$
ϕ_E	angle to electric field direction	ω_{pe}	electron plasma frequency
ϕ_i	angle of incidence	ω_{pj}	plasma frequency for particles of species j
Φ	radiant flux	ω_{Rk}	real part of complex ω corresponding to real wave number k
$\Phi_\lambda, \Phi_\omega$	spectral flux	ω_s	frequency of scattered light
$\Phi(x)$	$= \int_0^\infty \exp(y^2 - x^2) \, dy$	ω_{sn}	frequency of n'th Fourier component of scattered light
χ_H	ionization energy of ground state hydrogen atom $= 13 \cdot 60$ eV	ω_{trans}	frequency region of transition from black body to Bremsstrahlung continuum
χ_{max}	highest ionization energy of any atom or ion in plasma	ω_0	frequency of incident light
χ_s	ionization energy from state s	ω_1, ω_2	frequencies of waves 1, 2
χ_z	ionization energy of ground state atom or ion of charge Z $(Z = 0, 1 \ldots)$	Ω	solid angle

1

Small-Amplitude Waves
in a Plasma

1. INTRODUCTION

In this chapter we consider briefly the theory of wave propagation in a plasma. For our purposes a review of the more important results will suffice: the reader will find full treatments of the subject in the books by Bekefi (1966), Ginsburg (1970), Heald & Wharton (1965) and Stix (1962). We shall simplify the discussion by restricting it to plane waves, of small amplitude, in fully-ionized homogeneous plasmas.

We shall begin by investigating the conditions under which solutions to the wave equation exist. We shall then discuss the growth and decay of waves. The simple case of a cold unmagnetized plasma will be considered first, and then we shall take into account the anisotropy due to a magnetic field and also the effects of thermal and drift velocities of the plasma particles.

2. THE WAVE EQUATION

An electromagnetic field is characterized by the time and space dependence of the electric and magnetic field vectors **E** and **H**. Maxwell's equations for a plasma may be written

$$\nabla \times \mathbf{E} = -\frac{\partial \mathbf{B}}{\partial t} \tag{1.1}$$

$$\nabla \times \mathbf{H} = \varepsilon_0 \frac{\partial \mathbf{E}}{\partial t} + \mathbf{J} \tag{1.2}$$

$$\nabla \cdot \mathbf{E} = \frac{\rho_q}{\varepsilon_0} \tag{1.3}$$

$$\nabla \cdot \mathbf{B} = 0. \tag{1.4}$$

Here ε_0 is the permittivity of free space. All currents are included in the current density $\mathbf{J}$ and all charges in the charge density ρ_q. The magnetic flux density $\mathbf{B}$ is given by

$$\mathbf{B} = \mu_0 \mathbf{H} \tag{1.5}$$

where μ_0 is the permeability of free space.

The current density and the charge density are related by the continuity equation, obtained from (1.2) and (1.3):

$$\frac{\partial \rho_q}{\partial t} + \nabla \cdot \mathbf{J} = 0. \tag{1.6}$$

The current density is related to the electric field by the conductivity $\boldsymbol{\sigma}$, which is in general a tensor quantity. For a particular angular frequency ω the generalized Ohm's law may be written

$$\mathbf{J}(\omega) = \boldsymbol{\sigma}(\omega) \cdot \mathbf{E}(\omega). \tag{1.7}$$

For the small-amplitude waves discussed in this chapter we may assume that $\boldsymbol{\sigma}$, though dependent on the frequency, is independent of the amplitudes of $\mathbf{E}$ and $\mathbf{H}$. We shall also find it useful to introduce the *dielectric tensor*

$$\boldsymbol{\varepsilon} = 1 + \frac{\boldsymbol{\sigma}}{i\omega\varepsilon_0}. \tag{1.8}$$

Let us now suppose that both $\mathbf{E}$ and $\mathbf{H}$ vary in time and space as $\exp i(\omega t - \mathbf{k} \cdot \mathbf{r})$, $\mathbf{k}$ being the propagation vector ($k = 2\pi/\lambda$, where λ is the wavelength). Equations (1.1), (1.2) and (1.4) become

$$\nabla \times \mathbf{E} = -i\omega\mu_0 \mathbf{H} \tag{1.9}$$

$$\nabla \times \mathbf{H} = i\omega\varepsilon_0 \boldsymbol{\varepsilon} \cdot \mathbf{E} \tag{1.10}$$

$$\nabla \cdot \mathbf{B} = 0. \tag{1.11}$$

From equations (1.6) and (1.7), noting that $\partial \rho_q/\partial t = i\omega\rho_q$ from equation (1.3), we also have

$$\nabla \cdot (\boldsymbol{\varepsilon} \cdot \mathbf{E}) = 0. \tag{1.12}$$

The wave equation is obtained by taking the curl of equation (1.9) and substituting for $\mathbf{H}$ from equation (1.10): the result is

$$\nabla \times (\nabla \times \mathbf{E}) = \mu_0 \varepsilon_0 \omega^2 \boldsymbol{\varepsilon} . \mathbf{E}$$

$$= \left(\frac{\omega}{c}\right)^2 \boldsymbol{\varepsilon} . \mathbf{E}, \tag{1.13}$$

$c = (\mu_0 \varepsilon_0)^{-\frac{1}{2}}$ being the velocity of light in free space. Since both $\mathbf{E}$ and $\mathbf{H}$ vary as $\exp i(\omega t - \mathbf{k} . \mathbf{r})$, (1.13) may be written

$$\mathbf{k} \times (\mathbf{k} \times \mathbf{E}) + \left(\frac{\omega}{c}\right)^2 \boldsymbol{\varepsilon} . \mathbf{E} = 0. \tag{1.14}$$

This is the general wave equation for a plasma, and consists of three homogeneous equations in the components of $\mathbf{E}$. The determinant of the coefficients, $|\mathbf{D}|$, must vanish if there is a solution, that is, if $\mathbf{E} \neq 0$: the determinantal equation $|\mathbf{D}| = 0$ is known as the *dispersion equation*. There may be more than one solution to the dispersion equation, each representing a different family of waves, with different *dispersion relations* between ω and $\mathbf{k}$.

In general $\boldsymbol{\varepsilon}$ is a complex tensor, but in an isotropic medium it is a scalar and is known as the *dielectric constant*. For a transverse wave $\mathbf{k} . \mathbf{E} = 0$ and (1.14) becomes

$$\frac{c^2 k^2}{\omega^2} = \varepsilon_T. \tag{1.15}$$

The phase velocity of a wave is

$$V_p = \frac{\omega}{k}. \tag{1.16}$$

The refractive index $\tilde{\mu}$ is the ratio of the velocity of light in vacuum to the phase velocity, and is in general complex:

$$\tilde{\mu}^2 = \varepsilon_T = (\mu + i\kappa)^2. \tag{1.17}$$

It is convenient to take a real frequency ω and a complex propagation vector k, so that the *real refractive index*,

$$\mu = \frac{c}{\omega} \operatorname{Re}(k), \tag{1.18}$$

while the *absorption index*

$$\kappa = \frac{c}{\omega} \operatorname{Im}(k). \tag{1.19}$$

In describing the exponential spatial attenuation of a wave, the parameter normally used is the *linear absorption coefficient*, a_ω. This is defined in terms of the spectral intensity $\mathscr{I}_\omega$: for a wave travelling in the z direction,

$$\frac{\mathrm{d}\mathscr{I}_\omega}{\mathrm{d}z} = -a_\omega \mathscr{I}_\omega. \tag{1.20}$$

The intensity is proportional to the square of the amplitude, so

$$a_\omega = -2 \, \mathrm{Im} \, (k)$$

$$= -2\frac{\omega}{c} \, \kappa.$$

(1.21)

It follows from (1.17) that

$$\mu^2 - \kappa^2 = \mathrm{Re} \, (\varepsilon_T)$$

(1.22)

and

$$2\mu\kappa = \mathrm{Im} \, (\varepsilon_T).$$

(1.23)

We may also write

$$a_\omega = \frac{\omega}{c} \frac{\mathrm{Im} \, (\varepsilon_T)}{\mu}.$$

(1.24)

Thus a_ω depends upon the real part of the high-frequency conductivity: if the conductivity is zero, there is no absorption.

The *group velocity* of a wave in a medium

$$\mathbf{V}_g = \frac{\partial \omega}{\partial \mathbf{k}}$$

(1.25)

may be found from the dispersion relations. It is the velocity with which energy propagates in the medium. For a dispersionless wave $V_g = V_p$.

3. WAVES IN A COLD PLASMA WITH NO STATIC MAGNETIC FIELD

Let us consider first the very simple case of a plasma in which the electrons and ions are so cold that they have no motion other than the oscillations caused by the electromagnetic wave. We shall suppose that there is no static magnetic field in the plasma. This means that ε becomes a scalar quantity, the dielectric constant ε, which we shall now determine for a plasma in which the ion mass is very much larger than the electron mass, so that the contribution of the ion motion to the current may be ignored. This model is known as a Lorentz plasma.

We may write

$$\mathbf{J} = -n_e e \mathbf{v}_e$$

(1.26)

where n_e is the electron density, e the proton charge and $\mathbf{v}_e$ the electron velocity.

To begin with we shall ignore collisions between electrons and ions. In this case, to lowest order, the equation of motion of an electron of mass m_e in the electric field of the wave is

$$m_e \frac{\partial \mathbf{v}_e}{\partial t} = -e\mathbf{E} \tag{1.27}$$

so with E varying as $\exp[i(\omega t - \mathbf{k} \cdot \mathbf{r})]$ we find that

$$\mathbf{J} = \frac{n_e e^2}{m_e i \omega} \mathbf{E}, \tag{1.28}$$

so

$$\sigma = \frac{n_e e^2}{m_e i \omega}. \tag{1.29}$$

Thus from (1.8),

$$\varepsilon = 1 - \frac{\omega_{pe}^2}{\omega^2} \tag{1.30}$$

where

$$\omega_{pe}^2 = \frac{n_e e^2}{m_e \varepsilon_0}. \tag{1.31}$$

The frequency ω_{pe} is the *electron plasma frequency*. It is the natural frequency of electrostatic oscillation of the electrons resulting from a disturbance of the charge neutrality of the plasma, and is a most important parameter. Numerically, $\omega_{pe}/2\pi \simeq 9n_e^{\frac{1}{2}}$ Hz with n_e in m^{-3}.

The wave equation (1.14) may be split up into two separate equations, one for transverse waves ($\mathbf{k} \cdot \mathbf{E} = 0$) and the other for longitudinal waves ($\mathbf{k} \times \mathbf{E} = 0$). (Longitudinal waves are often described as *electrostatic* rather than electromagnetic waves, because no magnetic field is associated with them.) It follows that, in a cold unmagnetized plasma, transverse waves obey the dispersion relation

$$k^2 c^2 = \omega^2 \varepsilon = \omega^2 - \omega_{pe}^2, \tag{1.32}$$

so transverse waves of any frequency above the plasma frequency can propagate.

If $\mathbf{k} \times \mathbf{E} = 0$, the wave equation in a cold unmagnetized plasma reduces to

$$\varepsilon = 0, \tag{1.33}$$

so from (1.30) only oscillations at $\omega = \omega_{pe}$ are possible. The group velocity is zero, so these oscillations do not propagate as longitudinal waves.

Let us now return to the beginning of this section and include the effects of collisions between electrons and ions. We define an *electron–ion collision frequency for momentum transfer*, v_{ei}, as the frequency with which an electron loses all its momentum through collisions with the positive ions. The average collisional force acting on the electron is equal to the rate of change of momentum, $v_{ei} m_e \mathbf{v}_e$, so the equation of motion of an electron (1.27) must be modified by including this damping term:

$$m_e \frac{\partial \mathbf{v}_e}{\partial t} = -e\mathbf{E} - v_{ei} m_e \mathbf{v}_e. \tag{1.34}$$

If we make the assumption that v_{ei} is independent of the electron velocity, we find that the solution is

$$\mathbf{J} = n_e e \mathbf{v}_e = \frac{n_e e^2}{m_e (v_{ei} + i\omega)} \mathbf{E}, \tag{1.35}$$

so

$$\sigma = \frac{n_e e^2}{m_e} \frac{(v_{ei} - i\omega)}{(v_{ei}^2 + \omega^2)}. \tag{1.36}$$

Thus

$$\varepsilon = 1 + \frac{\sigma}{i\omega\varepsilon_0} = 1 - \frac{\omega_{pe}^2}{(v_{ei}^2 + \omega^2)} - \frac{i\omega_{pe}^2 v_{ei}}{\omega(v_{ei}^2 + \omega^2)}. \tag{1.37}$$

This is sometimes called the Lorentz dielectric constant: the corresponding value of σ is the Lorentz conductivity. The dispersion relation for transverse waves now becomes

$$k^2 c^2 = \omega^2 \varepsilon = \omega^2 - \frac{\omega^2 \omega_{pe}^2}{(v_{ei}^2 + \omega^2)} - \frac{i\omega\omega_{pe}^2 v_{ei}}{(v_{ei}^2 + \omega^2)}. \tag{1.38}$$

The real refractive index, given by (1.18), is

$$\mu = \left\{ \frac{1}{2}\left(1 - \frac{\omega_{pe}^2}{v_{ei}^2 + \omega^2}\right) + \frac{1}{2}\left[\left(1 - \frac{\omega_{pe}^2}{v_{ei}^2 + \omega^2}\right)^2 + \left(\frac{\omega_{pe}^2}{v_{ei}^2 + \omega^2} \frac{v_{ei}}{\omega}\right)^2\right]^{\frac{1}{2}} \right\}^{\frac{1}{2}}, \tag{1.39}$$

while the absorption coefficient, given by (1.24), is

$$a_\omega = \frac{v_{ei}}{\mu c} \frac{\omega_{pe}^2}{(v_{ei}^2 + \omega^2)}. \tag{1.40}$$

At frequencies well above the plasma frequency ($\omega \gg \omega_{pe}$) we may find approximate expressions for μ and a_ω. Provided $v_{ei}^2 \ll \omega^2 - \omega_{pe}^2$,

$$\mu \simeq \left(1 - \frac{\omega_{pe}^2}{\omega^2}\right)^{\frac{1}{2}} \tag{1.41}$$

and

$$a_\omega \simeq \frac{v_{ei}}{\mu c} \frac{\omega_{pe}^2}{\omega^2}. \tag{1.42}$$

Thus in this frequency region the real refractive index is essentially independent of the collision frequency, while the linear absorption coefficient is directly proportional to it.

In reality, the collision frequency is velocity dependent. The simple Lorentz expression for the absorption coefficient is useful, however, and is often retained for a velocity-dependent collision frequency, but with v_{ei} replaced by an appropriate average value $\langle v_{ei} \rangle$.

4. THE GENERAL DISPERSION EQUATION FOR A COLLISIONLESS PLASMA

We shall now deal with the more general case of waves in a hot plasma in a constant magnetic field. It is necessary to take into account the motion of the various plasma particles, in the absence of the wave, in our expressions for ρ_q and **J**.

We introduce the distribution function $f_j(\mathbf{r}, \mathbf{v}, t)$, such that $f_j d^3r \, d^3v$ is the number of particles of species j with mass m_j at time t in the six-dimensional element $d\mathbf{r} \, d\mathbf{v}$ centred at $(\mathbf{r}, \mathbf{v})$ in position and velocity space. The number density of particles per unit volume $n_j(\mathbf{r}, t) = \int f_j(\mathbf{r}, \mathbf{v}, t) \, d^3v$. The behaviour in time of the distribution function is described by the *Boltzmann equation*, which may be written

$$\frac{\partial f_j}{\partial t} + \mathbf{v} \cdot \frac{\partial f_j}{\partial \mathbf{r}} + \frac{\mathbf{F}_j}{m_j} \cdot \frac{\partial f_j}{\partial \mathbf{v}} = \left(\frac{\partial f_j}{\partial t} \right)_{\text{collisions}}. \tag{1.43}$$

In this general expression the external forces acting on the particles are represented by $\mathbf{F}_j$. In an electromagnetic field

$$\mathbf{F}_j = q_j(\mathbf{E} + \mathbf{v} \times \mathbf{B}), \tag{1.44}$$

where q_j is the charge on a particle of species j. The right hand term in equation (1.43) is the contribution to the rate of change of f_j due to close collisions between two particles. We shall assume that the frequency of the wave is sufficiently high for this term to be negligible. Then (1.43) reduces to the *collisionless Boltzmann*, or *Vlasov*, *equation*:

$$\frac{\partial f_j}{\partial t} + \mathbf{v} \cdot \frac{\partial f_j}{\partial \mathbf{r}} + \frac{q_j}{m_j}(\mathbf{E} + \mathbf{v} \times \mathbf{B}) \cdot \left(\frac{\partial f_j}{\partial \mathbf{v}} \right) = 0. \tag{1.45}$$

The current density is now given by

$$\mathbf{J} = \sum_j q_j \int \mathbf{v} f_j \, d^3v, \tag{1.46}$$

and the charge density

$$\rho_q = \sum_j q_j \int f_j \, d^3v. \tag{1.47}$$

We have restricted our discussion to waves of small amplitude, so in making use of the Vlasov equation only small departures from equilibrium need be considered.

Then we may write

$$f_j = f_{j0} + f_{j1}, \tag{1.48}$$

where f_{j0} is the unperturbed distribution function and f_{j1} is the small perturbation due to the wave. We shall suppose that there is no electric field in the absence of the wave, so

$$\mathbf{E} = \mathbf{E}_1, \tag{1.49}$$

The magnetic field may be written as the sum of the static field $\mathbf{B}_0$, which we shall assume to be parallel to the z direction, and the field $\mathbf{B}_1$ due to the wave:

$$\mathbf{B} = \mathbf{B}_0 + \mathbf{B}_1. \tag{1.50}$$

We now linearize the Vlasov equation, introducing (1.48), (1.49) and (1.50) into (1.45) and ignoring all products of f_{j1}, $\mathbf{E}_1$ and $\mathbf{B}_1$, with the result that

$$\mathbf{v} \cdot \frac{\partial f_{j0}}{\partial \mathbf{r}} + \frac{q_j}{m_j} (\mathbf{v} \times \mathbf{B}_0) \cdot \frac{\partial f_{j0}}{\partial \mathbf{v}} = 0 \tag{1.51}$$

and

$$\frac{\partial f_{j1}}{\partial t} + \mathbf{v} \cdot \frac{\partial f_{j1}}{\partial \mathbf{r}} + \frac{q_j}{m_j} (\mathbf{v} \times \mathbf{B}_0) \cdot \frac{\partial f_{j1}}{\partial \mathbf{v}} + \frac{q_j}{m_j} (\mathbf{E}_1 + \mathbf{v} \times \mathbf{B}_1) \cdot \frac{\partial f_{j0}}{\partial \mathbf{v}} = 0. \tag{1.52}$$

In a uniform plasma, equation (1.51) is satisfied provided

$$f_{j0} = f_{j0}(v_\parallel, v_\perp), \tag{1.53}$$

where $v_\parallel$ and $v_\perp$, the components of velocity parallel and perpendicular to the static magnetic field $\mathbf{B}_0$, are constants of the unperturbed motion. Some lengthy calculations (see for example Harris (1970)), lead to a solution of (1.52) for $f_{j1}(\mathbf{r}, \mathbf{v}, t)$ which may be substituted into expression (1.46) to give the current density $\mathbf{J}$ due to the wave. Writing

$$\mathbf{J} = \boldsymbol{\sigma} \cdot \mathbf{E}_1 \tag{1.54}$$

the result is an expression for the conductivity tensor $\boldsymbol{\sigma}$ of form

$$\boldsymbol{\sigma}(\mathbf{k}, \omega) = -i \sum_j \frac{q_j^2}{m_j \omega} \sum_{n=-\infty}^{+\infty} \int \frac{\mathbf{M}_j \, \mathrm{d}^3 v}{(\omega - k_\parallel v_\parallel - n\omega_{cj})}, \tag{1.55}$$

where

$$\mathbf{M}_j = \begin{bmatrix} v_\perp \mathscr{A} \left(\dfrac{n J_n}{l_j} \right)^2 & -i v_\perp \mathscr{A} \dfrac{n}{l_j} J_n J_n' & v_\perp \mathscr{B} \dfrac{n}{l_j} J_n^2 \\[2ex] i v_\perp \mathscr{A} \dfrac{n}{l_j} J_n J_n' & v_\perp \mathscr{A} (J_n')^2 & i v_\perp \mathscr{B} J_n J_n' \\[2ex] v_\parallel \mathscr{A} \dfrac{n}{l_j} J_n^2 & -i v_\parallel \mathscr{A} J_n J_n' & v_\parallel \mathscr{B} J_n^2 \end{bmatrix}. \tag{1.56}$$

Here

$$\mathcal{A} = (\omega - k_{\|}v_{\|})\frac{\partial f_{j0}}{\partial v_{\perp}} + k_{\|}v_{\perp}\frac{\partial f_{j0}}{\partial v_{\|}}, \tag{1.57}$$

$$\mathcal{B} = \frac{n\omega_{cj}v_{\|}}{v_{\perp}}\frac{\partial f_{j0}}{\partial v_{\perp}} + (\omega - n\omega_{cj})\frac{\partial f_{j0}}{\partial v_{\|}}, \tag{1.58}$$

$$\omega_{cj} = \frac{q_j B_0}{m_j}, \tag{1.59}$$

the cyclotron frequency, which is negative for electrons, and $k_{\|}$, $k_{\perp}$ are the components of $\mathbf{k}$ parallel and perpendicular to the magnetic field direction. The function J_n is a Bessel function, the argument being

$$l_j = \frac{k_{\perp}v_{\perp}}{\omega_{cj}}. \tag{1.60}$$

The function J'_n is the derivative of the Bessel function with respect to its argument.

The components of the dielectric tensor $\boldsymbol{\varepsilon}$ may now be obtained by inserting values of $\boldsymbol{\sigma}$ into expression (1.8). In the absence of a magnetic field, the dielectric tensor becomes a scalar quantity which depends on the orientation of the electric field of the wave relative to its propagation vector. For longitudinal waves, the *longitudinal dielectric constant* is given by

$$\varepsilon_L(k, \omega) = 1 + \sum_j \frac{q_j^2}{m_j k^2 \varepsilon_0} \int \frac{\mathbf{k} \cdot \partial f_{j0}/\partial \mathbf{v}}{(\omega - \mathbf{k} \cdot \mathbf{v})} d^3 v. \tag{1.61}$$

For transverse waves, the motion of the particles has no effect on the dielectric constant ε_T in the absence of a magnetic field, and

$$\varepsilon_T(\omega) = 1 - \sum_j \frac{\omega_{pj}^2}{\omega^2} \tag{1.62}$$

where

$$\omega_{pj}^2 = \frac{n_j q_j^2}{m_j \varepsilon_0}. \tag{1.63}$$

We may now define the general condition for a solution to exist for the wave equation (1.14). The condition is given (see for example Bekefi (1966)) by the determinantal dispersion equation

$$\begin{vmatrix} \left(\dfrac{\omega}{c}\right)^2 \varepsilon_{xx} - k_{\|}^2 & \left(\dfrac{\omega}{c}\right)^2 \varepsilon_{xy} & \left(\dfrac{\omega}{c}\right)^2 \varepsilon_{xz} + k_{\|}k_{\perp} \\[3mm] \left(\dfrac{\omega}{c}\right)^2 \varepsilon_{yx} & \left(\dfrac{\omega}{c}\right)^2 \varepsilon_{yy} - (k_{\|}^2 + k_{\perp}^2) & \left(\dfrac{\omega}{c}\right)^2 \varepsilon_{yz} \\[3mm] \left(\dfrac{\omega}{c}\right)^2 \varepsilon_{zx} + k_{\|}k_{\perp} & \left(\dfrac{\omega}{c}\right)^2 \varepsilon_{zy} & \left(\dfrac{\omega}{c}\right)^2 \varepsilon_{zz} - k_{\perp}^2 \end{vmatrix} = 0, \tag{1.64}$$

together with the conditions

$$
\begin{aligned}
\varepsilon_{xy} &= -\varepsilon_{yx}, \\
\varepsilon_{yz} &= -\varepsilon_{zy}, \\
\varepsilon_{zx} &= \varepsilon_{xz}.
\end{aligned}
\left.\rule{0pt}{52pt}\right\}
\tag{1.65}
$$

Here (x, y, z) are rectangular coordinates with z parallel to $\mathbf{B}_0$ and $k_\parallel$. The general solution of (1.64) and (1.65) presents some formidable problems. In certain circumstances, however, considerable simplification is possible.

If the direction of propagation of the wave is parallel to the magnetic field, or if the plasma is cold, we may set

$$
l_j = \frac{k_\perp v_\perp}{\omega_{cj}} = 0. \tag{1.66}
$$

In this case the dielectric tensor $\varepsilon(\mathbf{k}, \omega)$ reduces to the following components:

$$
\begin{aligned}
\varepsilon_{xx} = \varepsilon_{yy} = 1 + \sum_j \frac{q_j^2}{m_j \omega^2 \varepsilon_0} \int \frac{v_\perp}{4}\left[(\omega - k_\parallel v_\parallel)\frac{\partial f_{j0}}{\partial v_\perp} + k_\parallel v_\perp \frac{\partial f_{j0}}{\partial v_\parallel} \right] \\
\times \left[\frac{1}{(\omega - k_\parallel v_\parallel + \omega_{cj})} + \frac{1}{(\omega - k_\parallel v_\parallel - \omega_{cj})} \right] d^3 v,
\end{aligned}
\tag{1.67}
$$

$$
\varepsilon_{zz} = 1 + \sum_j \frac{q_j^2}{m_j \omega \varepsilon_0} \int v_\parallel \frac{\partial f_{j0}}{\partial v_\parallel}\left(\frac{1}{\omega - k_\parallel v_\parallel} \right) d^3 v \tag{1.68}
$$

and

$$
\begin{aligned}
\varepsilon_{xy} = -\varepsilon_{yx} = -i\sum_j \frac{q_j^2}{m_j \omega^2 \varepsilon_0} \int \frac{v_\perp}{4}\left[(\omega - k_\parallel v_\parallel)\frac{\partial f_{j0}}{\partial v_\perp} + k_\parallel v_\perp \frac{\partial f_{j0}}{\partial v_\parallel} \right] \\
\times \left[\frac{1}{(\omega - k_\parallel v_\parallel + \omega_{cj})} - \frac{1}{(\omega - k_\parallel v_\parallel - \omega_{cj})} \right] d^3 v,
\end{aligned}
\tag{1.69}
$$

the remaining components being zero. It should be noted that ε_{zz} is independent of the magnetic field.

5. WAVES IN A COLD PLASMA IN A MAGNETIC FIELD

Before discussing hot plasmas further, we shall consider the propagation of waves in cold magnetized plasmas. In this case both $v_\perp$ and $v_\parallel$ tend to zero, and $f_{j0} \to n_j \delta(\mathbf{v})$. Expressions (1.67), (1.68) and (1.69) now become

$$
\varepsilon_{xx} = \varepsilon_{yy} = 1 - \sum_j \frac{\omega_{pj}^2}{(\omega^2 - \omega_{cj}^2)}, \tag{1.70}
$$

$$\varepsilon_{zz} = 1 - \sum_j \frac{\omega_{pj}^2}{\omega^2} \tag{1.71}$$

and

$$\varepsilon_{xy} = -\varepsilon_{yx} = -i \sum_j \frac{\omega_{pj}^2 \omega_{cj}}{\omega(\omega^2 - \omega_{cj}^2)}. \tag{1.72}$$

We may also write

$$\varepsilon_{xx} \pm i\varepsilon_{xy} = 1 - \sum_j \frac{\omega_{pj}^2}{\omega(\omega \pm \omega_{cj})}. \tag{1.73}$$

Again,

$$\varepsilon_{xz} = \varepsilon_{zx} = \varepsilon_{yz} = \varepsilon_{zy} = 0.$$

5.1. $k_\perp = 0$

Let us now seek solutions to the determinantal equation for waves propagating parallel to the magnetic field, i.e. in the z direction. In this case $k_\perp = 0, k = k_\parallel$, so (1.64) becomes

$$\begin{vmatrix} \left(\frac{\omega}{c}\right)^2 \varepsilon_{xx} - k^2 & \left(\frac{\omega}{c}\right)^2 \varepsilon_{xy} & 0 \\ -\left(\frac{\omega}{c}\right)^2 \varepsilon_{xy} & \left(\frac{\omega}{c}\right)^2 \varepsilon_{xx} - k^2 & 0 \\ 0 & 0 & \left(\frac{\omega}{c}\right)^2 \varepsilon_{zz} \end{vmatrix} = 0, \tag{1.74}$$

i.e.

$$\left(\frac{\omega}{c}\right)^2 \varepsilon_{zz} \left\{ \left[\left(\frac{\omega}{c}\right)^2 \varepsilon_{xx} - k^2\right]^2 + \left[\left(\frac{\omega}{c}\right)^2 \varepsilon_{xy}\right]^2 \right\} = 0. \tag{1.75}$$

There are two solutions:

(i)
$$\varepsilon_{zz} = 0 \tag{1.76}$$

so from (1.71)

$$\omega^2 = \sum_j \omega_{pj}^2. \tag{1.77}$$

This solution is purely oscillatory; the group velocity is zero and the oscillations do not propagate.

(ii)
$$\left(\frac{\omega}{c}\right)^2 \varepsilon_{xx} - k^2 = \pm i \left(\frac{\omega}{c}\right)^2 \varepsilon_{xy}, \tag{1.78}$$

so from (1.73)

$$\frac{k^2c^2}{\omega^2} = 1 - \sum_j \frac{\omega_{pj}^2}{\omega(\omega \pm \omega_{cj})}. \tag{1.79}$$

Here we have two oppositely circularly polarized transverse waves ($\mathbf{k} \cdot \mathbf{E} = 0$), the positive sign being for a right-handed wave. If we ignore the motion of the ions, (1.79) becomes

$$\frac{k^2c^2}{\omega^2} = 1 - \frac{\omega_{pe}^2}{\omega(\omega \mp |\omega_{ce}|)}. \tag{1.80}$$

In this expression we have replaced ω_{ce}, which is negative, by $|\omega_{ce}|$ and so now the positive sign corresponds to a left-handed wave.

For the left-handed wave, k falls to zero when ω is the positive root of

$$\omega(\omega + |\omega_{ce}|) = \omega_{pe}^2, \tag{1.81}$$

i.e. when

$$\omega = \omega_l = \left[\left(\frac{\omega_{ce}}{2} \right)^2 + \omega_{pe}^2 \right]^{\frac{1}{2}} - \frac{|\omega_{ce}|}{2}. \tag{1.82}$$

Very large values of k only occur when

$$\omega^2 = k^2c^2 \tag{1.83}$$

which is the dispersion relation for free space. For the right-handed wave, $k = 0$ when $\omega = 0$ or when

$$\omega = \omega_r = \left[\left(\frac{\omega_{ce}}{2} \right)^2 + \omega_{pe}^2 \right]^{\frac{1}{2}} + \frac{|\omega_{ce}|}{2}, \tag{1.84}$$

and k becomes large either when $\omega^2 = k^2c^2$ again or when $\omega = |\omega_{ce}|$. We may sketch expressions (1.80) on an (ω, k) diagram (Figure 1.1). It is apparent that waves in certain frequency bands will not propagate parallel to the magnetic field in a cold plasma. At the frequencies defining the limits of these *stop bands*, k either goes to zero, or goes to infinity. If k goes to zero, we describe the limit as a *cut-off*: if k goes to infinity we call it a *resonance*. A wave encountering a region in the plasma where $k \to 0$ is strongly reflected: if it enters a region where $k \to \infty$ it is strongly absorbed, as a result of resonant energy transfer to some plasma motion (see for example, Stix (1962)).

Let us now return to the dispersion relation (1.70) and investigate the effects of including the ion motion in the transverse wave solution for low frequencies. If we include only one ion species i, (1.79) becomes

$$\frac{k^2c^2}{\omega^2} = 1 - \frac{\omega_{pe}^2}{\omega(\omega \pm \omega_{ce})} - \frac{\omega_{pi}^2}{\omega(\omega \pm \omega_{ci})}. \tag{1.85}$$

For very low frequencies, such that $\omega \ll |\omega_{ce}|, \omega_{ci}$, we may expand (1.85) in the form

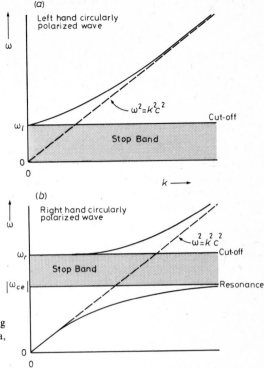

1.1 Dispersion relations for waves propagating parallel to the magnetic field in a cold plasma, neglecting ion motion

$$\frac{k^2 c^2}{\omega^2} \simeq 1 \pm \frac{1}{\omega}\left(\frac{\omega_{pe}^2}{\omega_{ce}} + \frac{\omega_{pi}^2}{\omega_{ci}}\right) + \left(\frac{\omega_{pe}^2}{\omega_{ce}^2} + \frac{\omega_{pi}^2}{\omega_{ci}^2}\right). \tag{1.86}$$

Equilibrium charge neutrality requires that the second term on the right be zero, and we find that

$$\frac{k^2 c^2}{\omega^2} \simeq 1 + \frac{n_e m_e + n_i m_i}{\varepsilon_0 B^2}$$

$$= 1 + \frac{\rho}{\varepsilon_0 B^2}, \tag{1.87}$$

where ρ is the mass of unit volume of the plasma. If we put

$$\frac{\rho}{\varepsilon_0 B^2} = \frac{c^2}{V_A^2} \tag{1.88}$$

in (1.87) we obtain

$$\omega = \frac{kV_A}{(1 + V_A^2/c^2)^{\frac{1}{2}}} \simeq kV_A. \tag{1.89}$$

This wave is known as an Alfvén wave, and V_A is the Alfvén velocity. It may be thought of as a vibration of a loaded string, the string being a magnetic line of force and the load the charged particles spiralling round it. It is a transverse wave when $k_\perp = 0$, in which case two independent orthogonal polarizations are possible.

At somewhat higher frequencies, waves of different polarizations behave differently. For the left-handed wave, (1.85) gives

$$\frac{k^2 c^2}{\omega^2} = 1 - \frac{\omega_{pe}^2}{\omega(\omega + |\omega_{cel}|)} - \frac{\omega_{pi}^2}{\omega(\omega - \omega_{ci})}, \tag{1.90}$$

while for the right-handed wave

$$\frac{k^2 c^2}{\omega^2} = 1 - \frac{\omega_{pe}^2}{\omega(\omega - |\omega_{cel}|)} - \frac{\omega_{pi}^2}{\omega(\omega + \omega_{ci})}. \tag{1.91}$$

Thus a resonance ($k \to \infty$) occurs for the left-handed wave when $\omega = \omega_{ci}$, and for the right-handed wave when $\omega = |\omega_{cel}|$. These are known as the *ion* and *electron cyclotron resonances*. In the frequency range $\omega_{ci} \ll \omega \ll |\omega_{cel}|$, equation (1.91) may be approximated by

$$\omega = \frac{k^2 c^2 |\omega_{cel}|}{\omega_{pe}^2}. \tag{1.92}$$

Such waves have a group velocity which increases with frequency. They occur in the ionosphere, being initiated simultaneously over a wide range of audio frequencies

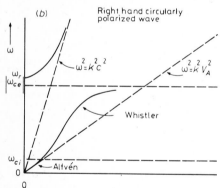

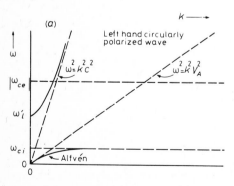

1.2 Dispersion relations for waves propagating parallel to the magnetic field in a cold plasma, including effects of ion motion

by a lightning flash. After propagating along the lines of the earth's magnetic field they become spread out. They are received as a descending glissando, and are known as 'whistlers'. (They also occur in solid state plasmas, where they are known as 'helicons').

If the large mass ratio between ions and electrons is taken into account, approximate cut-off frequencies may be found. Setting $k = 0$ in (1.90) and (1.91), for the right-handed wave the cut-off again occurs at

$$\omega_r = \left[\left(\frac{\omega_{ce}}{2}\right)^2 + \omega_{pe}^2\right]^{\frac{1}{2}} + \frac{|\omega_{ce}|}{2},$$

(1.93)

while for the left-handed wave the cut-off frequency becomes

$$\omega_{l'} = \left[\left(\frac{|\omega_{ce}| + \omega_{ci}}{2}\right)^2 + \omega_{pe}^2\right]^{\frac{1}{2}} + \frac{\omega_{ci} - |\omega_{ce}|}{2}.$$

(1.94)

The dispersion relations are shown in Figure 1.2.

5.2. $k_{\parallel} = 0$

We next look for solutions to the dispersion equation in the case of waves propagating across the static magnetic field, so that $k_{\parallel} = 0, k = k_{\perp}$. Equation (1.64) now becomes

$$\begin{vmatrix} \left(\frac{\omega}{c}\right)^2 \varepsilon_{xx} & \left(\frac{\omega}{c}\right)^2 \varepsilon_{xy} & 0 \\ -\left(\frac{\omega}{c}\right)^2 \varepsilon_{xy} & \left(\frac{\omega}{c}\right)^2 \varepsilon_{xx} - k^2 & 0 \\ 0 & 0 & \left(\frac{\omega}{c}\right)^2 \varepsilon_{zz} - k^2 \end{vmatrix} = 0,$$

(1.95)

i.e.

$$\left[\left(\frac{\omega}{c}\right)^2 \varepsilon_{zz} - k^2\right]\left\{\left(\frac{\omega}{c}\right)^2 \varepsilon_{xx}\left[\left(\frac{\omega}{c}\right)^2 \varepsilon_{xx} - k^2\right] + \left[\left(\frac{\omega}{c}\right)^2 \varepsilon_{xy}\right]^2\right\} = 0.$$

(1.96)

Again, there are two solutions:

(i)

$$\varepsilon_{zz} = \frac{k^2 c^2}{\omega^2}$$

(1.97)

so from (1.71)

$$\omega^2 = k^2 c^2 + \sum_j \omega_{pj}^2.$$

(1.98)

This is a transverse wave with **E** parallel to **B**$_0$ and is known as the *ordinary* wave. It is unaffected by the magnetic field.

(ii)   The other solution is

$$k^2 = \left(\frac{\omega}{c}\right)^2 \left[\varepsilon_{xx} + \frac{\varepsilon_{xy}^2}{\varepsilon_{xx}}\right],$$ (1.99)

or using (1.61) and (1.63),

$$\frac{k^2 c^2}{\omega^2} = 1 - \sum_j \frac{\omega_{pj}^2}{(\omega^2 - \omega_{cj}^2)} - \left\{\frac{\left[\sum_j \left[\frac{\omega_{pj}^2 \omega_{cj}}{\omega(\omega^2 - \omega_{cj}^2)}\right]\right]^2}{1 - \sum_j \frac{\omega_{pj}^2}{(\omega^2 - \omega_{cj}^2)}}\right\}.$$ (1.100)

This wave, known as the *extraordinary* wave, is neither purely longitudinal nor purely transverse since neither E_x nor E_y are zero. If we again ignore the motion of all the ions, we find that $k = 0$ when

$$\omega = \left[\left(\frac{\omega_{ce}}{2}\right)^2 + \omega_{pe}^2\right]^{\frac{1}{2}} \pm \frac{|\omega_{ce}|}{2},$$ (1.101)

i.e. $\omega = \omega_l$ or $\omega = \omega_r$. Also, k becomes very large either in the free space limit, when $\omega^2 = k^2 c^2$, or when $\varepsilon_{xx} = 0$, i.e. when

$$\omega^2 = \omega_{pe}^2 + \omega_{ce}^2 = \omega_{UH}^2.$$ (1.102)

The resonance frequency ω_{UH} is known as the *upper hybrid* frequency. These dispersion relations are sketched in Figure 1.3.

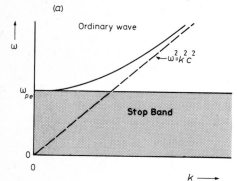

(a)

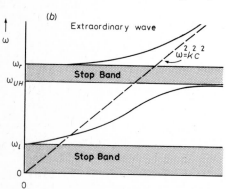

1.3   Dispersion relations for waves propagating across the magnetic field in a cold plasma, neglecting ion motion

The extraordinary wave exhibits additional resonances if the motion of the ions is taken into account. The condition for $\varepsilon_{xx} = 0$ is, from (1.61),

$$\sum_j \frac{\omega_{pj}^2}{(\omega^2 - \omega_{cj}^2)} = 1, \tag{1.103}$$

so there is a resonance for each species of particle. If we consider only electrons and a single species i of ion, and take into account the large ratio between the mass of an ion and that of an electron, (1.103) is approximated by

$$\omega^4 - \omega^2(\omega_{ce}^2 + \omega_{pe}^2) + \omega_{ci}^2 \omega_{ce}^2 + \omega_{pi}^2 \omega_{ce}^2 = 0 \tag{1.104}$$

with the approximate roots

$$\omega^2 = \omega_{pe}^2 + \omega_{ce}^2 = \omega_{UH}^2 \tag{1.105}$$

and

$$\omega^2 = |\omega_{ce}|\omega_{ci}\left(\frac{\omega_{pe}^2 + |\omega_{ce}|\omega_{ei}}{\omega_{pe}^2 + \omega_{ce}^2}\right), \tag{1.106}$$

which when $|\omega_{ce}| \ll \omega_{pe}$ reduces to

$$\omega^2 \simeq |\omega_{ce}|\omega_{ci} = \omega_{LH}^2, \tag{1.107}$$

The resonance frequency ω_{LH} is the *lower hybrid* frequency.

The cut-off frequencies are found similarly from equation (1.100) with $k = 0$ to be approximately

$$\omega = \left[\left(\frac{\omega_{ce}}{2}\right)^2 + \omega_{pe}^2\right]^{\frac{1}{2}} + \frac{|\omega_{ce}|}{2} = \omega_r \quad \text{(right-handed)} \tag{1.108}$$

and

$$\omega = \left[\left(\frac{|\omega_{ce}| + \omega_{ci}}{2}\right)^2 + \omega_{pe}^2\right]^{\frac{1}{2}} + \frac{\omega_{ci} - |\omega_{ce}|}{2} = \omega_{l'} \quad \text{(left-handed)}. \tag{1.109}$$

The cut-off frequencies are thus the same for waves propagating along and across the lines of force. A more general analysis shows that cut-off frequencies are independent of the direction of propagation (Aström (1950)).

5.3. PROPAGATION OF TRANSVERSE WAVES IN OVERDENSE PLASMAS

In an unmagnetized plasma, transverse waves can only propagate provided $\omega_{pe} < \omega$. From (1.31) this means that the electron density must be less than the *critical density*, $n_{ec} = \omega^2 m_e \varepsilon_0 / e^2$. The plasma is then described as '*underdense*'.

In a magnetized plasma, transverse waves can propagate in an '*overdense*' plasma $(n_e > n_{ec})$ provided $\omega_{ce} \cos \phi_B / \omega > 1$, where ϕ_B is the angle between the direction of propagation and **B**. Some practical considerations have been discussed by Bunkin *et al.* (1972).

6. GROWTH AND DECAY OF WAVES

Waves can act as effective means of energy transfer in a plasma. If the energy associated with the electric and magnetic fields of a wave increases at the expense of particle kinetic energy as the wave propagates through a plasma, or as time passes at some point in the plasma, the plasma is said to be *unstable* against the wave, or the wave is described as being unstable. The subject of plasma instability is a vast one, and will come up again in later chapters. For detailed discussion the reader is referred to books by Briggs (1964) and Akhiezer *et al.* (1967).

In general, solutions of the dispersion relations can only be obtained if ω or k are complex quantities. In Section (2) of this chapter we were interested in the spatial variation in amplitude of a wave. We solved for real ω and complex k. The imaginary part of k then gave a measure of the absorption or amplification per unit length. If, on the other hand, we are interested in the local temporal variation in amplitude, we solve for real k and complex ω, and the imaginary part of ω gives the temporal growth or decay rate. We shall fix our attention on a point in the plasma and consider only the growth or decay of a wave in time. For a given real value of k, we may then divide ω into real and imaginary parts, writing

$$\omega = \omega_{Rk} + i\gamma_k. \tag{1.110}$$

γ_k is the reciprocal of the exponential decay or growth time for the wave amplitude. The components ε of the dielectric tensor $\boldsymbol{\varepsilon}$ may also be divided into real and imaginary parts:

$$\varepsilon = \varepsilon_R(k, \omega_{Rk}, \gamma_k) + i\varepsilon_I(k, \omega_{Rk}, \gamma_k). \tag{1.111}$$

Very often we may treat ε_I as a small quantity compared with ε_R, and if the growth or decay coefficient γ_k is small compared with ω_{Rk} we may use a Taylor expansion for ε:

$$\varepsilon \simeq \varepsilon_R(k, \omega_{Rk}) + i\gamma_k \left[\frac{\partial \varepsilon_R(k, \omega_{Rk})}{\partial \omega_{Rk}} \right] + i\varepsilon_I(k, \omega_{Rk}). \tag{1.112}$$

In the next section we shall make use of this expression to find ω_{Rk} and γ_k for longitudinal waves. Here the dispersion relation is just $\varepsilon = 0$, and so we may solve

$$\varepsilon_R(k, \omega_{Rk}) = 0 \tag{1.113}$$

for ω_{Rk} and find γ_k from the relation

$$\gamma_k = \frac{-\varepsilon_I(k, \omega_{Rk})}{\left[\dfrac{\partial \varepsilon_R(k, \omega_{Rk})}{\partial \omega_{Rk}} \right]}. \tag{1.114}$$

7. LONGITUDINAL WAVES IN HOT PLASMAS WITH NO STATIC MAGNETIC FIELD

For longitudinal waves $(\mathbf{k} \times \mathbf{E} = 0)$ the dispersion relation in a hot unmagnetized plasma is

$$\varepsilon_L = 0, \tag{1.115}$$

ε_L being given by expression (1.61). Let us divide ε_L into its real and imaginary parts ε_{RL} and ε_{IL}.

A singularity occurs in the integral in ε_L, which is not well defined when both ω and k are real and $\omega = \mathbf{k} \cdot \mathbf{v}$. This may be dealt with (Landau (1946)) by adding a small positive imaginary part $i\xi$ to ω, integrating, and finding the limit as ξ goes to zero (for a discussion of this procedure see e.g. Krall & Trivelpiece (1973)). We make use of the Plemelj formula:

$$\frac{1}{\omega - kv + i\xi} \underset{\xi \to 0}{\longrightarrow} P\left(\frac{1}{\omega - kv}\right) - i\pi\delta(\omega - kv). \tag{1.116}$$

Here P denotes the principal value. Applying this result to (1.61) we find that

$$\varepsilon_L(\mathbf{k}, \omega) = 1 + \sum_j \frac{q_j^2}{m_j k^2 \varepsilon_0} P \int \frac{\mathbf{k} \cdot \partial f_{j0}/\partial \mathbf{v}}{\omega - \mathbf{k} \cdot \mathbf{v}} \, d^3 v$$

$$+ i\pi \sum_j \frac{q_j^2}{m_j k^2 \varepsilon_0} \int \delta(\omega - \mathbf{k} \cdot \mathbf{v}) \mathbf{k} \cdot \frac{\partial f_{j0}}{\partial \mathbf{v}} \, d^3 v. \tag{1.117}$$

It is now necessary to specify the distribution functions for every species of particle in the plasma. We shall suppose that each species has a Maxwellian distribution function with its own temperature T_j. It may then be shown (see for example Bekefi (1966)) that

$$\varepsilon_L(k, \omega) = 1 - \sum_j \frac{\omega_{pj}^2}{\omega^2} \mathcal{H}(y_j) \tag{1.118}$$

where

$$\mathcal{H}(y) = -2y^2 \left[1 - 2y \exp(-y^2) \int_0^y \exp(t^2) \, dt - i\pi^{\frac{1}{2}} y \exp(-y^2) \right] \tag{1.119}$$

and

$$y_j = \omega/Y_j,$$

with

$$Y_j = (2k^2 \kappa_B T_j/m_j)^{\frac{1}{2}}, \tag{1.120}$$

κ_B being Boltzmann's constant.

Asymptotic expressions are available for the error-function integral in (1.119)* for

---

* For a further discussion and numerical values of this function see Fried & Conte (1961).

large and small values of y. For large y an asymptotic series may be used:

$$2y \exp(-y^2) \int_0^y \exp(t^2) \, dt \simeq 1 + \frac{1}{2y^2} + \frac{3}{4y^4} + \ldots \qquad (y \gg 1), \qquad (1.121)$$

while for small y we have a Taylor series:

$$2y \exp(-y^2) \int_0^y \exp(t^2) \, dt \simeq 2y^2 - \frac{4y^4}{3} + \frac{8y^6}{15} + \ldots \qquad (y < 1). \qquad (1.122)$$

If we ignore the motion of the ions, we find that for high frequencies

$$\varepsilon_L(k, \omega) \simeq 1 - \left(\frac{\omega_{pe}}{\omega}\right)^2 \left[1 + \frac{3}{2}\left(\frac{Y_e}{\omega}\right)^2\right] - i\left(\frac{\omega_{pe}}{\omega}\right)^2 \pi^{\frac{1}{2}} 2\left(\frac{\omega}{Y_e}\right)^3 \exp\left[-\left(\frac{\omega}{Y_e}\right)^2\right],$$
$$(1.123)$$
$$(\omega \gg Y_e)$$

while for low frequencies,

$$\varepsilon_L(k, \omega) \simeq 1 + 2\left(\frac{\omega_{pe}}{Y_e}\right)^2 - i\left(\frac{\omega_{pe}}{\omega}\right)^2 \pi^{\frac{1}{2}} 2\left(\frac{\omega}{Y_e}\right)^3 \qquad (\omega \ll Y_e). \qquad (1.124)$$

The ion motion is most justifiably neglected in the high-frequency limit. Here equation (1.113) with (1.123) gives

$$\omega_{Rk}^2 \simeq \omega_{pe}^2 \left[1 + \frac{3Y_e^2}{2\omega_{Rk}^2}\right] \qquad (\omega \gg Y_e). \qquad (1.125)$$

Solving for ω_{Rk}^2, we find

$$\omega_{Rk}^2 \simeq \frac{\omega_{pe}^2}{2}\left\{1 + \left[1 + 6\left(\frac{Y_e}{\omega_{pe}}\right)^2\right]^{\frac{1}{2}}\right\}$$
$$(1.126)$$
$$\simeq \omega_{pe}^2\left[1 + \frac{3}{2}\left(\frac{Y_e}{\omega_{pe}}\right)^2\right],$$

since we have assumed that $\omega_{Rk} \gg Y_e$, i.e. that the phase velocity of the wave is much greater than the electron thermal velocity. This correction to the cold plasma result, $\omega = \omega_{pe}$, was derived by Bohm & Gross (1949). It may also be written

$$\omega_{Rk}^2 \simeq \omega_{BG}^2 = \omega_{pe}^2[1 + 3k^2\lambda_D^2], \qquad (1.127)$$

where the Debye shielding length (see for example Spitzer (1962))

$$\lambda_D = \left(\frac{\kappa_B T_e \varepsilon_0}{n_e e^2}\right)^{\frac{1}{2}} \qquad (1.128)$$

and

$$\left(\frac{Y_e}{\omega_{pe}}\right)^2 = 2k^2\lambda_D^2.$$

Let us now investigate the damping of these waves. We may write

$$\frac{\partial \varepsilon_{RL}(k, \omega_{Rk})}{\partial \omega_{Rk}} \simeq 2\frac{\omega_{pe}^2}{\omega_{Rk}^3}, \tag{1.129}$$

so from (1.114) and (1.123)

$$\frac{-\gamma_k}{\omega_{Rk}} = \pi^{\frac{1}{2}}\left(\frac{\omega_{Rk}}{Y_e}\right)^3 \exp\left[-\left(\frac{\omega_{Rk}}{Y_e}\right)^2\right]. \tag{1.130}$$

This expression has a maximum of about 0·7 when $(\omega_{Rk}/Y_e)^2 = 1\cdot5$ and then falls off rapidly as ω_{Rk}/Y_e increases. The value of $-\gamma_k/\omega_{Rk}$ is less than 0·1 when $(\omega_{Rk}/Y_e)^2 > 6$, so in this region (1.114) is valid. Our calculation of ω_{Rk} is only valid for $(\omega_{Rk}/Y_e) \gg 1$, so in the region of validity the damping is small. As the wavelength of the oscillations, $2\pi/k$, falls to values nearing the Debye length the damping becomes significant. This is not due to collisions, and is known as *Landau damping*. It arises because for a Maxwellian distribution of velocities the quantity $\mathbf{v}\,\partial f_0/\partial\mathbf{v}$ is always negative. Electrons travelling slower than the phase velocity of the wave, ω/k, are accelerated by it, while electrons travelling faster than ω/k are slowed down. When $\mathbf{v}\,\partial f_0/\partial\mathbf{v}$ is always negative, more electrons will be accelerated than slowed down, so the wave will lose energy to the electrons and will thus be damped.

The inverse process may occur if a non-Maxwellian distribution exists in a plasma, such that $\mathbf{v}\,\partial f_0/\partial\mathbf{v}$ is positive at the phase velocity of the wave. The wave will then increase in amplitude. This is known as a *two-stream instability*, because it can occur when there are two maxima in $f_0(v)$. For a discussion of stability criteria the reader is referred to Krall & Trivelpiece (1973) or Boyd & Sanderson (1969).

The effects of collisions on the dispersion relation have been discussed by Niimura *et al.* (1973) who have given numerical results. The Bohm–Gross frequency (1.127) is increased by a factor which, for $k\lambda_D < 0\cdot3$, is very approximately $(1+(1/118\lambda_D^3 n_e))^{\frac{1}{2}}$.

Next, let us return to expression (1.109) for the longitudinal dielectric constant and consider low-frequency longitudinal waves in a plasma in which ion thermal velocities are small, but electron thermal velocities are large. We shall suppose that only one species i of ion is present, whose charge is Z proton charges, and that $y_i \gg 1$ while $y_e \ll 1$. In this case the contribution of the electrons to ε_L is given by (1.124), while the contribution of the ions is found from (1.123) by replacing ω_{pe} and Y_e by ω_{pi} (from (1.63)) and Y_i (from (1.120)). Then

$$\varepsilon_L(k, \omega) \simeq 1 + 2\left(\frac{\omega_{pe}}{Y_e}\right)^2 - \left(\frac{\omega_{pi}}{\omega}\right)^2 - i\pi^{\frac{1}{2}}2\left(\frac{\omega_{pi}}{\omega}\right)^2 $$
$$\times \left\{\left(\frac{\omega}{Y_e}\right)^3\frac{m_i}{Zm_e} + \left(\frac{\omega}{Y_i}\right)^3 \exp\left[-\left(\frac{\omega}{Y_i}\right)^2\right]\right\}, \tag{1.131}$$

since

$$\frac{\omega_{pe}^2}{\omega_{pi}^2} = \frac{m_i}{Zm_e}.$$

From equation (1.113)

$$\omega_{Rk}^2 \simeq \omega_{pi}^2 \frac{k^2 \lambda_D^2}{1 + k^2 \lambda_D^2}$$

$$= \frac{Z \kappa_B T_e}{m_i} \frac{k^2}{1 + k^2 \lambda_D^2}.$$

(1.132)

These waves are known as *ion acoustic* waves and we shall denote their real frequency by ω_{ia}. The phase velocity V_{ia} is given by

$$V_{ia}^2 = \frac{\omega_{ia}^2}{k^2} = \frac{Z \kappa_B T_e}{m_i} \frac{1}{1 + k^2 \lambda_D^2}.$$

(1.133)

If $k\lambda_D \ll 1$, $V_{ia}^2 \simeq Z\kappa_B T_e/m_i$.

To investigate the damping we again use equation (1.114). From (1.131) and (1.132),

$$\frac{\partial \varepsilon_{RL}(k, \omega_{ia})}{\partial \omega_{ia}} = \frac{2}{\omega_{ia}} \left(1 + \frac{1}{k^2 \lambda_D^2}\right),$$

(1.134)

so

$$\frac{-\gamma_k}{\omega_{ia}} = \pi^{\frac{1}{2}} \left\{ \left(\frac{\omega_{ia}}{Y_e}\right)^3 \frac{m_i}{Zm_e} + \left(\frac{\omega_{ia}}{Y_i}\right)^3 \exp\left[-\left(\frac{\omega_{ia}}{Y_i}\right)^2\right] \right\}.$$

(1.135)

It is convenient to introduce the dimensionless parameter ζ:

$$\zeta = \frac{m_e}{m_i} \left(\frac{Y_e}{\omega_{ia}}\right)^2$$

$$= \frac{2}{Z}(1 + k^2 \lambda_D^2).$$

(1.136)

The minimum value of ζ is therefore $2/Z$.

Expression (1.135) may now be written

$$\frac{-\gamma_k}{\omega_{ia}} = \frac{\pi^{\frac{1}{2}}}{\zeta^{\frac{3}{2}}} \left[\frac{1}{Z}\left(\frac{m_e}{m_i}\right)^{\frac{1}{2}} + \left(\frac{T_e}{T_i}\right)^{\frac{3}{2}} \exp\left(\frac{-T_e}{\zeta T_i}\right)\right].$$

(1.137)

The first term in the square brackets may be neglected. Then (1.137) has the same structure as (1.130) with $(T_e/\zeta T_i)$ replacing (ω_{Rk}^2/Y_e): the damping is a maximum when $(T_e/\zeta T_i) = 1\cdot5$ and falls rapidly as $(T_e/\zeta T_i)$ increases, being less than $0\cdot1$ when $(T_e/\zeta T_i) > 6$. Since $V_{ia}^2 = 2\kappa_B T_e/m_i\zeta$, the condition for small damping is

$$V_{ia}^2 \gtrsim \frac{12\kappa_B T_i}{m_i},$$

(1.138)

i.e. the phase velocity of the wave must be substantially greater than the mean thermal velocity of the ions.

If a drift velocity exists between the electrons and the ions, which sufficiently exceeds the ion mean thermal velocity, ion acoustic waves can grow. This occurs at a phase velocity at which the positive contribution of $\partial f_{e0}/\partial v$ to γ_k exceeds the negative contribution of $\partial f_{i0}/\partial v$. The ion acoustic instability will be discussed further in Chapter 2.

8. WAVES IN A HOT MAGNETIZED PLASMA

The results of Section (5) for a cold magnetized plasma are modified by the motion of the plasma particles in a hot plasma. We shall simply indicate very briefly the more important results.

The most interesting effect is the occurrence of resonances and damping (or instability, given the appropriate non-Maxwellian distribution) of waves near the ion or electron cyclotron frequencies or their harmonics. As a result of the Doppler effect, a particle j moving along $\mathbf{B}$ with velocity $v_{j\parallel}$ will feel a modified wave frequency which will be close to the cyclotron frequency, or a harmonic, if $v_{j\parallel}$ satisfies the relation

$$v_{j\parallel}k_\parallel = \omega - n\omega_{cj} \qquad (n = 1, 2 \ldots). \tag{1.139}$$

If the velocity distribution is Maxwellian, the dispersion relation will only exhibit structure at harmonics of the cyclotron frequency of these particles if

$$\frac{2k_\parallel^2 \kappa_B T_j}{m_j} \lesssim \omega_{cj}^2. \tag{1.140}$$

At higher temperatures or larger values of $k_\parallel$ the effect will be smeared out and all frequencies between two harmonics will experience a similar damping.

For left-handed transverse waves propagating parallel to $\mathbf{B}$ there is a resonance at each ion plasma frequency (but not at harmonics thereof), while for right-handed waves a resonance occurs at the electron plasma frequency. Cyclotron damping occurs near these frequencies.

The ordinary transverse waves propagating perpendicular to $\mathbf{B}$ are unaffected by thermal motion, as are longitudinal waves propagating parallel to $\mathbf{B}$.

The extraordinary waves propagating perpendicular to $\mathbf{B}$ show a harmonic structure in their dispersion relation, as do largely longitudinal waves (defined by $(\mathbf{k} \cdot \mathbf{\varepsilon} \cdot \mathbf{k}/k^2) = 0$) propagating almost perpendicular to $\mathbf{B}$. The latter are known as *Bernstein modes* (Bernstein & Trehan (1960); Bernstein *et al.* (1964)). They are undamped when propagating exactly perpendicular to $\mathbf{B}$.

2

Radiative Energy Transfer

1. INTRODUCTION

When we described in Chapter 1 the propagation of small-amplitude waves in plasmas, we defined the absorption coefficient in terms of the plasma conductivity tensor, which we took to be constant. We took no account of radiation emitted by the plasma. In this chapter we shall consider the absorption and emission of light by a plasma in some detail.

At very high intensities the interaction of light with a plasma becomes non-linear. We shall discuss a number of the interesting new effects which then appear. In particular, an incident transverse wave can couple to other waves in the plasma, resulting in anomalous absorption or scattering. As we shall see in later chapters, it is probable that these and other non-linear processes will be important for the production of dense, high-temperature plasmas using high-power lasers.

2. SOME DEFINITIONS

The radiation power flowing in all directions through a surface is known as the *radiant flux*, Φ. The radiant flux per unit area is called the *irradiance*, I.* The radiant flux emitted per unit area by a surface is known as the *radiant exitance*, $\mathcal{M}$, and has the same dimensions as irradiance.

The radiation power traversing an element of area dA in directions within an

---

* In many papers dealing with the effects of focused laser light, the irradiance is called the 'flux', the 'power density' or the 'radiation intensity'.

element of solid angle $d\Omega$ which makes an angle ϕ_i with the outward normal is given by

$$d\Phi = \mathscr{I}(\phi_i) \cos \phi_i \, d\Omega \, dA, \qquad (2.1)$$

where $\mathscr{I}(\phi_i)$ is the *intensity* of the radiation in the direction of ϕ_i. Thus the intensity has the dimensions of irradiance per steradian. If a pencil of radiation traverses a medium of varying refractive index μ without losing or gaining energy, the quantity $\mathscr{I}/\mu^2$ is constant along the ray.

Very often we are interested in the wavelength or frequency distribution of these quantities. We then make use of the *spectral flux*, Φ_λ or Φ_ω, which is the flux per unit wavelength or angular frequency interval. Thus

$$\Phi = \int \Phi_\lambda \, d\lambda = \int \Phi_\omega \, d\omega. \qquad (2.2)$$

Similarly we define the *spectral irradiance* I_ω, the *spectral exitance* $\mathscr{M}_\lambda$ or $\mathscr{M}_\omega$ and the *spectral intensity* $\mathscr{I}_\lambda$ or $\mathscr{I}_\omega$.

The radiation energy per unit volume per unit angular frequency interval at frequency ω, or *spectral energy density*, w_ω, is given by

$$w_\omega = \int_{4\pi} \frac{\mathscr{I}_\omega}{V_g} \, d\Omega, \qquad (2.3)$$

where V_g is the magnitude of the group velocity $\mathbf{V}_g = \partial\omega/\partial\mathbf{k}$. For an isotropic, non-dispersive medium of refractive index μ,

$$w_\omega = \int \frac{\mathscr{I}_\omega \mu}{c} \, d\Omega. \qquad (2.4)$$

The total *radiant energy density*, w is given by

$$w = \int w_\omega \, d\omega. \qquad (2.5)$$

The radiation power emitted per unit solid angle from unit volume of a plasma is called the *emission coefficient* of the plasma, j. The emission coefficient per unit angular frequency interval at ω, j_ω, is the *spectral emission coefficient*.

We have already met the *linear absorption coefficient*, a_ω, in Chapter 1:

$$a_\omega = -\frac{1}{\mathscr{I}_\omega} \frac{d\mathscr{I}_\omega}{dz} \qquad (2.6)$$

in the absence of emission. It is often convenient to refer to the fractional loss of intensity on passing between two points z_1, z_2 in a plasma. This is known as the *optical thickness* or *optical depth*, $\mathscr{D}_\omega$ and is given by

$$\mathscr{D}_\omega = \int_{z_1}^{z_2} a_\omega \, dz. \qquad (2.7)$$

Finally, we define the *source function*

$$\mathscr{S}_\omega = \frac{1}{\mu^2}\frac{j_\omega}{a_\omega}.$$

(2.8)

The change in intensity on passing through an emitting and absorbing layer of thickness $\mathrm{d}z$ is given by

$$\mu^2\,\mathrm{d}\left(\frac{\mathscr{I}_\omega}{\mu^2}\right) = (-\mathscr{I}_\omega a_\omega + j_\omega)\mathrm{d}z,$$

(2.9)

which may be written in terms of the source function and the optical depth as

$$\frac{\mathrm{d}}{\mathrm{d}\mathscr{D}_\omega}\left(\frac{\mathscr{I}_\omega}{\mu^2}\right) = -\frac{\mathscr{I}_\omega}{\mu^2} + \mathscr{S}_\omega.$$

(2.10)

This is known as the *equation of radiative transfer*.

3. EMISSION PROCESSES IN A PLASMA

We shall next summarize very briefly the various processes through which radiation may be emitted from a plasma. For a full discussion the reader is referred to Griem's book (Griem (1964)) on plasma spectroscopy.

3.1. BOUND–BOUND TRANSITIONS

When an atom or ion makes a transition from one bound state to another of lower energy, the energy of the emitted photon is well defined. In the absence of perturbations the transition gives rise to a spectral line whose profile depends upon the spontaneous lifetime of the upper state, and on the distribution of velocities of the emitting atoms, which causes a distribution of Doppler shifts. Collisions, electric fields and magnetic fields may all perturb the initial and final states of the emitting atoms: the perturbations affect the spectral line profile, which thus contains much information about conditions existing in the plasma, though often in an indecipherable form. A further complication arises if the emitted radiation interacts with other particles, *en route* from the point of emission in the plasma to the observer, so as to distort the line profile.

Emission follows excitation, which may be caused by collisions with other particles or by the absorption of photons. Collisional excitation by energetic electrons is a very important process, and the spectrum of the line radiation depends on the electron temperature (or velocity distribution) and density. In a cool partially-ionized gas ($T_e = 1\,\mathrm{eV} \equiv 11\,605\,\mathrm{K}$) much of the line radiation is in the infrared visible regions of the spectrum. As the temperature rises, atoms can be raised to more energetic excited states and thus tend to emit lines of shorter wavelengths. When $T_e \sim 10\,\mathrm{eV}$ nearly all the atoms will be ionized and multiple ionization of many-electron atoms occurs. The reduced screening of the nuclear charge in multiply-ionized atoms leads to larger energy differences between bound states, so that line radiation is emitted mostly at shorter wavelengths in the far ultraviolet and X-ray regions of the spectrum. At sufficiently high temperatures, when virtually all the ions are reduced to bare nuclei, bound–bound line radiation ceases.

3.2. FREE–BOUND TRANSITIONS

When a free electron is captured by an n-times ionized atom and makes a transition to a bound state of the $(n-1)$-times ionized atom, the surplus energy may be emitted as a photon. The emission is known as *recombination* or *free–bound* radiation. For a particular final bound state, the emission spectrum is a continuum with a fairly sharp low-frequency cut-off. The cut-off is known as the recombination limit, and corresponds to the minimum energy required to ionize the atom from the bound state. The profile of the continuum depends on the free electron velocity distribution and on the (velocity dependent) capture cross section into the bound state. Each bound state has its own continuum, and these overlap: they may differ greatly in intensity.

The inverse process to recombination radiation is known as photo-ionization.

Recombination may also occur without the emission of radiation if two free electrons collide with the ion simultaneously. One electron is captured while the other carries away the surplus energy. This 'three-body' recombination is in competition with radiative recombination and becomes more probable as the density rises.

A complex ion may have two or more ionization potentials corresponding to different configurations of the bound electrons. A free electron may thus be captured, without the emission of radiation, into a bound state s of a configuration with a higher ionization potential. This transition is generally balanced by the inverse 'autoionization' process: however, the recombination may be stabilized by a radiative transition from s to another bound state of lower energy which is not subject to autoionization. This process is known as 'dielectronic' recombination. A review of all these processes has been given by Biberman & Norman (1967).

3.3. FREE–FREE TRANSITIONS

Free–free radiation, or Bremsstrahlung, occurs when a free electron collides with another particle and makes a transition to another free state of lower energy, with the emission of a photon. The spectrum is a continuum. Electron–electron collisions do not produce radiation except at relativistic velocities. In a hot plasma the important collisions are those between electrons and ions; in a slightly ionized gas the more numerous, though less effective, collisions between electrons and neutral atoms are more important.

In a plasma sufficiently hot for most of the ions to be stripped of all their orbital electrons, electron–ion Bremsstrahlung is the dominant radiation mechanism.

(This situation occurs typically at electron temperatures of order $10\,\text{eV}$ ($10^5\,\text{K}$) in hydrogen isotope plasmas, which are of particular interest to us.) The spectral emission coefficient (per unit wavelength) has a maximum at about $6{\cdot}2/T_e\,\text{Å}$, where T_e is in keV, and falls off very rapidly at shorter wavelengths. We shall discuss Bremsstrahlung emission and absorption processes in detail in Section (8).

3.4. CYCLOTRON RADIATION

In a magnetic field electrons spiral around the lines of force. The acceleration they experience gives rise to radiation at the cyclotron frequency and also (as a result of relativistic effects) at its harmonics. The process is sometimes known as magnetic Bremsstrahlung, but it is quite unlike collisional Bremsstrahlung. It is anisotropic, it is usually of much longer wavelength and it consists of lines, although these may be smeared out by perturbations.

For a Maxwellian electron velocity distribution, the total electron cyclotron radiation power per unit volume is (Trubnikov & Kudryavtsev (1958): see also Heald & Wharton (1965))

$$P_{ce} = \frac{n_e e^2 \omega_{ce}^2}{3\pi\varepsilon_0 c}\left(\frac{\kappa_B T_e}{m_e c^2}\right)\left[1 + \frac{5}{2}\left(\frac{\kappa_B T_e}{m_e c^2}\right) + \cdots\right]. \tag{2.11}$$

Numerically, $P_{ce} \simeq 5 \times 10^{-24}\, n_e B^2 T_e(1 + 4 \times 10^{-10}T_e)\,\text{Wm}^{-3}$, where T_e is in K, B in tesla ($1\,\text{T} = 10^4$ gauss) and n_e in m^{-3}.

The distribution over the angle ϕ_B between the direction of emission and·that of the magnetic field vector is as $(1 + \cos^2 \phi_B)$.

Positive ions also emit cyclotron radiation, but at equal electron and ion temperatures the total ion cyclotron radiative power P_{ci} is much less than P_{ce}.

$$\frac{P_{ci}}{P_{ce}} \simeq \left(\frac{Z m_e}{m_i}\right)^3. \tag{2.12}$$

In a very hot, tenuous and strongly magnetized plasma the electron cyclotron radiation power may exceed the Bremsstrahlung power. This occurs when

$$\left(\frac{\omega_{ce}^2}{Z\omega_{pe}^2}\right)^2 \frac{\kappa_B T_e}{\chi_H} \gtrsim 1, \tag{2.13}$$

$\chi_H = 13{\cdot}6\,\text{eV}$ being the ionization energy of a ground state hydrogen atom.

3.5. ČERENKOV RADIATION

Čerenkov radiation occurs when a charged particle travels through a medium faster than the local phase velocity of waves. Huyghens' construction shows that in the absence of dispersion, in an isotropic plasma, a pulse is generated which propagates in directions making an angle ϕ_C with the direction of motion of the particle, ϕ_C being given by the coherence condition

$$\cos \phi_C = \frac{v_{\text{phase}}}{v_{\text{particle}}} = \frac{\omega}{k v_{\text{particle}}}. \tag{2.14}$$

In a dispersive medium, ϕ_C depends on the frequency.

The refractive index of a fully-ionized plasma is always less than unity in the absence of a magnetic field. Transverse Čerenkov radiation can only occur in a plasma when a magnetic field is present. The plasma is then anisotropic, and ϕ_C will also depend on the azimuthal angle about the direction of motion of the particle, unless this coincides with the magnetic field vector. In a non-relativistic plasma, i.e. when $(v_{\text{particle}}/c) \ll 1$, the Čerenkov condition (2.14) can only be satisfied when $k \to \infty$. This is also the condition for the cyclotron resonance: the two processes of Čerenkov and cyclotron radiation are indistinguishable in a non-relativistic plasma.

The phase velocity of longitudinal electron plasma waves is approximately equal to the mean electron thermal velocity, and so high-energy 'suprathermal' electrons can generate longitudinal waves by a Čerenkov process (see for example Tsytovich (1970)). No magnetic field is necessary. This is another way of looking at the two-stream instability (Chapter 1, Section (7)): the inverse process is Landau damping.

4. BLACK BODY RADIATION

If a plasma is locally in thermodynamic equilibrium, the radiation within it has a well-defined frequency distribution of energy density which depends only on the temperature. The spectral energy density in vacuum at temperature T, for a single polarization, is given by the Planck function

$$w_{\omega B_0} = \frac{\hbar \omega^3}{2\pi^2 c^3} \left[\exp\left(\frac{\hbar \omega}{\kappa_B T}\right) - 1 \right]^{-1}. \tag{2.15}$$

In an isotropic, non-dispersive plasma the spectral energy density is

$$w_{\omega B} = \mu^3 w_{\omega B_0}. \tag{2.16}$$

The effects of dispersion and anisotropy have been considered by Rytov (1953) and by Mercier (1964): see also Bekefi (1966).

The spectral intensity of black body radiation at temperature T, which we shall call $\mathscr{I}_{\omega B}(T)$, is now given by expression (2.3): in an isotropic, non-dispersive medium

$$\mathscr{I}_{\omega B}(T) = \frac{\mu^2 \hbar \omega^3}{8\pi^3 c^2} \left[\exp\left(\frac{\hbar \omega}{\kappa_B T}\right) - 1 \right]^{-1}. \tag{2.17}$$

This expression is also valid for dispersive media. For long wavelengths, such that $\hbar \omega \ll \kappa_B T$, the classical Rayleigh–Jeans law gives a good approximation:

$$\mathscr{I}_{\omega B}(T) \simeq \frac{\mu^2 \omega^2 \kappa_B T}{8\pi^3 c^2}. \tag{2.18}$$

The spectral exitance from the surface of a black body in a vacuum, for a single polarization, is given in terms of the frequency, per unit angular frequency interval, by

$$\mathscr{M}_{\omega B_0} = \frac{c}{4} w_{\omega B_0}$$

$$= \frac{\hbar \omega^3}{8\pi^2 c^2} \left[\exp\left(\frac{\hbar \omega}{\kappa_B T}\right) - 1 \right]^{-1}, \tag{2.19}$$

or in terms of the wavelength λ, per unit wavelength interval,

$$\mathscr{M}_{\lambda B_0} = \frac{2\pi \hbar c^2}{\lambda^5} \left[\exp\left(\frac{hc}{\lambda \kappa_B T}\right) - 1 \right]^{-1}. \tag{2.20}$$

Numerically, if the wavelength is expressed in microns and the temperature in degrees absolute,

$$\mathscr{M}_{\lambda B_0} = \frac{3 \cdot 74 \times 10^8}{\lambda^5} \left[\exp\left(\frac{14388}{\lambda T}\right) - 1 \right]^{-1} \text{Wm}^{-2}\mu\text{m}^{-1}. \tag{2.21}$$

The wavelength λ_{max} of maximum spectral exitance per unit wavelength interval occurs when $d\mathscr{M}_{\lambda B_0}/d\lambda = 0$. The result is Wien's law:

$$T\lambda_{max} = 2898 \text{ K } \mu\text{m}. \tag{2.22}$$

The frequency of maximum spectral exitance per unit angular frequency interval is obtained from expression (2.19): $d\mathscr{M}_{\omega B_0}/d\omega = 0$ when $\omega = \omega_{max}$, given by

$$\omega_{max} = 3 \cdot 69 \times 10^{11} \text{ T rad s}^{-1}. \tag{2.23}$$

On integrating expression (2.19) over all frequencies (see for example Tolman (1938), p. 384) and multiplying by 2 to include both possible polarizations, we obtain Stefan's law for the total radiant exitance from a black body in vacuum:

$$\mathscr{M}_{B_0} = \frac{2\pi^5 \kappa_B^4 T^4}{15 c^2 h^3} = \sigma_S T^4, \tag{2.24}$$

where $\sigma_S = 5 \cdot 67 \times 10^{-8} \text{ Wm}^{-2} \text{ K}^{-4}$ is the Stefan–Boltzmann constant.

5. THE EINSTEIN RELATIONS

In thermodynamic equilibrium, each transition must occur at an equal rate in both directions. This very general principle is known as the *principle of detailed balance*. It enables us to relate the emission and absorption coefficients in a plasma.

Let us consider a collection of identical systems which can have two possible states of energies w_1 and w_2 such that $w_2 > w_1$, with populations n_1 and n_2 per unit volume. The frequency ω associated with the transition is given by

$$\hbar\omega = w_2 - w_1. \tag{2.25}$$

The number of spontaneous downward transitions which occur per unit time, volume and solid angle per unit bandwidth of the transition may be written as

$$\mathcal{N}_S = n_2 A_{21}, \tag{2.26}$$

where A_{21} is called the *spontaneous emission coefficient*.

Upward transitions involve the absorption of a photon from the radiation field. If the spectral intensity at ω is $\mathcal{I}_\omega$, then the number of upward transitions in unit time, volume and solid angle per unit bandwidth may be written as

$$\mathcal{N}_A = n_1 \mathcal{I}_\omega B_{12}. \tag{2.27}$$

Here B_{12} is the *induced absorption coefficient*.

In thermodynamic equilibrium, the populations n_1 and n_2 are related by the Maxwell–Boltzmann distribution:

$$\frac{n_1}{n_2} = \frac{g_1}{g_2} \exp\left(\frac{\hbar\omega}{\kappa_B T}\right). \tag{2.28}$$

The quantities g_1, g_2 are the statistical weights or degeneracies of the two states.

The spectral intensity in thermodynamic equilibrium is the black body value, which in an isotropic medium of refractive index μ at temperature T is given by expression (2.17). Thus the upward transition rate $\mathcal{N}_A$ may be written, using (2.27), (2.28) and (2.17),

$$\mathcal{N}_A = n_2 B_{12} \frac{g_1}{g_2} \exp\left(\frac{\hbar\omega}{\kappa_B T}\right) \frac{\mu^2 \hbar\omega^3}{8\pi^3 c^2 [\exp(\hbar\omega/\kappa_B T) - 1]}. \tag{2.29}$$

Clearly this temperature-dependent quantity cannot be equated to the spontaneous downward transition rate $\mathcal{N}_S$, because A_{21} is characteristic of the system, not its environment. It is necessary to invoke an additional emission process whose rate depends on the spectral intensity. The number of these transitions per unit bandwidth per unit time, volume and solid angle may be written

$$\mathcal{N}_I = n_2 \mathcal{I}_\omega B_{21}. \tag{2.30}$$

The quantity B_{21} is known as the *induced* or *stimulated emission coefficient*. Induced radiation is identical in frequency, direction and phase with the incident radiation which gives rise to it.

If we now equate the upward transition rate $\mathcal{N}_A$ with the total downward rate $\mathcal{N}_S + \mathcal{N}_I$ in thermodynamic equilibrium, we obtain the relation

$$\mathcal{I}_{\omega B}(T) = \frac{A_{21}}{B_{12}(g_1/g_2)\exp(\hbar\omega/\kappa_B T) - B_{21}}. \tag{2.31}$$

Since A_{21}, B_{12} and B_{21} must be independent of temperature, a comparison of (2.31) with (2.17) shows that

$$B_{21} = \frac{g_1}{g_2} B_{12} \tag{2.32}$$

and

$$A_{21} = \mu^2 \frac{\hbar\omega^3}{8\pi^3 c^2} B_{21}. \tag{2.33}$$

These relations were first derived (for a system in vacuum) by Einstein in 1917, and A_{21}, B_{12} and B_{21} are known as the Einstein coefficients. Although deduced from considerations of thermodynamic equilibrium, relations (2.32) and (2.33) apply to single-photon transitions in any two-level system in any isotropic medium (the generalization to multi-photon transitions has been discussed by Pert (1971)). We shall now make use of these relations to link the absorption and emission coefficients of a plasma.

The number of photons lost per unit bandwidth from a parallel beam of radiation of intensity $\mathscr{I}_\omega$ traversing a two-level medium is determined by the *net* upward transition rate after allowing for induced emission:

$$\begin{aligned}
\mathscr{N}_A - \mathscr{N}_I &= (n_1 B_{12} - n_2 B_{21})\mathscr{I}_\omega \\
&= \left(n_1 - \frac{g_1}{g_2} n_2\right) B_{12}\mathscr{I}_\omega.
\end{aligned} \tag{2.34}$$

This relation is valid in non-equilibrium situations. If $(g_1/g_2)n_2 > n_1$, amplification of the incident radiation occurs. Spontaneous emission has been ignored since it is distributed over all directions and only a tiny fraction of it lies within the parallel beam.

If the populations n_1 and n_2 are governed by a Maxwellian distribution, expression (2.28) is valid even though there may be no thermodynamic equilibrium. For the moment we assume this to be the case. Then

$$\mathscr{N}_A - \mathscr{N}_I = n_2 B_{12}\mathscr{I}_\omega \frac{g_1}{g_2}\left[\exp\left(\frac{\hbar\omega}{\kappa_B T}\right) - 1\right]. \tag{2.35}$$

Thus from (2.26), (2.32) and (2.33)

$$\mathscr{N}_A - \mathscr{N}_I = \mathscr{N}_S \mathscr{I}_\omega \frac{8\pi^3 c^2}{\mu^2 \hbar\omega^3}\left[\exp\left(\frac{\hbar\omega}{\kappa_B T}\right) - 1\right], \tag{2.36}$$

or in terms of the black body function $\mathscr{I}_{\omega B}(T)$,

$$\mathscr{N}_A - \mathscr{N}_I = \frac{\mathscr{I}_\omega \mathscr{N}_S}{\mathscr{I}_{\omega B}(T)}. \tag{2.37}$$

If several different pairs of states contribute to the transition at ω, their effects must be summed.

We may now find an expression for the linear absorption coefficient a_ω. Let us suppose that our beam of radiation of intensity $\mathscr{I}_\omega$ is travelling in the z direction.

For a volume element of plasma of unit area perpendicular to the beam, and of thickness dz, we may write

$$\hbar\omega(\mathcal{N}_A - \mathcal{N}_I)dz = d\mathcal{I}_\omega. \tag{2.38}$$

Thus from (2.37)

$$a_\omega = -\frac{1}{\mathcal{I}_\omega}\frac{d\mathcal{I}_\omega}{dz} = \frac{\hbar\omega\mathcal{N}_S}{\mathcal{I}_{\omega B}(T)} = \frac{j_\omega}{\mathcal{I}_{\omega B}(T)}, \tag{2.39}$$

where j_ω is the spontaneous spectral emission coefficient. This is Kirchhoff's law. It is important to note that (2.39) holds for any radiation field, provided the populations of the states involved in the transition are governed by a Maxwellian distribution at temperature T.

In the case of free–free transitions, with which we shall be mostly concerned, the electrons have a continuum of possible states and it is necessary to integrate over the velocity distribution function f. Instead of the transition rates $\mathcal{N}_S$, $\mathcal{N}_A$ and $\mathcal{N}_I$ we now use $\mathcal{P}_S$, $\mathcal{P}_A$ and $\mathcal{P}_I$, the powers per unit volume per steradian per unit angular frequency interval for a specified direction z and frequency ω. The Einstein coefficients A_{21}, B_{12} and B_{21} will be replaced by integrals over differential coefficients $\mathcal{A}_S$, $\mathcal{B}_A$, $\mathcal{B}_I$. In terms of the electron velocity, the spontaneous emission power d$\mathcal{P}_S$ due to the small group of electrons making transitions from states of velocity $\mathbf{v}'$ to states of velocity $\mathbf{v}$ with the emission of photons of frequency ω in the z direction is given by

$$d\mathcal{P}_S = \mathcal{A}_S(\mathbf{v}')f(\mathbf{v}')\,d^3\mathbf{v}'. \tag{2.40}$$

The total power absorbed from a beam of frequency ω and intensity $\mathcal{I}_\omega$ travelling in the z direction, by electrons making transitions from $\mathbf{v}$ to $\mathbf{v}'$, is

$$d\mathcal{P}_A = \mathcal{I}_\omega\mathcal{B}_A(\mathbf{v})f(\mathbf{v})\,d^3\mathbf{v}. \tag{2.41}$$

Similarly, the power radiated by induced emission from $\mathbf{v}'$ to $\mathbf{v}$ is

$$d\mathcal{P}_I = \mathcal{I}_\omega\mathcal{B}_I(\mathbf{v}')f(\mathbf{v}')\,d^3\mathbf{v}'. \tag{2.42}$$

In thermodynamic equilibrium $\mathcal{I}_\omega = \mathcal{I}_{\omega B}(T)$ and detailed balance requires that

$$d\mathcal{P}_S = d\mathcal{P}_A - d\mathcal{P}_I. \tag{2.43}$$

The distribution function is now Maxwellian and if the energies of the states with velocities $\mathbf{v}'$ and $\mathbf{v}$ are $w' = m_e v'^2/2$ and $w = m_e v^2/2$ respectively, with $w' - w = \hbar\omega$, we find that

$$\mathcal{I}_{\omega B}(T) = \frac{\mathcal{A}_S(\mathbf{v}')\exp(-w'/\kappa_B T)\,d^3\mathbf{v}'}{\mathcal{B}_A(\mathbf{v})\exp(-w/\kappa_B T)\,d^3\mathbf{v} - \mathcal{B}_I(\mathbf{v}')\exp(-w'/\kappa_B T)\,d^3\mathbf{v}'}. \tag{2.44}$$

A comparison with expression (2.17) shows that

$$\mathcal{A}_S(\mathbf{v}') = \mu^2\frac{\hbar\omega^3}{8\pi^3 c^2}\mathcal{B}_I(\mathbf{v}') \tag{2.45}$$

and

$$\mathcal{B}_A(\mathbf{v})\,d^3\mathbf{v} = \mathcal{B}_I(\mathbf{v}')\,d^3\mathbf{v}'. \tag{2.46}$$

The spectral emission coefficient j_ω, which is the power emitted spontaneously per unit volume, steradian and angular frequency interval, is obtained by integrating $\mathscr{A}_S$ over the distribution function. It may be written

$$j_\omega = \int \mathscr{A}_S(\mathbf{v}') f(\mathbf{v}') \, d^3\mathbf{v}'. \tag{2.47}$$

From (2.41), (2.42), (2.45) and (2.46) we find that the net power absorbed, $\mathscr{P}_A - \mathscr{P}_I$, is given by

$$\mathscr{P}_A - \mathscr{P}_I = \mathscr{I}_\omega \frac{8\pi^3 c^2}{\mu^2 \hbar \omega^3} \int \mathscr{A}_S(\mathbf{v}')[f(\mathbf{v}) - f(\mathbf{v}')] \, d^3\mathbf{v}'. \tag{2.48}$$

Thus the linear absorption coefficient

$$\begin{aligned}
a_\omega &= \frac{\mathscr{P}_A - \mathscr{P}_I}{\mathscr{I}_\omega} \\
&= \frac{8\pi^3 c^2}{\mu^2 \hbar \omega^3} \int \mathscr{A}_S(\mathbf{v}')[f(\mathbf{v}) - f(\mathbf{v}')] \, d^3\mathbf{v}'.
\end{aligned} \tag{2.49}$$

Expressions (2.47) to (2.49) are quite general and hold for any electron distribution function f. Thus

$$a_\omega = j_\omega \frac{8\pi^3 c^2}{\mu^2 \hbar \omega^3} \frac{\int \mathscr{A}_S(\mathbf{v}')[f(\mathbf{v}) - f(\mathbf{v}')] \, d^3\mathbf{v}'}{\int \mathscr{A}_S(\mathbf{v}') f(\mathbf{v}') \, d^3\mathbf{v}'}, \tag{2.50}$$

which is a generalized form of Kirchhoff's Law. If the electrons have a Maxwellian distribution with a temperature T_e, it may be shown that (2.50) reduces to

$$a_\omega = \frac{j_\omega}{\mathscr{I}_{\omega B}}. \tag{2.51}$$

6. RADIATION FROM A FINITE PLASMA

The emission spectrum observed from a finite, optically thin plasma is obtained by integrating the emission coefficients of elements of plasma along the line of sight. If the plasma is not optically thin it is necessary to take into account the reabsorption of radiation as it traverses the plasma.

Let us consider a plasma slab in which the spectral emission coefficient j_ω and the absorption coefficient a_ω are independent of the radiation density. The change in intensity of a ray on passing through a thin layer of the plasma of thickness dz is given by the equation of transfer (2.10):

$$\frac{d}{d\mathscr{D}_\omega} \left(\frac{\mathscr{I}_\omega}{\mu^2} \right) = -\frac{\mathscr{I}_\omega}{\mu^2} + \mathscr{S}_\omega. \tag{2.52}$$

We shall suppose that $\omega \gg \omega_{pe}$ so that the refractive index μ is real. The integrating factor $\exp(-\mathcal{D}_\omega)$ allows us to find the following solution for the intensity of a ray emitted by the plasma, ignoring reflections at the boundaries:

$$\mathcal{I}_\omega = \int_0^\mathcal{D} \mathcal{S}_\omega \exp(-\mathcal{D}_\omega)\,d\mathcal{D}_\omega, \tag{2.53}$$

the integral being taken along the ray.

In the case of Bremsstrahlung emission, provided the electrons in the plasma have a thermal velocity distribution, expressions (2.8) and (2.51) show that the source function $\mathcal{S}_\omega = j_\omega/\mu^2 a_\omega$ is equal to $\mathcal{I}_{\omega B}(T)$, the black body function (2.17). If the electron temperature is uniform in the plasma slab, it follows that $\mathcal{S}_\omega$ is independent of $\mathcal{D}_\omega$ so

$$\mathcal{I}_\omega = \mathcal{I}_{\omega B}[1 - \exp(-\mathcal{D}_{\omega tot})]. \tag{2.54}$$

where $\mathcal{D}_{\omega tot}$ is the total optical depth of the plasma along the ray. Thus if $\mathcal{D}_{\omega tot} \gg 1$ the slab emits like a black body at frequency ω. On the other hand if the optical depth is small, so that $\mathcal{D}_{\omega tot} \ll 1$, then $\mathcal{I}_\omega = \mathcal{I}_{\omega B} \cdot \mathcal{D}_{\omega tot}$. In this case, for a homogeneous plasma of thickness l, along the line of sight $\mathcal{D}_{\omega tot} = a_\omega l$ so $\mathcal{I}_\omega = j_\omega l$ as would be expected.

The transition from a black body spectrum to an optically thin continuum occurs in the vicinity of a frequency ω_{trans} which is determined by the thickness of the plasma slab and by the frequency dependence of a_ω, which will be found in Section (8) below.

At frequencies below the electron plasma frequency ω_{pe} the plasma is a good reflector and a poor emitter of radiation. In general, therefore, there are three distinct

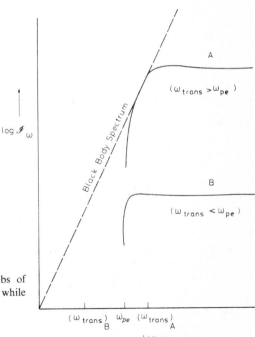

2.1 Emission spectra from two plasma slabs of different thickness: for slab (A) $(\omega_{trans})_A > \omega_{pe}$, while for the thinner slab (B) $(\omega_{trans})_B < \omega_{pe}$.

regions in the continuous emission spectrum from a finite plasma:

$$\omega < \omega_{pe} \qquad\qquad \text{virtually no emission}$$

$$\omega_{pe} < \omega < \omega_{trans} \qquad \text{black body}$$

$$\omega_{trans} < \omega \qquad\qquad \text{optically thin continuum}$$

However, it is often the case that $\omega_{trans} < \omega_{pe}$, so that as the frequency falls, the spectrum cuts off without touching the black body curve. Figure 2.1 shows these effects schematically.

7. PLASMA MODELS

The emission spectrum of a plasma is determined by the combined effects of many processes. Both collisional and radiative excitation, de-excitation, ionization and recombination occur, involving many energy levels and (in general) several stages of ionization. A knowledge of the rates of all these atomic processes would in principle allow us to predict the emission spectrum. More tractable approximate calculations make use of simplified models of the plasma. We shall consider briefly three plasma models. It is assumed throughout that the free electrons have a Maxwellian distribution at temperature T_e, determined by external conditions. Thus the number of electrons with speeds between v and $v+dv$ is

$$dn_e(v) = 4\pi n_e \left(\frac{m_e}{2\pi\kappa_B T_e}\right)^{\frac{3}{2}} \exp\left(\frac{-m_e v^2}{2\kappa_A T_e}\right) v^2 \, dv. \qquad (2.55)$$

The ion temperature may differ considerably from the electron temperature. It determines the Doppler breadth of emission and absorption lines, but has little effect otherwise on the processes discussed in this section.

7.1. LOCAL THERMAL EQUILIBRIUM (LTE)

Collisions become more frequent as the density of a plasma is increased. At sufficiently high densities, collisional processes become more important than radiative processes in determining excited state populations. The LTE model assumes that the populations of atoms and ions in the various bound states are controlled entirely by electron collisions, and may thus be determined from the principle of equipartition provided plasma conditions do not change too rapidly. Thus the populations n_1 and n_2 of two bound states 1 and 2 of an atom or ion, with energies w_1 and w_2, are related by the Boltzmann equation:

$$\frac{n_1}{n_2} = \frac{g_1}{g_2} \exp\left(\frac{w_2 - w_1}{\kappa_B T_e}\right). \qquad (2.56)$$

The densities n_Z, n_{Z+1} of ground state atoms or ions of charge Z, and ions of charge $Z+1$ ($Z = 0, 1 \ldots$), are related by Saha's equation (see for example Kennard (1938), p. 427):

$$\frac{n_e n_{Z+1}}{n_Z} = \frac{2(g_0)_{Z+1}}{(g_0)_Z} \left(\frac{2\pi m_e \kappa_B T_e}{h^2}\right)^{\frac{3}{2}} \exp\left(\frac{-\chi_Z}{\kappa_B T_e}\right). \qquad (2.57)$$

Here $(g_0)_Z$ is the statistical weight of the ground state of an atom or ion of charge Z, and χ_Z its ionization energy. The set of equations for all pairs of adjacent stages of ionization, together with the total density of nuclei, determine the ionization equilibrium populations completely.

For the LTE model to apply, the plasma electron density must be sufficiently high for collisional de-excitation to be at least, say, ten times more probable than radiative decay for all transitions. In an optically thin plasma it may be shown (see for example McWhirter (1965)) that this is equivalent to the requirement that

$$n_e \gtrsim 1 \cdot 6 \times 10^{12} T_e^{\frac{1}{2}} (\Delta w)^3_{\max} \text{ cm}^{-3}, \qquad (2.58)$$

where $(\Delta w)_{\max}$ is the largest interval between adjacent energy levels of the atoms and ions in the plasma, in eV, and T_e is in K. In an optically thick plasma the LTE model is valid at lower densities. Since $(\Delta w)_{\max} \lesssim 0 \cdot 8\chi$ for all atoms and ions, a safe lower limit to n_e for LTE to apply is

$$n_e \gtrsim 8 \times 10^{11} T_e^{\frac{1}{2}} \chi^3_{\max} \text{ cm}^{-3}, \qquad (2.59)$$

where $\chi_{\max}$ is the highest ionization energy of any of the atoms or ions present.

7.2. CORONAL EQUILIBRIUM

In very low density plasmas, such as are found in the solar corona, an excited atom will generally have time to radiate a photon before it can make a transition by a collision process. Assuming that the plasma is optically thin, the inverse process of excitation by absorption of a photon is unlikely to occur, so 'detailed balance' does not apply. Instead, in equilibrium a balance occurs between collisional excitation and spontaneous radiative de-excitation. Nearly all the atoms and ions are in their ground states, so in the coronal model it is assumed that collisional excitation from the ground state balances radiative decay from the upper level to all lower levels.

Collisional ionization is balanced by radiative recombination in the coronal model. Since the rates of both these processes are proportional to the electron density, the populations of the various ion species are independent of electron density. From approximate values of the ionization and recombination coefficients McWhirter (1965) gives the following general result due to Seaton:

$$\frac{n_Z}{n_{Z+1}} \simeq 8 \times 10^{-9} \chi_Z^{11/4} (\kappa_B T_e)^{-\frac{3}{4}} \exp\left(\frac{\chi_Z}{\kappa_B T_e}\right). \qquad (2.60)$$

Average values of the ion charge, $\overline{Z}$, have been calculated for a number of elements as functions of T_e using the coronal model, allowing for dielectronic recombination and autoionization. Shearer & Barnes (1971) give the following approximate empirical expression deduced from several detailed calculations:

$$\bar{Z} = 26\left[\frac{T_e}{1 + (26/Z_n)^2 T_e}\right]^{\frac{1}{4}}. \tag{2.61}$$

Here the electron temperature is in keV, and Z_n is the nuclear charge (the atomic number) of the element.

The coronal model is valid provided the electron density is sufficiently low for collisional transitions from an excited state to be slower than radiative decay. As the principal quantum number n increases, the probability of spontaneous radiation falls but the probability of a collisional transition rises. At sufficiently high n, therefore, the coronal model inevitably breaks down. If n is restricted to 6, McWhirter gives the following expression for the electron density below which the coronal model may be applied for hydrogen-like ions of charge Z:

$$n_e \lesssim 6 \times 10^8 (Z+1)^6 T_e^{\frac{1}{2}} \exp\left[\frac{1 \cdot 162 \times 10^3 (Z+1)^2}{T_e}\right] \text{ cm}^{-3}, \tag{2.62}$$

where T_e is in K.

7.3. COLLISIONAL–RADIATIVE EQUILIBRIUM

While LTE applies at high electron densities and coronal equilibrium at low densities, there is an important intermediate range of densities where neither model is valid. The collisional radiative model of Bates *et al.* (1962a, b) was developed to fill the gap by modifying the coronal model to take into account collisional transitions from the higher bound levels as well as radiative decay, and three-body recombination as well as radiative recombination. Detailed calculations of the equilibrium populations of the various ion species in a carbon plasma, as functions of temperature, have been made by Colombant & Tonon (1973), who have discussed the application of this model to laser-produced plasmas. Figure 2.2 shows their results for an electron density of 10^{21} cm^{-3}.

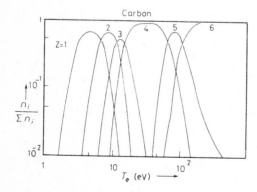

2.2 Equilibrium populations of carbon ions as functions of temperature for an electron density of 10^{21} cm^{-3}. (After Colombant & Tonon (1973))

7.4. TRANSIENT EFFECTS

At low densities the populations of the various ion species and energy levels take time to adjust to changes in plasma conditions. The slowest process, which determines the relaxation time, is the last stage of ionization (if the plasma is being heated) or recombination (if the plasma is being cooled). McWhirter (1965) has shown that the order of magnitude of the relaxation time for any coronal plasma is given by

$$t \sim \frac{10^{12}}{n_e} \text{ s,} \tag{2.63}$$

where n_e is in cm^{-3}.

More detailed calculations of ionization times have been made using the collisional–radiative model by Colombant & Tonon (1973) for iron and by Tonon & Rabeau (1973) for carbon (see Chapter 8, Section (2.2)).

8. BREMSSTRAHLUNG EMISSION AND ABSORPTION

We have already noted in Section (3) of this chapter that the Bremsstrahlung continuum dominates the emission spectrum of fully-ionized hydrogen isotope plasmas and is very important in other hot plasmas. The reverse process, in which an electron absorbs a photon as it moves from one free state to a more energetic one in the field of an ion, is of basic importance as a mechanism for plasma heating by laser light of all wavelengths. It is often known as 'inverse Bremsstrahlung' or simply as 'free–free absorption'.

At frequencies well above the electron plasma frequency, it is possible to obtain accurate theoretical expressions for the Bremsstrahlung emission and absorption processes by considering only binary collisions between an electron and an ion or atom. In the vicinity of the plasma frequency collective effects must be taken into account.

8.1. BINARY PROCESSES

We begin by discussing the emission and absorption of radiation in electron–ion and electron–atom collisions at low electron densities such that the radiation frequency $\omega \gg \omega_{pe}$. If the photon energy $\hbar\omega$ is much less than the electron kinetic energy $\frac{1}{2}m_e v_e^2$ a classical approach is appropriate provided the electron energy is not too great. On the other hand if $\hbar\omega \gtrsim \frac{1}{2}m_e v_e^2$ a quantum mechanical treatment is necessary. Quantum mechanical methods are needed also for high-energy electrons approaching very close to the scattering ion or atom, when the de Broglie wavelength of the electron $h/m_e v_e$ is comparable with the impact parameter.

8.1.1. Electron–ion Bremsstrahlung emission from Maxwellian plasmas

The emission spectrum due to the free–free transitions of non-relativistic electrons (i.e. electrons with energies less than about 10 keV) in the field of positive ions has been calculated by many authors since the early work of Kramers (1923). The theory was reviewed by Grant (1958), by Oster (1961) for radio frequencies, and more recently by Johnston (1967). The general results are somewhat intractable and many approximate expressions valid in different frequency and energy ranges have been found which have been collected together by Brussaard & Van de Hulst (1962).

Classical calculations

Kramers (1923) calculated, classically, the spectrum of power emitted spontaneously by an electron of speed v_e undergoing large-angle scattering in the pure Coulomb field of positive ions. The result, per unit bandwidth and solid angle, for a single polarization, may be written in the form

$$\mathscr{A}_{SK} = \frac{\mu n_i Z^2 e^6}{96\pi^4 \varepsilon_0^3 c^3 m_e^2} \frac{1}{v_e} \frac{\pi}{\sqrt{3}}. \tag{2.64}$$

The refractive index of the plasma μ has been inserted following Westfold (1950). The emission per unit bandwidth is independent of frequency according to this expression, which cannot be valid at high frequencies.

For small deflection angles, i.e. for low frequencies such that $\omega \ll 4\pi\varepsilon_0 m_e v_e^3/Ze^2$, Kramers found that a correction factor was necessary:

$$\mathscr{A}'_{SK}(\omega, v_e) = \mathscr{A}_{SK} g_c(\omega, v_e) \tag{2.65}$$

where

$$g_c(\omega, v_e) = \frac{\sqrt{3}}{\pi} \ln\left(\frac{8\pi\varepsilon_0 m_e v_e^3}{\gamma_g Z e^2 \omega}\right). \tag{2.66}$$

Here $\ln \gamma_g = 0.5772$, the Euler–Mascheroni constant ($\gamma_g = 1.781$). The factor $g_c(\omega, v_e)$ is known as the Gaunt factor after the author of some early quantum mechanical corrections to the Kramers results (Gaunt (1930)). The results of subsequent calculations are customarily expressed in the form of (2.65) with a variety of Gaunt factors.

Expression (2.65) with (2.66) is valid provided the electron kinetic energy $1/2m_e v_e^2$ is much less than the ionization energy of the ion, $Z^2 \chi_H$, and provided $\hbar\omega \ll \frac{1}{2}m_e v_e^2$.

For a thermal distribution of electron velocities, it is necessary to integrate (2.65) to obtain the spectral emission coefficient j_ω given by (2.47):

$$j_\omega(T_e) = \frac{\mu n_e n_i Z^2 e^6}{48\pi^4 \varepsilon_0^3 c^3 m_e^2}\left(\frac{m_e}{2\pi\kappa_B T_e}\right)^{\frac{1}{2}}$$

$$\times \ln\left[\left(\frac{2}{\gamma_g}\right)^{\frac{5}{2}}\left(\frac{\kappa_B T_e}{m_e}\right)^{\frac{3}{2}}\left(\frac{4\pi\varepsilon_0 m_e}{Ze^2\omega}\right)\right]. \tag{2.67}$$

It is necessary also to take into account the refractive index of the plasma. We may write (2.67) in the form

$$j_\omega(T_e) = \frac{\mu n_e n_i Z^2 e^6}{48\pi^4 \varepsilon_0^3 c^3 m_e^2} \left(\frac{m_e}{2\pi\kappa_B T_e}\right)^{\frac{1}{2}} \frac{\pi}{\sqrt{3}} \cdot \bar{g}_c(\omega, T_e),$$ (2.68)

where

$$\bar{g}_c(\omega, T_e) = \frac{\sqrt{3}}{\pi} \ln\left[\left(\frac{2}{\gamma_g}\right)^{\frac{5}{2}}\left(\frac{\kappa_B T_e}{m_e}\right)^{\frac{3}{2}}\left(\frac{4\pi\varepsilon_0 m_e}{Ze^2\omega}\right)\right]$$ (2.69)

is called the average Gaunt factor. Oster (1961) has shown that expression (2.67) is valid up to temperatures of about $50Z^2$ eV for low frequencies ($\hbar w \ll \kappa_B T_e$), but at higher temperatures quantum mechanical effects must be included.

Quantum-mechanical calculations

A non-relativistic treatment by Sommerfeld (1951) gave an exact expression for the differential spontaneous emission coefficient at frequency ω for an electron with initial speed v_{e1} making a transition to a final speed v_{e2} in a pure Coulomb field. The result may be written in the same form as expression (2.65) with a new Gaunt factor $g_S(\omega, v_{e1}, v_{e2})$ given by

$$g_S(\omega, v_{e1}, v_{e2}) = \frac{\sqrt{3} \cdot \pi x \dfrac{d}{dx}|\mathsf{F}(i\eta_1, i\eta_2, 1; x)|}{[\exp(2\pi\eta_1) - 1][1 - \exp(-2\pi\eta_2)]},$$ (2.70)

where $\eta_1 = Ze^2/4\pi\varepsilon_0 \hbar v_{e1}, \eta_2 = Ze^2/4\pi\varepsilon_0 \hbar v_{e2}, x = -4\eta_1\eta_2/(\eta_2 - \eta_1)^2$ and $\mathsf{F}(\alpha, \beta, \gamma; x)$ is the hypergeometric function:

$$\mathsf{F}(\alpha, \beta, \gamma; x) = 1 + \frac{\alpha\beta}{\gamma}\frac{x}{1!} + \frac{\alpha(\alpha+1)\beta(\beta+1)}{\gamma(\gamma+1)}\frac{x^2}{2!} + \cdots.$$ (2.71)

Numerical computations of g_S have been tabulated by Karzas & Latter (1961) and by Grant (1958). Several different analytical approximations have been found. Classical theory applies if $\eta_1, \eta_2 \gg 1$, that is, if the de Broglie wave number of the electron, $m_e v_e/\hbar$, is much greater than Z/a_0, a_0 being the Bohr radius. When $|\eta_2 - \eta_1| \gg 1$, the Sommerfeld expression reduces to Kramers' expression (2.64).

At low frequencies, when $\omega \ll 4\pi\varepsilon_0 m_e v_e^3/Ze^2$, the Sommerfeld Gaunt factor reduces to the classical expression (2.66) for low electron velocities ($\eta_1 \to \infty$).

At relatively high energies ($\eta_1 \ll 1$) an approximation due to Elwert (1939) to the Sommerfeld expression (2.70) may be used:

$$g_S(\omega, v_{e1}, v_{e2}) \simeq \frac{\sqrt{3}}{\pi} \frac{\eta_2}{\eta_1} \frac{(1 - e^{-2\pi\eta_1})}{(1 - e^{-2\pi\eta_2})} \ln\left(\frac{\eta_2 + \eta_1}{\eta_2 - \eta_1}\right).$$ (2.72)

If both $\eta_1 \ll 1$ and $\eta_2 \ll 1$, Elwert's result reduces to what is known as the Born approximation result:

$$g_B(\omega, v_{e1}, v_{e2}) = \frac{\sqrt{3}}{\pi} \ln\left(\frac{\eta_2 + \eta_1}{\eta_2 - \eta_1}\right).$$ (2.73)

This expression, which may also be written (within the approximation)

$$g_B(\omega, v_{e1}) = \frac{\sqrt{3}}{\pi} \ln \frac{2m_e v_{e1}^2}{\hbar \omega},$$ (2.74)

was first derived by Gaunt. It is appropriate for visible and infrared wavelengths at electron temperatures exceeding a few hundred eV.

Quantum-mechanically, it is apparent that the spectral emission coefficient (2.68) for a thermal distribution of electron velocities must be modified to allow for the fact that only electrons with energy $\frac{1}{2} m_e v_e^2 \geq \hbar \omega$ can emit photons of frequency ω. For a Maxwellian distribution of electron velocities this leads to a factor $\exp(-\hbar \omega / \kappa_B T_e)$ in expression (2.67), which is usually inserted also in (2.68) and is not included in $\bar{g}(\omega, T_e)$. This factor was first introduced by Cillie (1932). At low frequencies such that $\hbar \omega \ll \kappa_B T_e$ it may be ignored.

A quantum-mechanical average Gaunt factor at high temperatures may be obtained analytically (Greene (1959)) from expression (2.73):

$$\bar{g}_B(\omega, T_e) = \frac{\sqrt{3}}{\pi} \exp\left(\frac{\hbar \omega}{2\kappa_B T_e}\right) K_0\left(\frac{\hbar \omega}{2\kappa_B T_e}\right)$$ (2.75)

where $K_0(x)$ is the modified Bessel function. At high frequencies ($\hbar \omega \gg \kappa_B T_e$) this reduces to

$$\bar{g}_B(\omega, T_e) = \frac{\sqrt{3}}{\pi}\left(\frac{\pi \kappa_B T_e}{\hbar \omega}\right)^{\frac{1}{2}}.$$ (2.76)

At low frequencies ($\hbar \omega \ll \kappa_B T_e$) Elwert (1954) and Oster (1961) found that

$$\bar{g}_B(\omega, T_e) = \frac{\sqrt{3}}{\pi} \ln\left(\frac{4\kappa_B T_e}{\gamma_g \hbar \omega}\right).$$ (2.77)

Heald & Wharton (1965) have shown that the low frequency expressions (2.69) and (2.77) provide reasonably accurate approximations for temperatures below and above their point of intersection, which is at $T_e = 8.9 \times 10^5 Z^2$ K or $77Z^2$ eV. Numerically, at low temperatures ($T_e < 8.9 \times 10^5 Z^2$ K)

$$j_\omega(T_e) = 2.39 \times 10^{-47} \mu \frac{n_e n_i Z^2}{T_e^{\frac{1}{2}}}\left[19.56 + \ln\left(\frac{T_e^{\frac{3}{2}}}{\omega Z}\right)\right]$$ (2.78)

and at high temperatures ($T_e > 8.9 \times 10^5 Z^2$ K)

$$j_\omega(T_e) = 2.39 \times 10^{-47} \mu \frac{n_e n_i Z^2}{T_e^{\frac{1}{2}}}\left[26.41 + \ln\left(\frac{T_e}{\omega}\right)\right] \text{ W cm}^{-3} \text{ sterad}^{-1}(\text{rad s}^{-1})^{-1},$$ (2.79)

with n_e, n_i in cm^{-3} and T_e in K.

The wavelength and temperature ranges over which the classical and the Born approximations for $\bar{g}$ (expressions (2.69) and (2.77)) hold to 2% accuracy are shown in Figure 2.3. For wavelengths longer than $\sim 1/Z^2$ μm and temperatures above $\sim 130Z^2$ eV, the Born approximation is satisfactory.

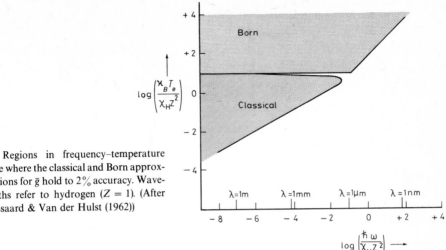

2.3 Regions in frequency–temperature space where the classical and Born approximations for $\bar{g}$ hold to 2% accuracy. Wavelengths refer to hydrogen $(Z = 1)$. (After Brussaard & Van der Hulst (1962))

It should be noted that at high frequencies and low temperatures the observed continuous radiation spectrum may contain important contributions from recombination (free–bound) transitions which have not been included in our discussion. Detailed calculations were made by Menzel & Pekeris (1935). Brussaard & Van de Hulst (1962) calculated the emission spectrum from a plasma containing a Maxwellian distribution of electrons, at temperature T_e, and hydrogen-like positive ions of charge Z. Figure 2.4 shows the product of the Cillie factor and the average Gaunt factor, with and without allowance for recombination, at a temperature $T_e = \chi_H Z^2/\kappa_B$ (for hydrogen this corresponds to $T = 1\cdot58 \times 10^5$ K). As may be seen, at this temperature the contribution from recombination becomes appreciable at wavelengths below 1 μm for hydrogen. Many references to work on continuous spectra are given by Biberman & Norman (1967). In the absence of accurate knowledge of the recombination radiation spectrum, Brussaard & Van de Hulst suggest that this contribution may be taken into account approximately by multiplying j_ω by $\exp(\hbar\omega/\kappa_B T_e)$.

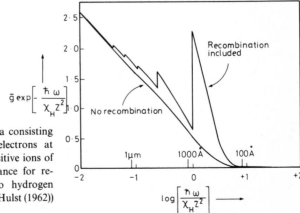

2.4 Emission spectra from a plasma consisting of a Maxwellian distribution of electrons at $T_e = \chi_H Z^2/\kappa_B$ and hydrogen-like positive ions of charge Z, with and without allowance for recombination. Wavelengths refer to hydrogen $(Z = 1)$. (After Brussaard & Van der Hulst (1962))

Total Bremsstrahlung emission

The total power emitted spontaneously as Bremsstrahlung over all frequencies, per unit solid angle and in a single polarization, $j(T_e)$, may be found by integrating the spectral emission coefficient $j_\omega(T_e)$. It is given (on dropping the refractive index) by

$$j(T_e) = \frac{n_e n_i Z^2 e^6}{96\pi^4 \varepsilon_0^3 \hbar m_e c^3} \left(\frac{2\pi\kappa_B T_e}{3m_e}\right)^{\frac{1}{2}} \bar{g},$$

(2.80)

where $\bar{g}$ is the Gaunt factor averaged over both electron velocity and frequency. The total power radiated per unit volume in all directions in both polarizations is given by

$$P_\beta = 1{\cdot}4 \times 10^{-34} n_e n_i Z^2 T_e^{\frac{1}{2}} \bar{g} \text{ W cm}^{-3},$$

(2.81)

where T_e is in K and n_e, n_i in cm^{-3}.

Greene (1959) calculated values of $\bar{g}$ from Sommerfeld's formula and found that $1{\cdot}45 > \bar{g} > 1{\cdot}10$ for all temperatures above 1 eV. In the Born approximation

$$\bar{g}_B = \frac{2}{\pi}\sqrt{3}.$$

(2.82)

This value may be used at temperatures above $130Z^2$ eV.

8.1.2. Electron–ion Bremsstrahlung absorption from weak radiation fields

The linear Bremsstrahlung absorption coefficient for radiation in a plasma with a Maxwellian distribution of electron velocities may be calculated immediately from expressions (2.17), (2.51) and (2.68) with the Cillie factor included:

$$a_\omega = \frac{j_\omega}{\mu^2 \mathscr{I}_{\omega B}}$$

$$= \frac{n_e n_i Z^2 e^6 [1 - \exp(-\hbar\omega/\kappa_B T_e)]}{\mu 6\pi\varepsilon_0^3 c\hbar\omega^3 m_e^2} \left(\frac{m_e}{2\pi\kappa_B T_e}\right)^{\frac{1}{2}} \frac{\pi}{\sqrt{3}} \bar{g}.$$

(2.83)

If recombination radiation contributes significantly to j_ω, the inverse process of photo-ionization will contribute significantly to the absorption coefficient α_ω.

At high temperatures and for low frequencies such that $\hbar\omega \ll \kappa_B T_e$, the term $[1 - \exp(-\hbar\omega/\kappa_B T_e)]$ is very nearly equal to $(\hbar\omega/\kappa_B T_e)$ so $a_\omega \simeq a_{\omega T}$, given by

$$a_{\omega T} = \frac{n_e n_i Z^2 e^6}{\mu 6\pi\varepsilon_0^3 c\omega^2} \left(\frac{1}{2\pi}\right)^{\frac{1}{2}} \left(\frac{1}{m_e \kappa_B T_e}\right)^{\frac{3}{2}} \frac{\pi}{\sqrt{3}} \bar{g}.$$

(2.84)

For laser light in a hot plasma this approximation for the weak-field binary absorption coefficient is generally used.

If we compare expression (2.84) with (1.42) we obtain a value for the effective electron–ion collision frequency

$$v_{ei}^* = a_{\omega T} c \mu \frac{\omega^2}{\omega_{pe}^2}$$

$$= \frac{1}{6\pi} \left(\frac{1}{2\pi m_e} \right)^{\frac{1}{2}} \left(\frac{1}{\kappa_B T_e} \right)^{\frac{3}{2}} \frac{n_e Z e^4}{\varepsilon_0^2} \frac{\pi}{\sqrt{3}} \bar{g}. \qquad (2.85)$$

From (2.85) and (1.36) we may find the Lorentz conductivity of the plasma at frequency ω.

Expressions (2.83), (2.84) and (2.85) are only valid provided the radiation field is sufficiently weak so as not to perturb the Maxwellian electron velocity distribution significantly. If the field is too strong, non-linear effects appear. These are discussed in the next section.

Hora & Wilhelm (1970) published calculated values of the absorption coefficient and the refractive index of fully-ionized hydrogen plasmas in the range $1 < T_e < 10^6$ eV at the wavelengths of the ruby, neodymium and CO_2 lasers (6943 Å, 1·06 μm and 10·6 μm) and their second harmonics. Only linear Bremsstrahlung absorption was considered. Absorption coefficients at 1·06 μm and 10·6 μm are shown in Figures 2.5(a) and (b).

Raizer (1959) described a simple approximate method of calculating the absorption length $1/a_\omega$ of radiation in ionized gases, and quoted values for ruby laser light (6943 Å) in air at normal density as a function of temperature (Raizer (1965a)). Effective average values of Z were used, assuming thermal equilibrium between the various ion species. The results are shown in Figure 2.6. Calculations by Key (1969)

2.5   Linear Bremmstrahlung absorption coefficient of fully-ionized hydrogen as function of temperature for various electron densities. (a) for neodymium laser light ($\lambda = 1·06 \mu$m), (b) for carbon dioxide laser light ($\lambda = 10·6 \mu$m). (After Hora & Wilhelm (1970))

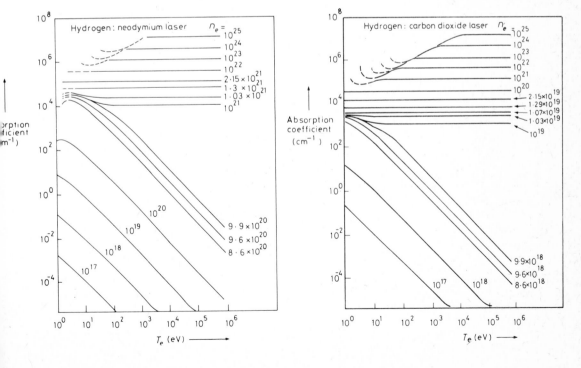

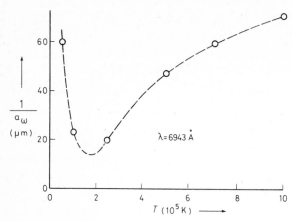

2.6   Absorption length for $\lambda = 6943$ Å in air at normal density as function of temperature. (Data from Raizer (1965a))

for helium at a density of $1\cdot7 \times 10^{20}$ nuclei cm^{-3} indicate a similar temperature dependence with absorption lengths about half those shown in the diagram.

8.1.3. Electron–ion Bremsstrahlung processes in strong radiation fields

An electron in an oscillating electric field of amplitude E_0, corresponding to an irradiance

$$I = \varepsilon_0 c \frac{E_0^2}{2},$$
(2.86)

experiences an acceleration $eE_0 \sin \omega t$. The maximum speed of the oscillating electron, which is sometimes known as the 'quivering' speed, is then

$$v_E = \left| \frac{eE_0}{m_e \omega} \right|$$

$$= 25\lambda\sqrt{I} \, \text{cm s}^{-1}$$
(2.87)

when λ is in microns and I in watts cm^{-2}.

If the plasma temperature is T_e, the thermal velocity of the electrons is of the order of

$$v_T = \sqrt{\frac{\kappa_B T_e}{m_e}}.$$
(2.88)

It is convenient to use the ratio of these two velocities as the *field strength parameter*, η:

$$\eta = \frac{v_E}{v_T}.$$
(2.89)

Provided $\eta \ll 1$, the radiation field has little direct effect on the electron velocity distribution, and expressions (2.83) and (2.84) for the Bremsstrahlung absorption coefficient are valid. If however $\eta \gtrsim 1$, the electron velocities are markedly affected by the radiation. It is necessary to find a new expression for a_ω which involves the radiation field and is therefore non-linear. Quantum mechanically, multiphoton processes must be taken into account at high photon densities.

In the strong field limit, when $\eta \gg 1$, it might be expected that $\kappa_B T_e = m_e v_T^2$ should be replaced by $m_e v_E^2 = e^2 E_0^2 / m_e \omega^2$, so instead of expression (2.84) for the absorption coefficient we would have

$$a_{\omega E} = \frac{n_e n_i Z^2 e^3 \omega}{\mu 6 \pi \varepsilon_0^3 c E_0^3} \left(\frac{1}{2\pi} \right)^{\frac{1}{2}} \frac{\pi}{\sqrt{3}} \bar{g}_E \tag{2.90}$$

where $\bar{g}_E$ is some new Gaunt factor. Detailed calculations have been made of the non-linear absorption coefficient by Rand (1964), Silin (1964), Bunkin & Fedorov (1965), Nicholson-Florence (1971), Brehme (1971), Osborn (1972a), Pert (1972) and Seely & Harris (1973). All agree on the general form of this expression but differ regarding the Gaunt factor. Another form derived by Hughes & Nicholson-Florence (1968) is incorrect because the wrong averaging procedure was used.

The power P absorbed by an element of plasma is proportional to $I a_\omega$. When $\eta \ll 1$, P increases as the irradiance is increased. However, when $\eta \gg 1$ the absorption coefficient varies as $1/E_0^3$ or $1/I^{\frac{3}{2}}$, and at very high irradiance, P will fall as the irradiance increases. There is therefore an optimum irradiance I_{opt} at which the power absorbed by the plasma element due to binary collisions is a maximum. This occurs approximately when $\eta = 1$. Thus for a laser beam of wavelength λ microns in a plasma at an electron temperature T_e eV,

$$I_{\text{opt}} \simeq 3 \times 10^{12} \frac{T_e}{\lambda^2} \text{ W cm}^{-2}. \tag{2.91}$$

The frequency dependence of the absorption coefficient also changes as the irradiance increases. Whereas at low fields $a_{\omega T}$ varies as $1/\omega^2$, at high fields $a_{\omega E}$ varies as ω. This does not necessarily mean that higher frequency light is more effectively absorbed by a plasma element: if we suppose that expression (2.84) is approximately valid up to the optimum irradiance, the product $I_{\text{opt}} a_{\omega T}$ is approximately independent of ω apart from the logarithmic Gaunt factor.

The modified electron velocity distribution due to the strong radiation field must also be manifested in the Bremsstrahlung emission spectrum. This effect was discussed by Pert (1973) who predicted a significant increase in the emitted power when $I \gtrsim I_{\text{opt}}$, but little change in the spectral distribution of the emission.

As we shall see later in this chapter, many other non-linear effects are also expected to occur in strong radiation fields. It may prove difficult experimentally to identify non-linear effects in Bremsstrahlung emission and absorption.

8.1.4. Electron–atom Bremsstrahlung processes

Interactions between electrons and neutral atoms have a much shorter range than the Coulomb interaction between electrons and ions. Electron–neutral atom Bremsstrahlung emission and absorption are therefore much weaker than the corresponding electron–ion processes discussed above.

The classical treatment valid at low frequencies ($\hbar\omega \ll \frac{1}{2}m_e v_e^2$) has been discussed by many authors (see for example Bekefi (1966)). The average power of spontaneous free–free emission due to an electron of speed v_e colliding with heavy particles, per unit solid angle in a single polarization, may be written as

$$\mathscr{A}_S(v_e) = \frac{e^2}{24\pi^3\varepsilon_0 c^3} v_e^2 v_{ea}(v_e). \tag{2.92}$$

Here $v_{ea}(v_e)$ is the collision frequency for momentum transfer between the electron and the atom. If the electron distribution function is $f(v_e)$, then the spectral emission coefficient is

$$j_\omega = \frac{e^2}{6\pi^2\varepsilon_0 c^3} \int v_e^4 v_{ea}(v_e) f(v_e)\, dv_e. \tag{2.93}$$

At low frequencies, therefore, the emission spectrum is independent of ω.

The corresponding absorption coefficient is obtained from expression (2.49): in the classical limit

$$a_\omega = \frac{4\pi e^2}{3c\varepsilon_0\omega^2 m_e} \int v_e^3 v_{ea}(v_e) \frac{\partial f}{\partial v_e}\, dv_e. \tag{2.94}$$

The velocity dependence of the collision frequency is well known for several gases (see for example Brown (1967)). For hydrogen and helium the collision frequency is almost independent of the velocity and is proportional to the gas density. For other gases numerical integrations are necessary.

Detailed quantum-mechanical calculations of transition probabilities were carried out by Breene & Nardone (1960, 1961, 1963) and by Kivel (1967a, b) for nitrogen and oxygen (see also Akcasu & Wald (1967)), and by Ashkin (1966) for argon. Simpler approximate methods which may be applied generally have been developed by Ohmura & Ohmura (1960), Firsov & Chibisov (1960), Kasyanov & Starostin (1965) and Dalgarno & Lane (1966). Phelps (1966) used a method due to Holstein to derive absorption coefficients for helium, argon and molecular nitrogen at $\lambda\,6943$ Å. The subject has been reviewed by Biberman & Norman (1967).

Electron–atom Bremsstrahlung coefficients are sometimes defined by imagining that the atom is an ion and quoting an effective charge Z' (usually $Z' \ll 1$). These results must be treated with caution because the electron–atom and electron–ion Bremsstrahlung coefficients depend differently on temperature and wavelength. Nevertheless, the use of an effective Z gives some idea of the relative importance of neutral atom and ion Bremsstrahlung processes in a partially ionized plasma. When more than a few percent of the atoms are ionized, neutral atom Bremsstrahlung may be ignored. On the other hand, as we shall see in Chapter 5, neutral atom Bremsstrahlung is very important in the early stages of gas breakdown by a laser pulse when only very few electrons are present.

Non-linear effects are unlikely to be important in connection with electron–atom Bremsstrahlung. This is because the radiation field strength necessary to cause significant non-linearity will produce very rapid multiphoton ionization of the atoms.

8.1.5. Ion–ion collisions in strong radiation fields

Let us consider the effect of a strong oscillating electric field of amplitude E_0 and frequency ω on an underdense plasma containing two species of ions with different charge-to-mass ratios m_1/Z_1, m_2/Z_2. The ion species move in the field with the same period but with different maximum streaming speeds $v_j = Z_j e E_0/m_j\omega$. Thus collisions occur which tend to thermalize the two ion velocity distributions.

This *driven collisional heating* was discussed by Mjolsness & Ruppel (1972), who concluded that at very high irradiances ($I \gtrsim 10^{16}$ W cm^{-2} for a CO_2 laser; 10^{18} W cm^{-2} for a neodymium laser) it might be more effective than electron–ion collisions in heating the ions in a kilovolt D–T plasma.

8.2. COLLECTIVE EFFECTS

The motion of the constituent particles in a plasma causes fluctuations in the local particle densities and currents. The spectra of these fluctuations may be deduced for thermal plasmas with the help of the fluctuation–dissipation theorem (Callen & Welton (1951)). For stable, non-thermal distributions an extension of the theorem due to Hubbard (1961) and Rostoker (1961) may be used. The general result for the spectral power density $P(\mathbf{k}, \omega)$ of fluctuations with wavenumber $\mathbf{k}$ and frequency ω is

$$P(\mathbf{k}, \omega) = \sum_j \frac{n_j q_j^2}{|k|^2 \varepsilon_0^2 |\varepsilon_L(k, \omega)|^2} \int f_j(\mathbf{v})\delta(\omega - \mathbf{k} \cdot \mathbf{v}) \, d^3v. \tag{2.95}$$

Here ε_L, the longitudinal dielectric constant, is given by expression (1.61) for an unmagnetized plasma. The sum is over all the charged particle species in the plasma. In thermal equilibrium the spectral power density becomes

$$P(\mathbf{k}, \omega) = \sum_j \frac{\kappa_B T_j}{\pi \omega \varepsilon_0} \operatorname{Im} \frac{1}{\varepsilon_L(k, \omega)}. \tag{2.96}$$

The fluctuations are essentially electrostatic, longitudinal oscillations in the plasma. However, if an appropriate conversion mechanism is available their energy can appear as transverse waves. One such process is scattering of the longitudinal electron waves from the grainy distribution of ions, which leads to an enhanced emission of Bremsstrahlung near the plasma frequency. The inverse process enhances the absorption coefficient for transverse radiation near ω_{pe}. Birmingham et al. (1965) found that the spectral emission coefficient allowing for these effects is

$$j_\omega = \frac{\mu n_i Z^2 e^6}{24\pi^2 \varepsilon_0^3 m_e^2 c^3} \frac{1}{(2\pi)^3} \int \int \frac{f(\mathbf{v}_e)\delta(\omega - \mathbf{k} \cdot \mathbf{v}_e)}{|k|^2 |\varepsilon_L(k, \omega)|^2} \, d^3k \, d^3v_e. \tag{2.97}$$

The corresponding absorption coefficient is

$$a_\omega = \frac{n_i \pi Z^2 e^6}{\mu 3\varepsilon_0^3 m_e^3 c\omega^3} \frac{1}{(2\pi)^3} \int \int \left(\frac{\partial f(\mathbf{v}_e)}{\partial v_e}\right) \frac{\delta(\omega - \mathbf{k} \cdot \mathbf{v}_e)}{|k|^2 |\varepsilon_L(k, \omega)|^2} \, d^3k \, d^3v_e \tag{2.98}$$

In the case of a thermal plasma, these expressions may be written in the same form as (2.68) and (2.83) or (2.84) with a new average Gaunt factor $\bar{g}$ (Dawson & Oberman (1962), Johnston & Dawson (1973)). The Gaunt factor is of the form

$$\bar{g} = \frac{\sqrt{3}}{\pi} \ln \Lambda, \qquad (2.99)$$

Λ being the smaller of $v_T/\omega_{pe}b_{min}$ and $v_T/\omega b_{min}$, where $v_T = (\kappa_B T_e/m_e)^{\frac{1}{2}}$ and b_{min} is the larger of $Ze^2/4\pi\varepsilon_0\kappa_B T_e$ and $\hbar/(m_e\kappa_B T_e)^{\frac{1}{2}}$. These values of $\bar{g}$ give good approximations to numerical computations.

The effects of ion correlations have been considered by Dawson & Oberman (1963), who showed that they may be neglected in a thermal plasma for frequencies well above ω_{pe}, and that near ω_{pe} they reduce Λ by a factor $1/(Z+1)$.

In a non-thermal plasma, both ion and electron correlations may have major effects on emission and absorption coefficients. Here we are getting very close to the more general question of the effects of wave coupling processes in plasmas, which is discussed in Section (9).

Collective effects are also manifest in the behaviour of the refractive index of the plasma near the electron plasma frequency. For a detailed discussion of this topic the reader is referred to Heald & Wharton (1965).

8.3. RELATIVISTIC EFFECTS: ELECTRON–ELECTRON BREMSSTRAHLUNG

At temperatures above about 30 keV, relativistic effects should be taken into account. The dipole approximation for the electron–ion Bremsstrahlung spectrum is no longer satisfactory and multipole contributions must be included. Corrections to the dipole spectrum have been given by Quigg (1968).

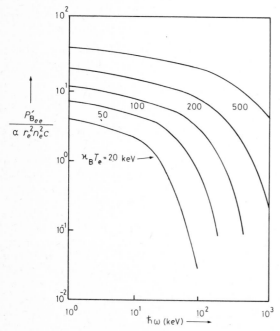

2.7   Electron–electron Bremsstrahlung emission spectra for five values of T_e (see text). (After Maxon (1972))

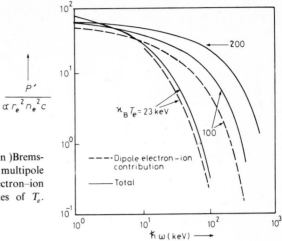

2.8   Total (electron–electron + electron–ion)Bremsstrahlung emission spectra, including multipole contributions, compared with dipole electron–ion Bremsstrahlung emission for three values of T_e. (After Maxon (1972))

Electron–electron Bremsstrahlung, which consists only of quadrupole and higher even-order multipole radiation, also becomes important at high temperatures. Maxon (1972) has interpolated between the quadrupole and the extreme relativistic spectra for five values of $\kappa_B T_e$ from 20 keV to 500 keV. The power emitted per unit volume per unit energy interval, $P'_{\beta_{ee}}$ (in units of $\alpha r_e^2 n_e^2 c$, where $\alpha = e^2/2hc\varepsilon_0$ is the fine structure constant and $r_e = e^2/4\pi c^2 m_e \varepsilon_0$ is the classical electron radius), is plotted as a function of the photon energy $\hbar\omega$ in Figure 2.7. The sum of the electron–electron spectrum and the electron–ion spectrum including multipole contributions, for ions with $Z = 1$, are shown in Figure 2.8, with the dipole contribution to the electron–ion spectrum for comparison. Figure 2.9 shows the total Bremsstrahlung emission power $P_{\beta_{rel}}$ integrated over all frequencies. The departure from the dipole emission

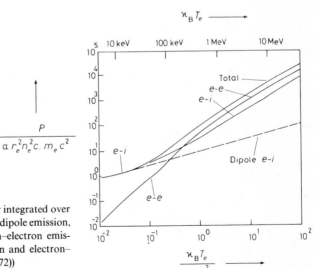

2.9   Bremsstrahlung emission power integrated over all frequencies, showing electron–ion dipole emission, total electron–ion emission, electron–electron emission and the total of all electron–ion and electron–electron emission. (After Maxon (1972))

curve above 30 keV is apparent. For higher values of Z the electron–ion contributions should be multiplied by Z: the electron–electron contributions then become less important.

8.4. BREMSSTRAHLUNG EMISSION FROM A FINITE PLASMA

We saw in Section (6) that the emission spectrum of a uniform plasma slab of thickness l changes from optically thin Bremsstrahlung to black body radiation when $a_\omega l \simeq 1$. The frequency ω_{trans} near which this transition occurs may be found from (2.84) for low frequencies and high (but non-relativistic) temperatures:

$$\omega^2_{\text{trans}} = \frac{n_e n_i Z^2 e^6}{\mu 6 \pi \varepsilon_0^3 c} \left(\frac{1}{2\pi}\right)^{\frac{1}{2}} \left(\frac{1}{m_e \kappa_B T_e}\right)^{\frac{3}{2}} \frac{\pi}{\sqrt{3}} \bar{g} l. \tag{2.100}$$

For a single species of ion $n_i = n_e/Z$, so in terms of the plasma frequency $\omega^2_{pe} = n_e e^2/m_e \varepsilon_0$ we find that with $n = 1$

$$\frac{\omega^2_{\text{trans}}}{\omega^2_{pe}} = \frac{e^4}{6 \varepsilon_0^2 c} \left(\frac{1}{6 \pi m_e}\right)^{\frac{1}{2}} \frac{n_e Z \bar{g} l}{(\kappa_B T_e)^{\frac{3}{2}}}. \tag{2.101}$$

Numerically

$$\frac{\omega^2_{\text{trans}}}{\omega^2_{pe}} = 1 \cdot 75 \times 10^{-16} \frac{n_e Z \bar{g} l}{T_e^{\frac{3}{2}}}, \tag{2.102}$$

where n_e is in cm^{-3}, l in cm and T_e in eV. As we shall see in later chapters, in many laser-heated plasmas $\omega^2_{\text{trans}}/\omega^2_{pe} < 1$, so the plasma frequency cut-off intervenes to prevent these plasmas radiating like black bodies at any frequency.

8.5. EXPERIMENTAL MEASUREMENTS OF BREMSSTRAHLUNG COEFFICIENTS

8.5.1. Fully-ionized plasmas

Experimental measurements of the Bremsstrahlung emission spectrum from a fully-ionized hydrogen θ-pinch* plasma were made by Bogen & Rusbüldt (1968) over a wide range of wavelengths, from $0 \cdot 25 \, \mu\text{m}$ to $25 \, \mu\text{m}$. The electron temperature and density were $\sim 100 \, \text{eV}$ and $\sim 3 \times 10^{17} \, \text{cm}^{-3}$. The emission spectrum agreed well with values calculated using expression (2.76): the Gaunt factor was found to vary by a factor of 2 over the wavelength range investigated.

Measurements of far-infrared emission spectra above $\lambda \, 70 \, \mu\text{m}$ have been made for a variety of plasmas by Harding et al. (1961, 1962) and by Kimmitt et al. (1961, 1963). Where independent values of n_e and T_e are available, the results agree well with theory. The transition from the Bremsstrahlung continuum to the black body continuum and the cut-off in the vicinity of the plasma frequency have been clearly observed.

The absorption coefficient of a fully-ionized argon plasma for CO_2 laser light at $\lambda \, 10 \cdot 6 \, \mu\text{m}$ was measured by Weisbach & Ahlstrom (1973). The electron temperature

---

* For a description of the θ-pinch and other magnetized plasma configurations see e.g. Krall & Trivelpiece (1973).

and density were in the ranges 1·6 to 2·5 eV and $(2\cdot3$ to $6\cdot4)\times 10^{16}$ cm^{-3}. The measured linear absorption coefficient was found to be approximately twice the calculated value.

8.5.2. Partially-ionized gases

Experimental measurements of continuous emission from shock-heated gases have been reported by Taylor & Caledonia for neon, argon and xenon (1969a) and for atomic oxygen and nitrogen and molecular nitrogen (1969b). The results, which were expressed in terms of the 'effective Z^2' (see Section 8.1.4)) lay in all cases in the range $4\times 10^{-3} < (Z')^2 < 4\times 10^{-2}$. For oxygen and nitrogen they did not agree well with the theoretical results of Kivel (1967b) and Mjolsness & Ruppel (1967).

Batenin & Chinnov (1971) measured the continuous emission spectra from arcs in high-pressure inert gases seeded with potassium. With argon and helium in the wavelength range from 3000 Å to 1 μm, good agreement was found with theoretical values obtained using the theory of Kasyanov & Starostin (1965), after allowing for recombination. Batenin & Chinnov also measured the absorption at 4500 Å, which agreed well with calculated values: in the region of 6000 K the absorption cross-section per atom, per unit electron density, for helium at a density $n_a \sim 2\times 10^{18}$ cm^{-3} was about 6×10^{-40} cm^5, whilst for argon at $n_a \sim 4\times 10^{18}$ cm^{-3} the cross-section was about 2×10^{-40} cm^5. These cross-sections, multiplied by $n_a n_e$, give the linear absorption coefficients.

The absorption of ruby laser light at 6943 Å by a dense, partially-ionized laser produced plasma was measured by Litvak & Edwards (1966) who found that it was as much as 100 times greater than the value calculated on the basis of electron–ion Bremsstrahlung and photo-ionization alone. Tsai et al. (1971) gave numerical values for

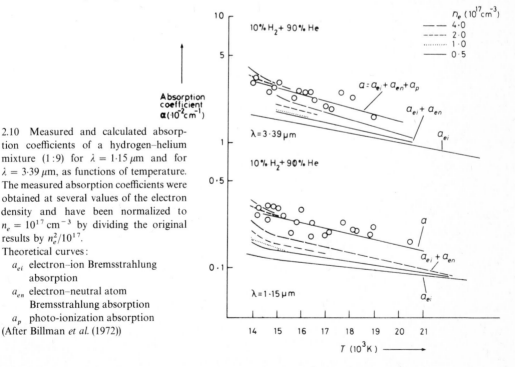

2.10 Measured and calculated absorption coefficients of a hydrogen–helium mixture (1:9) for $\lambda = 1\cdot15\,\mu$m and for $\lambda = 3\cdot39\,\mu$m, as functions of temperature. The measured absorption coefficients were obtained at several values of the electron density and have been normalized to $n_e = 10^{17}$ cm^{-3} by dividing the original results by $n_e^2/10^{17}$.

Theoretical curves:

a_{ei} electron–ion Bremsstrahlung absorption

a_{en} electron–neutral atom Bremsstrahlung absorption

a_p photo-ionization absorption

(After Billman et al. (1972))

the absorption coefficient of hydrogen at 6943 Å, allowing also for electron–neutral atom Bremsstrahlung, over a range of electron temperatures ($\sim 1 < T_e < 100$ eV) and densities ($5 \times 10^{17} \leqslant n_e \leqslant 10^{19}$ cm^{-3}) and found that these data gave better agreement with Litvak & Edwards' experimental observations.

Billman *et al.* (1972) measured the absorption coefficients, at $1\cdot 15$ μm and $3\cdot 39$ μm, of a partially-ionized shock-heated hydrogen–helium mixture (90% He, 10% H_2) with electron temperatures in the range $(1\cdot 4$ to $2\cdot 0) \times 10^4$ K and electron densities between 8×10^{16} and 5×10^{17} cm^{-3}. Their results, shown in Figure 2.10, agree well with theoretical calculations.

9. PARAMETRIC INTERACTIONS

Non-linear effects arise when large-amplitude disturbances propagate in a plasma: high-power lasers can give irradiances which are sufficiently great to produce strongly non-linear behaviour. The non-linear behaviour of a solid dielectric is described in terms of the non-linear susceptibility (see for example Yariv (1967)). The non-linear properties of a plasma may be described in terms of higher-order correlation functions for the charge density fluctuations of the plasma particles: a generalization of the fluctuation–dissipation theorem for non-linear media has been given by Sitenko (1973).

In a non-linear medium coupling can occur between different modes of oscillation, resulting in a transfer of energy from one mode to another. An immense and very significant new region of plasma physics is implied by the existence of these 'parametric interactions'; it has only recently begun to be explored theoretically and has only been touched on experimentally. Coupling between waves is strongest for frequencies in the vicinity of natural resonances of the medium, so it is to be expected that electron plasma waves and ion acoustic waves will be involved in many of the processes. We shall ignore the further complexities which arise in the presence of a magnetic field.

Processes in which an intense beam of transverse radiation is converted into other forms of wave energy are of special interest. Quantum-mechanically, we may think of photons of transverse radiation T with energy $\hbar\omega_0$ and wave vector $\mathbf{k}_0$ being converted into a combination of 'plasmons' (or quanta of electron plasma waves) E, ion acoustic phonons I, and other photons T'. In order to satisfy energy and momentum balance more than one mode of oscillation must be excited in the plasma: for the simplest cases, in which two modes are excited at frequencies ω_1, ω_2 with wave vectors $\mathbf{k}_1$ and $\mathbf{k}_2$, the energy and momentum conservation equations are

$$\omega_0 = \omega_1 + \omega_2 \tag{2.103}$$

$$\mathbf{k}_0 = \mathbf{k}_1 + \mathbf{k}_2. \tag{2.104}$$

(These 'selection rules' need not be satisfied exactly: some degree of 'mismatch' may occur.)

Several processes in this category have been studied theoretically. They include the following:

(a) $T \rightarrow E+I$ the 'parametric instability'.
(b) $T \rightarrow E+E$ the 'two-plasmon instability'.
(c) $T \rightarrow T'+E$ stimulated Raman scattering.
(d) $T \rightarrow T'+I$ stimulated Brillouin scattering.

As a result of processes (a) and (b), laser light energy may under certain conditions be efficiently converted into electron and ion waves in a plasma and thence into particle kinetic energy. On the other hand, processes (c) and (d) may result in a plasma boundary becoming highly reflecting to intense laser light. It is important to know the thresholds and the growth rates for the various competing processes to find out which will dominate a given experimental situation, and much of the theoretical work has been concerned with these questions, being restricted to relatively small-amplitude waves.

When large-amplitude electron and ion waves are generated in a plasma, these waves too may couple non-linearly, producing new waves at other frequencies. The new waves may again couple to other waves and so, in principle at least, *ad infinitum*. The result of this cascade of non-linear interactions is to cause a rapid redistribution of energy in the plasma. A further effect of large-amplitude electrostatic waves is to cause significant changes in the shape of the particle distribution functions. In order to estimate the absorption coefficient of a plasma for intense light due to non-linear processes, it is thus necessary to follow its development in considerable detail. This very difficult task has been done in a limited way by computer simulations which have given useful insights into the saturation of some of the parametric processes.

In order to explain the general principles of parametric interactions we shall give one form of the basic theory which underlies many theoretical treatments (for a full discussion see for example Tsytovich (1970)). We shall then quote a number of results for specific processes.

9.1. BASIC THEORY

In this section we follow closely the general theory of Nishikawa (1968a). (For another general treatment see Lashmore-Davies (1973)). We consider a homogeneous plasma in which a 'pump' wave of very long wavelength at frequency ω_0 is externally maintained at a large constant amplitude A_0, so that

$$A_0(t) = 2A_0 \cos \omega_0 t. \tag{2.105}$$

Let us suppose that in the absence of the pump, and in the absence of any coupling, two 'normal modes' of oscillation can exist in the plasma which satisfy the wave equations

$$\frac{d^2 A_1}{dt^2} + 2\Gamma_1 \frac{dA_1}{dt} + \omega_1^2 A_1 \equiv \mathcal{W}_1[A_1(t)] = 0, \tag{2.106}$$

$$\frac{d^2 A_2}{dt^2} + 2\Gamma_2 \frac{dA_2}{dt} + \omega_2^2 A_2 \equiv \mathcal{W}_2[A_2(t)] = 0. \tag{2.107}$$

Here Γ_1, Γ_2 are the damping coefficients for the uncoupled waves at frequencies ω_1 and ω_2. We shall assume that $|\omega_1| < |\omega_2|$. If we take coupling into account we may write

$$\mathcal{W}_1[A_1(t)] = c_{120} A_2(t) A_0(t) \tag{2.108}$$

$$\mathcal{W}_2[A_2(t)] = c_{201} A_0(t) A_1(t). \tag{2.109}$$

Here c_{120}, c_{201} are coupling coefficients, which we suppose to be real and positive. On taking Fourier transforms of (2.108) and (2.109) with

$$A_j(t) = \int_{-\infty}^{+\infty} e^{i\omega t} A_j(\omega) \, d\omega, \tag{2.110}$$

we have

$$(\omega^2 - \omega_1^2 + 2i\Gamma_1\omega) A_1(\omega) + c_{120} A_0 [A_2(\omega - \omega_0) + A_2(\omega + \omega_0)] = 0 \tag{2.111}$$

and

$$(\omega^2 - \omega_2^2 + 2i\Gamma_2\omega) A_2(\omega) + c_{201} A_0 [A_1(\omega - \omega_0) + A_1(\omega + \omega_0)] = 0. \tag{2.112}$$

According to (2.111), A_1 at frequency ω couples with A_2 at frequencies $(\omega \pm \omega_0)$. But according to (2.112), A_2 at frequencies $(\omega \pm \omega_0)$ will couple not only with A_1 at ω but also with A_1 at $(\omega \pm 2\omega_0)$. A whole hierarchy of coupling equations appears. However, we are concerned with resonant effects and we may assume that $A_1(\omega \pm 2\omega_0)$ is sufficiently far off resonance to be ignored. In deriving (2.111) and (2.112) it was assumed that Γ_1 and Γ_2 were independent of frequency. This is not unreasonable very near resonances.

Thus equations (2.111) and (2.112) give a set of three homogeneous equations in $A_1(\omega)$, $A_2(\omega - \omega_0)$ and $A_2(\omega + \omega_0)$:

$$(\omega^2 - \omega_1^2 + 2i\Gamma_1\omega) A_1(\omega) + c_{120} A_0 [A_2(\omega - \omega_0) + A_2(\omega + \omega_0)] = 0 \tag{2.113}$$

$$[(\omega - \omega_0)^2 - \omega_2^2 + 2i\Gamma_2(\omega - \omega_0)] A_2(\omega - \omega_0) + c_{201} A_0 A_1(\omega) = 0 \tag{2.114}$$

$$[(\omega - \omega_0)^2 - \omega_2^2 + 2i\Gamma_2(\omega + \omega_0)] A_2(\omega + \omega_0) + c_{201} A_0 A_1(\omega) = 0. \tag{2.115}$$

These may be written in matrix form:

$$\begin{bmatrix} \omega^2 - \omega_1^2 + 2i\Gamma_1\omega & c_{120} A_0 & c_{120} A_0 \\ c_{201} A_0 & (\omega - \omega_0)^2 - \omega_2^2 + 2i\Gamma_2(\omega - \omega_0) & 0 \\ c_{201} A_0 & 0 & (\omega + \omega_0)^2 - \omega_2^2 + 2i\Gamma_2(\omega + \omega_0) \end{bmatrix}$$

$$\times \begin{bmatrix} A_1(\omega) \\ A_2(\omega - \omega_0) \\ A_2(\omega + \omega_0) \end{bmatrix} = 0. \tag{2.116}$$

If the determinant of the matrix of coefficients is set equal to zero we obtain an expression for the complex frequency ω, which we may write in the usual form:

$$\omega = \omega_R + i\gamma. \tag{2.117}$$

Thus the coupled system consists of a wave of amplitude A_1 at frequency ω_R near ω_1, and waves of amplitude A_2 at $(\omega_R \pm \omega_0)$ all of which grow, at the expense of pump wave energy, at the same rate γ.

To simplify further, let us suppose that the wave at $(\omega_R + \omega_0)$ is far from resonance and may be neglected, and that ω_R is close to ω_1. Then if Γ_1 and Γ_2 are small the determinantal equation may be approximated by

$$\begin{vmatrix} 2\omega_1(\omega - \omega_1 + i\Gamma_1) & c_{120}A_0 \\ c_{201}A_0 & -2\omega_2(\omega - \omega_0 + \omega_2 + i\Gamma_2) \end{vmatrix} = 0. \tag{2.118}$$

It is convenient now to define two new parameters. One measures the deviation from perfect frequency matching:

$$\Delta = \omega_0 - \omega_1 - \omega_2. \tag{2.119}$$

The other is a measure of the intensity of the incident wave:

$$\Psi = \frac{c_{120}c_{201}}{\omega_1 \omega_2} A_0^2. \tag{2.120}$$

The real and imaginary parts of equation (2.118) yield the following two equations:

$$(\omega_R - \omega_1)(\omega_R - \omega_1 - \Delta) + \frac{\Psi}{4} = (\gamma + \Gamma_1)(\gamma + \Gamma_2) \tag{2.121}$$

and

$$\omega_R = \omega_1 + \left(\frac{\gamma + \Gamma_1}{2\gamma + \Gamma_1 + \Gamma_2}\right)\Delta. \tag{2.122}$$

On substituting expression (2.122) for ω_R into (2.121) we find that

$$\Psi = 4(\gamma + \Gamma_1)(\gamma + \Gamma_2)\left[1 + \frac{\Delta^2}{(2\gamma + \Gamma_1 + \Gamma_2)^2}\right]. \tag{2.123}$$

In order that the instability should grow, γ must be positive: at the threshold $\gamma = 0$. The value of Ψ at the threshold, Ψ_{th}, is least when $\Delta = 0$:

$$\Psi_{th}(\Delta) = \Psi_{th}(0)\left[1 + \frac{\Delta^2}{(\Gamma_1 + \Gamma_2)^2}\right] \tag{2.124}$$

where

$$\Psi_{th}(0) = 4\Gamma_1\Gamma_2. \tag{2.125}$$

Thus if either Γ_1 or Γ_2 are zero, the instability is excited even at very low pump intensity. At the threshold, the frequencies of A_1 and A_2 are $\omega_R = \omega_1 + [\Gamma_1/(\Gamma_1 + \Gamma_2)]\Delta$

and $\omega_R - \omega_0 = -\omega_2 - [\Gamma_2/(\Gamma_1 + \Gamma_2)]\Delta$ respectively, which reduce to ω_1 and $-\omega_2$ when $\Delta = 0$.

For given ω_1 and ω_2, expression (2.124) may be used to determine the range of frequencies over which the instability will grow with a given pump intensity Ψ:

$$|\omega_0 - \omega_1 - \omega_2| < \Gamma_1 + \left[\frac{\Psi}{\Psi_{th}(0)} - 1 \right]^{\frac{1}{2}} \Gamma_2. \tag{2.126}$$

Let us now find the maximum growth rate for a given pump intensity, i.e. for a given value of Ψ, above the threshold. Equation (2.123) shows that γ has a maximum value γ_{max} when $\Delta = 0$, given by

$$\gamma_{max} = \tfrac{1}{2}\{-(\Gamma_1 + \Gamma_2) + [(\Gamma_1 + \Gamma_2)^2 + \Psi]^{\frac{1}{2}}\}. \tag{2.127}$$

Thus when $\Psi \gg \Gamma_1, \Gamma_2$ the maximum growth rate is simply

$$\gamma_{max} = \frac{\Psi^{\frac{1}{2}}}{2}. \tag{2.128}$$

If the full determinantal equation of (2.116) is solved, it is found that in addition to a solution of the form (2.117) there is also a solution of form $\omega = i\gamma$, implying that in this case A_1 grows at zero frequency, while A_2 grows at the frequency of the pump wave. The purely growing instability only occurs when $\omega_0 < \omega_2$: the threshold is determined by the parameter $\delta = \omega_0 - \omega_2$ and occurs at

$$\Psi_{th}(\delta) = -\omega_1 \frac{(\Gamma_2^2 + \delta^2)}{\delta}. \tag{2.129}$$

It is independent of Γ_1. The minimum value of the threshold occurs when $\delta = -\Gamma_2$, and equals

$$\Psi_{th} = 2\omega_1 \Gamma_2. \tag{2.130}$$

The maximum growth rate, well above threshold, is

$$\gamma_{max} = \left(\frac{\Psi \omega_1}{2} \right)^{\frac{1}{2}}. \tag{2.131}$$

For the other solution, of form $\omega = \omega_R + i\gamma$ the minimum threshold for small damping ($\omega_1 \gg \Gamma_1, \omega \gg \Gamma_2$) is again given (as in (2.125)) by

$$\Psi_{th} = 4\Gamma_1 \Gamma_2 \tag{2.132}$$

and is therefore lower than that for the purely growing instability (2.130). The maximum growth rate is now

$$\gamma_{max} = \frac{\sqrt{3}}{2} \left(\frac{\Psi \omega_1}{4} \right)^{\frac{1}{2}}. \tag{2.133}$$

These results are very general and may be applied to several different sets of oscillations if the appropriate coupling coefficients can be found.

9.2. PARAMETRIC INSTABILITIES

9.2.1. Onset of instability

Homogeneous plasma

As an example of the application of the general theory let us consider (still following Nishikawa's treatment (1968b)) the interaction

$$T \rightarrow E + I.$$

In this case the first of the uncoupled equations (2.106) describes the behaviour of low-frequency ion density fluctuations, of amplitude n'_i, near the ion acoustic frequency ω_{ia}. The second uncoupled equation (2.107) describes high-frequency electron density fluctuations of amplitude n'_e near the Bohm–Gross frequency ω_{BG} (expression (1.127)), which is close to the electron plasma frequency, ω_{pe}. Since $\omega_0 = \omega_1 + \omega_2$, the transverse pump wave will also be in the region of the electron plasma frequency. The wavelength of the pump radiation may be assumed to be infinite, so that $k_0 = 0$. Thus $k_1 = -k_2$ and the wave numbers of the two coupled oscillations will be equal and will be called k.

A proper derivation, from the hydrodynamic equations, of the coupling of the electron and ion density fluctuations by the pump radiation is lengthy. However, the interaction may be understood from a very simple model (Kruer & Dawson (1972)). A low-frequency ion fluctuation of a given wave number causes low-frequency variations in the electron density with the same wavenumber, because the electron density follows the ion density (we shall assume $Z = 1$). An electron in the high-frequency electric field of the pump radiation of amplitude E_0 oscillates back and forth with a spatial amplitude $x_0 = eE_0/m\omega_0^2$. Thus the effect of the pump field is to move the ion-induced electron density fluctuations rapidly past a given point, resulting in an additional high-frequency contribution to the electron density fluctuations, of amplitude

$$\delta n'_{ek} = n'_{ik}(x + x_0) - n'_{ik}(x). \tag{2.134}$$

The x-dependence of n'_{ik} is of the form

$$n'_{ik}(x) = n'_{ik} \exp{(ikx)} \tag{2.135}$$

so

$$\delta n'_{ek} = x_0 \frac{dn'_{ik}}{dx} = \frac{eE_0}{m_e \omega_0^2} ikn'_{ik}. \tag{2.136}$$

A full analysis of this coupling gives a very similar result: the second of the coupled equations (2.109) then becomes

$$\frac{\partial^2 n'_{ek}}{\partial t^2} + \omega_{BG}^2 n'_{ek} + v_{ek} \frac{\partial n'_{ek}}{\partial t} = \frac{ikeE_0}{m_e} n'_{ik}. \tag{2.137}$$

Here v_{ek} is the effective electron collision frequency, allowing for Landau damping.

The corresponding equation for the ions is

$$\frac{\partial^2 n'_{ik}}{\partial t^2} + \omega_{ia}^2 n'_{ik} + v_{ik}\frac{\partial n'_{ik}}{\partial t} = -\frac{ikeE_0}{m_i}n'_{ek}. \qquad (2.138)$$

Thus the results of the previous section for thresholds and growth rates may be applied by substituting n'_{ik} for A_1, n'_{ek} for A_2, ω_{ia} for ω_1, ω_{BG} for ω_2, kE_0 for A_0, $-ie/m_i$ for c_{120}, ie/m_e for c_{201}, v_{ik} for $2\Gamma_1$ and v_{ek} for $2\Gamma_2$.

The solution of form $\omega = i\gamma$ which occurs when the ion oscillations have zero frequency is known as the 'oscillating two-stream' instability and was first noted by Nishikawa (1968a) and by Kaw & Dawson (1969). The other solution, of form $\omega = \omega_R + i\gamma$, in which a pump photon decays to a plasmon and an ion acoustic phonon, is known as the 'parametric decay' instability. It was first studied by Oraevski & Sagdeev (1962), who considered a longitudinal pumping wave (the theory is the same for a transverse or a longitudinal wave if $k_0 = 0$). Silin (1965), using the hydrodynamic equations, studied the growth of electron plasma oscillations when a transverse wave propagates through a cold plasma. The growth rates he obtained, ignoring damping, are valid only when the pump field is well above threshold. Dubois & Goldman (1965), and Goldman (1966) used a Green's function method to investigate the case of a weak pump field ($\eta \ll 1$, see (2.89)) and found the threshold for growth of the instability, allowing for Landau damping. Lee & Su (1966) obtained similar results using a hydrodynamic model. Jackson (1967) used the linearized Vlasov equation to investigate the instability without restriction on the pump field amplitude. His treatment has been criticized by Nishikawa (1968b) on the grounds that the derivation of the dispersion relation was incorrect.

We have already described Nishikawa's analysis: his results were verified and extended to arbitrary pump fields by Sanmartin (1970) using the Vlasov equation. The case of finite pump wavelength ($k_0 \neq 0$) has been discussed by Tsytovich (1967) and Makhankov & Tsytovich (1973) and by Ott (1971). The theory has also been extended to deal with some of the many complex situations occurring in magnetized plasmas (Forslund et al. (1972), Gorbunov & Silin (1969), Kindel et al. (1972), Kovrizhnykh et al. (1966)).

The stability properties of a plasma against both these instabilities for $k_0 = 0$ were computed by Freidberg & Marder (1971). Some of their results are shown in Figures 2.11(a) to (e). Figure 2.11(a) shows the region of instability in $(\omega_p/\omega_0, kx'_0)$ space for a cold hydrogen plasma. Here $\omega_p = (\omega_{pe}^2 + \omega_{pi}^2)^{\frac{1}{2}}$, k is the wavenumber of the electron oscillation, ω_0 is the incident light wave frequency and

$$x'_0 = \frac{eE_0}{\omega_0^2}\left(\frac{m_i + m_e}{m_i m_e}\right) \simeq \frac{eE_0}{\omega_0^2 m_e}$$

is a scale length approximately equal to the amplitude of electron excursions in the pump field. As the plasma temperature rises the unstable regions shrink. The field strength parameters for Figures 2.11(b) and (c) are $\eta = 10$ and 2 respectively, with $T_e = 10T_i$ in each case. Instabilities occur only at long wavelengths, cutting off at about $k\lambda_D = 1$. As kx'_0 increases, instabilities occur at lower densities. The Bohm–Gross dispersion relation, expression (1.127), separates the decay instability (on the

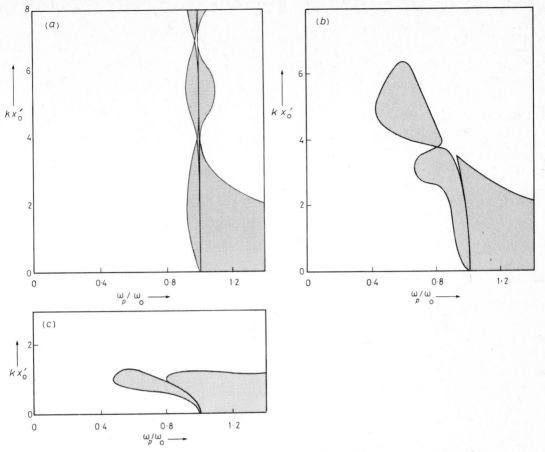

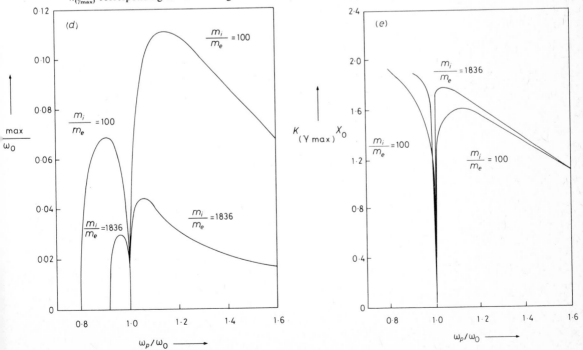

2.11 Parametric instability of plasmas. (a) to (c): regions of instability in $(\omega_p/\omega_0, kx_0')$-space for hydrogen plasmas $(m_i/m_e = 1836)$: (a) cold plasma, (b) $T_e = 10T_i$, $\eta = 10$, (c) $T_e = 10T_i$, $\eta = 2$. (d) maximum growth rates as functions of ω_p/ω_0 for $m_i/m_e = 1836$ and 100. (e) wave numbers $k_{(\gamma\max)}$ corresponding to maximum growth rates. (After Freidberg & Marder (1971))

left, in the lower density region) and the two-stream instability. At high temperatures or in weak fields the latter extends not only into the overdense region but also into regions of lower density. The maximum growth rates are greatest in the strong field limit and are then independent of temperature: they are shown in Figure 2.11(d). The wavenumbers corresponding to maximum growth rates are shown in Figure 2.11(e). The maximum possible growth rate depends on the mass ratio of ions and electrons: for the two-stream instability it is given approximately by

$$\frac{\gamma_{max}}{\omega_{pe}} \simeq \frac{1}{2}\left(\frac{m_e}{m_i}\right)^{\frac{1}{3}},$$ (2.139)

while for the decay instability it is given by

$$\frac{\gamma_{max}}{\omega_{pe}} \simeq \frac{1}{3}\left(\frac{m_e}{m_i}\right)^{\frac{1}{3}}.$$ (2.140)

Provided the pump wave is not too strong, i.e. if $\eta \lesssim 7$, the wave vectors of the modes with maximum growth rates are parallel to the electric field $\mathbf{E}_0$ of the pump. For stronger fields, according to De Groot & Katz (1973), the modes with maximum growth rates propagate at an angle ϕ_E to $\mathbf{E}_0$, such that $\cos \phi_E \simeq 7/\eta$.

The threshold pump irradiance for the onset of the instabilities in homogeneous plasmas has been estimated, assuming $\eta \ll 1$, by several authors.* For the oscillating two-stream mode it is found to be given by

$$\eta^2 = 4\left(\frac{v_{ek}}{\omega_{pe}}\right)\left(1 + \frac{T_i}{T_e}\right),$$ (2.141)

while for the parametric decay mode it is given at high temperatures (such that $v_{ek} < \omega_i$) by

$$\eta^2 = 4\frac{v_{ek}}{\omega_{pe}}\frac{v_{ik}}{\omega_i}$$ (2.142)

and at low temperatures ($v_{ek} > \omega_i$) by

$$\eta^2 = \frac{8\sqrt{3}}{9}\frac{v_{ek}^2 v_{ik}}{\omega_{pe}\omega_i^2}.$$ (2.143)

Here ω_i is the ion wave frequency and v_{ek}, v_{ik} are the effective electron–ion and ion–ion collision frequencies respectively, taking Landau damping into account. Yamanaka et al. (1972c) have calculated the threshold irradiances for neodymium laser light incident on a deuterium plasma (Figure 2.12). At low temperatures ($T_e \lesssim 200\,\text{eV}$) the two-stream instability has the lower threshold, but at higher temperatures the parametric decay instability is more readily excited, especially when $T_e \gg T_i$. The threshold is always below $\eta = 1$ so the binary collision frequency included in v_{ek} remains linear.

---

* Goldman (1966), Sanmartin (1970), Nishikawa (1968b), Kaw & Dawson (1969), Kaw et al. (1970), Andreev et al. (1969), Pustovalov & Silin (1970), Dubois & Goldman (1967), Yamanaka et al. (1972c), Meyer & Shatas (1972): see also the review by Galeev & Sagdeev (1973).

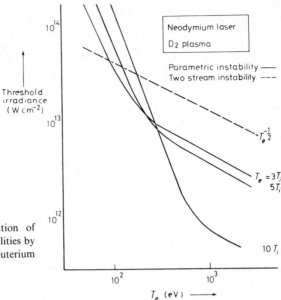

2.12 Threshold irradiances for excitation of parametric decay and two-stream instabilities by neodymium laser light incident on a deuterium plasma. (After Yamanaka *et al.* (1972c))

Inhomogeneous plasma

In an inhomogeneous plasma these threshold conditions must be modified. The unstable region has a finite spatial extent so that energy is carried away by electron plasma waves and the threshold pump irradiance rises. Perkins & Flick (1971) considered a plasma with a density gradient scale length L and found that when $v_{ek} \ll \omega_i$ the two-stream instability threshold is given by

$$\eta^2 = \left(1 + \frac{T_i}{T_e}\right)\left(\frac{4v_{ek}}{\omega_{pe}} + \frac{2}{k_{\|}L}\right)$$

$$\geqslant \left(1 + \frac{T_i}{T_e}\right)\left(\frac{4v_{ek}}{\omega_{pe}} + \frac{8\lambda_D}{L}\right), \tag{2.144}$$

the second expression being due to $k_{\|}$, the component of the electron plasma wave vector parallel to the electric field, having an upper limit of $1/4\lambda_D$ to avoid strong Landau damping.

The decay mode becomes a convective instability (see for example Briggs (1964) or Akhiezer *et al.* (1967)) in the presence of a density gradient. Although the amplitude of the instability at a fixed point does not grow indefinitely, the limited unstable region acts as an amplifying medium which increases the intensity of waves propagating through it. The threshold may then be defined as the electric field which causes an intensity amplification of e^5, and was found by Perkins and Flick to be given by

$$\eta^2 = \frac{8}{k_{\|}L}\left(1 + \frac{3T_i}{T_e}\right)\left(\frac{v_{ik}}{\omega_i}\right)^{\frac{1}{2}} + 3 \cdot 2\left(\frac{v_{ik}}{\omega_i}\right)\left(\frac{v_{ek}}{\omega_{pe}}\right). \tag{2.145}$$

When the density gradient is important, unless $T_e \gg T_i$ this threshold is not very different from that for the two-stream instability.

The effects of inhomogeneities have also been discussed by Aliev et al. (1972) and by Amano & Okamoto (1969), whose results have been criticized by Perkins & Flick (1971).

Effects of incoherent pump wave

The theoretical calculations described above assumed a perfectly monochromatic, coherent pump wave, whereas the output from a real laser has a finite bandwidth. The effect of limited pump wave coherence on the growth rate for the parametric decay instability was considered by Tsytovich & Shvartsburg (1967) and by Valeo & Oberman (1973), who showed how the growth rate falls as the coherence deteriorates. The effect of incoherence on the threshold irradiance was discussed by Tamor (1973). If the number of phase jumps in the pump wave per unit time is v_ϕ, and all phase jumps are equally probable, the effect on the parametric equations is simply to replace v_{ek} by $v_{ek} + 2v_\phi$. In terms of pump wave bandwidth $\Delta\omega_0$, $v_\phi \sim \Delta\omega_0/2$. For a hydrogen plasma at $T_e = 1$ keV, pumped by neodymium laser light at λ 1·06 μm, the threshold for the parametric instability is greater by a factor of 2·7 for light with a bandwidth of $10^{-3}\omega_0$ than for perfectly coherent light.

9.2.2. Development of the instability

The development of these instabilities must be followed in order to estimate their contribution to absorption of incident radiation energy and to find out how the energy absorbed is ultimately distributed among the plasma particles. Most of the theoretical studies made so far refer to homogeneous plasmas.

A rough estimate of the non-linear absorption coefficient due to the two-stream instability for pump fields not too far above threshold ($\eta \ll 1$) was obtained by Kaw & Dawson (1969) by means of the following argument. The threshold pump field (given by expression (2.141)) varies as the square root of the electron–ion collision frequency v_{ek}. As the instability develops, strong ion fluctuations occur which interact with the pump wave and as a result (Dawson & Oberman (1962)) the effective collision frequency increases. Eventually the collision frequency rises to a value $v_{ek_{max}}$ at which the pump field is no longer above the threshold value. The instability then ceases to grow. Here $v_{ek_{max}} \simeq \eta^2 \omega_{pe}/4$. The absorption coefficient for the pump wave is related to the effective collision frequency (expression (1.42)) and thus the maximum absorption coefficient due to this instability is given by

$$a_{max} \simeq \frac{30I}{n_e^{\frac{1}{2}} T_e} \text{ cm}^{-1}, \tag{2.146}$$

where I is in watts cm^{-2}, n_e in cm^{-3} and T_e in eV. The normal Bremsstrahlung absorption coefficient (expression (2.84)) may be written approximately, for ω not too far from ω_{pe} and $\bar{g} \sim 5$, as

$$a_{\omega_{pe}} \simeq 2 \times 10^{-15} \frac{n_e}{T_e^{\frac{3}{2}}} \text{ cm}^{-1}, \tag{2.147}$$

and is therefore exceeded by the parametric absorption when

$$I \gtrsim 10^{-16} \frac{n_e^{\frac{3}{2}}}{T_e^{\frac{1}{2}}} \, \text{W cm}^{-2}. \qquad (2.148)$$

The non-linear evolution of a parametrically unstable plasma has been treated analytically for weak pump fields using weak turbulence theory* and also for moderate pump fields using a theory involving the broadening of the wave-particle resonance by plasma wave turbulence (Bezzerides & Weinstock (1972)) (for a discussion of wave propagation in turbulent plasmas see Kadomtsev (1965)). For strong pump fields computer simulations of plasmas (Morse & Nielson (1971)) have provided some interesting results (De Groot and Katz (1973), Kruer *et al.* (1970), Kruer & Dawson (1972)). The computer is used to follow the self-consistent response of a large number of charged particles (in one dimension, usually) to a driving electric field. Figure 2.13 shows the energy of the electron plasma wave and the total energy of the plasma computed by Kruer & Dawson (1972) as functions of the duration of a constant oscillatory pump field of amplitude E_0. In this one-dimensional example $\eta = 0.5$,

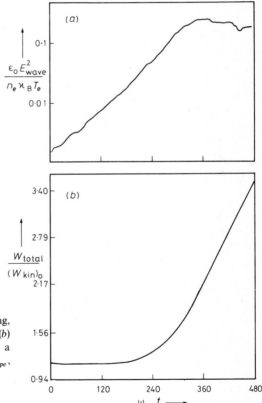

2.13 Computer simulation of plasma heating, showing (a) self-consistent wave energy and (b) total plasma kinetic energy, averaged over a plasma oscillation. $m_i/m_e = 100$, $\omega_0 = 1.04 \, \omega_{pe}$, $\eta = 0.5$. (After Kruer & Dawson (1972))

* See Pustovalov & Silin (1970), Dubois & Goldman (1967, 1972a, b), Kuo & Fejer (1972), Valeo *et al.* (1972), Kruer & Valeo (1973) and Fejer & Kuo (1973).

$\omega = 1.04\omega_{pe}$, $T_e/T_i = 30$ and (to facilitate computing) $m_i/m_e = 100$. Initially, the plasma heats slowly as the result of binary collisions, while the wave energy rises exponentially. After a time of about $t = 250\omega_{pe}^{-1}$, the wave energy begins to saturate and the plasma energy immediately begins to rise rapidly as anomalous heating sets in. The pump wave energy is efficiently converted into plasma oscillations which at large amplitude are rapidly Landau damped and can trap a significant number of electrons (Dawson & Shanny (1968)), creating a high-energy tail on the electron distribution function. As plasma heating continues it is found that energy is being fed into plasma oscillations of progressively longer wavelengths which, because of their higher phase velocity, experience less Landau damping and become the modes of maximum growth rate.

For large values of η the wave field at saturation becomes strong enough to trap a large proportion of the electrons. One-dimensional computations of the time variation of the wave energy and the electron and ion 'temperatures' by De Groot & Katz (1973) are shown in Figure 2.14 for a plasma with $m_i/m_e = 100$, $\omega_0 = \omega_{pe}$, when $\eta = 7$. The 'temperatures' are defined by

$$T_{e,i} \equiv m_{e,i}[\langle v^2 \rangle_{e,i} - \langle v \rangle_{e,i}^2]. \tag{2.149}$$

In this case the development of the plasma is more rapid, and saturation occurs at

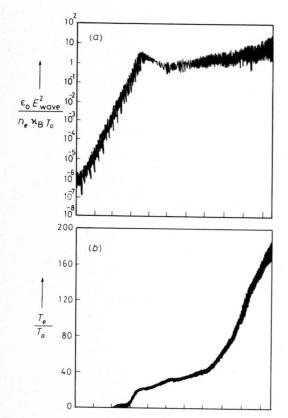

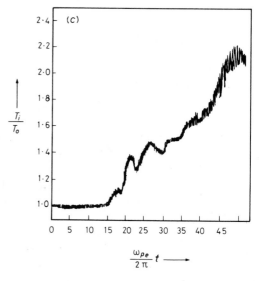

2.14 Computer simulation of plasma heating, showing (a) self-consistent field energy, (b) the electron temperature and (c) the ion temperature as functions of time. $m_i/m_e = 100$, $\omega_0 = \omega_{pe}$, $T_{i0} = T_{e0} = T_0$, $\eta = 7$. (After De Groot & Katz (1973))

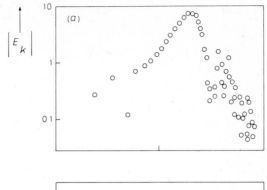

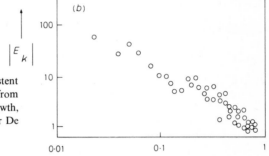

2.15 Wavenumber spectrum of self-consistent electric field as function of $k\lambda_{D_o}$, obtained from computer simulation, (a) during linear growth, (b) when turbulence is well developed. (After De Groot & Katz (1973))

a time of about $17 \times 2\pi/\omega_{pe}$. The wavenumber spectrum of the self-consistent electric field is shown in Figure 2.15 for two times, the first during the steady growth phase and the second at $t = 40 \times 2\pi/\omega_{pe}$, when turbulence is well developed. The maximum in the early spectrum occurs at a wavenumber very close to $k\lambda_{D_o} = 0.23$ where the most unstable electron plasma mode is predicted by linear theory (λ_{D_o} is the initial value of λ_D). At the later time the maximum has moved to the mode with the lowest available wavenumber. The most unstable modes have trapped electrons and decayed. Strongly non-linear coupling occurs between modes and a turbulent state has developed. The wave energy has ceased to grow. The rapid coupling of energy between modes is probably connected with the presence of strong ion density fluctuations (Kruer & Dawson (1970), Kainer et al. (1972), Kruer (1972)) which will also enhance the rate of energy transfer from the wave field to particles.

While the electron temperature is low the entire electron distribution function gains energy in Landau damping the large-amplitude waves. As a result the electron 'temperature' rises rapidly, the electron velocity distribution remaining quasi-Maxwellian. This, however, causes η to fall at constant E_0, so that when $t = 48 \times 2\pi/\omega_{pe}$ the value of η is only 0.52. The situation is then very similar to that of the weak-field case of Figure 2.14, though at a higher temperature, and a high-energy tail again appears on the electron velocity distribution.

The ion temperature rises much more slowly than the electron temperature.

Although the external pump frequency is set equal to the electron plasma frequency, so that only the two-stream instability is excited, a frequency analysis of the turbulent plasma shows a maximum in the spectrum close to the ion plasma frequency. The phase velocity of these waves is much greater than the average ion velocity, so no Landau damping is expected unless a very large amplitude is attained, when ion trapping may occur (Sleeper et al. (1972)).

Effective collision frequencies v_{ei}^* for the rate of energy transfer from the laser field to the plasma have been derived from the simulations. A composite estimate for a deuterium plasma suggests that v_{ei}^*/ω_{pe} rises (initially approximately quadratically) with increasing η to a maximum of about 0·04 when η is about 0·5, and then falls linearly to about 4×10^{-3} as η rises to 10. For $\eta > 1$, therefore,

$$\frac{v_{ei}^*}{\omega_{pe}} \simeq \frac{4 \times 10^{-2}}{\eta}. \qquad (2.150)$$

The absorption coefficient a then becomes proportional to $I^{-\frac{1}{2}}$.

Shearer & Duderstadt (1973) calculated the efficiency with which a semi-infinite deuterium plasma absorbs normally incident light, allowing for both collisional and parametric processes, and concluded that the overall absorption efficiency increases as the wavelength is decreased.

When a solid target is being heated by a laser pulse, the interaction region is strongly inhomogeneous. Hot electrons flow out of the unstable layer and are replaced by cold ones. As a result the electron temperature saturates in the unstable layer, and the energy absorbed from the laser ultimately produces an extended region of hot electrons, together with hydrodynamic expansion. Ions move through the unstable layer much more slowly, with approximately the acoustic velocity, and are heated en route by turbulence. Bodner et al. (1973) showed by a computer simulation that under these conditions the ion temperature can approach the electron temperature. The flow of electrons through the unstable layer was simulated by resetting the velocities of a proportion of the electrons, selected randomly, to their initial values. It was found that the ion heating rate was not very sensitive to the rate at which electrons are reset. The results shown in Figure 2.16 are for a plasma of density 10^{21} cm^{-3}, with $m_i/m_e = 100$, $\eta = 0.36$, $\omega_0 = \omega_{pe}$ and $T_{e0} = 1$ keV. As may be seen, the electron temperatures saturates when $t \simeq 130 \times 2\pi/\omega_{pe}$ and then rises only slowly (the resetting procedure does not properly reduce the high-energy tail of the electron distribution). The ion temperature rises almost linearly with time long after the electron temperature has saturated: the ion distribution remains roughly Maxwellian. The ion heating rate varies with the ion mass as $m_i^{-\frac{1}{2}}$ and is directly proportional to the incident laser irradiance: for a deuterium plasma, with I in watts cm^{-2}.

$$\frac{dT_i}{dt} \simeq 10^{-16} I \text{ keV per picosecond.} \qquad (2.151)$$

The time spent by an ion in the heating region varies roughly as $I^{-\frac{1}{2}}$, so the final temperature should vary as $I^{\frac{2}{3}}$.

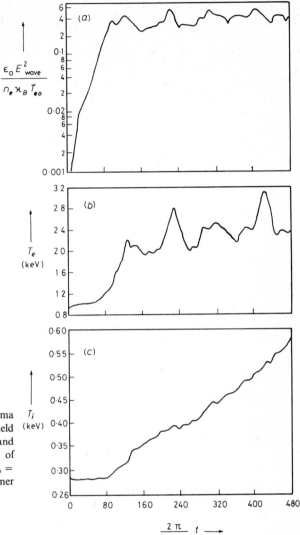

2.16 Computer simulation of plasma heating showing (a) self-consistent field energy, (b) the electron temperature and (c) the ion temperature as functions of time. $n_e = 10^{21}$ cm^{-3}, $T_{e0} = 1$ keV, $\omega_0 = \omega_{pe}$, $m_i/m_e = 100$, $\eta = 0.36$. (After Bodner et al. (1973))

9.3. 'TWO PLASMON' INSTABILITY

Another parametric process which may be important in plasma heating may occur when $\omega_{pe} \simeq \omega_0/2$. It involves the decay of a phonon into two plasmons and has been discussed by Goldman (1966), Jackson (1967), Galeev et al. (1972), Amano & Okamoto (1969) and Rosenbluth (1972). Only the electrons are involved, so in a homogeneous plasma there can be no instability in the dipole approximation ($k_0 = 0$). Jackson (1967) found that the threshold irradiance for this instability with $k_0 = \omega_0/c$ is

$$I \simeq 8 \times 10^{-2} \left(\frac{v_{ek}}{\omega_{pe}}\right)^2 n_e \ \text{W cm}^{-2}, \tag{2.152}$$

and that the most unstable modes are at 45° to $\mathbf{E}_0$ and $\mathbf{k}_0$.

Amano & Okamoto (1969) suggested that in a plasma with a density gradient of scale length L, k_0 could be replaced by $1/L$. Another treatment by Rosenbluth (1972) suggests that the threshold irradiance for significant spatial amplification for this instability is

$$I \gg 1 \cdot 7 \times 10^{15} \frac{T_e}{L\lambda_0} \ \text{W cm}^{-2}, \tag{2.153}$$

where T_e is in keV and L and λ_0 are in microns.

9.4. COUPLING OF TWO LASER BEAMS TO A PLASMA WAVE

Two laser beams with frequencies ω_1 and ω_2, such that $\omega_1, \omega_2 \gg \omega_{pe}$ and $\omega_1 - \omega_2 = \omega_{pe}$, can couple parametrically to a plasma wave (Kroll et al. (1964)). The heating due to this process has been discussed by a number of authors.* According to Rosenbluth & Liu (1972), the heating rate is greater for opposing (antiparallel) beams than for parallel beams by a factor $4(\omega_{1,2}/\omega_{pe})^2$. The Manley–Rowe relations (Manley & Rowe (1956, 1959)) show that the maximum proportion of the incident energy that can be converted to plasma wave energy is $\sim \omega_{pe}/\omega_1$. The proportion of incident energy transferred to an ion wave when $\omega_1 - \omega_2 = \omega_{ia}$ will be even smaller. However, a cascade of similar processes is possible (Cohen et al. (1972), Kaufman & Cohen (1973)) which might lead to more effective heating.

The coupled mode equations have also been derived using a Lagrangian formulation by Boyd & Turner (1972). Coherence effects have been discussed by Fisher & Hirschfield (1973).

9.5. STIMULATED RAMAN AND BRILLOUIN SCATTERING

The previous section discussed the excitation of a plasma wave by the interaction of two light waves whose frequencies are separated by ω_{pe}. A single incident light wave scattered from plasma density fluctuations at ω_{pe} is shifted in frequency by ω_{pe} (a process discussed by Bornatici et al. (1969)). Scattered light can then interact with incident light to enhance both the plasma fluctuations and the scattered wave. The incident beam thus generates a plasma wave from which it is itself scattered. This process is known as stimulated Raman scattering (or stimulated Thomson scattering) when an electron plasma wave is involved. A similar process can occur involving an ion acoustic wave, and is known as stimulated Brillouin scattering (or stimulated Mandelshtam–Brillouin scattering).

Goldman & Dubois (1965) showed that for Raman scattering the gain G_s, which

---

* Montgomery (1965), Sjölund & Stenflo (1967), Cohen et al. (1972), Meyer (1972), Rosenbluth & Liu (1972), Beaudry & Martineau (1973), Chang et al. (1973), Lee et al. (1973), Fuchs et al. (1973).

equals the fraction dN_s/N_s of scattered photons in mode s produced in a distance dz along the incident pump beam of irradiance I_0, is given by

$$G_s \simeq 10^{-4} \frac{I_0}{k_0^3} \frac{(n_e \kappa_B T_e)^{\frac{1}{2}}}{\ln (\kappa_B T_e/\hbar\omega_{pe})} \left(\frac{k}{k_D}\right)^2 \frac{1+\cos^2\theta}{2} \text{ cm}^{-1}. \qquad (2.154)$$

Here θ is the scattering angle, $k_D = 1/\lambda_D$, n_e is in cm^{-3}, $\kappa_B T_e$ and $\hbar\omega_{pe}$ are in eV and I is in W cm^{-2}. In order that Landau damping shall be small, $k/k_D \lesssim \frac{1}{6}$. Provided $\theta \gtrsim (v_T/c)(k_D/k_0)$,

$$\frac{k}{k_D} \simeq 2\frac{k_0}{k_D} \sin\left(\frac{\theta}{2}\right). \qquad (2.155)$$

Optimum gain then occurs when $k/k_D \simeq \frac{1}{6}$ for back-scattered light ($\theta = \pi$) at $k_D \simeq 12k_0$. The electron density for optimum gain therefore depends on temperature and on the laser wavelength: the condition is

$$\left(\frac{n_e}{T_e}\right)^{\frac{1}{2}} \lambda \simeq 5 \times 10^8, \qquad (2.156)$$

where T_e is in eV, n_e in cm^{-3} and λ in microns. Subject to this condition, the optimum gain occurs at the highest density. For a neodymium laser with $n_e = 10^{20}$ cm^{-3}, $T_e = 400$ eV, $G_s \simeq 3 \times 10^{-10} I$, and could be as high as 3×10^5 cm^{-1} for $I = 10^{15}$ W cm^{-2}.

When light is incident on a plasma boundary with a steep density gradient the matching conditions hold only over a limited distance. The propagation of the parametrically generated waves out of this region of the plasma restricts the growth of the instability (Rosenbluth (1972), Galeev et al. (1973)). This raises the threshold for backscattering of light incident in a direction parallel to the density gradient (Pesme et al. (1973)). Liu et al. (1973) considered sidescattering at 90° to the direction of the incident light. In this case the scattered light stays much longer in the unstable region before refracting out (or reaching the edge of the laser beam) and may again represent a significant reduction in the fraction of incident light energy that reaches the strongly absorbing critical density layer. The threshold power for Raman sidescattering is lower by a factor $18\kappa_B T_e/m_e c^2$ than that for backscattering, and is comparable to that for Brillouin backscattering. Computer simulations by Klein et al. (1973) confirmed these results. Arnush & Kennel (1973) have pointed out that an intense pump wave will affect the real refractive index and may trap an electrostatic wave.

Detailed one-dimensional analyses of backscattering in plasmas with two different density gradients were given by Tsytovich et al. (1973). Saturation must be considered (Litvak & Trakhtengerts (1971), Vinogradov et al. (1973)). A computer simulation study has been made by Forslund et al. (1973) of backscattering of a light wave of fixed amplitude incident from the right on a plasma with the density profile shown in Figure 2.17(a). The density is in units of n_{ec}, where $\omega_{pe} = \omega_0$ at n_{ec}. The transmitted wave amplitude is shown in Figure 2.17(b) as a function of time, t being in units of $1/\omega_0$. Electron and ion phase space plots show that the first transmission minimum, which occurs at $t \simeq 500/\omega_0$, is due to reflection caused by Raman scattering. This

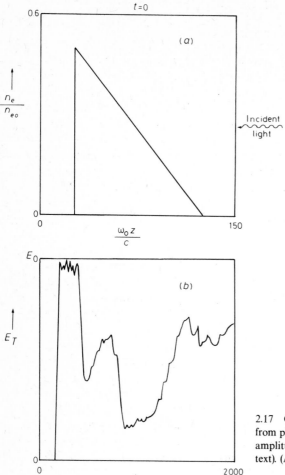

$\dfrac{n_e}{n_{eo}}$

$t=0$

(a)

Incident
light

$\dfrac{\omega_o z}{c}$

E_0

(b)

E_T

$\omega_o t$ ⟶

2.17   Computer simulation of effects of backscattering
from plasma with density profile shown in (a): (b) shows
amplitude of transmitted wave as function of time (see
text). (After Forslund *et al.* (1973))

saturates as a result of electron trapping and heating. The second transmission
minimum, which occurs at $t \simeq 1000/\omega_0$, is due to reflection caused by Brillouin
scattering. By $t = 2000/\omega_0$, the intensity transmission coefficient has settled down
to about 0·4.

Though much more information is needed about these processes, it seems clear
that they may deflect away a significant fraction of intense light energy incident on
a dense plasma surface.

10. OTHER NON-LINEAR INTERACTIONS

10.1. STIMULATED COMPTON SCATTERING

Scattering of a photon from a light beam by an electron may be regarded as absorption followed by emission at a new frequency in a new direction. If a second intense beam coincides in frequency and direction with the scattered light the emission will be stimulated (Dreicer (1964)). The scattering rate into the second beam becomes proportional to the product of the two beam intensities. If the two beams have frequencies ω_1 and $\omega_2 < \omega_1$, when scattering occurs from beam 1 into beam 2 the energy difference $\hbar(\omega_1 - \omega_2)$ is delivered to the scattering electron. The heating rate per particle is independent of density and varies as the product of the two beam intensities. This process was discussed in detail by Peyraud (1969). Bunkin & Kazakhov (1970a) suggested that the effect could take place in a single laser beam of finite bandwidth $\Delta\omega_0$ around ω_0, converging in a plasma. The heating rate now varies as the square of the converging beam intensity, so this process might be helpful for heating very hot, relatively low density plasmas when inverse Bremsstrahlung is ineffective (Beaudry et al. (1971), Martineau & Pepin (1972)). However, the proportion of the incident energy given to the electrons is limited by the bandwidth of the radiation to $\Delta\omega_0/\omega_0$. For lasers, even mode-locked lasers, this is unlikely to exceed 1%. Further calculations were made by Zeldovich et al. (1972) and by Vinogradov & Pustovalov (1972a: see also 1972b).

Stimulated Compton scattering may, however, play another more significant role in dense plasma heating experiments, as was pointed out by Krasyuk et al. (1973). When a laser beam is incident on an isolated high-density target some of the light will be reflected from the region of critical plasma density. The incident and reflected waves can then interact through stimulated Compton scattering in the less dense outer layers of the plasma. The reflected wave is amplified, and if the irradiances are sufficiently great a significant part of the incident wave energy may be converted into reflected wave energy. The spatial amplifications of the incident and reflected irradiances $I_1(z)$ and $I_2(z)$ are related by the expression

$$\frac{dI_1(z)}{dz} = \frac{dI_2(z)}{dz} = G'_C I_1(z) I_2(z), \tag{2.157}$$

the positive direction of z being towards the laser. Provided the incident bandwidth $\Delta\omega_0$ satisfies $v_T/c \gg \Delta\omega_0/\omega_0$, the parameter G'_C is given* by

$$G'_C = \frac{8\pi^2 r_e^2 c^2}{\omega_0^3} \frac{v_p}{\kappa_B T_e} n_e \left(\frac{m_e}{2\pi\kappa_B T_e}\right)^{\frac{1}{2}} \exp\left(\frac{-m_e v_p^2}{2\kappa_B T_e}\right). \tag{2.158}$$

Here r_e is the classical electron radius and v_p is the plasma velocity along z. The overall intensity reflection coefficient, $\mathcal{R}$, for the incident beam is given by

$$\mathcal{R} = \mathcal{R}_0 \exp G'_C I_{10} l(1 - \mathcal{R}), \tag{2.159}$$

---

* In the paper by Krasyuk et al. (1973) the expression for G'_C should contain c^2, not c (see Kazakov et al. (1971)).

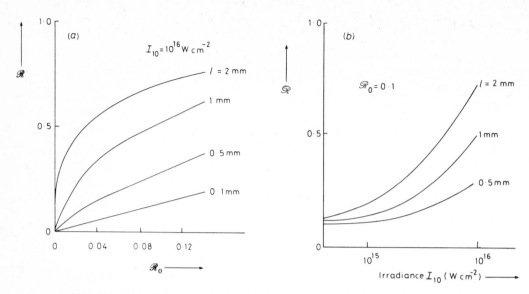

2.18   Enhanced reflectivity due to stimulated Compton scattering in the outer regions of a plasma boundary. (a) total reflection coefficient $\mathscr{R}$ as function of reflection coefficient of critical density layer, $\mathscr{R}_0$ for four plasma thicknesses l, with incident laser irradiance $I_{10} = 10^{16}$ W cm^{-2}: (b) total reflection coefficient $\mathscr{R}$ as function of I_{10}, for $\mathscr{R}_0 = 0.1$ and three values of l. (After Krasyuk et al. (1973))

where $\mathscr{R}_0$ is the initial reflection coefficient at $z = 0$, l is the length of the plasma and I_{10} is the original incident laser irradiance.

Krasyuk et al. solved equation (2.159) for conditions approximating those envisaged in an isentropic compression proposal (see Chapter 7) using a neodymium laser: $\kappa_B T_e \simeq 10$ keV, $n_e \simeq (0.5$ to $1) \times 10^{21}$ cm^{-3}, $v_p = 10^8$ cm s^{-1}, so that $G'_C = 3.14 \times 10^{-17}$ cm W^{-1}. The overall reflectivity $\mathscr{R}$ for $I_{10} = 10^{16}$ W cm^{-2} is shown as a function of the initial target reflectivity $\mathscr{R}_0$ for four values of l in Figure 2.18(a), while in Figure 2.18(b) the overall reflectivity is shown as a function of the incident irradiance for three values of l at $\mathscr{R}_0 = 0.1$. The dependence of $\mathscr{R}_0$ on the local irradiance at the critical density layer has not been taken into account.

10.2. OPTICAL RESONANCE ABSORPTION

A plasma whose density rises monotonically in the z direction from zero to more than the critical density will partially reflect an incident transverse wave. The depth of penetration in the geometrical optics approximation depends on the angle of incidence ϕ_i of the wave at the zero density surface: at oblique incidence the wave will be refracted, reaching the turning point at a depth in the plasma where $\omega_{pe}^2 = \omega_0^2 \cos \phi_i$. If the obliquely incident wave has an electric field component in the plane of incidence, then provided it approaches reasonably close to the critical density layer when it reaches the turning point, it can tunnel through to drive electron plasma oscillations along the direction of the density gradient and thus loses energy. The effect does not occur at normal incidence or near grazing incidence. The theory has been worked out in considerable detail (see for example Ginzburg (1970) pages 252–277) and has been discussed in the context of laser-heated plasmas by several authors.* Some results will be discussed in Chapter 7, Section (3).

Optical resonant absorption is essentially a linear process, in the sense that the absorption coefficient is normally independent of the incident irradiance, and there is no threshold. However, at high irradiances non-linear parametric coupling to plasma waves must be expected. These waves will propagate parallel to the density gradient into regions of lower density, so that k/k_D increases and Landau damping occurs. Parametric decay at oblique incidence has been discussed by White *et al.* (1973), who concluded that for a density scale length of 1 mm and $T_e = 10$ keV, the growth rate would be substantial over a range of angles of incidence of about $10°$ for λ 1·06 μm and $15°$ for λ 10·6 μm. The maximum ratio of power absorbed to incident power was calculated to be about $\frac{1}{2}$.

10.3. RELATIVISTIC EFFECTS

In a laser beam of irradiance exceeding $I_{rel} \simeq 10^{18}/\lambda^2$ W cm^{-2}, (λ in microns), an electron oscillates with a velocity approaching that of light, and its effective mass is increased so that the plasma frequency is reduced. The light can now propagate in a somewhat overdense plasma (Kaw and Dawson (1970)). It is no longer reflected completely, because the plasma current is limited to $n_e ec$ and is not large enough to cancel the displacement current (Max & Perkins (1971)). Self-focusing may occur, leading to the breakup of a laser wavefront into filaments (Kaw (1969)). Bunkin & Kazakov (1970a, b) have discussed the possibility of multiphoton Compton scattering: a significant rate of heating requires a strongly relativistic irradiance ($I \gg I_{rel}$). Max (1973) has considered the parametric instabilty of a relativistically strong, circularly polarized pump wave, and has found that the growth rate of instability may be as great as $\omega_0/2$. Some particles may be accelerated to very high energies, and considerable backscatter may occur.

These effects are slightly beyond the reach of present-day laser techniques, but may be important later.

11. EXPERIMENTAL STUDIES OF NON-LINEAR INTERACTIONS OF RADIATION WITH PLASMAS

11.1. MICROWAVE EXPERIMENTS

Parametric processes are much more readily studied with microwaves in low-density plasmas than with lasers. Indeed up to the present time, the only detailed experimental investigations of wave–wave interactions in plasmas have been made at microwave frequencies.

---

* Dawson *et al.* (1968), Vinogradov & Pustovalov (1971), Freidberg *et al.* (1972), Godwin (1972), Burnett (1972), Mueller (1973).

Gekker & Sizukhin (1969) observed anomalously strong absorption of 10 cm waves, in a collisionless plasma ($\omega_{pe} \gg \omega \gg \nu_{ei}$) with $T_e = 4\,\mathrm{eV}$, above a threshold power corresponding to $\eta = 0.1$, and suggested that this was caused by the excitation of some instability. Kaw et al. (1970) showed that the effect could be explained as being due to one of the two parametric instabilities (Section (8.2)): parametric coupling had already been observed between electron plasma and ion acoustic waves by Stern & Tzoar (1966). Dreicer et al. (1971), using a potassium plasma ($T_e \simeq T_i \simeq 0.25\,\mathrm{eV}$), and Eubank (1971) with rare gas plasmas ($T_e = 6$ to $8\,\mathrm{eV}$, $T_i < 1\,\mathrm{eV}$) also observed a well-defined threshold for anomalous absorption consistent with the onset of parametric instabilities. Phelps et al. (1971, 1973) demonstrated non-linear coupling between electromagnetic, electron plasma and ion acoustic waves in a potassium plasma and verified the frequency and wave number selection rules and the magnitudes of the coupling coefficients.

Detailed studies of the parametric decay instability were made by Stenzel & Wong (1972) in an argon plasma ($T_e = 2\,\mathrm{eV} = 10T_i$, $n_e = 10^9\,\mathrm{cm}^{-3}$). Studies of the ion acoustic wave showed that the threshold pump intensity varied as ν_{ik}/ω_i. At higher pump intensities the oscillations saturated, and their amplitude then fluctuated at a frequency, below ν_{ik}, which increased linearly with the pump intensity. When an additional small amplitude plasma wave was excited at the difference frequency between the pump and the ion wave frequencies, the ion wave amplitude was greatly enhanced. Experiments by Chu & Hendel (1972) with a potassium plasma also confirmed the selection rules and showed that, above the instability threshold, the power absorbed by the plasma varied as E_0^4, while the effective collision frequency varied as E_0^2 as is expected theoretically (Dubois & Goldman (1972a), Valeo et al. (1972)). Rapid heating of the ions at a rate proportional to the energy density of the ion acoustic instability was also observed in similar experiments by Hendel & Flick (1973). The purely growing two-stream instability was not observed.

The parametric decay of a longitudinal electron plasma wave into another plasma wave and an ion acoustic wave, predicted by Oraevski & Sagdeev (1962) was experimentally observed in a sodium plasma by Franklin et al. (1971).

Evidence for the production of a high-energy, non-thermal tail on the electron distribution in a potassium plasma was found by Dreicer et al. (1973). The electron energy distribution depended sensitively on the electron density and the microwave pump power, the high-energy tail being strongest when the driving frequency was very close to the plasma frequency.

Related effects have been observed in magnetized laboratory plasmas* and (at lower frequencies) in the ionosphere (Wong & Taylor (1971), see also Kaw et al. (1971)). Parametric coupling between two electron plasma waves and an ion wave has also been investigated (Franklin et al. (1971, 1972)). All the available experimental evidence is consistent with theory in the microwave region.

---

* Arkhipenko et al. (1971), Porkolab et al. (1972), Sugaya (1972), Chang et al. (1972), Keen & Fletcher (1971, 1973).

11.2. LASER EXPERIMENTS

The proportion of the incident light reflected when a high-power laser pulse produces a plasma from a solid target has been measured in a number of experiments,* though the geometry and the condition of the target surface are not always specified. Surface irregularities are likely to be particularly significant with picosecond pulses, because of the short distance travelled by plasma particles during the laser pulse. Caruso et al. (1970a) measured the reflectivity of a nickel target for 100 ps neodymium laser pulses and found that it fell rapidly if the irradiance was increased above about 10^{13} W cm^{-2}. The dependence of the reflected energy on the angle of incidence of the target surface to the laser beam was consistent with specular reflection.

Andrews et al. (1972) measured the specular reflectivity of polished copper, bismuth, boron and lithium fluoride targets as a function of the incident pulse energy with single transverse mode, mode-locked 10 ps neodymium laser pulses. Measurements for copper are shown in Figure 2.19: above a clearly defined threshold the reflectivity fell steadily. The results were independent of the polarization of the incident light.

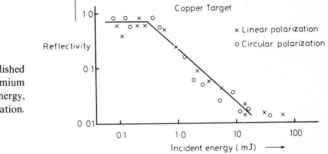

2.19 Specular reflectivity of polished copper target for 10 ps neodymium laser pulses as function of pulse energy, for linear and circular polarization. (After Andrews et al. (1972))

The threshold was at very nearly the same incident pulse energy, corresponding to an irradiance of about 10^{13} W cm^{-2}, for all three conductors. The very small reflectivity of the dielectric, lithium fluoride, showed no perceptible change over the range of observation. The diffusely reflected light energy was also measured and found to be approximately proportional to the incident pulse energy. The fall in specular reflection above threshold corresponded, therefore, to a non-linear increase in absorption. It seems possible that this was due to the excitation of the two-stream parametric instability. Ion waves may not have had time to develop during the 10 ps pulse.

The spectrum of the reflected light was examined by Caruso et al. (1970b), who found that the second harmonic of the incident neodymium light wavelength was generated from perspex and from solid deuterium targets. More complex harmonics were recorded by Bobin et al. (1973) in light emitted from the vicinity of a solid hydrogen target at the focus of a 15 ns, 1 GW neodymium laser. The peak irradiance was about 10^{12} W cm^{-2}. The emitted light was collected at an angle of 135° to the incident

---

*Yamanaka et al. (1972c), Büchl et al. (1971), Salzmann (1973), Olsen et al. (1972), Caruso et al. (1970a), Andrews et al. (1972), Shearer et al. (1972).

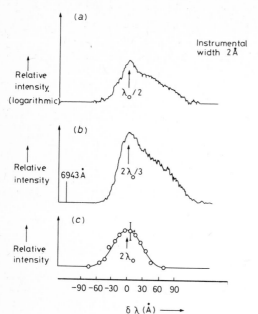

2.20   Profiles of harmonics of the incident neodymium laser wavelength (λ_0 = 10 600 Å) emitted from vicinity of a solid hydrogen target: (a) $\lambda_0/2$, (b) $2\lambda_0/3$, (c) $2\lambda_0$. (After Bobin et al. (1973))

beam axis. Three broad lines were observed, at $3\omega_0/2$, $2\omega_0$ and $\omega_0/2$ in order of decreasing intensity: their profiles are shown in Figure 2.20. Bobin et al. suggested that the harmonics and subharmonics are generated in the region of the strong density gradient as the result of parametric interactions with electron and ion waves, the line profile being characteristic of the ion wave spectrum.

The experiment proposed by Kroll et al. (1964), in which light is scattered from plasma oscillations enhanced by optical mixing of two laser beams, was carried out successfully by Stansfield et al. (1971) and is described in Chapter 3 (Section (8.2)).

There is evidence also for the occurrence of stimulated Compton scattering. Krasyuk et al. (1970a) focused a 50 ps mode-locked ruby laser pulse in helium at a pressure of 3 atmospheres and examined the incident and transmitted spectra. The transmitted line profile was found to be slightly distorted towards longer wavelengths. When a thin aluminized polyester film was placed at the focus of the laser the transmitted line profile was again slightly displaced to the red, and a secondary maximum appeared on the long wavelength side. These changes in the line profile were ascribed to stimulated Compton scattering: from the redistribution of the energy spectrum, the fraction of the incident light energy absorbed by the electrons was found to be 1 to 2×10^{-5}.

Further evidence for Compton scattering was obtained by Decroisette et al. (1970, 1972). A 1 GW, 15 to 20 ns neodymium laser pulse was focused on to a 2 mm thick solid D_2 target with an $f/1$ lens to give an irradiance of about 10^{12} W cm^{-2} and the incident and transmitted light spectra were recorded with time resolution. The transmitted line profiles (see Chapter 8, Figure 8.92) showed a number of maxima superimposed on the smooth incident light profile on the long wavelength side only. Both the incident and the transmitted light beams were almost completely polarized. Doppler shifts as the result of plasma motion are not seen in forward scattering, and Decroisette *et al.* suggested that if the effect was due to a parametric interaction with ion acoustic waves the modified spectrum would have shown some depolarization. The results were also discussed by Babuel-Peyrissac (1972): it was concluded that stimulated Compton scattering was involved.

Kazakov *et al.* (1971) divided a ruby laser pulse (20 to 200 ps) into two beams and arranged for them to collide, at the coincident foci of two coaxial lenses, in argon at a pressure of 0·4 atm. The stronger beam produced an irradiance, I_1, of 3×10^{14} W cm^{-2} at the focus. After allowing for Bremsstrahlung absorption it was found that the weaker beam was amplified by a factor of 1·32 when $I_2 = 0·2I_1$, and by a smaller factor as I_2 was increased. The laser light was strongly broadened on passing through the interacting region. Again, the effect was ascribed to stimulated Compton scattering. The gain calculated from expression (2.158) for the conditions of the experiment was reported to be in good agreement with observation.

Stimulated Compton scattering has also been reported for carbon dioxide laser light in a hydrogen plasma (Offenberger & Burnett (1973)).

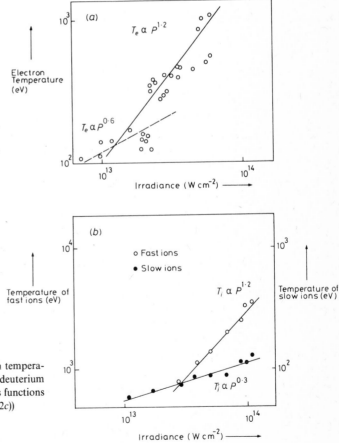

2.21   (a) Electron temperature and (b) ion temperatures in plasmas produced from solid deuterium targets by 2 ns neodymium laser pulses, as functions of irradiance. (After Yamanaka *et al.* (1972c))

Less direct evidence for non-linear effects comes from studies of the behaviour of laser-heated plasmas. Yamanaka *et al.* (1972b, c) produced plasmas from solid deuterium targets using 2 ns neodymium laser pulses of up to 40 J. Figures 2.21(a) and (b) show the electron temperature, measured by the X-ray absorption method (see Chapter 6, Section (8.5)), and the ion temperature derived from time-of-flight velocity profiles. At irradiances above about 3×10^{13} W cm^{-2} a fast group of ions was observed: their appearance coincided with the appearance of neutrons from the plasma, and with a sudden increase in the plasma reflectivity from 4% to 20% of the pulse energy. Here the electron temperature was about 300 eV. Figure 2.12 shows that these conditions coincide with the threshold for the oscillating two-stream instability. At a slightly lower laser irradiance, about 2×10^{12} W cm^{-2}, when $T_e \simeq 200$ eV, the reflected laser power often oscillated at about 10^9 Hz (Waki *et al.* (1972)). Yamanaka *et al.* suggested that the onset of the instability produces a sharp boundary of very dense plasma which increase the effective reflectivity.

Shearer *et al.* (1972) produced a plasma with neodymium laser pulses of 2 to 5 ns duration and 10 GW peak power, focused on a polyvinylchloride target by an $f/7$ lens. The peak irradiance at the target was about 2×10^{14} W cm^{-2}. Electron temperatures deduced from X-ray absorption ratios using two nickel foils did not agree with

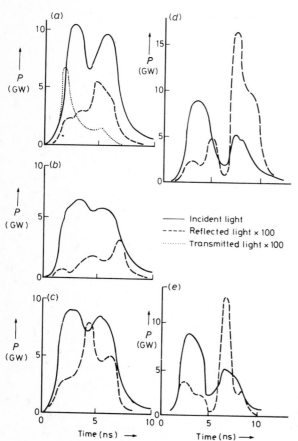

——— Incident light
---- Reflected light × 100
·········· Transmitted light × 100

2.22 Incident and reflected light power as functions of time during five experiments with polyvinylchloride targets. In (a) the transmitted light is also shown. The reflected and transmitted light powers are shown multiplied by 100. (After Shearer *et al.* (1972))

results using aluminium foils : for aluminium $T_e \simeq 2$ keV, but for nickel $T_e \simeq 40$ keV, suggesting that the electron velocity distribution was strongly non-thermal and that anomalous heating was taking place. The expected value from hydrodynamic theory is about 500 eV. The incident and reflected light intensities were recorded and are shown in Figure 2.22 for five different shots. In the first diagram the transmitted light is also shown. Shearer *et al.* interpreted the anomalous heating as being due to the onset of the parametric decay instability (the two-stream instability has a higher threshold at 500 eV) and suggested that the bursts of reflected light are due to stimulated Brillouin scattering.

It is apparent that much experimental work remains to be done before the processes responsible for the observed non-linear interactions between laser light and plasma can be unambiguously identified.

3

Incoherent Scattering of Light by Plasmas

1. INTRODUCTION

In Chapter 2 we discussed the processes through which light is absorbed by and emitted from a plasma. We considered several parametric interactions, including stimulated coherent scattering processes which can deflect a large fraction of the incident light energy. We now examine the incoherent Thomson scattering of light by a plasma, a process involving such a minute fraction of the incident light energy that it was entirely neglected in the previous chapter. It is nevertheless important because the spectrum of light scattered by a plasma from a monochromatic beam carries a great deal of information: as we shall see below, the electron density and temperature, the ion temperature, the magnetic field and information about the spectrum of plasma waves can all be deduced in appropriate circumstances.

As a diagnostic technique, scattering has the advantages of providing excellent spatial and temporal resolution. Only light from a short length of a narrow incident beam need be collected and with high-power laser pulses observations may be made in nanoseconds or even picoseconds. Unlike measurements involving the introduction of probes, scattering studies do not in general perturb the plasma under investigation significantly, unless a very powerful and strongly-focused laser beam is used. The technique has become a basic (albeit expensive) tool in experimental plasma physics.

The first observations of scattering from electrons were made by Bowles (1958), not with light but with radar waves, the plasma investigated being the ionosphere.

However, the theory developed for this situation is also valid for optical wavelengths and denser laboratory plasmas.

We begin with an outline of scattering theory, making use of some standard results of electromagnetic theory and plasma kinetic theory. We shall then discuss some experimental problems. Finally, we review the results of scattering observations made on a wide variety of plasmas.

2. SCATTERING BY A SINGLE FREE PARTICLE

Let us consider a plane polarized wave of irradiance I propagating in the z direction, incident on a stationary electron of mass m_e and charge e. Consider an element of solid angle $d\Omega$ about the direction making an angle ϕ_E with the electric vector of the

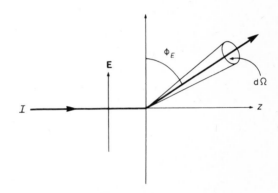

3.1   Scattering geometry

incident wave (Figure 3.1). The power dP scattered from the wave into $d\Omega$ is given (see for example Sommerfeld (1964)) by

$$dP = r_e^2 \sin^2 \phi_E I \, d\Omega \tag{3.1}$$

where

$$r_e = \frac{e^2}{4\pi\varepsilon_0 m_e c^2} = 2 \cdot 82 \times 10^{-15} \, \text{m} \tag{3.2}$$

is the classical electron radius. Figure 3.2 is a polar diagram of this distribution in space.

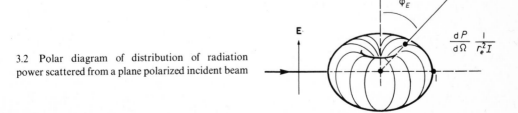

3.2   Polar diagram of distribution of radiation power scattered from a plane polarized incident beam

If the incident wave is unpolarized, the distribution of scattered light must be averaged over all possible directions of the electric field vector in the plane perpendicular to z. The resulting distribution is then symmetrical about the direction of propagation of the incident wave, and for light deflected through an 'angle of scatter' θ,

$$dP = \frac{r_e^2}{2}(1+\cos^2\theta)I\,d\Omega \tag{3.3}$$

Figure 3.3 shows this distribution.

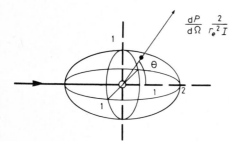

3.3 Polar diagram of distribution of radiation power scattered from an unpolarized incident beam

On integrating over all solid angles, the total scattered power is found to be

$$P = \frac{8\pi}{3}r_e^2 I. \tag{3.4}$$

The total scattering cross-section is known as the *Thomson scattering cross-section*,

$$\sigma_T = \frac{8\pi r_e^2}{3} = 6\cdot 65\times 10^{-29}\,\mathrm{m}^2. \tag{3.5}$$

The quantities $r_e^2\sin^2\phi_E$ and $r_e^2(1+\cos^2\theta)/2$ in expressions (3.1) and (3.3) are differential Thomson cross-sections, and will be written as $\sigma_T'(\phi_E)$ and $\sigma_T'(\theta)$ respectively.

The exceedingly small value of the cross-section explains why Thomson scattering may be neglected so far as the transfer of energy from the incident beam to the plasma is concerned. It also explains why Thomson scattering is difficult to detect unless a very powerful source is available. On the other hand the small cross-section ensures that further scattering of scattered light on its way out of the plasma, which would greatly complicate the interpretation of observations, is very unlikely and may be ignored. It should be noted that the cross-section is independent of the wavelength of the incident light.

The scattering cross-section for a single proton is found by multiplying the Thomson cross-section for an electron by $(m_e/m_p)^2 \simeq 3\times 10^{-7}$. For any other atom stripped of all its electrons the cross-section is even lower. The contribution of bare nuclei to the scattered light intensity is therefore normally negligible. On the other hand atoms with bound electrons, and molecules, can have quite large polarizabilities and correspondingly large Rayleigh scattering cross-sections, which vary as λ^{-4}. The alkali metal atoms Li, Na, K, Sr, Rb and Cs, which have only a single valence electron outside an inert gas configuration, have particularly large cross-sections, within a factor of 3 of σ_T at λ 6943 Å. For most other atoms, ions and molecules for which data are available, the cross-section is smaller than σ_T by a factor of at least a hundred (see e.g. Allen (1962), Dalgarno & Kingston (1960), Pauling (1927) and Sternheimer (1954)).

3. EFFECT OF MOTION OF A SINGLE SCATTERING PARTICLE ON THE WAVELENGTH OF SCATTERED LIGHT

Consider a stationary source emitting light of angular frequency ω_0 and wavelength λ_0 towards a scattering particle, and a stationary observer receiving scattered light. If the particle is moving, and its velocity has a component in the direction of propagation of the incident light wave, a Doppler shift will occur. The particle will oscillate in the field of the wave at a different frequency,

$$\omega_d = \omega_0 - \mathbf{k}_0 \cdot \mathbf{v}, \tag{3.6}$$

$\mathbf{k}_0$ being the propagation vector of the incident wave. The scattered light emitted by the particle will also have the frequency ω_d in the moving coordinate frame of the particle. However, if the particle velocity has a component in the line of sight of the observer a further Doppler shift will occur and the scattered light will appear to have the frequency ω_s, such that

$$\begin{aligned} \omega_s &= \omega_d + \mathbf{k}_s \cdot \mathbf{v} \\ &= \omega_0 - (\mathbf{k}_0 - \mathbf{k}_s) \cdot \mathbf{v} \\ &= \omega_0 - \mathbf{k} \cdot \mathbf{v}, \end{aligned} \tag{3.7}$$

where $\mathbf{k}_s$ is the propagation vector of the scattered wave, and

$$\mathbf{k} = \mathbf{k}_0 - \mathbf{k}_s. \tag{3.8}$$

The geometry of the situation is shown in Figure 3.4. Provided $k_s \simeq k_0$, as is the case

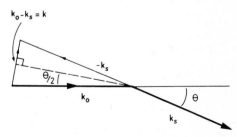

3.4 Diagram showing relations between $\mathbf{k}_0$, $\mathbf{k}_s$, $\mathbf{k}$ and θ

for non-relativistic velocities,

$$k = 2k_0 \sin\left(\frac{\theta}{2}\right)$$

$$= 2\frac{\omega_0}{c} \sin\left(\frac{\theta}{2}\right).$$

(3.9)

Thus the frequency shift observed, $\omega_s - \omega_0 = (2\omega_0 v_k/c) \sin(\theta/2)$, v_k being the component of the particle velocity along the direction of $(\mathbf{k}_0 - \mathbf{k}_s)$. For a given direction of the incident beam and a given direction of observation, therefore, information is only available about the component of the particle velocity in a single direction. Under certain conditions, as will be seen later, the contribution of all the particles in the scattering volume to the observed spectrum of scattered light may be simply added. In this case, a Maxwellian distribution of electron velocities will yield a Gaussian profile for the scattered light, with a full width at half intensity $\Delta\omega$ given by

$$\frac{\Delta\omega}{\omega_0} = 4 \sin\left(\frac{\theta}{2}\right)\left(\frac{2\kappa_B T_e}{m_e c^2} \ln 2\right)^{\frac{1}{2}}.$$

(3.10)

A full discussion of the classical theory of scattering by a single free electron has been given by Sarachik & Schappert (1970).

4. EFFECTS OF DENSITY FLUCTUATIONS

The total power scattered from a uniform distribution of particles in the scattering volume would be almost zero, as a result of interference between contributions of opposite phase. Scattering can only be observed when variations in density are present (Rayleigh (1871)). In a plasma, electrons are the only important scatterers, and electron density fluctuations determine the observed spectrum of scattered light. Quantum mechanically, the process may be described as a non-linear Raman interaction between an incident photon, of frequency ω_0 and propagation vector $\mathbf{k}_0$, and a plasmon of frequency ω and propagation vector $\mathbf{k}$, resulting in a scattered photon of frequency ω_s and propagation vector $\mathbf{k}_s$. This process can only occur provided the usual energy and momentum conservation conditions are satisfied, i.e.

$$\left.\begin{aligned}\omega &= \omega_0 - \omega_s\\ \mathbf{k} &= \mathbf{k}_0 - \mathbf{k}_s\end{aligned}\right\}.$$

(3.11)

The observed spectrum of light scattered along $\mathbf{k}_s$ thus represents the spectrum of those electron density fluctuations whose propagation vector $\mathbf{k} = \mathbf{k}_0 - \mathbf{k}_s$. The intensity of the scattered light depends on the product of the intensity of the incident wave and that of the fluctuations. When the corresponding wavelength of the density fluctuation is short compared to the Debye shielding distance λ_D, i.e.

$$\frac{1}{k\lambda_D} = \frac{\lambda_0}{4\pi\lambda_D \sin (\theta/2)} \ll 1,$$

the effects of screening are small, and for a Maxwellian distribution of electron velocities only thermal fluctuations are represented in the scattered spectrum. When $1/k\lambda_D \gg 1$, on the other hand, collective effects dominate the scattered spectrum.

Instead of expressions (3.1) or (3.3) we now seek a general expression for the frequency spectrum of light scattered in a given direction. If the scattering volume is V and the mean electron density is n_e we may write for the power scattered into an element of solid angle $d\Omega$

$$P_T(\mathbf{k}, \omega) = \sigma'(\mathbf{k}, \omega)In_eV\,d\omega\,d\Omega, \tag{3.12}$$

where the angle of scattering θ is defined for non-relativistic electron velocities by $k = (2\omega_0/c) \sin (\theta/2)$.

It is convenient to write the differential cross-section $\sigma'(\mathbf{k}, \omega)$ in the form

$$\sigma'(\mathbf{k}, \omega) = \sigma'_T S(\mathbf{k}, \omega), \tag{3.13}$$

where σ'_T is the appropriate differential Thomson cross section, and $S(\mathbf{k}, \omega)$, which is known as the *dynamic form factor*, represents the spectrum of electron density fluctuations.

Scattering by plasma density fluctuations was first investigated by Akhiezer et al. (1957). Early work by Salpeter (1960), Dougherty & Farley (1960) and Fejer (1960) was done to explain the spectrum observed in radar waves scattered back to earth from the ionosphere, and was mainly concerned with plasmas in which the ions and electron each have Maxwellian velocity distributions, although at different temperatures. The electron and ion distribution functions are assumed to satisfy separate Vlasov equations, which are coupled by a Poisson equation. The generalization to arbitrary stable, non-relativistic velocity distributions by Rosenbluth & Rostoker (1962) yields the following result (Bernstein et al. (1964)) for a plasma consisting of electrons and a single species of ion of charge Z, in which binary collisions may be neglected:

$$S(\mathbf{k}, \omega) = \left| \frac{1-\mathscr{G}_i}{1-\mathscr{G}_e-\mathscr{G}_i} \right|^2 \mathscr{F}_e(\mathbf{k}, \omega) + Z\left| \frac{\mathscr{G}_e}{1-\mathscr{G}_e-\mathscr{G}_i} \right|^2 \mathscr{F}_i(\mathbf{k}, \omega). \tag{3.14}$$

Here $\mathscr{F}_e(\mathbf{k}, \omega)$, $\mathscr{F}_i(\mathbf{k}, \omega)$ are the unperturbed distribution functions for the components of velocity along $\mathbf{k}$ of the electrons and ions respectively, normalized to unity. The screening integrals $\mathscr{G}_e$, $\mathscr{G}_i$ are given by

$$\mathscr{G}_e(\mathbf{k}, \omega) = \frac{e^2}{m_ek^2\varepsilon_0} \int \frac{\mathbf{k} \cdot \partial f_{e0}(\mathbf{v})/\partial \mathbf{v}}{\mathbf{k} \cdot \mathbf{v} - \omega}\,d\mathbf{v}, \tag{3.15}$$

$$\mathscr{G}_i(\mathbf{k}, \omega) = \frac{Z^2e^2}{m_ik^2\varepsilon_0} \int \frac{\mathbf{k} \cdot \partial f_{i0}(\mathbf{v})/\partial \mathbf{v}}{\mathbf{k} \cdot \mathbf{v} - \omega}\,d\mathbf{v}. \tag{3.16}$$

where f_{e0}, f_{i0} are the unperturbed electron and ion distribution functions respectively. For details of the derivation the reader is referred to Bernstein *et al.* (1964), Evans & Katzenstein (1969), or Bekefi (1966).

The dynamic form factor is closely related to the longitudinal dielectric constant. From (1.118),

$$\varepsilon_L = 1 - \mathcal{G}_e - \mathcal{G}_i. \tag{3.17}$$

Thus both terms in (3.14) may become large in the vicinity of a resonance, when $\varepsilon_L \rightarrow 0$.

In the next section we shall consider the scattering spectrum in detail for a number of special conditions.

5. SOME CALCULATIONS OF SCATTERED LIGHT SPECTRA

5.1. MAXWELLIAN VELOCITY DISTRIBUTIONS

When the electrons and ions each have Maxwellian velocity distributions, at temperatures T_e and T_i, we may write

$$f_{e0}(v) = n_e \left(\frac{m_e}{2\pi \kappa_B T_e} \right)^{\frac{1}{2}} \exp \left[- \left(\frac{v^2 m_e}{2\kappa_B T_e} \right) \right] \tag{3.18}$$

and

$$f_{i0}(v) = n_i \left(\frac{m_i}{2\pi \kappa_B T_i} \right)^{\frac{1}{2}} \exp \left[- \left(\frac{v^2 m_i}{2\kappa_B T_i} \right) \right]. \tag{3.19}$$

The screening integrals are then given (Salpeter (1960)) by

$$\mathcal{G}_e = -\alpha^2 \mathcal{W}(y_e) \tag{3.20}$$

$$\mathcal{G}_i = -Z\alpha^2 \frac{T_e}{T_i} \mathcal{W}(y_i) \tag{3.21}$$

where (as in Chapter 1, (1.120))

$$y_e = \frac{\omega}{Y_e}, \qquad y_i = \frac{\omega}{Y_i},$$

$$Y_e = \left(\frac{2k^2 \kappa_B T_e}{m_e} \right)^{\frac{1}{2}}, \qquad Y_i = \left(\frac{2k^2 \kappa_B T_i}{m_i} \right)^{\frac{1}{2}} \tag{3.22}$$

and

$$\alpha = \frac{1}{k\lambda_D} = \frac{\lambda_0}{4\pi\lambda_D \sin(\theta/2)}. \tag{3.23}$$

The parameter α is very important in determining the nature of the scattering spectrum, as will be seen below. It has the numerical value

$$\alpha = 1\cdot58 \times 10^{-12} \lambda_0 \left[\frac{n_e}{T_e(1-\cos\theta)} \right]^{\frac{1}{2}}, \tag{3.24}$$

where λ is in Å, n_e in cm^{-3} and T_e in eV. The dependence of α on n_e, T_e and θ is shown, for $\lambda_0 = 6943$ Å, in Figure 3.5.

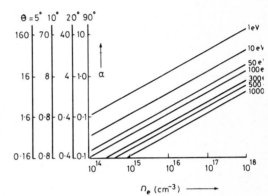

3.5 Scattering parameter α as function of electron density n_e for several values of the electron temperature T_e and of the scattering angle θ. (After John (1972))

The function $\mathscr{W}(y)$ is given by

$$\mathscr{W}(y) = 1 - 2y \exp(-y^2) \int_0^y \exp(t^2)\,dt - i\pi^{\frac{1}{2}} y \exp(-y^2), \tag{3.25}$$

which may be written in the form

$$\mathscr{W}(y) = -\frac{\mathscr{H}(y)}{2y^2}. \tag{3.26}$$

The function $\mathscr{H}(y)$ has already been encountered (expression (1.119)): asymptotic forms were given for the integral in (1.121) and (1.122).

Inserting these expressions for f_{e0}, f_{i0} and $\mathscr{G}_e, \mathscr{G}_i$ into (3.14) yields

$$S(\mathbf{k},\omega)\,d\omega = \left| \frac{1+\alpha^2 Z(T_e/T_i)\mathscr{W}(y_i)}{1+\alpha^2\mathscr{W}(y_e)+\alpha^2 Z(T_e/T_i)\mathscr{W}(y_i)} \right|^2 \frac{\exp(-y_e^2)\,dy_e}{\pi^{\frac{1}{2}}}$$

$$+ Z \left| \frac{-\alpha^2\mathscr{W}(y_e)}{1+\alpha^2\mathscr{W}(y_e)+\alpha^2 Z(T_e/T_i)\mathscr{W}(y_i)} \right|^2 \frac{\exp(-y_i^2)\,dy_i}{\pi^{\frac{1}{2}}}. \tag{3.27}$$

5.2. THE SALPETER APPROXIMATION

Salpeter (1960) found that under the conditions of Section (5.1) and provided that T_e is neither less than, nor very much greater than, T_i, expression (3.27) is conveniently and closely* approximated by the sum of two analytically similar terms:

$$S(\mathbf{k}, \omega)\,d\omega = \left[\Gamma_\alpha(y_e)\,dy_e + Z\left(\frac{\alpha^2}{1+\alpha^2}\right)^2 \Gamma_\beta(y_i)\,dy_i \right] \pi^{-\frac{1}{2}} \qquad (3.28)$$

where

and

$$\left.\begin{aligned} \beta^2 &= Z\frac{T_e}{T_i}\left(\frac{\alpha^2}{1+\alpha^2}\right) \\[2mm] \Gamma_\alpha(y) &= \frac{\exp(-y^2)}{|1+\alpha^2\mathscr{W}(y)|^2} \end{aligned}\right\} \qquad (3.29)$$

The two components of $S(\mathbf{k}, \omega)$ are known as the electron spectrum and the ion spectrum respectively. Their integrals over all frequencies are (see Salpeter (1963))

$$S(\mathbf{k})_e = \frac{1}{1+\alpha^2} \qquad (3.30)$$

for the electron component, and

$$S(\mathbf{k})_i = \frac{Z\alpha^4}{(1+\alpha^2)[1+\alpha^2+Z\alpha^2(T_e/T_i)]} \qquad (3.31)$$

for the ion component. Thus for $\alpha \ll 1$ only the electron component need be considered and the integral is almost unity, while for $\alpha \gg 1$ the ion component is dominant and the integral is approximately $1/[1+(T_e/T_i)]$.

The sum $S(\mathbf{k})$ of the integrals of the two terms in the full expression (3.27) has been evaluated numerically by Moorcroft (1963), and is shown in Figure 3.6. As may be

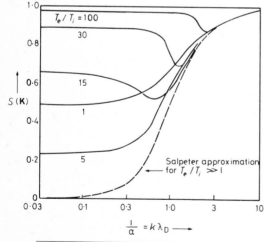

3.6   Total scattered power from a hydrogen plasma as function of $k\lambda_D = 1/\alpha$, calculated from expression (3.27), for several values of T_e/T_i. The Salpeter approximation gives the dashed curve for $T_e \gg T_i$. (After Moorcroft (1963))

---

* A discussion of the validity of the Salpeter approximation is given by Evans & Katzenstein (1969, p. 221).

seen, for large values of T_e/T_i the Salpeter approximation breaks down when $\alpha \gtrsim \frac{1}{2}$.

The function $\Gamma_\alpha(y)$ has been calculated by Salpeter for several values of α and is shown in Figure 3.7. Only half the profile is shown: the function is even in y. A logarithmic plot of some scattered light spectra is shown in Figure 3.8.

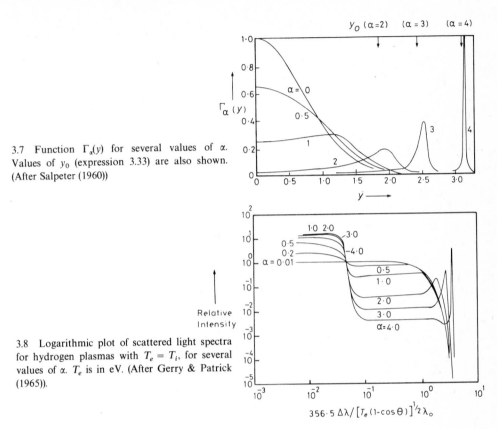

3.7 Function $\Gamma_\alpha(y)$ for several values of α. Values of y_0 (expression 3.33) are also shown. (After Salpeter (1960))

3.8 Logarithmic plot of scattered light spectra for hydrogen plasmas with $T_e = T_i$, for several values of α. T_e is in eV. (After Gerry & Patrick (1965)).

We are now in a position to investigate expression (3.27) for various values of α and T_e/T_i.

We consider first the electron component. For $\alpha \ll 1$ only this component makes any contribution, and the scattered light spectrum reduces to the Gaussian profile whose half-intensity width, given in expression (3.10), is determined by the thermal motion of the electrons. The conditions of observation ensure that only electron fluctuations of wavelength small compared to the Debye length, i.e. only random thermal fluctuations, are represented in the scattered spectrum.

For $\alpha \gg 1$, the electron component is very small, except for a resonance which occurs when $|1 + \alpha^2 \mathscr{W}(y_e)| \to 0$. This condition is satisfied when y_e is the larger root of the equation

$$-\mathrm{Re}\ \mathscr{W}(y_e) = \frac{1}{\alpha^2} \tag{3.32}$$

(for $\alpha \gg 1$ the imaginary part of $\mathscr{W}'(y_e)$ is small for this large value of y_e). Introducing the first three terms of the asymptotic series (expression (1.121)) into the integral in $\mathscr{W}'(y)$ (expression (3.25)) then gives the result that for $\alpha \gg 1$ the resonance in the electron component occurs in the region of $y_e = y_0$, where

$$y_0^2 = \tfrac{1}{2}(\alpha^2 + 3), \tag{3.33}$$

i.e. at frequencies displaced from that of the incident beam by $(\delta\omega)_e$, given by

$$(\delta\omega)_e^2 = (y_0 Y_e)^2 = \omega_{pe}^2 + \frac{3\kappa_B T_e}{m_e}k^2 = \omega_{BG}^2. \tag{3.34}$$

This result is just the dispersion relation obtained by Bohm & Gross for longitudinal electron plasma waves of wavelength much greater than the Debye length, derived in Chapter 1 (expression (1.127)). These waves thus give rise to the sharp spikes in the wings of the spectral profiles in Figure 3.7 for $\alpha \gg 1$. The profile of one of the spikes may be approximated in the region of y_0 by the Lorentzian shape:

$$\Gamma_a(y) = \frac{(\alpha^2/2)\exp(-y_0^2)}{4(y - y_0)^2 + [\tfrac{1}{2}\pi^{\frac{1}{2}}\alpha^4 \exp(-y_0^2)]^2}. \tag{3.35}$$

For $\alpha \gg 1$, $y_0^2 \simeq \alpha^2/2$ and the half-intensity width of the spike is given approximately by

$$(\Delta\omega)_e = \frac{\pi^{\frac{1}{2}}\alpha^4}{2}\exp\left(-\frac{\alpha^2}{2}\right). \tag{3.36}$$

As α increases the width decreases, corresponding to a reduction in the Landau damping of the plasma waves. Although the peak height of the spike increases as α increases, the total scattered energy in the two spikes (which now contain nearly all the contribution of the electron component) decreases as $1/\alpha^2$, as is to be expected from expression (3.30).

Figure 3.9 shows how the scattered light spectrum, calculated by Gerry & Patrick (1965) on the Salpeter approximation for a typical low-density plasma ($n_e = 10^{15}$ cm^{-3}), varies with the scattering angle. As the angle is reduced from $\theta = 90°$ to $\theta = 5°$, α varies from 0·16 to 2·6. The spectrum is dominated at 90° by a Gaussian

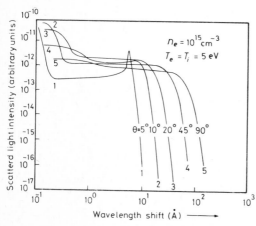

3.9   Scattered light spectra calculated on the Salpeter approximation for a plasma with $T_i = T_e = 5$ eV, $n_e = 10^{15}$ cm^{-3}, for several scattering angles θ. (After Gerry & Patrick (1965))

which is characteristic of the electron thermal velocity, but as α increases the central feature becomes narrow, characteristic of the ion thermal profile, and is accompanied by spikes displaced by a frequency close to the electron plasma frequency. By making observations at two different angles corresponding to large and small values of α, the electron and ion temperatures and the electron density can all be determined simultaneously. (In principle all this information can be obtained with moderate accuracy from a single profile provided $\alpha \simeq 1$).

For the limiting case of $\alpha \gg 1$ the ion spectrum shows very similar features to those of the electron component, but it is much narrower, and its shape is determined by the parameter β defined in (3.29):

$$\beta = \left[Z\frac{T_e}{T_i}\left(\frac{\alpha^2}{1+\alpha^2}\right) \right]^{\frac{1}{2}}$$

$$\simeq \left[Z\frac{T_e}{T_i} \right]^{\frac{1}{2}} \quad \text{for } \alpha \gg 1. \tag{3.37}$$

The Salpeter approximation only holds for a limited range of values of T_e/T_i greater than unity, within which no resonant behaviour can be discerned in the ion component.

Summing up the results of this section, for $\alpha \ll 1$ the total scattered spectrum has a Gaussian profile whose breadth is determined by the electron temperature. For $\alpha \gg 1$ the spectrum consists of a narrow central feature, whose shape is determined by both the ion and the electron temperatures, together with a symmetrical pair of sharp spikes displaced from the centre by the plasma frequency, which is determined by the electron density.

In practice the detector used to analyse light scattered from the plasma will receive light over a range of scattering angles θ. Both the breadth and the position of the electron spike vary with θ and an apparent broadening of the feature will be observed (mainly due to the shift) if too large an aperture is used in the detector system. Pyatnitskii et al. (1968) have calculated acceptable apertures $\Delta\Omega$ for the spurious broadening not to exceed 10% of the true breadth. The result for various angles of observation and values of α are shown in Figure 3.10.

3.10 Relations between scattering parameter α and the maximum detector solid angle $\Delta\Omega$ for spurious broadening not to exceed 10% of true breadth of scattered light spectrum, for various scattering angles θ. (After Pyatnitskii et al. (1968))

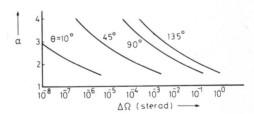

5.3. THE ION–ACOUSTIC WAVE RESONANCE

As has been explained above, the Salpeter approximation breaks down when $T_e \gg T_i$ and $\alpha \gg 1$. Under these conditions ion acoustic waves may be expected to make their presence known in the scattered spectrum. The dispersion relation for the ion acoustic wave is given by (1.133): for $k\lambda_D = 1/\alpha \ll 1$,

$$\omega_{ia} \simeq \left(\frac{Z\kappa_B T_e k^2}{m_i}\right)^{\frac{1}{2}}$$ (3.38)

so that when ω takes this value

and

$$y_e = \left(\frac{Z}{2}\frac{m_e}{m_i}\right)^{\frac{1}{2}}$$
$$y_i = \left(\frac{Z}{2}\frac{T_e}{T_i}\right)^{\frac{1}{2}}.$$
(3.39)

Thus $y_e \ll 1$ and $y_i \gg 1$. In order to investigate the effect of this resonance, it is necessary to return to expression (3.27) and to make the appropriate approximations in these limits.

Only the first term makes a significant contribution. The result (Evans & Katzenstein (1969) p. 225) is that in the region of ω_{ia}

$$S(\mathbf{k}, \omega) = \frac{1}{\pi^{\frac{1}{2}}} \frac{\left(\frac{\omega_{ia}}{\omega}\right)^4 \left(\frac{Zm_e}{2m_i}\right)^{\frac{1}{2}}\frac{d\omega}{\omega_{ia}}}{\left[1 - \left(\frac{\omega_{ia}}{\omega}\right)^2\right]^2 + \pi\left(\frac{Zm_e}{2m_i}\right)}$$ (3.40)

– a pair of Lorentzian spikes at $\omega = \pm\omega_{ia}$. Under these conditions they form the main contribution to the total scattering cross section integrated over all frequencies,

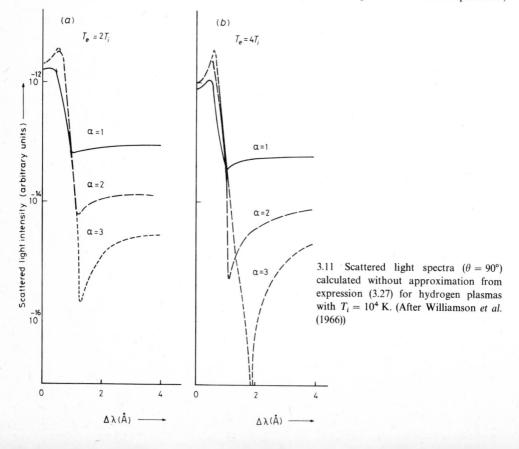

(a) $T_e = 2T_i$

(b) $T_e = 4T_i$

$\alpha = 1$

$\alpha = 2$

$\alpha = 3$

Scattered light intensity (arbitrary units)

10^{-12}

10^{-14}

10^{-16}

$\Delta\lambda(\text{Å})$

3.11 Scattered light spectra $(\theta = 90°)$ calculated without approximation from expression (3.27) for hydrogen plasmas with $T_i = 10^4$ K. (After Williamson et al. (1966))

$S(\mathbf{k})$, which amounts to unity. Some scattered light spectra have been computed from expression (3.27) without the Salpeter approximation by Williamson *et al.* (1966) for ruby light scattered from hydrogen plasmas at an ion temperature of 10 000 K under various conditions. The central region of the spectrum scattered at 90° is shown in Figure 3.11 for $T_e = 2T_i$ and $T_e = 4T_i$, for three electron densities, corresponding to $\alpha = 1, 2$ and 3. The ion acoustic peak is clearly seen when $\alpha \gtrsim 2$ and $T_e = 4T_i$ (a situation where according to Section (6) of Chapter 1 the ion acoustic wave should be only moderately damped), and is accompanied by an interesting dip in the region of the *ion* plasma frequency, when

$$\omega = \left[\frac{n_e Z e^2}{m_i \varepsilon_0} + \frac{3\kappa_B T_i k^2}{m_i} \right]^{\frac{1}{2}}. \tag{3.41}$$

This is the Bohm & Gross frequency (see Chapter 1, Section (7)) for electrostatic ion waves.

5.4. EFFECTS OF A DRIFT VELOCITY OF ELECTRONS RELATIVE TO IONS

Calculations similar to those described above have been made for the spectrum of light scattered from a plasma in which both the ions and the electrons have Maxwellian velocity distributions in their individual centre-of-mass frames, but in which the electron centre-of-mass frame has a drift velocity $\mathbf{u}$ while that of the ions is stationary (Rosenbluth & Rostoker (1962), Evans *et al.* (1966c)). While such calculations must be treated with caution, since a large drift velocity may invalidate the basis of the theoretical model used, they produce some very interesting results. The electron component of the scattered spectrum is displaced by a Doppler shift corresponding to the component of $\mathbf{u}$ along $\mathbf{k}$. The ion component becomes strongly asymmetrical. This may be explained on the basis that an ion acoustic wave travelling with a phase velocity component along $\mathbf{k}$ which is in the same direction as $\mathbf{u}$ suffers less Landau damping, whereas the wave whose velocity component along $\mathbf{k}$ is in the opposite direction to that of $\mathbf{u}$ will be more heavily damped. Some results of numerical calculations by Evans *et al.* (1966c) are given in Figure 3.12. The drift velocity parameter y_u is equal to $(\mathbf{u} \cdot \mathbf{k})/Y_e$.

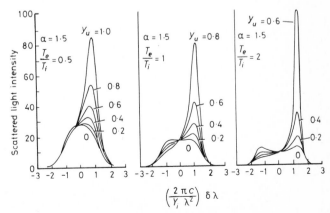

3.12 Scattered light spectra calculated for plasmas in which the electrons have a drift velocity $\mathbf{u}$ relative to the ions. $y_u = (\mathbf{u} \cdot \mathbf{k})/Y_e$. (After Evans *et al.* (1966c))

5.5. EFFECTS OF A MAGNETIC FIELD

A uniform magnetic field curves the paths of electrons in the plane perpendicular to the field lines, but has no effect on their motion parallel to the field. Thus the effect of a magnetic field **B** on the scattered light spectrum may be expected to depend on its orientation relative to $\mathbf{k} = \mathbf{k}_s - \mathbf{k}_0$ and to vanish when **B** is parallel to **k**.

Let us first consider a single electron in a magnetic field **B** whose direction is at an angle ψ to the normal to **k** (Lehner and Pohl (1970)). Let the electron velocity components parallel and perpendicular to the magnetic field be $v_{\parallel}$ and $v_{\perp}$ respectively (see Figure 3.13). The frequency of gyration of the electron around **B**, or electron

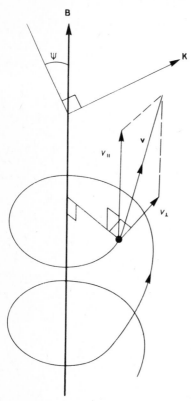

3.13   Scattering geometry in a magnetic field

cyclotron frequency, is $\omega_{ce} = eB/m_e$. Then the component of electron velocity parallel to **k** is given by

$$v_k = v_{\parallel} \sin \psi + v_{\perp} \cos \psi \sin (\omega_{ce} t). \qquad (3.42)$$

The frequency shift observed, $\mathbf{k} \cdot \mathbf{v}$ is therefore time-dependent, and on Fourier analysis the scattered light spectrum consists of a number of lines of frequency ω_{sn} given by

$$\omega_{sn} - \omega_0 = k v_{\parallel} \sin \psi + n \omega_{ce}, \qquad (3.43)$$

where the integer n can take positive and negative values.

Detailed analyses of the scattering spectrum from homogeneous thermal plasmas in magnetic fields have been given by Fejer (1961), Farley *et al.* (1961), Hagfors (1961) and Salpeter (1961) (see also Platzman *et al.* (1968)). For $\alpha \ll 1$, it is found that with moderate magnetic fields ($\omega_{ce} \ll Y_e$) the spectrum consists essentially of that given by expression (3.27), with a modulation superimposed at harmonics of the electron cyclotron frequency. The depth of modulation depends strongly on the alignment of the magnetic field relative to **k**. If $kr_{ge} \gg 1$, r_{ge} being the electron gyroradius, the spectrum may be approximated (Carolan & Evans (1971*a*)) by the expression

$$S(\mathbf{k}, \omega, \psi) = \left[\frac{\Gamma_\alpha(y_e)}{Y_e} + \left(\frac{\alpha^2}{1+\alpha^2} \right)^2 \frac{\Gamma_\beta(y_i)}{Y_i} \right] \pi^{-\frac{1}{2}} \sum_n \frac{\exp - \left(\frac{\omega - n\omega_{ce}}{Y_e \sin \psi} \right)^2}{Y_e \sin \psi}. \qquad (3.44)$$

The width of a single line is

$$\Delta\omega(T_e, \psi) = 2 Y_e \sin \psi. \qquad (3.45)$$

For the modulation to be well defined, $\Delta\omega(T_e, \psi) \lesssim \omega_{ce}$. This is equivalent to requiring that the electron gyroradius be less than $(k \sin \psi)^{-1}$, and in general it is necessary that $\psi \simeq 0$. This raises interesting practical problems, because if scattered light is to be detected a non-zero solid angle of acceptance is necessary, corresponding to a finite range of ψ. The effects of a finite cone of acceptance angles, and of mis-alignment of the axis of the cone, were discussed by Carolan & Evans (1971*b*). These authors pointed out that misalignment in *azimuth* relative to **B** has no effect on the scattered light spectrum. Figures 3.14(*a*) and (*b*) show scattering spectra calculated for two cone semi-angles, 0·425° and 0·85°, for various angles of misalignment ζ in *latitude* with respect to **B**. The results show that the degree of modulation observed in scattered light collected within a cone is much greater than might be expected if the effects of symmetry about **B** were ignored.

For $\alpha \gg 1$, the ion component dominates the scattered light spectrum, and in moderate magnetic fields ($\omega_{ci} \ll Y_i$) is modulated at the ion cyclotron frequency.

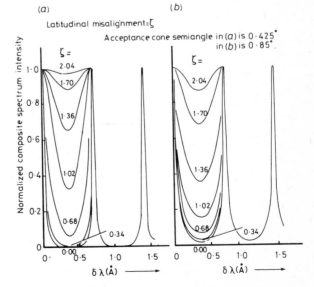

3.14 Scattered light spectra for a plasma in a magnetic field, calculated for several values of the angle ζ of misalignment in latitude w.r.t. **B**, and for two values of the acceptance cone semi-angle (*a*) 0·425°, (*b*) 0·85°. $T_e = 20\,\mathrm{eV}$, $n_e = 3 \times 10^{15}\,\mathrm{cm}^{-3}$, $B = 16\,\mathrm{kG}$ and $k = 4·7 \times 10^4\,\mathrm{cm}^{-1}$. (After Carolan & Evans (1971*b*))

The alignment of $\mathbf{B}$ perpendicular to $\mathbf{k}$ is now even more critical than for the electron component, and this makes observation of modulation very difficult. Evans & Carolan (1970), making use of the general theory given by Hagfors (1961), found that if $\omega_{ce} < Y_i = k(2\kappa_B T_i/m_i)^{\frac{1}{2}}$ it is also possible for the ion component to be modulated at the electron cyclotron frequency. Modulation at frequencies corresponding to a number of hybrid resonances has been predicted (Salpeter (1961)).

In very strong magnetic fields, when $\omega_{ce} \gg Y_e$ for the electron spectrum or when $\omega_{ci} \gg Y_i$ for the ion spectrum, the magnetic field causes a narrowing of the corresponding spectrum by a factor $\sin \psi$.

A general discussion of scattering from turbulent, inhomogeneous, magnetized plasmas has been given by Sakhokiya & Tsytovich (1968).

Sheffield (1972a) suggested that, by arranging for $\mathbf{k}$ and $\mathbf{B}$ to be almost orthogonal, the depth of modulation of the electron spectrum ($\alpha \ll 1$) may be used to determine accurately the direction of the magnetic field in a 'Tokamak' device. He pointed out that all the peaks of the modulated spectrum can be superimposed if a Fabry–Perot étalon with the free spectral range equal to the electron cyclotron frequency is used to analyse the scattered light. Related calculations were published by Meyer & Leclert (1972a, b) and Meyer et al. (1973). However, Boyd et al. (1971) have found that a drift velocity of electrons relative to ions can enhance the modulation and increase the total scattered power. This effect may have to be taken into account.

5.6. EFFECTS OF COLLISIONS

The effects of Coulomb collisions on the scattered light spectrum, which have been neglected so far in this chapter, were investigated by Dubois & Gilinsky (1964) for plasmas in which $\lambda_D^3 n_e \gg 1$.

The important parameter in the case of Coulomb collisions is the electron–electron collision frequency, which may be written

$$v_{ee} = \frac{3}{2^{\frac{1}{2}}} \omega_{pe} \frac{\ln \Lambda_e}{\Lambda_e}, \tag{3.46}$$

where $\Lambda_e = 1{\cdot}55 \times 10^{10} \, T_e^{\frac{3}{2}} n_e^{-\frac{1}{2}}$, n_e being in cm^{-3} and T_e in eV. At high frequencies, features in the scattered spectrum of smaller scale than v_{ee} tend to be blurred. This effect is likely to be particularly important in the case of the cyclotron modulation discussed in the previous section. On the other hand, the low-frequency ion acoustic resonances are *sharpened* by collisions if their frequency is much lower than the electron–electron and ion–ion collision frequencies, because ion acoustic waves suffer less damping as the collision frequency increases.

Linnebur & Duderstadt (1973a) used projection operator techniques to find $S(\mathbf{k}, \omega)$ allowing for collisions in plasmas with arbitrary values of $\lambda_D^3 n_e$. Figure 3.15 shows the computed profile of the electron feature for 90° scattering with $n_e = 10^{19} \, \text{cm}^{-3}$, $T_e = T_i = 4{\cdot}5 \, \text{eV}$, according to this method, compared to calculations based on the Fokker–Planck equation (Grewal (1964)). The Vlasov equation would give very

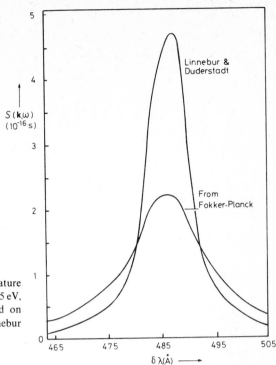

3.15 $S(\mathbf{k}, \omega)$ in vicinity of the electron feature calculated for $n_e = 10^{19}$ cm^{-3}, $T_e = T_i = 4.5$ eV, $\theta = 90°$, compared with calculations based on the Fokker–Planck equation. (After Linnebur & Duderstadt (1973a))

nearly a delta function under these conditions. Figure 3.16 shows the calculated shape of the central ion feature under the same conditions, compared to other calculations.

The effects of collisions between charged particles and neutral particles have been considered by Dougherty & Farley (1963), Lepechinski (1967), Alexandrov et al. (1968) and by Linnebur & Duderstadt (1973b).

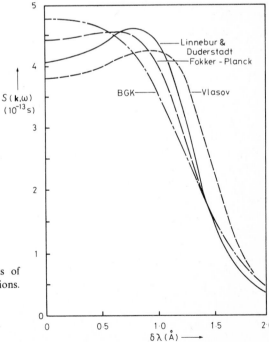

3.16   Ion feature of $S(\mathbf{k}, \omega)$ for conditions of Figure 3.15, compared with other calculations. (After Linnebur & Duderstadt (1973a))

5.7. EFFECTS OF IMPURITIES

So far we have considered only plasmas containing electrons and a single species of ion. The presence of even quite small proportions of impurities can have important effects on the spectrum of scattered light which were studied by Evans (1970) for the case of a hydrogen plasma (see also Fejer (1961)). Evans derived a generalized form of expression (3.14) for the spectrum of electron density fluctuations:

$$S(\mathbf{k}, \omega) = \frac{|1 - \sum_j \mathscr{G}_j|^2 \mathscr{F}_e + |\mathscr{G}_e|^2 \sum_j b_j \mathscr{F}_j}{|1 - \mathscr{G}_e - \sum_j \mathscr{G}_j|^2} \tag{3.47}$$

where $b_j = Z_j^2 n_j / (\sum_j n_j Z_j)$ and n_j, Z_j are the number density and charge of the jth ion species. As both the numerator and the denominator of (3.47) are changed by the presence of additional ion species, a linear superposition of contributions from the various species is not possible.

The electron component of the scattered light spectrum is not much affected by the presence of impurities. The ion component was computed without approximation by Evans for a large number of different situations. The effects of varying the concentration of fully-ionized oxygen atoms (O^{8+}) in a hydrogen plasma are shown in Figure 3.17, and the effects of varying the stage of ionization of 1% of iron atoms are shown in Figure 3.18. In each case $\alpha = 1$. The integrals under the curves of Figure 3.17 are shown in Figure 3.19: for $T_e/T_i = 1$ the enhancement of the total scattered light is greater than would be expected from a linear superposition of the contributions

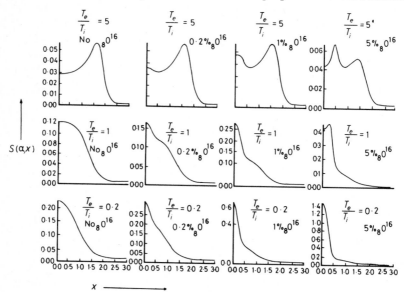

3.17   Effect on scattered light spectra of varying concentration of fully ionized oxygen atoms O^{8+} in a hydrogen plasma. $x = (\omega/k)(2\kappa_B T_i/m_p)^{1/2}$, $\alpha = 1$. (After Evans (1970))

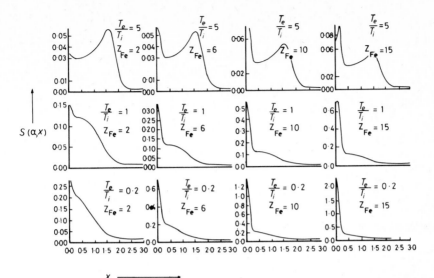

3.18 Effect on scattered light spectra of adding 1% of iron atoms of charge Z_{Fe} to a hydrogen plasma. $x = (\omega/k)(2\kappa_B T_i/m_p)^{1/2}$, $\alpha = 1$. (After Evans (1970))

from hydrogen and oxygen plasmas. However, the total scattering is unchanged by the presence of the impurity if $T_e/T_i \simeq 10$.

In a plasma containing heavy atoms several stages of ionization usually coexist in significant proportions within the scattering volume. Interpretation of the ion component scattering spectrum requires a knowledge of the ionization equilibrium, which may sometimes be deduced from a study of the electron component which is unaffected by impurities. Often, however, there is no thermal equilibrium, and ions of different species may have different temperatures. This possibility too was investigated by Evans. At low temperatures the contribution of Rayleigh scattering can be important. The limits of discrimination under these conditions have been discussed by Pyatnitskii et al. (1971) (see also Vriens (1973)).

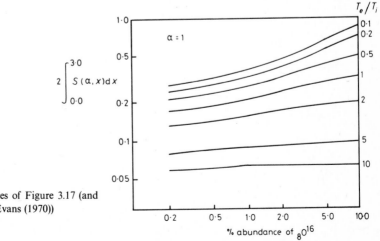

3.19 Integrals of curves of Figure 3.17 (and similar curves). (After Evans (1970))

5.8. EFFECTS OF INHOMOGENEITIES IN THE PLASMA

In the preceding sections it has been assumed that the plasma within the scattering volume is homogeneous. It is often possible to make the scattering volume sufficiently small for this assumption to be valid, but in some cases extremely large density and temperature gradients may occur, as for example in plasmas produced by a laser focused on a solid target (see Chapters 7 & 8). The interpretation of scattering spectra under these conditions presents some formidable problems.

5.9. EFFECTS OF LARGE-AMPLITUDE FLUCTUATIONS

The theory of Rosenbluth & Rostoker given in Section (4) holds for any stable particle distribution functions, and may be applied to scattering from enhanced fluctuations associated with stable non-Maxwellian distributions. Thus, for example, Kroll et al. (1964) have calculated the enhanced scattering to be expected when strong plasma oscillations are stimulated by the parametric interaction of two laser beams whose frequency difference is near the electron plasma frequency (see Chapter 2, Section (9.4)). However, the theory is not applicable to plasmas in which an instability is growing. Scattering from marginally stable plasmas has been discussed by several authors: Joyce & Salat (1971) have extended the general theory using the method of Rogister & Oberman (1968). The time history of the distribution functions must be known in order to find $S(\mathbf{k}, \omega, t)$ under these conditions.

5.10. EFFECTS OF HIGH TEMPERATURES

The theory given above is valid only for non-relativistic electrons. At plasma electron temperatures exceeding a few hundred eV, a significant proportion of the electrons have velocities approaching the speed of light, and relativistic effects must be taken into account.

Calculations by Pappert (1963), by Pechacek & Trivelpiece (1967) and by Theimer & Sollid (1968a, b) show that at high temperatures the scattered light spectrum for uncorrelated electrons becomes asymmetrical and the peak is displaced, with respect to the incident light frequency, to higher frequencies. The spectrum given by Pechacek & Trivelpiece for a scattering angle of 90° in the plane perpendicular to the incident electric vector is shown in Figure 3.20. The effects of a finite scattering volume are also discussed in this paper and it is concluded that the scattering spectrum for a bounded high-temperature plasma is significantly different from that calculated for an unbounded plasma. The effect may be illustrated by a simple example.

Figure 3.21 shows space–time diagrams for two situations. In (a) a single emitting particle travels continuously at velocity v for a time t towards a detector at X, so that the detector receives radiation energy W in a period $t_2 - t_1$. In (b) the particle travels at the same speed from the same origin over a distance d, but at this point leaves the scattering volume and is replaced by a second particle at the origin. This sequence is repeated indefinitely. The energy received by the detector in the interval $t_2 - t_1$ is

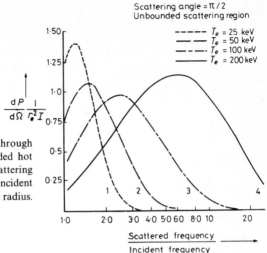

Scattering angle $= \pi/2$
Unbounded scattering region

- - - - - $T_e = 25$ keV
——— $T_e = 50$ keV
—·—· $T_e = 100$ keV
——— $T_e = 200$ keV

$$\frac{dP}{d\Omega}\frac{1}{r_e^2 I}$$

Scattered frequency
———————————
Incident frequency

3.20 Theoretical spectra of light scattered through 90° from uncorrelated electrons in unbounded hot plasmas, allowing for relativistic effects. Scattering plane is perpendicular to electric vector of incident beam (of irradiance I): r_e = classical electron radius. (After Pechacek & Trivelpiece (1967))

now only $[1-(v/c)]W$. Yet on average the scattering region contains one electron in each case. The effect has been taken into account in Figure 3.22. The transit time of scattering particles across the scattering volume may also necessitate a further correction to the scattered light spectrum. The spectrum is broadened as a result of the limited duration of the wave train emitted by a single particle, as in collision broadening. The effect is small unless the effective scattering volume has dimensions of the order of a few wavelengths or less.

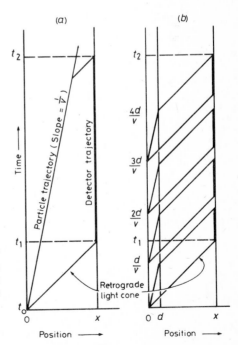

3.21 Effects of finite scattering volume on energy received by a detector. In each case the scattering region contains only one particle at any time. Heavy line indicates non-zero instantaneous scattered power at the detector. (After Pechacek & Trivelpiece (1967))

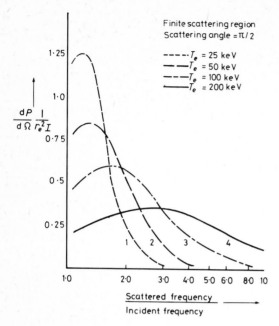

3.22 As for Figure 3.20, but for a bounded scattering volume. (After Pechacek & Trivelpiece (1967))

Sheffield (1972*b*) calculated spectra for incoherent ($\alpha \ll 1$) scattering from hot plasmas, including the effects of finite volume. He concluded that the shift $(\delta\lambda)_{\text{rel}}$ in the spectrum of light scattered through an angle θ from an incident beam at λ_0 is given by

$$\frac{(\delta\lambda)_{\text{rel}}}{\lambda_0} \simeq -2 \cdot 8 \times 10^{-5} T_e \sin^2\left(\frac{\theta}{2}\right), \tag{3.48}$$

where the electron temperature T_e is in eV. The apparent electron temperature, deduced from measurements at 90° of the spectrum on the short wavelength side of the laser line without making relativistic corrections, would be higher than the true temperature by a factor of about $(1 + 6 \times 10^{-3} T_e^{\frac{1}{2}})$, where T_e is in eV.

6. POWER AND ENERGY REQUIREMENTS FOR THE INCIDENT BEAM: THE PROBLEM OF SELF-LUMINOSITY

As we have seen, the scattering cross-section σ_T is extremely small. Even with very dense plasmas only a tiny fraction of the incident light is scattered. At the same time, plasmas are intensely luminous, emitting continuous radiation from free–free and free–bound transitions as well as line radiation. In order to be able to detect and analyse the scattered light it is necessary to have a very intense monochromatic incident beam, which produces sufficient scattered photons to provide a measurable signal which may be distinguished from the self-luminosity of the plasma. We shall follow the discussion of beam power and energy requirements given by Evans & Katzenstein (1969).

For scattering studies to be possible a plasma must be optically thin in the wavelength region of the incident light. We shall consider for simplicity a hot, fully-ionized hydrogen plasma ($T_e \approx 100\,\text{eV}$), in which the principal emission mechanism will be Bremsstrahlung. Making use of expression (2.79) for the emission coefficient, the power $P'_\beta(\lambda_0)$ emitted per unit volume of plasma into solid angle $\delta\Omega$ in the wavelength range $\delta\lambda$ at λ_0 is given by

$$P'_\beta(\lambda_0) = 2j_\omega \frac{d\omega}{d\lambda} \delta\lambda \, \delta\Omega$$

$$\simeq 4.3 \times 10^{-27} \frac{n_e^2}{\lambda_0^2 T_e^{\frac{1}{2}}} \delta\lambda \, \delta\Omega.$$

(3.49)

Here n_e is in cm^{-3}, λ_0 in Å and T_e in K.

The power scattered per unit volume of plasma into solid angle $\delta\Omega$ in the wavelength range $\delta\lambda$ at λ from a parallel beam of irradiance I is

$$P'_s(\lambda, \alpha) = I\sigma_T n_e \frac{\delta\Omega}{4\pi} S(\lambda, \alpha) \, \delta\lambda.$$

(3.50)

Here $S(\lambda, \alpha)$ is normalized so that the integral over all wavelengths is unity: it is obtained by dividing expression (3.27) for $S(\mathbf{k}, \omega)$ by the integral $[S(\mathbf{k})_e + S(\mathbf{k})_i]$ from expressions (3.30) and (3.31) and changing the variable. We assume $Z = 1$ and $T_e = T_i$. When $\alpha \ll 1$ the value of $S(\lambda, \alpha)$ at the centre of the scattered spectrum, λ_0, is

$$S(\lambda_0, 0) = \frac{c}{2\pi^{\frac{1}{2}}} \left[\frac{1}{\lambda_0 \sin(\theta/2)} \right] \left(\frac{m_e}{2\kappa_B T_e} \right)^{\frac{1}{2}},$$

(3.51)

and when $\alpha \gg 1$

$$S(\lambda_0, \infty) = \frac{c}{4\pi^{\frac{1}{2}}} \left[\frac{1}{\lambda_0 \sin(\theta/2)} \right] \left(\frac{m_i}{2\kappa_B T_e} \right)^{\frac{1}{2}}.$$

(3.52)

The laser beam irradiance I_1 required in order that the scattered light power and the self-luminosity power received by the detector from the same element of plasma at the wavelength of the incident light should be equal is found by equating P'_β and P'_s from (3.49) and (3.50):

$$I_1 \simeq 5.3 \times 10^{-6} \frac{n_e}{\lambda_0} \sin\left(\frac{\theta}{2}\right) \text{W cm}^{-2}$$

(3.53)

when $\alpha \ll 1$, or

$$I_1 \simeq 2.5 \times 10^{-7} \frac{n_e}{\lambda_0} \sin\left(\frac{\theta}{2}\right) \text{W cm}^{-2}$$

(3.54)

when $\alpha \gg 1$. Here n_e is in cm^{-3} and λ_0 in Å.

What matters from the point of view of recording the scattered light spectrum, however, is the signal-to-noise ratio in the output from the detector. The laser pulse duration τ_L is generally much shorter than the time constant characterizing significant variations in the self-luminous plasma light. Thus if the detector bandwidth is matched to the reciprocal of the laser pulse duration, in addition to the scattered light signal only random statistical fluctuations in the self-luminous signal will affect the detector. In a photomultiplier shot noise predominates, so the overall signal-to-noise ratio will be given by the ratio

$$\Sigma = \frac{\text{number of photoelectrons produced by scattered light in time } \tau_L}{(\text{total number of photoelectrons produced in time } \tau_L)^{\frac{1}{2}}}. \quad (3.55)$$

If the fraction of the light directed towards the photomultiplier which actually reaches the photocathode is f, and the quantum efficiency of the photocathode is q ($q < 1$),

$$\Sigma = \frac{P'_s \text{D}}{[(P'_s + P'_\beta)\text{D}]^{\frac{1}{2}}} \quad (3.56)$$

where $\text{D} = Vfq\tau_L\lambda_0/hc$. Here we have assumed for simplicity that the volume of plasma, V, from which Bremsstrahlung is received equals the scattering volume.

We may distinguish two limiting cases:
1.   The Bremsstrahlung power received by the detector, P'_β, is much less than the scattered power received, P'_s ($I \gg I_1$).
2.   The Bremsstrahlung power greatly exceeds the scattered power ($I \ll I_1$).

In the first case, the signal-to-noise ratio is limited simply by a shortage of scattered photons. This situation is most likely to occur in a low-density plasma. Here

$$\Sigma = \left[I\sigma_T n_e \frac{\delta\Omega}{4\pi} S(\lambda, \alpha)\text{D}\delta\lambda \right]^{\frac{1}{2}}. \quad (3.57)$$

For a given laser pulse energy, if scattered light is accepted from a fixed length of the laser beam, no improvement is possible by varying the laser beam diameter or power. The signal-to-noise ratio varies as the square root of both the laser pulse energy and the electron density, and as $T_e^{-\frac{1}{4}}$ since the half-width of the scattered light spectrum varies as $T_e^{\frac{1}{2}}$. For a given experimental arrangement, this sets a lower limit to the density at which useful scattering measurements may be made. Many factors enter into Σ and it must be worked out for each experimental situation. As a rough guide, for a $90°$ scattering experiment with $\alpha \ll 1$ it should be possible to determine the broad profile of the electron component with reasonable resolution using normal techniques provided the electron density exceeds $10^{13}l^{-1}W_L^{-1}$ cm^{-3}, W_L being the laser pulse energy in joules and l (cm) the beam length from which scattered light is received. At very low densities, however, stray light entering the detector may cause severe problems, which are discussed below.

In the second case, when self-luminosity predominates,

$$\Sigma \propto I\sigma_T S(\lambda, \alpha)T_e^{\frac{1}{4}}\lambda_0 \left[\text{D}\delta\lambda\frac{\delta\Omega}{4\pi} \right]^{\frac{1}{2}}, \quad (3.58)$$

and for a given laser pulse energy it is now advantageous to focus the laser beam into a small area and to concentrate the energy into a short, high-power pulse. The signal-to-noise ratio again varies as $T_e^{-\frac{1}{4}}$, but in this case it is independent of the electron density.

It is sometimes helpful to make use of the fact that the light scattered from a linearly polarized beam is itself fully polarized, whereas the self-luminosity of the plasma is normally only slightly polarized. Suppose a polarizer, placed before the detector, has a fractional power transmittance of $\mathcal{T}$ for the allowed polarization, which coincides with that of the scattered light. Then the resulting signal-to-noise ratio will be

$$\Sigma_p = \frac{P'_s \mathrm{D}\mathcal{T}}{\left[\left[\left(\frac{P'_\beta}{2}+P'_s\right)\mathrm{D}\mathcal{T}\right]\right]^{\frac{1}{2}}} \tag{3.59}$$

so

$$\frac{\Sigma_p}{\Sigma} = \left[\frac{1+\left(\frac{P'_s}{P'_\beta}\right)\mathcal{T}}{\frac{1}{2}+\frac{P'_s}{P'_\beta}}\right]^{\frac{1}{2}}. \tag{3.60}$$

The polarizer is most effective when $P'_\beta \gg P'_s$, in which case it can produce a maximum improvement of $\sqrt{2}$ in the signal-to-noise ratio if $\mathcal{T} = 1$. However, it will always do more harm than good if $\mathcal{T} < \frac{1}{2}$.

If a very long laser pulse is used, during which significant variations in the self-luminous intensity occur, the situation is somewhat different. It is necessary to measure the self-luminous signal and deduct it from the total to obtain the scattered light signal. This can be done by repeated experiments with and without the laser pulse, but the shot-to-shot variation in plasma luminosity is often rather large. Alternatively a differential method may be used, as suggested in an early feasibility study (Hughes (1962)), making use of the polarization properties mentioned above. The scattered light is divided into two beams at a polarizing beam-splitter, one containing the scattered light and half the self-luminosity, the other containing only half the self-luminosity. These two beams pass to two detectors A and B respectively with balanced outputs, and the signal from B is subtracted from that from A. Then the signal-to-noise ratio in the resulting differential output is given by

$$(\Sigma'_p)_{\text{diff}} = \frac{P'_s \mathrm{D}\mathcal{T}}{\left[\frac{P'_\beta}{2}\mathrm{D}\mathcal{T}+\left(\frac{P'_\beta}{2}+P'_s\right)\mathrm{D}\mathcal{T}\right]^{\frac{1}{2}}}$$

$$= \frac{P'_s \mathrm{D}\mathcal{T}}{[(P'_\beta+P'_s)\mathrm{D}\mathcal{T}]^{\frac{1}{2}}}, \tag{3.61}$$

if we assume that the beam-splitter transmission coefficient is the same for both polarizations. If, instead, two successive measurements are made with the polarizer, one being made without the laser pulse, then ignoring the shot-to-shot variation, the result of combining the two gives

$$\Sigma'_p = \frac{P'_s \mathrm{D}\mathscr{T}}{\left[\left(\frac{P'_\beta}{2} + P'_s\right)\mathrm{D}\mathscr{T} + \left(\frac{P'_\beta}{2}\right)\mathrm{D}\mathscr{T}\right]^{\frac{1}{2}}}$$

$$= \frac{P'_s \mathrm{D}\mathscr{T}}{[(P'_\beta + P'_s)\mathrm{D}\mathscr{T}]^{\frac{1}{2}}}. \tag{3.62}$$

In this case, the ratio

$$\frac{(\Sigma'_p)_{\mathrm{diff}}}{\Sigma'_p} = 1. \tag{3.63}$$

However, if two shots are allowed for both techniques, the differential method will give a value of Σ which is $\sqrt{2}$ times greater. The differential, single-shot method is preferable anyway because it eliminates the often considerable variation from shot to shot.

The differential method is no use for a short laser pulse in a long plasma pulse. One channel then collects the scattered light, P'_s, plus the *fluctuations* in the self-luminous radiation of one polarization, assuming that the detector bandwidth is matched to the laser pulse duration; the other channel collects the fluctuations in the self-luminous radiation of the other polarization. The fluctuations in the two polarizations are uncorrelated. In this case the differential method again gives the signal-to-noise ratio $(\Sigma'_p)_{\mathrm{diff}}$ of expression (3.61) while the direct measurement using a polarizer gives Σ_p of expression (3.59), which is higher.

In all the above calculations it has been assumed that the radiation received by the detector is emitted only from the scattering volume. One of the great merits of the scattering technique is that excellent spatial resolution can be achieved even in a large plasma. However, the cone of observation which contains the small scattering volume may also contain a much larger volume of self-luminous plasma, and different values of V should be used in expressions for P_s and P_β. The problem of discrimination then becomes more difficult.

7. EXPERIMENTAL METHODS

7.1. THE LIGHT SOURCE

The light source most conveniently satisfying the requirements of brightness and monochromaticity for scattering studies is the ruby laser in its various forms. The spectrum of ruby laser emission has been extensively studied by many authors (see especially Röss (1969)). Mode selection techniques restricting the longitudinal mode number in the laser oscillator to a single value can reduce the bandwidth to as little as 10^{-3} Å (Hercher (1965)), which is narrow enough for most scattering experiments. Mode-hopping must be avoided. Q-switching and 'cavity dumping' or 'pulse transmission mode' operation of the oscillator allow the laser energy to be concentrated into pulses with a duration of a few nanoseconds—as short as the resolving time of a fast photomultiplier. If very high spectral resolution is not required, mode-locking techniques may be used to reduce the laser pulse duration to about 10^{-10} s. Amplifiers can readily bring the energy of the longer pulses up to some tens of joules in an almost perfectly parallel beam. The wavelength of the ruby laser light (6943 Å at room temperature) is sufficiently short to allow it to penetrate most plasmas (the critical electron density at which a plasma normally reflects light at 6943 Å is about 2×10^{21} cm^{-3}), and photocathodes with quantum efficiencies of a few per cent are available to detect it. If for any reason a shorter wavelength should be needed, efficient frequency-doubling crystals are available which provide an output at 3472 Å. The high-power ruby laser is, in fact, an almost perfect source for scattering experiments in laboratory plasmas, apart from its high cost and rather low pulse repetition frequency. There are, however, certain circumstances in which longer wavelengths are desirable, and carbon dioxide lasers at 10·6 μm have been used in some scattering experiments.

7.2. METHODS OF REDUCING STRAY LIGHT

The total light power P_s scattered from a length l of a laser beam of power P_0 in a plasma of electron density n_e is given approximately by

$$\frac{P_s}{P_0} = \sigma_T n_e l. \tag{3.64}$$

Because the Thomson scattering cross-section σ_T is extremely small, only 10^{-12} to 10^{-10} of the incident light power is scattered per cm from plasmas with densities of 10^{12} to 10^{14} cm^{-3}. A much larger fraction of the incident beam power can find its way into the detector as stray light unless great care is taken to prevent it. Scattering from optical components in the laser beam, and diffuse reflection from the walls of the plasma chamber are the main problems. Gerry & Rose (1966) discussed methods of reducing stray light in scattering experiments with ruby laser light and quoted some diffuse reflectivities given by Kachen for several materials, which are reproduced in Table (3.1).

It is essential for the laser beam to be absorbed as completely as possible after it has passed through the scattering volume. This is normally done by means of a tapered absorber, often consisting of black or strongly-absorbing filter glass so that light not absorbed is specularly reflected. The absorption is most effective if the laser beam (which is usually linearly polarized) meets the first surface at Brewster's angle.

It is important, too, that the solid angle within which scattered light is collected by the detector should have as black as background as possible.

A focal plane stop should be placed at an intermediate focus in the path of light

TABLE 3.1: Diffuse reflectivities

Material	Fraction of 45° incident light reflected per steradian at 0°
Black polished glass	1×10^{-5}
Fused quartz	2×10^{-5}
Acetylene soot	$1\text{--}2 \times 10^{-3}$
Kodak dull black paint	7×10^{-3}
Black velvet	2×10^{-3}
Brass, polished, ebanol-C black, buffed	6×10^{-3}
Aluminium, hard, polished, black, anodized	2×10^{-3}
Stainless steel, 304 passivate black, polished	1×10^{-2}

After Kachen, from Gerry & Rose (1966).

entering the detector, with the aperture just fitting the image of the scattering volume, so as to limit the volume of the plasma and the area of the walls contributing to the input to the detector.

The use of a focal plane stop with auxiliary lenses to clean up the spatial distribution of light in the laser beam is also very helpful, particularly in small-angle forward scattering experiments. Care must be taken not to spread the beam by diffraction from too small an aperture, nor to cause air breakdown at intermediate foci.

Dust is disastrous, causing very large spurious signals, and must be eliminated.

Stray light has the same spectrum as the incident laser beam and for a monochromatic incident beam will appear in the output of the dispersing and detecting system with the instrumental profile. One approach to the problem of eliminating its effects when measuring the broader profile of scattered light is to concentrate on filtering out light in the immediate vicinity of the laser wavelength, using a dispersing system of high contrast (Daehler & Ribe (1967)).

7.3. METHODS OF ANALYSING AND DETECTING THE SCATTERED LIGHT

In order to determine the spectrum of the low-intensity scattered light, efficient dispersing instruments are necessary. Early work relied on isolating a single wavelength band at a time, and repeating the observation in successive bands to build up the complete spectrum. This method is rather slow, and depends heavily on the reproducibility of the plasma (and in some cases the laser wavelength) from one shot to the next. As the technique of scattering improved, and more powerful ruby laser beams became available, several workers developed methods of recording complete spectra from a single laser shot using multi-channel detectors.

At the low resolution ($\delta\lambda \gtrsim 1$ Å) normally used to study the spectrum for $\alpha \ll 1$, monochromators and interference filters with narrow pass bands have been used successfully for single-channel recording, the latter being tuned to different wavelength ranges by tilting. For multi-channel recording several output slits are used to convert a monochromator into a polychromator: fibre optics have been used to carry the output in each of several wavelength elements to separate detectors. A flexible system has been described by Glock (1966). An array of three interference filters, each with its own detector, was used by Bottoms & Eisner (1966).

At the higher resolution ($\delta\lambda \lesssim 0.1$ Å) necessary, for example, to record the profile of the narrow central ion feature with $\alpha \gg 1$, measurements have usually been made with a Fabry–Perot interferometer, although a high-resolution grating polychromator has also been used (Glock (1966)). For single-channel recording, the central transmission wavelength may be varied smoothly and reproducibility by changing the pressure of a gas between and surrounding the étalon plates, whose spacing remains constant. The use of two identical étalons in series (Daehler & Ribe (1967)) improves contrast and thus helps to reduce stray light. Image dissection methods have been developed to record the fringe pattern in a single shot:* as many as 14 channels have been used. Examples of the use of all these techniques will be found in Section (8) of this chapter.

For carbon dioxide laser scattering from steady-state plasmas, a piezo-electrically scanned Fabry-Perot étalon with coated germanium plates has been used (Offenberger & Kerr (1971)).

Scattered ruby laser light is almost always detected by means of a photomultiplier. For the laser wavelength (6943 Å) an S20 photocathode has a quantum efficiency of about 5%: Röhr (1968) has stated that a device allowing tangential illumination of the photocathode (developed with Kronast) can improve this quantum efficiency by a factor of 3 to 5.

A shuttered image converter placed with its photocathode at the exit focal plane of a spectrograph has also been used as a detector (Dolgov-Savelev et al. (1967)). It was necessary to follow this with an image intensifier stage in order to obtain photographic recording of the spectrum, and the technique needs further development. However, it is potentially very useful, because it allows spatial distributions to be studied.

At the carbon dioxide laser wavelength, helium-cooled Ge:Hg (Offenberger & Kerr (1971): Kornherr et al. (1972)) and Ge:Cu (Yokoyama et al. (1971)) detectors have been used behind cooled filters (see for example Kimmitt (1970)). When a steady-state plasma is being investigated, a continuously-operating carbon dioxide laser may be used with a chopper and a phase-sensitive detection system (Yokoyama et al. (1971)).

---

* Hirschberg (1960), Katzenstein (1965), Hirschberg & Platz (1965), Evans et al. (1966a), Daehler et al. (1969).

7.4. CALIBRATION OF THE SCATTERING SYSTEM

It is possible in principle to calculate what fraction of the incident laser power should appear at the detector for a given plasma electron density from the Thomson cross-section, the geometry of the laser beam, the scattering angle, the angle relative to the electric vector of the incident light, the solid angle accepted by the detector, the transmittance of the dispersing elements, and several other factors. It is very much easier, and more accurate, to carry out a subsidiary experiment to calibrate the system, using a cold gas at relatively high pressure in place of the plasma and observing the Rayleigh scattered light. Several gases with known Rayleigh cross-sections have been used successfully by different authors. Pure nitrogen was used by Daehler & Ribe (1967), Baconnet et al. (1969), De Silva et al. (1964) and by Gerry & Rose (1966), who found that a two hour settling period was necessary for reproducible results to be obtained. Gerry & Rose also noted that atmospheric air produced signals that saturated the detector and remained anomalously high for 24 hours, but this may have been a local difficulty, because Patrick (1965) reported no problems with atmospheric air. Propane has been used at the Garching Laboratory (Röhr (1968), Kronast et al. (1966)). It has the advantage of a large Rayleigh cross-section, so that only relatively low pressures are necessary, which are more convenient to handle in a vacuum system. Helium was used by Malyshev et.al. (1966b): argon by Schwarz (1965) and by Consoli et al. (1966).

It is important to check that the output of the detector system increases linearly with the pressure of gas in the scattering volume. Solid particles carried into the plasma chamber or dislodged by the gas flow will produce inconsistent readings. A further check is obtained if two different gases are used: Evans et al. (1966a) used CO_2 and N_2, whose cross-sections differ by a factor of 2·3.

There was at one time uncertainty as to whether the usual Rayleigh theory for incoherent scattering applied to coherent laser light (George et al. (1965)). This was resolved by Watson & Clark (1965) who found that the normal incoherent theory agreed accurately with experimental measurements of the angular distribution of ruby light scattered from nitrogen.

8. EXPERIMENTAL OBSERVATIONS OF SCATTERED LIGHT FROM PLASMAS

The first reported observation of Thomson scattering from a laser beam was by Fiocco & Thomson (1963). In this experiment a 74 mA, 2 kV electron beam ($n_e = 5 \times 10^9$ cm^{-3}) scattered the light from a 20 J, 800 μs laser pulse and only a very few scattered photons were detected. Another beam experiment was described more recently by Ward et al. (1971) (see also Ward & Pechacek (1972)) using 50 keV electrons, which confirmed the high-energy theory described in Section (5.10) above. Thomson scattering from a plasma was probably first observed by Fünfer et al.

(1963): some doubt has been cast (Malyshev (1965)) on the interpretation of an early result by Schwarz (1963). Fünfer *et al.* used a 1 ms, 20 J ruby laser pulse and a differential polarization recording technique. A low-resolution monochromator was used to reduce the effect of stray light at the laser wavelength, but the scattering spectrum was not recorded.

Subsequent work has been concerned with measurements of the scattered light spectrum and is conveniently divided into three categories:

(1) experiments in which the electrons may be regarded as more or less independent scatterers ($\alpha < 1$),

(2) experiments in which collective effects become important ($\alpha \gtrsim 1$) and

(3) experiments in which the modulation of the spectrum due to magnetic fields is investigated.

8.1. EXPERIMENTS IN WHICH $\alpha < 1$

The first experimental measurement of the scattered light spectrum was reported by Davies & Ramsden (1964) on light from a 10 MW Q-switched ruby laser passed through the afterglow of a hydrogen θ-pinch plasma.* The electron density was 5×10^{15} cm^{-3} and the electron temperature 3.3 eV. A double monochromator dispersed the light scattered at 90° to the laser beam: the arrangement is shown in Figure 3.23. Shortly afterwards De Silva *et al.* (1964) reported similar experiments on a hydrogen arc using a 2.5 MW laser. These authors were the first to describe the use of a conical lens (or 'axicon') to collect all the light scattered through a given angle from the laser beam, which in this experiment was 170°. An interference filter isolated a narrow range of wavelengths which was varied by tilting the filter. The electron density and temperature were 9×10^{14} cm^{-3} and 2.2 eV respectively.

Patrick (1965) used a scattering method to find the electron density and temperature in plasmas behind high-speed shock waves in a magnetic annular shock tube. In these experiments $10^{15} \leqslant n_e < 10^{16}$ and 10 eV $\leqslant T_e \leqslant 100$ eV, and the scattering angle was 90°. Tilted interference filters were used to isolate various wavelength ranges. The Q-switched ruby laser pulse power was between 10 and 30 MW.

An arc plasma with $n_e = 2.5 \times 10^{13}$ cm^{-3}, $T_e = 1.8$ eV was studied using a 25 J, 0.5 ms ruby laser by Malyshev *et al.* (1966b). The light scattered through 90° was dispersed by a double monochromator with a pass bandwidth of 10 Å. Hollow-cathode arc plasmas with electron densities in the range $10^{13} < n_e < 10^{14}$ cm^{-3} and electron temperatures between 2 and 10 eV were investigated in careful experiments by Gerry & Rose (1966), using a 100 J, 1 ms laser pulse and observing the light scattered through 45° with tilted interference filters of about 3 Å pass bandwidth. The hollow-cathode arc ran steadily for long periods, which was fortunate since a set of 20 observations at different wavelengths took three hours. Some measurements are shown in Figure 3.24 for an argon plasma with an axial magnetic field of 455 G perpendicular to the scattering plane. The small bumps appearing at about 7 Å and 14 Å from the laser wave-length were noted by the authors, and similar bumps at

---

* For descriptions of the θ-pinch, z-pinch and other magnetized plasma configurations see e.g. Krall & Trivelpiece (1973).

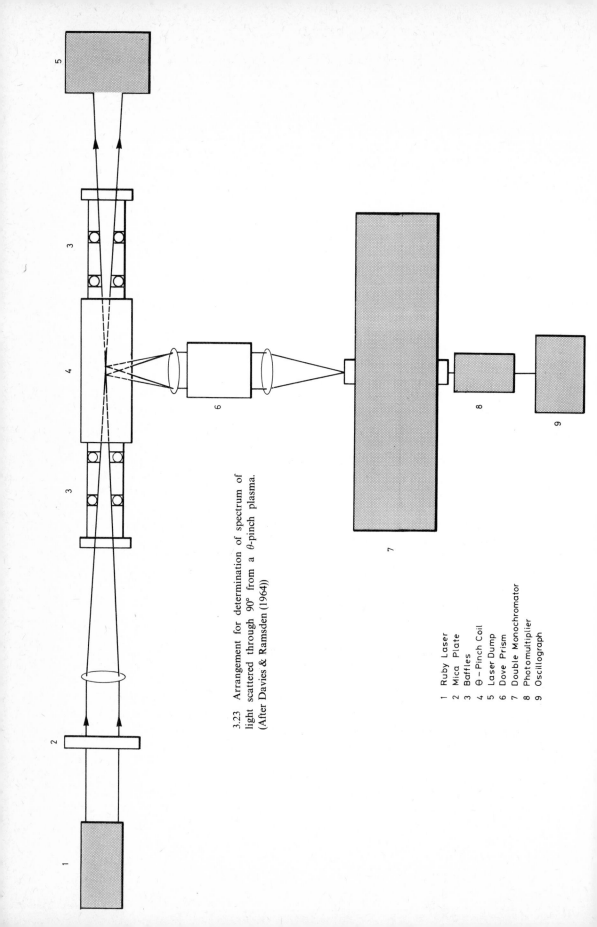

3.23 Arrangement for determination of spectrum of light scattered through 90° from a θ-pinch plasma. (After Davies & Ramsden (1964))

1 Ruby Laser
2 Mica Plate
3 Baffles
4 θ – Pinch Coil
5 Laser Dump
6 Dove Prism
7 Double Monochromator
8 Photomultiplier
9 Oscillograph

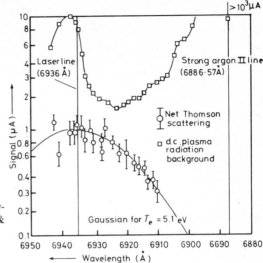

3.24 Spectrum of light scattered from a hollow-cathode argon arc plasma. (After Gerry & Rose (1966))

other wavelengths were observed under other conditions, but no good explanation was offered. Similar anomalies observed more recently are described below.

Consoli *et al.* (1966) measured the scattered light spectrum from a dense θ-pinch afterglow plasma ($n_e \sim 10^{16}\,\mathrm{cm^{-3}}$, $T_e \sim 2 \cdot 7\,\mathrm{eV}$) with a 0·6 J, 30 ns Q-switched laser. A grating monochromator resolved light scattered at 90°.

Bottoms & Eisner (1966), using a plasma produced by a coaxial plasma accelerator, and Andelfinger *et al.* (1966), using a θ-pinch plasma, determined scattered light spectra from single laser shots. Bottoms & Eisner used three interference filters, with 5 Å pass bands centred on different wavelengths, to define the spectrum, while Andelfinger *et al.* used an 8-channel grating polychromator with glass fibre exit slits.

Dolgov-Savelev (1967) imaged the light scattered from a relatively long section of the laser beam on to the slit of a grating spectrograph. The photocathode of an image convertor camera was placed in the output focal plane, and the output of the image convertor was passed through an image intensifier and finally recorded photographically. In this way the spatial variation of the spectrum of scattered light from a plasma whose properties varied along the length of the laser beam (in this case across a shock front) could be recorded from a single shot. The technical difficulties are considerable.

Dimock & Mazzucato (1968) succeeded in obtaining diametral electron density and temperature profiles for the large Stellarator C plasma although the electron density was less than $10^{13}\,\mathrm{cm^{-3}}$. The duration of the plasma was many milliseconds, so it was possible to use a 1 ms laser pulse of 50 to 100 J and yet obtain adequate time resolution. A 7-channel polychromator was used to record the spectrum of light scattered through 90°. Scattering measurements were made at several points across a diameter of the plasma from successive discharges at several different times during the discharge period.

Time- and space-resolved studies of the electron density and temperature of a plasma were also made by Watson & Beach (1969: see also Beach *et al.* (1969)). Light scattered from plasmas in both large and small θ-pinch discharge tubes was analysed with a 12-channel photomultiplier spectrometer with channels spaced at intervals of about 24 Å (Beach (1967)). Discrimination against plasma self-luminosity was facilitated by pulsing the photomultipliers for 70 ns: the outputs were delayed successively by intervals of 75 ns and displayed on a single oscilloscope trace. Temperatures of 100 to 300 eV and densities of 10^{15} to 10^{17} cm^{-3} were measured with an accuracy of about 15 %.

The experiments described above yielded measurements of plasma parameters which could often be compared with values obtained by other methods. When a comparison was possible good agreement was found, so for small values of α the theory of Thomson scattering appears to be generally satisfactory. However, anomalous enhancement of the scattered light spectrum similar to that noted by Gerry & Rose (1966) was also reported by Ringler & Nodwell (1969a, b) in light scattered at 90° from a magnetically-stabilized, high-current arc discharge. As in the experiments of Gerry & Rose, the plane of scattering was perpendicular to the axis of the arc. An axial magnetic field of 10 kG was used to stabilize the arc. The electric vector of the incident light was parallel to the magnetic field. The results are shown in Figure 3.25. The value of α in this experiment was about 0·5, so according to the Salpeter theory collective effects should be weak. Nevertheless, a narrow peak occurred in the central region of the scattering spectrum, suggesting the existence of enhanced ion oscillations, and a further maximum may be seen in the vicinity of the electron plasma frequency (14 Å) and its harmonics. Similar anomalies have been observed in hot θ-pinch discharges: a slight enhancement of the central region and a small maximum near the plasma frequency were observed with $\alpha = 0·2$ by Gondhalekar et al. (1970), and an enhancement near the plasma frequency with $\alpha = 0·45$ was

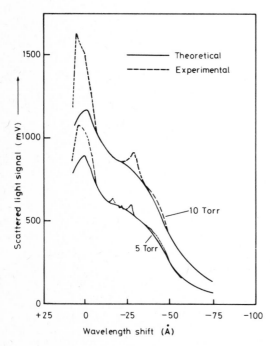

3.25   Spectrum of light scattered from a magnetically stabilized hydrogen arc plasma, for two initial filling pressures. (After Ringler & Nodwell (1969a))

noted by John *et al.* (1971). In contrast to these results, Berney (1973) found very close agreement with the Salpeter theory for scattering from a pulsed helium arc in an hour-glass shaped discharge tube without magnetic stabilization. In these experiments $0.9 \leqslant \alpha \leqslant 1.3$. The anomalies seem to occur when strong magnetic fields are present.

An asymmetry, in the form of a general displacement to shorter wavelengths, was noted by Gondhalekar & Kronast (1973) in the spectrum of light scattered from a θ-pinch plasma at an electron temperature of about 100 eV. The shift was consistent with the predictions of the relativistic theory discussed in Section (5.10) above. Shifts ascribed to electron drift velocities have been noted in θ-pinch studies by Gowers *et al.* (1971) and by Kronast & Pietrzyk (1971).

8.2. Experiments in which $\alpha \gtrsim 1$

In this section we describe experiments in which collective effects play an important part in determining the scattered light spectrum.

In their early scattering observations Fünfer *et al.* (1963) noted that the spectrum of scattered light was rather narrower than that corresponding to $\alpha = 0$ with the known electron thermal velocity distribution. Subsequently Kunze *et al.* (1964) (see also Kunze (1965)) succeeded in determining the scattered light spectra from θ-pinch plasmas with $\alpha = 0.53$ and $\alpha = 0.97$, and observed collective scattering effects unambiguously. They used a 30 ns, 5 to 10 MW ruby laser pulse. A polychromator with seven fibre optics output channels and photomultipliers was used to record the spectrum of light scattered through 90° from a single laser pulse. The measurements are shown with the theoretical profile for a plasma with $T = 55\,000$ K and $n_e = 4 \times 10^{16}$ cm^{-3} ($\alpha = 0.97$) in Figure 3.26.

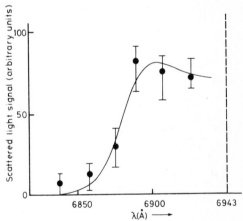

3.26 Spectrum of light scattered from a θ-pinch plasma. Theoretical curve is for $T_e = 5.5 \times 10^4$ K, $n_e = 4 \times 10^{16}$ cm^{-3}, $\alpha = 0.97$. (After Kunze *et al.* (1964))

Collective scattering was also observed by De Silva *et al.* (1964). The axicon used for their scattering measurements at 170°, described in the preceding section, was used in this experiment to collect light scattered through 10°, giving a value of $\alpha = 1.7$ with $T_e = 2.2$ eV and $n_e = 9 \times 10^{14}$ cm^{-3}. The Q-switched ruby laser had a line breadth of only 0·05 Å. A pressure-scanned Fabry–Perot interferometer provided an instrumental resolution which had to be restricted, because of the limited amount of light available, to 2 Å, within which the structure of the narrow central ion feature could not be resolved.

Ascoli-Bartoli et al. (1964) studied light scattered through only 3° from a Q-switched laser beam by a θ-pinch discharge in hydrogen, using a Fabry-Perot étalon and multiple-shot, single-channel recording. The observations at high resolution with an étalon free spectral range of 0·8 Å were inconclusive: in these measurements the laser was refrigerated to produce a narrow line and the output power was only 1·5 MW. One problem was a variation in the laser emission wavelength due to temperature changes from shot to shot. At lower resolution, with an étalon free spectral range of 30 Å, and a room temperature laser output power of 15 MW, the electron plasma frequency lines were observed on each side of the laser wavelength, the short wavelength line being much stronger than the long wavelength one, possibly as a result of an electron drift velocity relative to the ions. The displacements implied an electron density of about $3·2 \times 10^{14}$ cm^{-3}, consistent with the value of $2·4 \times 10^{14}$ cm^{-3} obtained from the total scattered intensity. Ascoli-Bartoli et al. (1965) subsequently obtained the spectrum of the narrow ion component by using a 14-channel analyser with the Fabry–Perot étalon (see Katzenstein (1965)). The outputs of all the channels were displayed on a single oscilloscope by delaying their photomultiplier output signals by successively longer times. A Kerr cell was placed before the étalon and pulsed open only for 50 ns at the time of the laser pulse. This prevented a large noise signal, the sum of the signals from all the photomultipliers, from appearing. An ion temperature of about 50 eV was deduced from the observations.

The electron plasma frequency lines were resolved in experiments by Chan & Nodwell (1966) on an argon plasma jet, with an electron density of 10^{16}–10^{17} cm^{-3} and an electron temperature of 1 or 2 eV. Here a large value of α (about 4) could be obtained at a scattering angle of 45°, which is much more convenient than small angles of 5° or 10°. A monochromator with 8 Å bandwidth and a single photomultiplier detector were used. The measured line breadth was much greater than was expected theoretically from a uniform plasma. However, the plasma temperature was known to vary rapidly over a distance of 1 mm, implying a consequent variation in electron density by a factor of 2, which would lead to a spread in the peak wavelength of the scattered lines over the scattering volume. This interpretation was later confirmed by experiments with spatial resolution (Nodwell & van der Kamp (1968)) and by a comparison of results for two values of the plasma current (Chan (1971a)).

Ramsden & Davies (1966) extended their earlier work on scattering from a θ-pinch afterglow plasma by making measurements at a scattering angle of 13° with a 5 Å bandwidth interference filter. A value of $9·0 \pm 1·5$ was obtained for the ratio of the energy in the central ion peak to the energy in a single electron plasma frequency line. This corresponds to a value of α of $3·0 \pm 0·25$ if $T_e = T_i$. The wavelength shifts of the electron plasma frequency lines were determined, using a pressure-scanned Fabry–Perot étalon of 12 Å free spectral range, to be ± 8 Å, corresponding to an electron density of $2·4 \times 10^{15}$ cm^{-3}. For $\alpha = 3$ this gave an electron temperature of

1·1 eV, in good agreement with that deduced from 90° scattering observations. The central ion peak was not resolved.

Kronast *et al.* (1966) succeeded in measuring the profile of the ion line scattered through 3° from a deuterium θ-pinch plasma. A powerful laser pulse of 500 MW was used, and the angle of observation was limited to $\pm 0.3°$. A high-resolution grating spectrograph was used with eight fibre optics output slits, each covering a wavelength range of 0·07 Å, adjacent ranges being centred 0·09 Å apart. The slits were disposed asymmetrically about the laser wavelength, only two being on the short wavelength side. Spectra were obtained with and without a plasma to give the effect of stray light, and the line breadth of the incident laser light (about 0·1 Å) was deconvoluted from the resulting profile to give the true ion line shape. One experiment gave the result shown in Figure 3.27, while another with a larger θ-pinch condenser bank

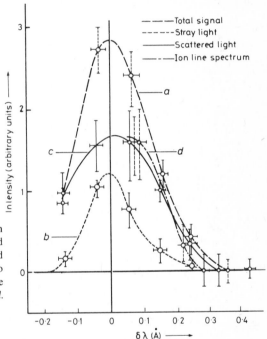

3.27 Spectrum of light scattered from a θ-pinch plasma. The stray light signal (*b*) is subtracted from the measured signal (*a*) to give the scattered light signal (*c*). Deconvolution of (*b*) from (*c*) to correct for the finite laser line breadth gives the true ion line spectrum (*d*). (After Kronast *et al.* (1966))

yielded Figure 3.28 in which a distinct central minimum occurs. On the assumption of Maxwellian electron and ion velocity distributions, the former result indicates $T_i \simeq 108$ eV, $\beta \simeq 0.65$, so for $\alpha \gg 1$, $T_e \simeq 45$ eV while the intensity of the line gives $n_e \simeq 6.3 \times 10^{16}$ cm^{-3}. The latter indicates strong longitudinal plasma waves with $k = 5 \times 10^3$ cm^{-1} and $\omega = 6 \times 10^{10}$ rad s^{-1}.

Evans *et al.* (1966*b*) used a curve-fitting technique to find the electron temperature and density of a θ-pinch plasma from the profile of the electron component with α in the region of 1. The profiles observed were very similar to those obtained by Kunze *et al.* (1964). The scattering angle was 5°. In one experiment the profile was best

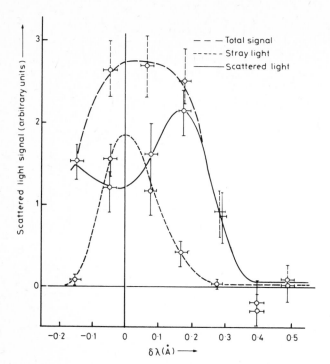

3.28   Spectrum of light scattered from a θ-pinch plasma produced with a larger condenser bank than that of Figure 3.27. (After Kronast *et al.* (1966))

fitted by $\alpha = 1\cdot3$, with the maximum scattered intensity 16 Å from the laser wavelength. From these parameters the plasma electron density and temperature were found to be $n_e = 6 \times 10^{15}\,\mathrm{cm}^{-3}$, $T_e = 103\,\mathrm{eV}$. In another experiment at a different time in the discharge the profile gave $\alpha = 1\cdot2$ and the maximum occurred 22 Å from the laser wavelength, from which $n_e = 1\cdot16 \times 10^{16}\,\mathrm{cm}^{-3}$ and $T_e = 233\,\mathrm{eV}$. A similar curve-fitting method was used later by Siemon & Benford (1969) for a low-temperature θ-pinch plasma.

The ion line from similar plasmas was also investigated by Evans *et al.* (1966c) using a pressure-scanned Fabry–Perot interferometer. A near-central minimum, similar to that reported by Kronast *et al.* (Figure 3.28), was observed. However, as may be seen in Figure 3.29, the spectrum showed a marked asymmetry which the detecting system of Kronast *et al.*, with only two channels (out of eight) on the short wavelength side of the laser wavelength, would not have recorded. The enhancement of the short wavelength wing of the spectrum was interpreted as arising from a drift velocity of the electrons parallel to **k**. In this experiment the laser beam passed axially through the θ-pinch plasma, and an axicon collected light scattered at $5°$. However, a quarter of the azimuthal aperture was obscured by the θ-pinch current leads, producing the asymmetry necessary to observe an asymmetrical spectrum due to the drift velocity.

Ramsden *et al.* continued their studies of θ-pinch plasmas using first a pressure-scanned Fabry–Perot interferometer (Ramsden & Davies (1966), Ramsden *et al.* (1966)) and subsequently a Fabry–Perot interferometer with a 7-channel image

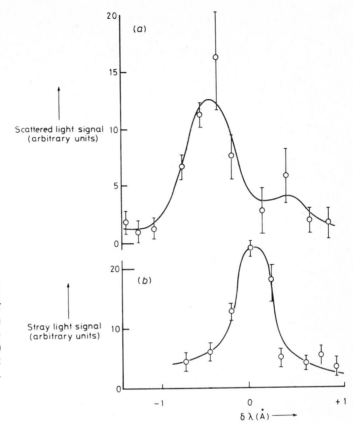

3.29 (a) Spectrum of light scattered from a θ-pinch plasma: (b) spectrum of stray laser light on the same wavelength scale. (a) is obtained by subtracting (b) from the total scattered light signal. (After Evans et al. (1966c))

dissector (Ramsden et al. (1967)) to study the ion spectrum scattered at 5° over all azimuthal angles. The θ-pinch was filled with deuterium. In later experiments, described in detail by John (1972), simultaneous 90° scattering spectra were also obtained with an 8-channel polychromator to give a measure of T_e from the electron component. The arrangement is shown in Figure 3.30. The ion spectrum recorded soon after the peak of the first half cycle of the θ-pinch current pulse (at the time of peak neutron emission) showed no sign of any departure from thermal equilibrium. The ion temperature (about 300 eV) was deduced from the breadth of the ion spectrum, making use of the measured electron temperature (between about 60 and 80 eV), and was found to be in good agreement with that obtained from the number of neutrons produced by D–D reactions.

The Scylla III θ-pinch plasma was investigated by Daehler & Ribe (1967) using a pressure-scanned pair of Fabry–Perot étalons in series and later by Daehler et al. (1969) using a 12-channel image dissector with the étalons. Scattered light was collected at a mean scattering angle of 6·25° over all azimuthal angles, using the arrangement shown in Figure 3.31. The ion feature with no bias magnetic field (Figure 3.32) was found to consist of a central peak whose breadth was only about half of that

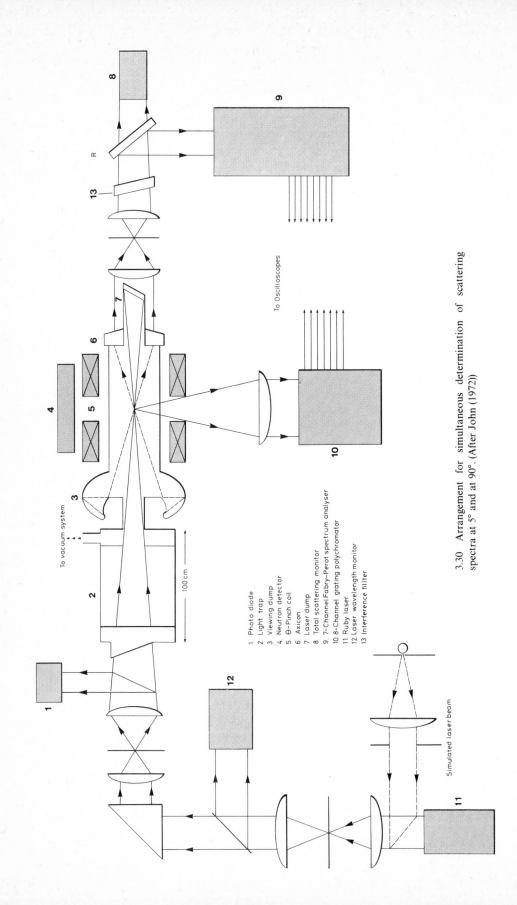

To vacuum system

100 cm

To Oscilloscopes

Simulated laser beam

R

1. Photo diode
2. Light trap
3. Viewing dump
4. Neutron detector
5. θ–Pinch coil
6. Axicon
7. Laser dump
8. Total scattering monitor
9. 7–Channel Fabry–Perot spectrum analyser
10. 8–Channel grating polychromator
11. Ruby laser
12. Laser wavelength monitor
13. Interference filter

3.30 Arrangement for simultaneous determination of scattering spectra at 5° and at 90°. (After John (1972))

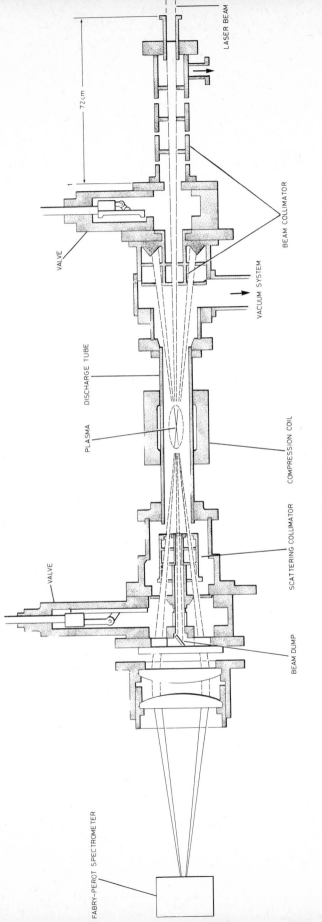

LASER BEAM

72 cm

VALVE

BEAM COLLIMATOR

VACUUM SYSTEM

DISCHARGE TUBE

PLASMA

COMPRESSION COIL

SCATTERING COLLIMATOR

VALVE

BEAM DUMP

FABRY-PEROT SPECTROMETER

3.31 Arrangement for determination of spectrum of light scattered through 6·25° from a θ-pinch plasma. (After Daehler et al. (1969))

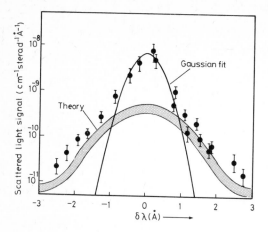

3.32 Measured scattered light spectrum from Scylla III at scattering angle of 6·25°, showing enhanced ion feature. Theoretical curve is for $T_e = 345$ eV, $T_i = 2$ keV, $n_e = 2·8 \times 10^{16}$ cm^{-3} ($\alpha = 1·23$), allowing for density variations. (After Daehler & Ribe (1967))

expected from the known properties of the discharge, but whose intensity was up to 15 times greater, superimposed on a broader distribution which fitted the theoretical predictions well. A further enhancement of the central intensity by a factor of 2 or 3 was observed when a bias field of 750–800 gauss was applied. It was found that the enhanced scattering cross-section decreased with increasing ion temperatures as determined from the neutron yield. The large observed cross-section may have been due to turbulence or microinstabilities. In these experiments the value of k observed was 10^4 cm^{-1}, so only waves or turbulent scale lengths very much shorter than the plasma column diameter of 1 cm were involved. Further experiments with other values of k might provide interesting information about the anomalous diffusion coefficient.

Röhr (1967, 1968) made careful studies of the profiles of the ion line and the electron plasma frequency lines in light scattered from a cool, dense θ-pinch plasma ($T_e \sim T_i \sim 3$ to 5 eV, $n_e \sim 5 \times 10^{17}$ cm^{-3}). The plasma parameters gave a large value of α at a scattering angle of 90°, which meant that the central ion line was relatively broad even at low temperatures and could be resolved by a high-resolution 7-channel grating polychromator. A further advantage of using 90° scattering is that because of the symmetry the light available for scattering measurements within a given solid angle may be doubled by the device shown in Figure 3.33. The ion spectra were found to be symmetric relative to the incident laser light wavelength. The experimental results for the ion line profiles were closely fitted by theoretical profiles for $T_e = T_i$, despite the fact that under these conditions the number of electrons in a Debye volume, $n_e\lambda_D^3$, was between about 10 and 2, so that the theory, which assumes $n_e\lambda_D^3 \gg 1$, might well not be valid. The electron plasma frequency lines were found to be anomalously broad, a result which was ascribed to variations in electron density over the scattering volume. When determining the electron line profile the total intensity of the ion line was measured simultaneously. The position of the electron line maximum yielded a value of the electron density in good agreement with that determined from the ion line intensity.

Scattered light spectra from a similar cool θ-pinch discharge in hydrogen were investigated by Kato (1972) under conditions where $n_e\lambda_D^3 \simeq 2$, with $\alpha \simeq 4$, and also

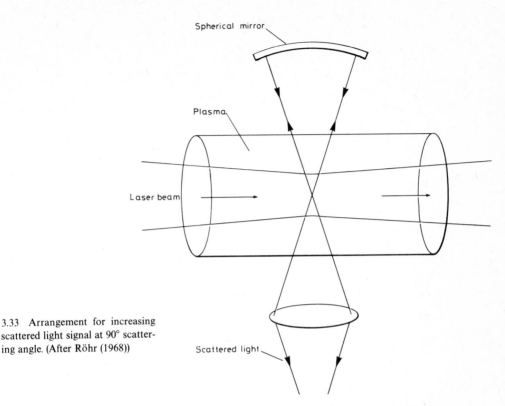

3.33 Arrangement for increasing scattered light signal at 90° scattering angle. (After Röhr (1968))

where $n_e\lambda_D^3 \simeq 12$, with $\alpha \simeq 2$. The plasma electron temperature and density were determined independently by spectroscopic measurements on the H_β line, and used to calculate theoretical scattering profiles. These were in excellent agreement with the experimentally observed scattering spectra, as may be seen from Figure 3.34 for the case of $n_e\lambda_D^3 \simeq 2$. Scattering theory appears to be valid even when only a very small number of particles is in a Debye sphere.

Röhr & Decker (1968) made a study very similar to that of Ramsden & Davies (1966) on a hot deuterium θ-pinch plasma. The ion line profile was studied at a scattering

3.34 Scattered light spectrum from a cool hydrogen θ-pinch plasma, compared with theoretical profile constructed using n_e and T_e from spectroscopic observations on H_β. (After Kato (1972))

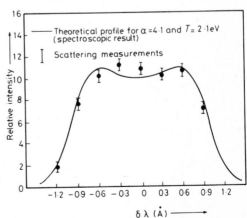

angle of 7°, and the electron component at small α at 90°. No enhancement of the ion line intensity was observed: the results were in satisfactory agreement with theory assuming thermal velocity distributions with $T_e \gg T_i$. Somewhat higher ion temperatures were deduced from neutron counting measurements than were obtained from the ion line profile, possibly as a result of spatial variations in T_i.

Martone & Segre (1969, see also Segre & Martone (1971)), observed light scattered through 90° with $\alpha = 1$ from a dense, cool plasma ($n_e = 2.4 \times 10^{16}$ cm^{-3}, T_e a few eV). The electron plasma lines were recorded with 5 Å resolution using a 10-channel grating polychromator, one channel of which measured the total intensity in the central ion line. From the relative intensity of the electron and ion features a value of 1.7 was obtained for α, which allowed a determination of the electron density from the position of the electron plasma line. The electron density deduced in this way was in good agreement with a value obtained from Stark broadening measurements.

The effects of high collision frequencies on the spectrum of light scattered from a plasma were investigated by Offenberger & Kerr (1971). In order to create collision-dominated conditions carbon dioxide laser light was scattered from a dense, low-temperature plasma jet ($n_e \sim 2 \times 10^{17}$ cm^{-3}, $T_e \sim T_i \sim 1$ eV). With $\alpha \gtrsim 40$, the ion feature was found to be considerably narrowed compared to the predictions of collisionless theory, as is to be expected (see Section (5.6)).

In experiments by Leonard & Bach (1973) ruby laser light was scattered from the dense plasma produced by exploding lithium wires. Although occasionally a broadened scattered light spectrum was observed, in general no broadening or shift was observed and the intensity of the scattered light was anomalously high. The results were ascribed to Fresnel reflection of the incident laser light at the plasma boundary.

Paul et al. (see Paul (1970) and Daughney et al. (1970)) studied cylindrical collision-less shock waves produced in a plasma by a fast linear z-pinch. Electron temperatures between 40 and 150 eV were measured from 90° scattering observations using interference filters. The profiles were closely Gaussian. In another experiment the intensity of the narrow central ion feature in forward scattered light ($\theta = 4.5°$) was compared with the intensity over a broader bandwidth which included the electron component. At the same time the backward scattered intensity ($\theta = 170°$) was measured in a 30 Å wavelength region offset by 50 Å from the laser line so as to give a signal when hot electrons were present in the scattering volume. The forward scattered signals were very brief, with a rise time of only 5 ns, and occurred as the shock front passed through the scattering volume. It was found that the forward scattered signal was enhanced by a factor of 16 during the passage of the shock, the factor being the same for the narrow-band and for the broad-band filters. This factor had to be corrected for the fact that the shock front did not fill the scattering volume: the true enhancement was by a factor of 37. According to the usual theory, over a wide range of ion- to electron-temperature ratios and α values (see Figure 3.6 above) the total scattered light should not vary much for plasmas in thermal equilibrium. The electron contribution would be expected to increase across the shock, so that the ion contribution should fall. Thus the observed enhancement implies an even greater enhancement of the ion feature. It was concluded that a micro-instability must be present in the shock front. The spectrum of scattered light as a function of the wave vector k was measured in more detail by Daughney *et al.* (1970) and compared with the predictions of non-linear theory. The data fitted a Kadomtsev spectrum (Kadomtsev (1965)) well. It was concluded that ion wave turbulence was present in the shock. Machalek & Nielsen (1973) obtained very similar experimental results, but could not identify the instability responsible for the turbulence. Measurements by Keilhacker & Steuer (1971) on shocks in deuterium, using a ruby laser, were extended to lower values of $k\lambda_D$ by Kornherr *et al.* (1972) who used a carbon dioxide laser. Density fluctuations more than 10^4 times the thermal level were observed. The results were found to be in good agreement with computer simulations of current-driven turbulence perpendicular to a magnetic field (Forslund *et al.* (1970)).

The spectrum of light scattered from a shock in a low-density θ-pinch plasma was observed by De Silva & Stamper (1967), who also found enhancement of the central region of the spectrum.

Two high-power monochromatic transverse waves of different variable frequencies were mixed in a plasma by Stansfield *et al.* (1971) to produce longitudinal waves at the difference frequency, as suggested by Kroll *et al.* (1964: see Chapter 2, Section (9)). A third wave was scattered from the interaction volume and showed enhanced scattering in the electron plasma line when the difference frequency equalled the plasma frequency. The two interacting beams were obtained from a single dye laser, pumped longitudinally by a Q-switched ruby laser. The arrangement is shown in Figure 3.35. Different parts of the dye cell were used for the two cavities, which were independently tunable over 80 Å in the region of 7400 Å by rotating the gratings.

3.35   Arrangement for observing scattering from two beams of different frequencies mixed in a plasma jet (see text). (After Stansfield *et al.* (1971))

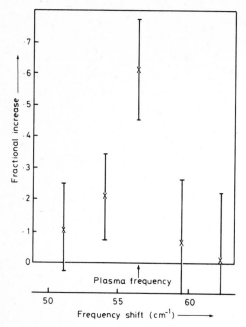

3.36 Increase in intensity of light scattered into the electron plasma line, as a function of difference frequency between two incident beams, showing enhancement when difference frequency equals plasma frequency. (After Stansfield *et al.* (1971))

An output power of 2 MW in a bandwidth of 2 Å was obtained in each beam. The plasma jet had a density of 2.2×10^{16} cm^{-3} and a temperature of 17 000 K. The enhancement of the electron plasma line as a function of the difference in frequency between the dye lasers is shown in Figure 3.36. The arrow shows the resonant frequency of the normal electron plasma line.

Several scattering observations have been made on laser-produced plasmas. These experiments are discussed in Chapters 6 and 8.

8.3. STUDIES OF MAGNETIC FIELD EFFECTS

Observations of magnetic field modulation of the scattered light spectrum at the electron cyclotron frequency were first reported by Kellerer (1970*a, b*). A hydrogen arc plasma ($n_e = 1.2 \times 10^{16}$ cm^{-3}, $T_e = 3.2$ eV) produced in a magnetic field of about 100 kG was used to scatter the light. A well-collimated 100 MW laser beam contained

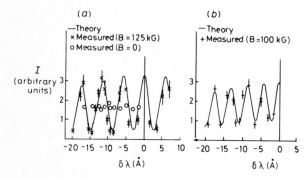

3.37 Experimental observations of scattered light from a hydrogen arc plasma in a magnetic field: (*a*) $B = 125$ kG and $B = 0$, (*b*) $B = 100$ kG (see text). (After Kellerer (1970*b*))

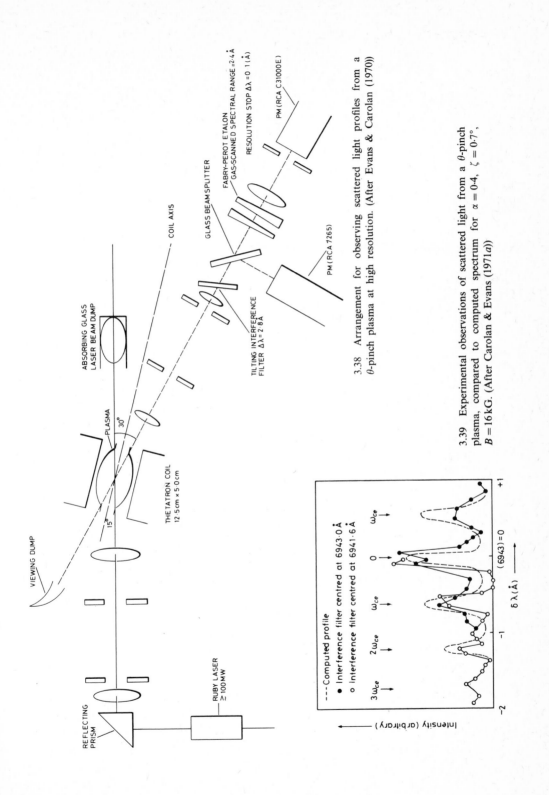

3.38 Arrangement for observing scattered light profiles from a θ-pinch plasma at high resolution. (After Evans & Carolan (1970))

3.39 Experimental observations of scattered light from a θ-pinch plasma, compared to computed spectrum for $\alpha = 0.4$, $\zeta = 0.7°$, $B = 16\,\text{kG}$. (After Carolan & Evans (1971a))

in an aperture angle of 30 mrad was scattered into a cone of divergence 40 mrad at 90° to the laser beam. Care was taken to ensure that **k** was exactly perpendicular to **B**. The value of α was 0·6. A 7-channel polychromator with 1·7 Å resolution was used to measure the profile of the scattered light. The results for magnetic fields of 125 and 100 kG are shown in Figures 3.37(a) and (b). The results were in good agreement with theoretical profiles.

Evans & Carolan (1970) succeeded in resolving modulation by the much lower magnetic field (~ 10 kG) occurring on the axis of a small θ-pinch discharge. The apparatus is shown in Figure 3.38. The tiltable interference filter provided low resolution of the scattered light profile, recorded by a photomultiplier receiving a fraction of the transmitted light. The rest of the light passed by the interference filter was further analysed by the high-resolution pressure-scanned Fabry–Perot étalon. In this way both the broad profile and the fine structure of the scattered light could be determined. The results are shown in Figure 3.39 for a magnetic field which was independently measured by the Faraday rotation method to be 16 kG. The computed spectrum (Carolan & Evans (1971a)) is also shown, allowing for the finite acceptance angle of the detector, with $\alpha = 0.4$ and with $\zeta = 0.7°$ (see Section (5.5)). The intervals between the peaks correspond approximately in each case to the electron gyro-

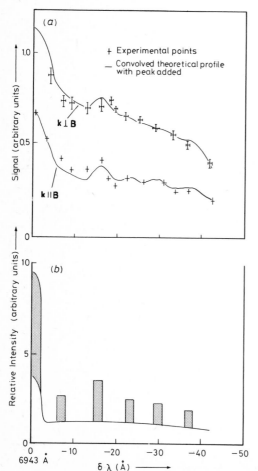

3.40 (a) Experimental observations of scattered light profiles for a hydrogen arc plasma in a magnetic field of 10 kG: (b) theoretical profile for $T_e = 7 \times 10^4$ K, $n_e = 1.2 \times 10^{16}$ cm^{-3}, $\alpha = 0.47$, with hatched areas superimposed at wavelengths corresponding to displacements of multiples of half the plasma frequency. The profile in (b), convoluted with instrumental profiles, is also shown in (a). (After Ludwig & Mahn (1971))

frequency. Magnetic structure in the light scattered from a θ-pinch plasma was also reported by Kronast & Benesch (1971), who used a 10-channel Fabry–Perot system to record the spectrum near an electron plasma frequency line.

Following the work of Ringler & Nodwell described in Section (8.1), Ludwig & Mahn (1971) investigated the spectra of light scattered with **k** parallel and also with **k** perpendicular to the magnetic field in a hydrogen arc plasma. The experimental results shown in Figure 3.40(*a*) were consistent with the convolution of the profile shown in Figure 3.40(*b*) with the instrumental profiles, and suggest that enhanced scattering at shifts of multiples of half the plasma frequency occurs for *both* orientations of **k** relative to the magnetic field. A study of the central maximum with higher resolution (using a pressure-scanned Fabry–Perot interferometer) for **k** perpendicular to the magnetic field showed a fine structure whose peaks were separated by a wavelength interval corresponding to the electron cyclotron frequency for the magnetic field used (0·45 Å). Further work is needed to elucidate these results, which may also explain other anomalies noted in Sections (8.1) and (8.2).

4

Thermonuclear Reactions

1. INTRODUCTION

The very great concentration of energy which can now be achieved at the focus of an intense laser pulse can be used to raise matter to very high temperatures. The possibility of making use of laser light to initiate thermonuclear reactions has stimulated much of the present extensive research on high-power lasers and on the interaction of intense light with plasmas. Recently, a number of developments have raised hopes that a useful yield of nuclear energy might be released in this way. At the same time, the rising costs of fossil fuels have given a new urgency to the development of alternative energy sources.

In this chapter we consider the conditions which must be satisfied if useful energy is to be obtained from thermonuclear reactions, and describe (in very general terms) some experimental approaches.

2. NUCLEAR REACTION RATES

In order to cause a fusion reaction, two suitable light nuclei must be brought close enough together for the short-range attractive nuclear forces to overcome the electrostatic repulsion. The necessary energy increases with the atomic number. The cross-sections of many fusion reactions are known, and it is found that the easiest

132

to initiate are those involving the nuclei of the heavy isotopes of hydrogen, deuterium and tritium. These may be written (see for example Rose & Clark (1961))

$$\left.\begin{array}{l} {}_1D^2 + {}_1D^2 \rightarrow {}_1T^3_{(1\cdot01)} + {}_1H^1_{(3\cdot03)} \\ {}_1D^2 + {}_1D^2 \rightarrow {}_2He^3_{(0\cdot82)} + {}_0n^1_{(2\cdot45)} \end{array}\right\} \tag{4.1}$$

and

$$_1D^2 + {}_1T^3 \rightarrow {}_2He^4_{(3\cdot52)} + {}_0n^1_{(14\cdot06)}. \tag{4.2}$$

The figures in brackets show the energy carried away by each of the reaction product particles in MeV. A further reaction can also take place (relatively slowly):

$$_1D^2 + {}_2He^3 \rightarrow {}_2He^4_{(3\cdot67)} + {}_1H^1_{(14\cdot67)}. \tag{4.3}$$

Deuterium occurs naturally in water in the proportion of one atom to 6500 atoms of hydrogen and is easily and cheaply extracted. Tritium, on the other hand, is scarce and very radioactive, and would be an expensive fuel. However, it is possible to produce tritium from lithium by a further reaction involving a neutron:

$$_3Li^6 + {}_0n^1 \rightarrow {}_2He^4 + {}_1T^3 + 4\cdot8 \text{ MeV}. \tag{4.4}$$

Thus the neutrons produced in the D–T and D–D reactions could be used to re-generate tritium in a lithium blanket surrounding the reaction region. If the yield of tritium from this reaction is too low, an intermediate reaction can be used to multiply the number of neutrons.

The only known method of obtaining a net yield of nuclear fusion energy is to heat up the reacting nuclei in a plasma to a very high temperature and to hold them there for long enough for a sufficient number of reactions to take place. Although the reaction cross-sections reach their maximum values at energies of 70 keV or more, if the nuclei have a near-Maxwellian velocity distribution in the plasma useful reaction rates can be obtained at lower mean energies: the important contributions are made by particles in the high-energy tail. The number of reactions per unit time and volume between two different kinds of nuclei with number densities n_{i1} and n_{i2} per unit volume is given by

$$R = n_{i1}n_{i2}\langle\sigma v\rangle. \tag{4.5}$$

Here $\langle\sigma v\rangle$ is the product of the relative velocity and the (velocity dependent) reaction cross-section, averaged over the velocity distribution. If only one species of particles is involved, with density n_i,

$$R = \frac{n_i^2}{2}\langle\sigma v\rangle. \tag{4.6}$$

The rate of nuclear energy release per unit volume is given by

$$P_N = RQ, \tag{4.7}$$

where Q is the average nuclear energy released per reaction.

Values of $\langle\sigma v\rangle$ are shown as functions of temperature in Figure 4.1 for D–T

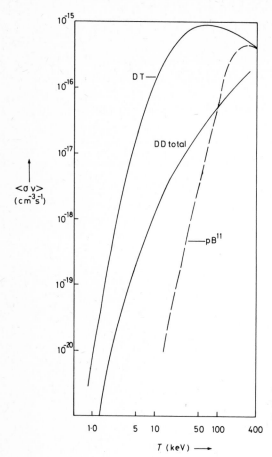

4.1   Values of $\langle \sigma v \rangle$ for D–T, D–D and p–B^{11} reactions for Maxwellian velocity distributions at temperature T. (After Tuck (1961) and Weaver *et al.* (1972))

and D–D reactions (Tuck, 1961). The D–D curve is the sum for the two reactions described by equations (4.1): the two branches are roughly equally probable. Between 0·5 and 1 keV $\langle \sigma v \rangle$ varies approximately as $T_i^{5·5}$: this is relevant to discussions of observations of neutron yield in Chapter 7.

Another reaction which may prove to be useful (Weaver *et al.* (1972)) takes place in a mixture of boron and hydrogen:

$$_1H^1 + {}_5B^{11} \rightarrow 3({}_2He^4) + 8·68 \text{ MeV}. \tag{4.8}$$

Boron is considerably more abundant than deuterium and can be extracted more cheaply. The cross-sections for the p–B reaction are not yet accurately known, but $\langle \sigma v \rangle$ probably reaches a maximum comparable with that of the D–T reaction, though at higher temperatures, in the region of 200 to 300 keV (see Figure 4.1).

3. INITIATION OF REACTIONS

Because of the very high temperatures involved, a thermonuclear reactor will almost certainly operate only in very short pulses. It is necessary to supply energy to the plasma initially from an external source to heat it to a sufficiently high temperature for nuclear reactions to begin. It may also be necessary to continue to supply energy during the remainder of the pulse in order to keep the plasma hot, compensating for Bremsstrahlung radiation and other cooling processes.

The nuclear energy released when reactions take place is in the form of kinetic energy of the reaction product particles. As may be seen from equations (4.1) and (4.2) most of the energy from D–T and D–D reactions goes to the lighter particles. The neutrons will probably escape from the plasma and will eventually be slowed down in the walls of the reactor, causing a rise in temperature. Charged reaction products, particularly α-particles, may be trapped and thermalized in the plasma if the dimensions and the density are sufficiently great, in which case they will contribute to the heating required to sustain the reaction. For a plasma with equal numbers of D and T nuclei, about $\frac{1}{5}$ of the total nuclear reaction energy is carried by the α-particles, while for a pure deuterium plasma (allowing one subsequent D–T reaction for each pair of D–D reactions) charged particles will carry about $\frac{1}{3}$ of the reaction energy. The three α-particles from the p–B^{11} reaction carry all the energy.

The charged reaction products have initial energies of several MeV. It is important to know how quickly and over what distance this energy is given to the plasma as the result of collisions. Butler & Buckingham (1962, see also Krokhin & Rozanov (1972)) found expressions for the rates of energy loss by charged particles to electrons and to ions due to Coulomb collisions: for α-particles with energies w_α of a few MeV in a D–D or D–T plasma the electrons receive a greater proportion of the energy than the deuterons and tritons if $\kappa_B T_e \sim \kappa_B T_i \lesssim w_\alpha/40$.* If the initial velocity of the α-particle is $v_{\alpha 0}$, its velocity at time t may be written (Dawson et al. (1971)) in the form

$$v_\alpha(t) = v_{\alpha 0} \exp -(t/\tau_{\alpha e}) \tag{4.9}$$

where

$$\tau_{\alpha e} = 4.5 \times 10^7 \frac{T_e^{\frac{3}{2}}}{n_e}, \tag{4.10}$$

T_e being in eV and n_e in cm^{-3}. The stopping distance $l_{\alpha e}$ is given by

$$l_{\alpha e} = v_{\alpha 0}\tau_{\alpha e}. \tag{4.11}$$

For an initial energy $w_{\alpha 0} = 3.5$ MeV, $v_{\alpha 0} = 1.3 \times 10^9$ cm s^{-1}, so $l_{\alpha e} = 6 \times 10^{16}$ $(T_e^{\frac{3}{2}}/n_e)$ cm. In terms of the mass density ρ of the D–T mixture, the stopping distance, for an electron temperature of 10 keV, is then given by

$$l_{\alpha e} \simeq 1.2\frac{\rho_s}{\rho} \text{ cm}, \tag{4.12}$$

---

* The origin of an expression given by Chu (1972, expression (10)) is not clear.

where $\rho_s = 0.2 \text{ gm cm}^{-3}$ is the density of uncompressed solid DT ($n_e \simeq 4.5 \times 10^{22}$ cm^{-3}).

The energy of the α-particle after travelling a distance x through the plasma, $w_\alpha(x)$, is given by

$$\frac{w_\alpha(x)}{w_{\alpha 0}} = \left(1 - \frac{x}{l_{\alpha e}}\right)^2. \tag{4.13}$$

Thus the particle loses half its energy in a distance of $0.3\, l_{\alpha e}$.

Although most of the energy of the α-particle goes to the plasma electrons, collisions with fuel ions may increase the reaction rate considerably by enhancing the high-energy tail of the ion distribution, even though only relatively small numbers of ions are involved. A few fuel ions may also be given high energies in collisions with energetic neutrons from nuclear reactions. If, on average, more than one fuel ion reacts as the result of gaining energy from the reaction products of a single reaction, a chain reaction will grow exponentially (Gryziński (1958)). This possibility has been considered for α-particle heating of protons in a p–B^{11} plasma (Weaver et al. (1972)).

Let us now consider a hypothetical pulsed reactor in which energy is supplied to an isolated, homogeneous D–T plasma so as to heat it rapidly to a temperature $T = T_e = T_i$, at a density of nuclei $n = 2n_D = 2n_T = n_e$, and to hold it there for a containment time t_c. We shall assume that only a small proportion of the nuclear fuel is 'burnt' during this time. We shall also assume that the only loss mechanism is Bremsstrahlung radiation, to which the plasma is optically thin.

The nuclear energy released, plus all the energy supplied from the external source, will eventually reach the walls of the reactor vessel and heat them up (we ignore the possibility of direct conversion of plasma energy to electrical energy for the moment). For the reactor to be useful, this energy must be at least sufficient, after conversion at an overall efficiency H into some suitable form, to heat up the plasma for the next pulse. Then if the fraction of the nuclear reaction energy retained as thermal energy in the plasma is C, the condition for there to be a net output of useful energy is

$$H[3n\kappa_B T + t_c P_\beta + (1 - C)t_c P_N] > 3n\kappa_B T + t_c P_\beta - Ct_c P_N, \tag{4.14}$$

where P_β is the total Bremsstrahlung power radiated per unit volume. We shall assume for the moment that $CP_N \leqslant P_\beta$: otherwise, the plasma temperature would rise.

The condition (4.14) may be written in the form

$$nt_c > \frac{3n^2 \kappa_B T}{P_\beta}\left[\frac{1}{(KP_N/P_\beta) - 1}\right], \tag{4.15}$$

where

$$K = \frac{H + C(1 - H)}{1 - H}. \tag{4.16}$$

From expressions (2.81) and (2.82), if T is in K and n in cm^{-3}

$$P_\beta(T) \simeq 1.6 \times 10^{-34} n^2 T^{\frac{1}{2}} \text{ W cm}^{-3} \tag{4.17}$$

According to expressions (4.6) and (4.7), P_N also varies as n^2, so the right hand side of expression (4.15) depends only on the temperature and on K. Thus the condition for a useful yield of energy is that the product of the density and the lifetime of the plasma should exceed a value determined only by the temperature, the efficiency of energy conversion and the proportion of nuclear energy retained in the plasma. This condition is known as the *Lawson criterion* (Lawson, (1957)).

Figure 4.2 shows the values of nt_c required to satisfy the Lawson criterion for the D–T, D–D and p–B^{11} reactions, as functions of temperature, when $K = \frac{1}{2}$ (with $C = 0$ this corresponds to the optimistic assumption that $H = \frac{1}{3}$: with $C = \frac{1}{3}, H = \frac{1}{7}$).

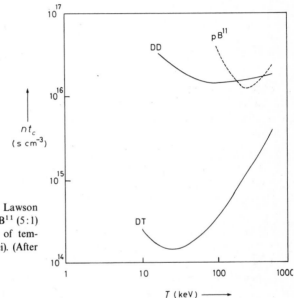

4.2 Values of nt_c satisfying the Lawson criterion for D–T (1:1), D–D and p–B^{11} (5:1) reactions with $K = \frac{1}{2}$ as functions of temperature (n is total density of nuclei). (After Weaver *et al.* (1972))

For the D–T reaction, the minimum value of nt_c is about 10^{14} s cm^{-3} and occurs at a temperature of about 25 keV. For the D–D and p–B^{11} reactions the minima are about a hundred times greater and occur at temperatures of more than 100 keV.

Figure 4.3 shows the ratio P_β/P_N as a function of the temperature for D–T and D–D reactions. This ratio is equal to the minimum allowable value of K if any net yield of nuclear energy is to be achieved, however large nt_c may be. If K is less than 7×10^{-3} a net yield cannot be achieved from D–T reactions at any temperature. For D–D reactions at temperatures below 100 keV, K must be as large as 0·18.

If $P_\beta/P_N < C$, the plasma is self-heating due to the retention and thermalization of nuclear energy. When self-heating occurs, expressions (4.14) and (4.15) are no longer valid. External energy is now only necessary to heat the plasma to the ignition temperature where $P_\beta/P_N = C$. The maximum value of C corresponds to the retention of all charged reaction products in the plasma. The temperature at which $P_\beta/P_N = C_{max}$ is known as the *ideal ignition temperature*, which (as may be seen from Figure 4.3) is about 4 keV for D–T reactions and about 36 keV for D–D reactions. The ideal

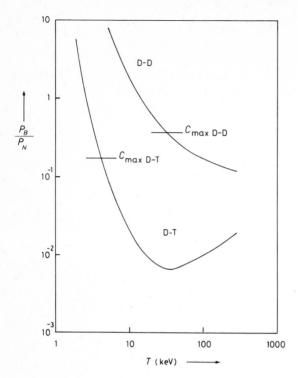

4.3 Ratio of Bremsstrahlung power lost to nuclear power generated per unit volume for D–T and D–D reactions as functions of temperature

ignition temperature is lowered if neutrons are slowed down significantly before they leave the plasma.

Our derivation of the Lawson criterion was based on the assumption that only a small fraction of the nuclear fuel was burnt. If in a D–T mixture the nuclear reaction goes to completion, known as 'burn-out', the nuclear energy released is 17·6 MeV per D–T pair. To this may be added a further 4·8 MeV if lithium is used to regenerate tritium in the walls, so that a total of 22·4 MeV is released per D–T pair. If we suppose that the plasma must be heated to $\kappa_B T = 10$ keV to initiate the reaction, the energy supplied is $6\kappa_B T = 60$ keV per D–T pair, neglecting all losses. Thus the energy released is at most about 370 times the input energy. This sets a lower limit of 3×10^{-3} to the acceptable overall efficiency of conversion from thermal energy to plasma energy in a homogeneously reacting D–T plasma. In the case of the p–B^{11} reaction the minimum acceptable efficiency is very much lower: the temperature might have to be ten times higher, the nuclear yield is only 8·7 MeV per reaction, and five electrons must be heated instead of two, so that $10 \cdot 5\kappa_B T$ must be supplied per reaction. The overall efficiency must now be at least 0·12, which is unlikely to be attained. However, so far we have explored only the simple concept of a homogeneously reacting plasma, and other possibilities must be considered. Before we do so, we shall discuss the problem of containing a plasma.

4. CONTAINMENT

In the preceding discussion the plasma was assumed to be held at constant density and isolated from the walls of the reactor vessel for the required duration of the reaction. Such containment is very difficult to achieve. Two main approaches to the problem have been investigated, which may be called *magnetic* and *inertial* methods.

4.1. MAGNETIC CONTAINMENT

Charged particles move in helical orbits around magnetic lines of force. Thus although in a uniform field they can move freely along the lines of force they are restrained from moving across them. We may consider the hypothetical example of a perfectly-conducting cylindrical plasma surrounded by a uniform magnetic field parallel to its axis. The field exerts a pressure at right angles to the lines of force equal to $B^2/2\mu_0$, which may be arranged to balance the kinetic pressure of the plasma

$$n_e \kappa_B T_e + \sum_i n_i \kappa_B T_i.$$

Because of the very high temperatures necessary for thermonuclear reactions to take place at a useful rate, only relatively low density plasmas ($n \lesssim 10^{17} \, \text{cm}^{-3}$) can be restrained in this way with currently available fields ($\sim 10^5 \, \text{G}$, although 10^6 to $10^7 \, \text{G}$ has been achieved in very small volumes). At these densities very little of the charged reaction product energy will be retained in a plasma of reasonable dimensions. Long containment times (milliseconds or longer) are necessary to satisfy the Lawson criterion. It has been found that the containment time is limited by diffusion of particles across the magnetic field lines, and by the growth of instabilities. Much attention has been given to these effects: for a review of recent progress the reader is referred to a paper by Pease (1972).

Until recently the main interest in high-power lasers in connection with magnetically-contained plasmas was for the purpose of Thomson scattering diagnosis, which as we have seen in Chapter 3 is a very useful technique. However, the development of very efficient far infrared lasers has prompted a number of workers to consider whether it will be worthwhile to use lasers to heat magnetically-confined plasmas. Practical limitations on the magnetic field strength restricting the plasma density to about $10^{17} \, \text{cm}^{-3}$ imply that a long-wavelength laser is necessary if the laser energy is to be absorbed in a reasonable distance. A plasma a few hundred metres long with a diameter of less than 1 cm has been suggested, contained by a magnetic field of 300 kG (Dawson *et al.* (1971)) and heated by a carbon dioxide laser ($\lambda = 10 \cdot 6 \, \mu\text{m}$). The tendency of the laser beam to diffract out of the plasma over such a long distance can perhaps be overcome by arranging for the electron density in the cylindrical plasma column to increase with increasing radius, thereby refracting the beam back towards the axis. The effects of the laser energy on the plasma density and temperature and on the confining field strength have been considered by Vlases (1971): in a θ-pinch configuration about 70% of the laser energy might remain in the plasma as thermal energy.

In contrast to conventional magnetically-contained plasmas, with laser heating the magnetic field is used solely to confine the plasma and not to compress and heat it. Thus greater freedom is available for optimizing the magnetic field as a function of space and time so as to prevent the growth of instabilities.

Some preliminary experiments on laser heating of magnetically-contained plasmas will be described in Chapter 6.

4.2. INERTIAL CONTAINMENT

Another approach to the problem of containment relies simply on making use of the inertia of the plasma. If a spherical plasma of radius r is formed rapidly in vacuum, by heating a small pellet of solid material, it will expand freely with a time constant

$$\tau_c \sim \frac{r}{V_s} \tag{4.18}$$

where V_s is the speed of sound in the plasma, which for ions of average mass $\bar{m}_i$ is given by

$$V_s = \left(\frac{2\gamma\kappa_B T}{\bar{m}_i}\right)^{\frac{1}{2}}. \tag{4.19}$$

Thus the Lawson criterion (4.15) may be approximately satisfied provided

$$nr \gtrsim nt_c V_s. \tag{4.20}$$

If we consider a D–T mixture, and suppose that $\kappa_B T = 10^4 \, \text{eV}$ and $\gamma = \frac{5}{3}$ so that $V_s = 1\cdot1 \times 10^8 \, \text{cm s}^{-1}$, then for $nt_c = 3 \times 10^{14} \, \text{s cm}^{-3}$ expression (4.20) becomes

$$nr \gtrsim 3\cdot3 \times 10^{22} \, \text{cm}^{-2}. \tag{4.21}$$

The energy W that must be delivered to the plasma to heat it to a temperature T, ignoring losses, is given by

$$W = 4\pi n\kappa_B T r^3. \tag{4.22}$$

Table 4.1 shows values of τ_c, n and W for D–T plasmas with several values of r, when $T = 10 \, \text{keV}$. These results are of course only rough estimates. More detailed analysis of the dynamics of isolated, laser-heated plasmas is deferred to Chapter 7.

The inertial containment time could be increased by 'tamping', that is, by surrounding the plasma with material of high atomic mass (Winterberg (1968)). There would be a number of practical problems in supplying energy to the central plasma (Eden & Saunders (1971)) and the presence of high-Z material would increase the Bremsstrahlung losses.

A hybrid scheme has been considered by Spalding (1972) in which small pellets of solid fuel are heated by lasers, making use of inertial containment to maintain high density and efficient absorption, within a magnetic containment device such as Stellarator. The nuclear reactions would then take place over a much longer time at relatively low density after the plasma has expanded to fill the magnetic 'bottle'.

TABLE 4.1: *Containment times, densities and input energies required to satisfy the Lawson criterion for inertially-contained D–T plasmas of radius r at $\kappa_B T = 10\,keV$.*

r	$1\,\mu m$	$10\,\mu m$	$100\,\mu m$	$1\,mm$	$10\,mm$	$100\,mm$
τ_c	$1\,ps$	$10\,ps$	$100\,ps$	$1\,ns$	$10\,ns$	$100\,ns$
n	$3.3 \times 10^{26}\,cm^{-3}$	$3.3 \times 10^{25}\,cm^{-3}$	$3.3 \times 10^{24}\,cm^{-3}$	$3.3 \times 10^{23}\,cm^{-3}$	$3.3 \times 10^{22}\,cm^{-3}$	$3.3 \times 10^{21}\,cm^{-3}$
W	$6.7\,J$	$670\,J$	$67\,kJ$	$6.7\,MJ$	$670\,MJ$	$67\,GJ$

It is apparent from Table 4.1 that if a useful yield of nuclear energy is to be obtained from a laser-heated homogeneous, untamped, inertially-confined plasma the initial density of the plasma must be very high. Even at solid density, the amount of laser energy required to satisfy the Lawson criterion would be very large, and the amount of nuclear energy released would be so large as to constitute a quite substantial explosion: $10^{10}\,J$ corresponds to the energy of about 1000 kg of high explosive, although the impulse of the nuclear reaction products would be much less than that of chemical reaction products with the same energy because of their smaller mass. If a 'superdense' plasma can be produced, whose density is greater than that of the cold solid, the input and output energies become smaller, although the plasma must be heated in a shorter time. Also, at higher densities the charged reaction products are more likely to be retained in the plasma. At very high densities, neutrons may be significantly slowed down before leaving the plasma: the range of 14 MeV neutrons is uncompressed solid DT is 22 cm, while in Li(DT) it is 3.5 cm (Fraley *et al.* (1974)).

5. TRIGGERED REACTIONS

So far we have considered homogeneous plasmas, uniformly heated. It may be possible to raise a small volume of plasma to the ignition temperature, within a larger mass of fuel (Winterberg (1968), Linhart (1968)). Provided certain conditions are satisfied, a thermonuclear reaction wave may then propagate through the medium, analogous to the detonation wave in a chemical explosive.

Approximate criteria for triggering a D–T mixture were derived by Linhart (1970). A small spherical region within an infinite cool D–T plasma at an initial density n_0 nuclei per unit volume is suddenly supplied with energy W, which raises it to a temperature above 10 keV. As a result, nuclear reactions are initiated. At the same time a spherical blast wave moves outwards and heats fresh material, while the hot plasma behind the wave cools. Linhart found that the heating due to the nuclear reactions compensates for the cooling due to expansion, provided

$$n_0 r > 2.6 \times 10^5 \frac{T^{\frac{3}{2}}}{\langle \sigma v \rangle}\,cm^{-2}, \tag{4.23}$$

where T is in keV. The right hand side of this expression has a minimum value when

$$\frac{d \ln \langle \sigma v \rangle}{d \ln T} = \frac{3}{2},$$ (4.24)

which occurs when $T = 18.2 \, \text{keV}$ for D–T reactions (and when $T = 36.8 \, \text{keV}$ for D–D reactions) (Babykin & Starykh (1971)). Thus for D–T reactions

$$n_0 r > 5 \times 10^{22} \, \text{cm}^{-2},$$ (4.25)

which is about one and a half times the Lawson criterion given in (4.21) for inertial containment. (Linhart obtained a somewhat lower value, using a temperature of $1.5 \times 10^8 \, \text{K} \equiv 13 \, \text{keV}$). The density of nuclei in a solid D–T mixture, n_s, is about $4.5 \times 10^{22} \, \text{cm}^{-3}$, so we may express (4.25) in the form

$$r > r_{\text{min}} \simeq 1.0 \frac{n_s}{n_0} \, \text{cm}.$$ (4.26)

The energy W_0 required to trigger a nuclear reaction wave is proportional to $n_0 r_{\text{min}}^3 T$. Since r_{min} varies as $T^{\frac{3}{2}}/\langle \sigma v \rangle$, W_0 has a minimum value when

$$\frac{d \ln \langle \sigma v \rangle}{d \ln T} = \frac{11}{6}$$ (4.27)

which occurs at $T = 15.4 \, \text{keV}$ for D–T and $20.5 \, \text{keV}$ for D–D reactions (Babykin & Starykh (1971)). Linhart found that for a D–T mixture

$$(W_0)_{\text{min}} \simeq 10^9 \left(\frac{n_s}{n_0} \right)^2 \, \text{J}.$$ (4.28)

The value of r corresponding to the minimum value of W_0 is slightly greater than r_{min}.

The energy W_0 must be supplied in a time t_h which is short compared to the initial radius divided by the initial velocity of the blast wave. If the average ion mass is $\bar{m}_i$,

$$t_h \lesssim \frac{r_0}{(32 \kappa_B T / 3 \bar{m}_i)^{\frac{1}{2}}}.$$ (4.29)

For a D–T mixture at a temperature of 16 keV, if r_0 is in centimetres,

$$t_h \lesssim 4 \times 10^{-9} r_0.$$ (4.30)

These estimates neglect the effects of a number of additional processes which tend to make triggering more difficult, including electron conduction, radiation and charged reaction product losses to regions outside the reaction zone, and the relatively slow development of the high-energy tails of the ion energy distributions. Detailed calculations for *plane* waves have been given for low-density plasmas ($n \sim 10^{15} \, \text{cm}^{-3}$) by Fuller & Gross (1968) and for solid-density plasmas by Chu (1972) and by Ahlborn (1971).

If a nuclear reaction wave can be initiated in a sufficiently large volume of fuel, there is in principle no limit to the amount of energy released. The efficiency limits discussed for homogeneous reactors at the end of Section (3) no longer apply. However, if the energy released in a pulse is to be converted into electrical energy it must still be confined within a reactor vessel, which will only be able to withstand a limited wall loading. It is still necessary to minimize the amount of energy needed for triggering.

When a laser pulse is incident on a solid surface the *ablation*, or evaporation of material from the surface, is accompanied by a recoil pressure. With sufficiently powerful lasers and spherical symmetry it is believed that densities of $10^4 \rho_s$ at thermonuclear temperatures can be achieved by ablation-compression in a region at the centre of a small pellet of nuclear fuel. It should then be possible to initiate a fusion reaction wave with a relatively small amount of energy. A net gain of nuclear energy may be possible. These possibilities will be discussed in detail in Chapter 7: they appear to be within the capabilities of current experimental techniques.

6. FISSION REACTIONS

In order to initiate a fission chain reaction it is necessary to assemble a certain critical mass of fissile material. If it is possible to compress the material the critical mass is reduced. Winterberg (1973a, b, c), and Askaryan et al. (1973) have suggested that laser ablation-compression could be used to initiate a chain reaction in a small pellet of fissile material. This, of course, would not be a thermonuclear reaction. However, if the pellet has a shell of neutron-reflecting material the critical mass becomes even smaller, and such a shell could consist of a D–T mixture, which might be heated by the fission reaction energy to the thermonuclear ignition temperature. The extra neutrons thus generated would increase the rate of the fission reactions, which would in turn raise the temperature more rapidly, and so on (Winterberg (1973c)).

It is not clear that it would be easier to achieve ignition in this way than with the pure fusion scheme discussed previously, and the technological problems of disposing of the reaction products would be more difficult.

Another hybrid proposal (see Linhart (1968)) involves placing fissile material in the walls of a pulsed fusion reactor to make use of the neutron flux. Again, additional difficulties would be encountered.

7. DESIGN STUDIES FOR FUSION REACTORS

Several design studies have been made for hypothetical fusion reactors, including a number of cost estimates. Much attention has been paid to the design of the walls of the reaction vessel, which must withstand large pulsed fluxes of high-energy particles over a long period.

If tritium is to be regenerated using neutrons from the reaction zone, the walls of the reactor vessel must contain lithium. Methods have been proposed for lining the inside of the reactor wall with liquid lithium, so as to reduce the radiation damage to the outer structural materials (see Hancox & Spalding (1973)). Alternatively, liquid lithium may be circulated behind a thin inner wall, thus acting also as a cooling medium. Instead of lithium, a eutectic mixture of beryllium, fluorine and lithium, 'flibe', has been proposed (Mitchell & Hancox (1972)). Sputtering of the wall material by the high-energy particles may prove to be an important factor limiting the life of the reactor (Laegreid & Dahlgren (1973)).

In magnetic containment devices the design of the reactor vessel is complicated by the proximity of the magnetic windings, which experience very large forces and may have toroidal geometry. The presence of large magnetic fields will also cause eddy-current braking of the flow of an electrically-conducting coolant such as liquid lithium, thus restricting the thermal power which can be carried away from unit area of the walls (Mitchell & Hancox (1972)). The magnet coils, which are generally assumed to be superconducting, necessitating cooling to liquid helium temperatures, may indeed be the most expensive parts of the reactor.

In laser heated reactors using only inertial containment, the design of the walls is complicated by the need to admit laser light. Windows or reflectors able to withstand the particle and electromagnetic radiation fluxes over very large numbers of pulses (perhaps 10^{10}) without deterioration are necessary. Magnet coils may again be needed if direct conversion of plasma energy to electrical energy is to be incorporated in the design.

The economics of pulsed laser fusion reactors have been discussed by Hancox & Spalding (1973), who conclude that electricity from a laser heated D–T system might be very roughly comparable in cost with that from a fast breeder fission reactor. No clear economic advantage in developing fusion reactors is apparent at present. Also, the amount of radioactivity produced in a D–T fusion reactor would probably be comparable with that in a fission reactor: within three months the activity in the walls might rise to 100 k curies/MW (Hancox (1972)). However, if it proves possible to use the p–B^{11} reaction instead, the radioactivity generated would be very much less (Weaver et al. (1972)).

There is, however, one overwhelmingly important advantage in using fusion reactors rather than fission reactors: only a very small amount of fuel is present in the reactor at any time. Whereas a fast fission reactor core may contain several thousand kilograms of plutonium and uranium, and represents a potentially disastrous explosion hazard, a D–T fusion reactor would probably contain considerably less than a gram of fuel within the reaction zone, whose energy could always be contained inside the reactor. Perhaps the most serious hazard to be anticipated with a D–T reactor would be a fire in the lithium circulating system, which might release a small amount of tritium. If p–B^{11} fuel can be used this risk does not of course arise.

5

Ionization and Breakdown of Gases by Light

1. INTRODUCTION

At the third International Conference on Quantum Electronics, held in Paris in February, 1963, Maker *et al.* (1964) reported that when a ruby laser pulse of sufficiently high power (a few megawatts) was focused by a lens in atmospheric air, a bright spark occurred at the focus, accompanied by a loud click. The mechanism of this spark was subsequently investigated by several authors. It is in many ways similar to breakdown by microwave radiation, and is often referred to as optical frequency breakdown. The essential feature is that the gas, which is normally a transparent, non-conducting medium, becomes ionized by the laser light and absorbs energy strongly from the beam. Rapid local heating occurs, leading to the production of a dense, strongly ionized plasma.

Provided a few free electrons are present in the focal volume, it is not difficult to explain the growth of ionization at moderate gas pressures. In intense light an electron can gain sufficient energy, by absorbing photons from the radiation field in collisions with neutral atoms (see Chapter 2), to produce ionization of an atom. The number of electrons thus doubles in a certain characteristic time, and a cascade process develops causing an exponential growth of ionization. Breakdown will occur if the electron density can reach a certain critical value despite losses due to diffusion, attachment, etc., within the duration of the laser pulse.

145

However, the origin of the first electron must still be explained. The focal volume usually used in breakdown studies is very small—of order 10^{-6} cm^3—and at sea level natural ionization by cosmic rays, radioactivity, etc., will normally produce only 10–100 ion pairs per second per cm^3 in air at S.T.P. The equilibrium ion density is only of the order of 10^3 to 10^4 ion pairs per cm^3 in air at S.T.P., and is lower at lower pressures. Thus, as was remarked by Tozer (1965), the probability of finding a naturally-occurring free electron within the focal volume during a laser pulse is small. Nevertheless, experiments show that many free electrons are produced within the focal volume even during laser pulses lasting for less than a nanosecond. It seems certain that at least the small number of electrons needed to initiate breakdown must be produced by the direct action of the radiation field on individual atoms or molecules. The normal photoelectric effect is ruled out, because the energy of a laser photon is small compared to the ionization energy of a gas atom. Ionization via intermediate bound levels by single photon steps is unlikely to be important. Thus we are led to conclude that the initial ionization is caused by a multi-photon process in which several laser photons are absorbed simultaneously by an atom or molecule, resulting in the liberation of an electron. At very low pressures, at which collisions are infrequent, this may be the only important ionization mechanism, but at higher pressures, as will be seen below, other processes control the subsequent development of high ion and electron densities.

In this chapter we first consider multi-photon ionization theory and see how this agrees with experimental studies of the initiation of ionization, mainly at very low gas pressures. We then look at the theory of the development of the electron cascade, and with this also in mind survey experimental measurements of the dependence of the breakdown threshold in various gases on parameters such as pressure and pulse duration. Finally we consider the possibility of 'self-focusing' of intense light in gases during breakdown.

2. MULTI-PHOTON IONIZATION THEORY

The development of this rather complex subject will be outlined briefly here, and the main results will be quoted. For details the reader is referred to the original papers.

The problem to be solved is a particular case of the more general one of calculating the probability that, under the influence of an electromagnetic field, an atomic system will make a transition from one state to another of higher energy by absorbing more than one photon. In multi-photon ionization the electron is initially bound, while in the final state it has been freed. A similar problem, but involving transitions between two states in both of which the electron is free, is that of calculating the transition probability for multi-photon inverse Bremsstrahlung discussed in Chapter 2.

Ionization of an atom by a strong d.c. electric field, the 'tunnel' effect, may be regarded as the limiting case of multi-photon ionization when the frequency of the electromagnetic field is allowed to go to zero, so that an infinite number of photons are necessary to produce ionization. The tunnel effect was first treated by Oppenheimer (1928) who calculated the ionization probability from the ground state, and then by Lanczos (1931), who obtained the probabilities of ionization of excited levels with principal quantum number $n = 5, 6, 7$ and 8 in hydrogen. (Lanczos was interested in the quenching of hydrogen emission lines by a strong electric field. An excited atom in a given state has a finite lifetime: if this is longer than the reciprocal of the ionization probability due to the field, the atom will be ionized before it can emit a photon, so emission lines corresponding to transitions from the upper state will be extinguished.) Further calculations for levels with $n = 5, 6$ and 7 were made by Rice & Good (1962). Experimental measurements of ionization rates by a d.c. field in hydrogen for $n = 9$ to 20 by Riviere & Sweetman (1964) were in good agreement with theory based on a general formula, given by Bethe & Salpeter (1957), which makes use of second-order perturbation theory to obtain level energies as a function of the electric field.

A useful approximate treatment of the ionization of a ground state hydrogen atom by a d.c. field E is given by Landau & Lifshitz (1965), which yields for the ionization probability

$$v_{I\text{d.c.}} = \frac{8\chi_H}{h} \frac{E_H}{E} \exp\left(-\frac{2}{3}\frac{E_H}{E}\right). \tag{5.1}$$

Here $E_H = \pi m_e^2 e^5/4\varepsilon_0^3 h^4 = 5\cdot14 \times 10^{11}\ \text{Vm}^{-1}$ is the electric field in the first Bohr orbit and $\chi_H = m_e e^4/8\varepsilon_0^2 h^2 = 13\cdot6\ \text{eV}$ is the ground state ionization energy. For future use we may note that an alternative form is

$$v_{I\text{d.c.}} = \frac{\pi^2 m_e^3 e^9}{2\varepsilon_0^5 h^7 E} \exp\left[-\frac{8\pi(2m_e)^{\frac{1}{2}}\chi_H^{\frac{3}{2}}}{3eEh}\right]. \tag{5.2}$$

A more general calculation of the probability of ionization by a d.c. field was made by Smirnov & Chibisov (1965) for an arbitrary atom in a state S whose ionization energy is χ_S, with orbital angular momentum l whose component in the direction of the field is m. The corrected result of their calculation was given by Perelomov et al. (1966) and may be written

$$v_{I\text{d.c.}} = \frac{\chi_S}{h}\Theta(l, m, \chi_S, Z)$$

$$\times \left[\frac{2E_H}{E}\left(\frac{\chi_S}{\chi_H}\right)^{\frac{3}{2}}\right]^{2\zeta-|m|-1} \exp\left[-\frac{2}{3}\frac{E_H}{E}\left(\frac{\chi_S}{\chi_H}\right)^{\frac{3}{2}}\right], \tag{5.3}$$

where $\zeta = Z(\chi_H/\chi_S)^{\frac{1}{2}}$ and Z is the charge on the atomic core. For the hydrogen atom in the ground state the dimensionless quantity Θ is equal to 4, $\zeta = 1$, $m = 0$ and the result is in agreement with expression (5.1).

In an alternating field, provided the electron can emerge through the potential barrier in a time short compared to the reciprocal of the frequency of the field, the field may be regarded as constant during ionization. The condition for this *adiabatic* approximation to be valid is that

$$\gamma_t = \omega \tau_t \ll 1, \tag{5.4}$$

where $\tau_t = (2\chi m_e)^{\frac{1}{2}}/eE$ is the 'tunnelling time', and ω is the frequency of the field. Since χ is generally between 5 and 25 eV, this condition is only satisfied for ruby or neodymium lasers if $E \gtrsim 10^9 \text{ V cm}^{-1}$, corresponding to irradiances I exceeding $10^{15} \text{ watts cm}^2$, while for carbon dioxide lasers, $E \gtrsim 10^8 \text{ V cm}^{-1}$ or $I \gtrsim 10^{13} \text{ W cm}^{-2}$. So far as most breakdown studies are concerned, therefore, the adiabatic approximation does not hold, and the ionization rate will be frequency dependent. However, with focused mode-locked lasers, γ_t can be less than 1.

Provided the adiabatic approximation is valid, the transition probability may be calculated by averaging the expression for $v_{I\text{d.c.}}$ over a cycle of the field. Thus, using the general expression (5.3) for $v_{I\text{d.c.}}$,

$$v_{I_{(\gamma_t \ll 1)}} = \frac{\omega}{2\pi} \int_0^{2\pi/\omega} v_{I\text{d.c.}}(E) \, dt. \tag{5.5}$$

The polarization form of the light is important. For an elliptically-polarized, monochromatic wave of ellipticity $\mathscr{E}$ ($\mathscr{E} = 0$ for linear polarization and ± 1 for circular polarization) Perelomov et al. (1966) found that for positive values of $\mathscr{E}$

$$v_{I_{(\gamma_t \ll 1)}} = \left[\frac{\mathscr{E}(1+\mathscr{E})}{2}\right]^{-\frac{1}{2}} g \left[\frac{(1-\zeta)}{3\mathscr{E}} \frac{E_H}{E}\left(\frac{\chi_S}{\chi_H}\right)^{\frac{3}{2}}\right] v_{I\text{d.c.}}. \tag{5.6}$$

Here $g(a) = e^{-a} I_0(a)$, $I_0(a)$ being the modified Bessel function of the first kind, so that $g(a)$ falls monotonically from unity when $a = 0$ to approximately $(2\pi a)^{-\frac{1}{2}}$ when $a \gg 1$. For small values of $\mathscr{E}$, then,

$$v_{I_{(\gamma_t \ll 1)}} = \left[\frac{3}{\pi(1-\mathscr{E}^2)}\left(\frac{\chi_H}{\chi_S}\right)^{\frac{3}{2}} \frac{E}{E_H}\right]^{\frac{1}{2}} v_{I\text{d.c.}}, \tag{5.7}$$

and for a linearly-polarized wave

$$v_{I_{(\gamma_t \ll 1)}} = \left[\frac{3}{\pi}\left(\frac{\chi_H}{\chi_S}\right)^{\frac{3}{2}} \frac{E}{E_H}\right]^{\frac{1}{2}} v_{I\text{d.c.}}. \tag{5.8}$$

For a circularly-polarized wave, on the other hand,

$$v_{I_{(\gamma_t \ll 1)}} = v_{I\text{d.c.}}. \tag{5.9}$$

However, small departures from perfect circular polarization produce large changes in $v_{I_{(\gamma_t \ll 1)}}/v_{I\text{d.c.}}$ and in general $v_{I_{(\gamma_t \ll 1)}} < v_{I\text{d.c.}}$ except at fields so large that the theory is invalid.

For the hydrogen atom in the ground state, subjected to linearly-polarized light, we have

$$v_{I_{(\gamma_t \ll 1)}} = \left(\frac{3}{\pi}\right)^{\frac{1}{2}} \frac{8\chi_H}{\hbar} \left(\frac{E_H}{E}\right)^{\frac{1}{2}} \exp\left(-\frac{2}{3}\frac{E_H}{E}\right)$$

$$= \left(\frac{48}{\pi}\right)^{\frac{1}{2}} 2^{\frac{7}{4}} \left(\frac{m_e \chi_H^7}{e^2 E^2 \hbar^6}\right)^{\frac{1}{4}} \exp\left[-\frac{4}{3}\frac{(2m_e)^{\frac{1}{2}}\chi_H^{\frac{3}{2}}}{eE\hbar}\right].$$

(5.10)

In the usual breakdown situation, when $\gamma_t \gg 1$, the electric field changes markedly during the tunnelling time and a more elaborate treatment is required. Keldysh (1964) used perturbation theory to calculate the transition probability from a bound state to a final state that takes into account the acceleration of the free electron by the electric field, but neglects the Coulomb interaction with the remainder of the atom. For the direct transition from the ground state to the ionized continuum the transition probability v_I' obtained in this way is given* by

$$v_I' = \left(\frac{2\chi_0}{\hbar\omega}\right)^{\frac{1}{2}} \omega \left[\frac{\gamma_t}{(1+\gamma_t^2)^{\frac{1}{2}}}\right]^{\frac{3}{2}} \mathscr{K}\left(\gamma_t, \frac{\tilde{\chi}_0}{\hbar\omega}\right)$$

$$\times \exp\left\{-\frac{2\tilde{\chi}_0}{\hbar\omega}\left[\sinh^{-1}\gamma_t - \frac{\gamma(1+\gamma_t^2)}{1+2\gamma_t^2}\right]\right\}.$$

(5.11)

Here $\tilde{\chi}_0$ is the effective ionization energy, which is greater than the normal ionization energy from the ground state, χ_0, by the average energy of oscillation of an electron in the field. For a linearly-polarized wave

$$\tilde{\chi}_0 = \chi_0 + \frac{e^2 E^2}{4m_e\omega^2} = \chi_0\left(1 + \frac{1}{2\gamma_t^2}\right),$$

(5.12)

while for a circularly-polarized wave

$$\tilde{\chi}_0 = \chi_0\left(1 + \frac{1}{\gamma_t^2}\right).$$

(5.13)

The function

$$\mathscr{K}\left(\gamma_t, \frac{\tilde{\chi}_0}{\hbar\omega}\right) = \sum_{n=0}^{\infty} \exp\left\{-2\left[N_0 - \frac{\tilde{\chi}_0}{\hbar\omega} - n\right]\left[\sinh^{-1}\gamma_t - \frac{\gamma_t}{(1+\gamma_t^2)^{\frac{1}{2}}}\right]\right\}$$

$$\times \Phi\left\{\left[\frac{2\gamma_t}{(1+\gamma_t^2)^{\frac{1}{2}}}\left(N_0 - \frac{\tilde{\chi}_0}{\hbar\omega} - n\right)\right]^{\frac{1}{2}}\right\}.$$

(5.14)

The symbol N_0 denotes the smallest whole number of photons whose combined energy exceeds the effective ionization energy, and

$$\Phi(x) = \int_0^{\infty} \exp(y^2 - x^2)\,dy.$$

In the adiabatic approximation ($\gamma_t \ll 1$), the main contribution to $\mathscr{K}$ comes from

---

* Keldysh's expression (16) is incorrectly printed and should be multiplied by $\omega^{-\frac{1}{2}}$.

large values of n. Summation may thus be replaced by integration, and it is found that $\mathscr{K}(\gamma_t, \tilde{\chi}_0/\hbar\omega) \simeq (3\pi)^{\frac{1}{2}}/4\gamma_t^2$. The adiabatic ionization probability is then given for a linearly-polarized wave by

$$
v'_{I_{(\gamma_t \ll 1)}} = \frac{(6\pi)^{\frac{1}{2}}}{4} \frac{\chi_0}{\hbar} \left(\frac{eE\hbar}{m_e^{\frac{1}{2}}\chi_0^{\frac{3}{2}}} \right)^{\frac{1}{2}}
$$

$$
\times \exp \left[-\frac{4}{3} \frac{(2m_e)^{\frac{1}{2}}\chi_0^{\frac{3}{2}}}{eE\hbar} \left(1 - \frac{m_e\omega^2\chi_0}{5e^2E^2} \right) \right].
$$

(5.15)

(For a circularly-polarized wave the exponent must be multiplied by $(1+\gamma_t^2)/(1+2\gamma_t^2)$). Since $\gamma_t \ll 1$ the exponential term becomes, in this approximation, just

$$
\exp \left[-\frac{4}{3} \frac{(2m_e)^{\frac{1}{2}}\chi_0^{\frac{3}{2}}}{eE\hbar} \right]
$$

and is thus identical with that derived previously from calculations for tunnelling in a d.c. field. However, the term in front of the exponential is different. Keldysh ascribes this to the neglect of the Coulomb interaction in the final state (see Perelomov et al. (1968) and Brodskii & Gurevich (1971) for a discussion of this point) and suggests a correction factor $\chi_0\gamma_t/[\hbar\omega(1+\gamma_t^2)^{\frac{1}{2}}]$. When this is included the general result becomes

$$
v_I = A\omega \left(\frac{\chi_0}{\hbar\omega} \right)^{\frac{3}{2}} \left[\frac{\gamma_t}{(1+\gamma_t^2)^{\frac{1}{2}}} \right]^{\frac{1}{2}}
$$

$$
\times \mathscr{K}\left(\gamma_t, \frac{\tilde{\chi}_0}{\hbar\omega} \right) \exp \left\{ \frac{-2\tilde{\chi}_0}{\hbar\omega} \left[\sinh^{-1}\gamma_t - \frac{\gamma_t(1+\gamma_t^2)^{\frac{1}{2}}}{1+2\gamma_t^2} \right] \right\},
$$

(5.16)

where A is a numerical factor. In the adiabatic approximation this gives

$$
v_{I_{(\gamma_t \ll 1)}} = A \left(\frac{m_e\chi_0^7}{e^2E^2\hbar^6} \right)^{\frac{1}{4}} \exp \left[-\frac{4}{3} \frac{(2m_e)^{\frac{1}{2}}\chi_0^{\frac{3}{2}}}{eE\hbar} \right],
$$

(5.17)

which agrees well with expression (5.10) for a hydrogen atom if $A \simeq 13$*.

In the other limiting case, $\gamma_t \gg 1$, corresponding to the usual laser breakdown situation, the important term in $\mathscr{K}(\gamma_t, \chi_0/\hbar\omega)$ is the zero order term. Then from expression (5.16)

$$
v_{I_{(\gamma_t \gg 1)}} = A\omega \left(\frac{\chi_0}{\hbar\omega} \right)^{\frac{3}{2}} \exp \left[2N_0 - \frac{\tilde{\chi}_0}{\hbar\omega}\left(1 + \frac{1}{\gamma_t^2} \right) \right]
$$

$$
\times \left(\frac{1}{4\gamma_t^2} \right)^{(\tilde{\chi}_0/\hbar\omega)+1} \Phi \left[\left(2N_0 - 2\frac{\tilde{\chi}_0}{\hbar\omega} \right)^{\frac{1}{2}} \right].
$$

(5.18)

---

* Peressini (1966) has suggested on the basis of calculations by Hartman (1962) that the tunnelling times usually quoted are much too long, so that the adiabatic approximation should be valid for fields as low as 10^8 V cm^{-1}. He has obtained an expression for the ionization rate $v_{I_{(\gamma_t \ll 1)}}$ which differs considerably from that given by Keldysh.

This result may be approximated (Raizer (1965b)) by the expression

$$v_{I_{(\gamma_t \gg 1)}} \simeq b\omega N_0^{\frac{3}{2}} \left(\frac{e^2 E^2}{8m_e \omega^2 \chi_0} \right)^{N_0}, \tag{5.19}$$

where b is a numerical factor of order unity. It may also be written in terms of F, the flux of photons per unit area per unit time:

$$v_{I_{(\gamma_t \gg 1)}} \simeq b\omega N_0^{\frac{3}{2}} \left(\frac{\pi e^2 \hbar F}{m_e c \omega \chi_0} \right)^{N_0}. \tag{5.20}$$

For hydrogen in the ground state $\chi_H = 13 \cdot 6$ eV, so for the ruby laser $N_0 = 8$. Taking b as unity, we find

$$v_I \simeq 10^{-250} F^8 \text{ s}^{-1}, \tag{5.21}$$

F being in units of $cm^{-2} s^{-1}$.* Generally this and similar expressions are used to obtain the value of the flux required to produce a given ionization rate, not vice versa, so an uncertainty in b is of no consequence.

Keldysh went on to consider the more difficult case of an atom first being excited to a bound state of higher energy and then being ionized. This problem was also investigated by Kotova & Terentev (1967). Such a process is important when there is a 'near-resonant' or 'quasi-resonant' state, that is to say when the energy of a stationary state is close to an integral multiple of the photon energy, $\hbar\omega$, above the ground state. In a strong radiation field the energy levels are shifted by the quadratic Stark effect, with the result that resonances are both frequency and intensity dependent. Keldysh (1964: see also Voronov (1966)) showed that the probability $v_I^{(N_0)}$ of N_0-photon ionization of an atom in the presence of an N_s-photon resonance with a bound state s is given by

$$v_I^{(N_0)} = v_{0s}^{(N_s)} v_{sI}^{(N_0-N_s)} \tag{5.22}$$

$$\simeq \text{constant} \cdot \frac{F^{N_s} F^{N_0-N_s}}{\Delta_{N_s}^2 + \hbar^2 (\Gamma_s)^2},$$

Here $v_{0s}^{(N_s)}$, $v_{sI}^{(N_0-N_s)}$ are the probabilities for excitation from the ground state to state s by N_s photons and for ionization from state s by $(N_0 - N_s)$ photons respectively, and Δ_{N_s} is the energy deviation from exact N_s-photon resonance with the Stark-displaced level, whose total breadth is Γ_s. This expression is only valid for relatively small values of F: at sufficiently high irradiances saturation of the ionizing second transition may occur and the overall ionization probability will be determined only by the first stage (Delone et al. (1971, 1972)). Experimental results must be interpreted carefully: Keldysh pointed out that in a strong radiation field whose amplitude varies in time the changing perturbation of the atomic energy levels due to the Stark effect may bring a state into temporary resonance. Similar considerations must apply to spatial variations in the distribution of the radiation field in a focused laser beam.

Quantum-mechanical methods using Nth-order perturbation theory have been applied, taking intermediate states into account but ignoring the intensity-dependent

---

* Some conversion factors are given in Appendix B.

Stark shifts. Zernik (1964) and Zernik & Klopfenstein (1965) carried out precise calculations of the probability of two-photon ionization from the metastable (2s) state in the hydrogen atom, using a technique due to Schwartz & Tieman (Schwartz, (1959), Schwarz & Tieman (1959)) to evaluate implicitly the sum over intermediate states. Subsequently Bebb & Gold (1966a, b) made detailed calculations of the rate of N-photon ionization from the ground state in hydrogen, from N = 2 to N = 12, but included only those intermediate states whose energies were near resonance with an integral number of photons. For ruby laser light they found that $v_I \simeq 10^{-244} F^8 \text{ s}^{-1}$. With these results to guide them they obtained improved values for the ionization rates in the rare gases, which they had previously estimated (Gold & Bebb (1965)) using approximate wavefunctions and matrix elements. Calculations of three-photon ionization rates for the alkali atoms (other than lithium) were made along similar lines by Bebb (1967). Related calculations for the rare gases and alkali metals, mercury and hydrogen were also carried out by Morton (1967). Gontier & Trahin (1968a, b; 1971a, b) refined Bebb & Gold's computations for the ground state hydrogen atom, taking all intermediate states from N = 2 to N = 8 into account. This degree of precision is only attainable in calculations for atoms whose wavefunctions are accurately known. Though the discrepancies in the ionization probabilities obtained by these two treatments appear large, sometimes exceeding a factor of 100, the corresponding ratios of the calculated breakdown threshold field strengths for a given experimental situation are of course much smaller, since $v_I^{(N)} \propto E^{2N}$.

Dynamic polarizabilities and Stark shifts have been calculated for a number of atomic states by Davydkin et al. (1971) by the quantum defect method: the method has been used to obtain parameters for multi-photon excitation of alkali metals and helium (Zon et al. (1971), Manakov et al. (1971a, b)).

A quantum electrodynamic approach has been described by Perel'man & Arutyunyan (1972) which allows for re-emission and for resonances with excited states but has not yet yielded numerical results. Another treatment by Chang & Stehle (1973) has been applied to the detailed calculation of the variation with laser frequency of the 4-photon ionization probability of caesium, near 3-photon resonance with the (6f) state (see Section (3)).

Nelson (1964, 1965, see also Nelson et al. (1964)) sought an approximate treatment of multi-photon ionization, subjecting a diagram technique to 'simplification brutale': theoretical objections have been raised to the procedure (Bunkin & Fedorov (1965)).

Perturbation theory fails at very high irradiances, when the perturbation Hamiltonian due to the electromagnetic field becomes comparable with or exceeds the energy of the atomic system. Several authors have used semi-classical methods to calculate multi-photon transition probabilities. Bunkin & Prokhorov (1964) considered the effect of circularly-polarized light, reducing the problem to that of ionization in a constant field by transforming to a rotating frame of reference. Henneberger (1968) made use of a unitary transformation to an accelerated reference frame. Reiss (1970a, 1971), developed what is known as the 'momentum translation' method, in which the electromagnetic field is approximately eliminated by a unitary transformation. The same transformation has also been used by Osborn (1972b). The approximation is satisfactory provided the wavelength of the radiation is long compared to the dimensions of the atomic system, and provided many photons are involved in the transition. The method takes into account the possibility of absorbing more than the minimum number of photons, with re-emission of the balance. It has also been applied to Raman-type processes involving many photons from the strong field and one (Reiss (1970a, b)) or two (Mohan & Thareja (1973a)) photons of another frequency. In the low-intensity limit the method gives results similar to perturbation theory (Reiss (1972a)): in the high-intensity limit the results are similar to those of tunnelling theory (Choudhury (1973b)). Multi-photon ionization probabilities from the ground state have been calculated using momentum translation theory for the hydrogen atom* and for the helium atom (Mohan (1973)): the effects of intermediate states have not, however, been taken into account in these calculations. As the intensity increases, processes involving more than the minimum number of photons become more important: the index of the power-law dependence of the overall transition probability, including all processes, on the flux is then reduced.

The polarization dependence of two- and three-photon ionization cross-sections was discussed by Lambropoulos (1972a), who obtained expressions for the ratio σ_{IC}/σ_{IL} of the cross-sections for circularly- and linearly-polarized light. For the two-photon case σ_{IC}/σ_{IL} has a maximum value, which occurs at frequencies near resonance with intermediate states, of 1·5, while for the three-photon case the maximum is about 2·5. Klarsfeld & Maquet (1972) investigated higher-order processes on the basis of Nth-order perturbation theory, and found an upper limit to σ_{IC}/σ_{IL} which increased exponentially with N, the number of photons absorbed. A similar result was obtained with momentum translation theory by Faisal (1972b). However, Reiss (1972b) pointed out that Klarsfeld & Maquet considered only the transition from an initial s state to a single final state of angular momentum $l = N$, whereas for large N many other final states are possible (and more important) for linear polarization. Reiss used the momentum translation method to show that in the limit where the ionized electron has zero momentum, when final states of low l are taken into account, linear polarization should be more effective than circular polarization above N = 4 or 5. De Witt (1973a) carried out more detailed calculations for the case of hydrogen ionized by neodymium laser light ($\lambda = 1·06 \mu m$) and considered the higher-order processes in which more photons are absorbed than the minimum number (12) required to produce ionization. Above N = 22 circularly-polarized light becomes more effective than linearly-polarized light; however, these very high-order processes are much less important than the dominant (N = 12) process, and De Witt predicted that the summed cross-sections for all N (which are observed experimentally) will show that linearly-polarized light is much more effective.

The calculation of the multi-photon ionization cross-section for unpolarized light is difficult, and has been discussed by Lambropoulos (1972b).

Bebb & Gold (1966a) took as their criterion for the initiation of breakdown by multi-photon ionization the liberation of one electron in a focal volume of 10^{-8} cm^3 in 10 ns. The results of their calculations for direct multi-photon ionization by ruby laser light are shown in Figure 5.1. It is apparent that very low concentrations of the

---

* De Witt (1973b), Mohan & Thareja (1972, 1973b), Tewari (1972), Faisal (1972a): see also Choudhury (1973a).

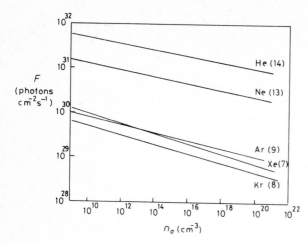

5.1 Calculated ruby laser photon flux F required to liberate 1 electron in $10^{-8}\,\text{cm}^3$ in 10 ns, as a function of density of atoms n_a. The minimum number of photons required to ionize each atom directly is shown in brackets. (After Bebb & Gold (1966a))

more easily ionized gases will have a pronounced effect on the breakdown threshold. Pure helium at S.T.P. has a calculated threshold field of $3 \times 10^7\,\text{V cm}^{-1}$, but if as few as ten argon, krypton or xenon atoms are present in the focal volume, i.e. less than 1 part in 10^{10}, the threshold according to this criterion will be reduced by a factor of 3 or more. Small concentrations of other impurities will have similar effects.

Small variations in the electric field will also produce large effects. The theoretical thresholds discussed above are calculated on the assumption that the radiation field is very coherent, as is the case for a single-mode laser beam. In multimode operation, fluctuations occur in the photon flux so that the light is statistically some-what similar to that from a thermal source. If thermal (Gaussian) statistics applied, the average photon flux required to produce breakdown by N-photon ionization would be lower, being equal to that needed for single-mode laser light multiplied by a factor $(\text{N}!)^{-1/\text{N}}$ which varies from about 0·7 for N = 2 to 0·12 for N = 10. The appropriate factor in a given experimental situation is probably somewhere between $(\text{N}!)^{-1/\text{N}}$ and unity. [We may note that theoretical discussions of this point (see, for example Carusotto (1968), Guccioni–Gush (1967), Gardner (1966a, b)) are not entirely in agreement.] Very large fluctuations have been observed experimentally: their effects have been discussed by Chin (1972) and by Lambropoulos (1972c).

Finally in this section it should be mentioned that the theory of multi-photon dissociation of molecules has been studied by several authors (Askaryan (1964), Bunkin et al. (1964), Nikishov & Ritus (1966), Lu Van et al. (1972, 1973)).

3. EXPERIMENTAL STUDIES OF IONIZATION OF GASES

The first reported attempt to observe ionization in gases by laser light was made in 1962 by Damon & Tomlinson (1963). They used a ruby laser capable of producing a peak power of 210 kW, and focused the beam by means of a lens in air, helium, or argon at a pressure of 2×10^{-2} Torr. Charged particles were collected by two electrodes 1 cm apart, in the focal plane of the lens, to which a d.c. potential of 3 V was applied. The low gas pressure ensured that recombination would not occur before the ions were collected, and that collisional ionization could be neglected. No ionization was detected when the peak laser power was less than 50 kW: the diameter of the focal spot was not stated. The number of ions generated at higher powers increased approximately exponentially with the laser pulse energy. However, the results obtained were disconcertingly similar with all three gases.

Voronov & Delone (1965, 1966) began a series of careful ionization measurements in 1965. Using xenon at a pressure below 10^{-2} Torr they were able to show that in apparatus similar to that of Damon & Tomlinson, the ions detected were produced outside the focal volume, and almost certainly came from the surface of the focusing lens. They were able to eliminate this contribution to the ionization in the vacuum chamber by using the arrangement shown in Figure 5.2, in which a d.c. potential between two sets of apertures drew out positive ions from the focal region alone on

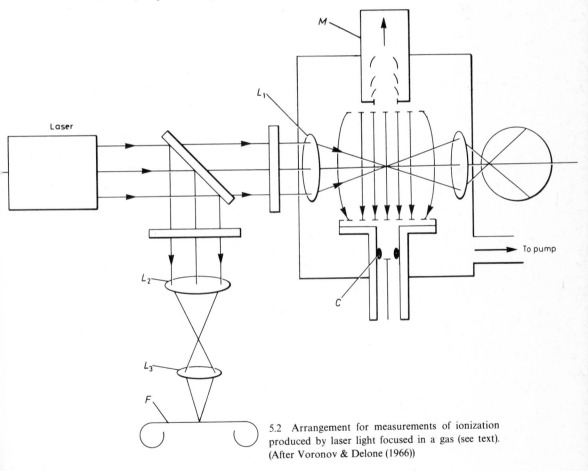

5.2   Arrangement for measurements of ionization produced by laser light focused in a gas (see text). (After Voronov & Delone (1966))

to the Faraday cup collector (C) and the electron multiplier (M). The minimum numbers of ions that could be detected by these were 10^3 and 10 respectively. The beam splitter in the laser beam deflected some of the light to a lens (L_2) identical with the focusing lens (L_1), so that an enlarged photograph of the averaged spatial distribution of the laser pulse could be recorded by means of the objective (L_3) and film (F). A fast photocell was used to measure the temporal behaviour of the laser pulse. The analysis of the results deserves a brief discussion.

Let us consider a pulsed laser beam in which the photon flux per unit area and time is a function $F(x, y, z; t)$ of position and time. The beam passes through a medium in which the density of neutral atoms is n_a. We suppose that the probability of ionization of an atom in a flux F is given by

$$v_I = CF^N, \qquad (5.23)$$

N being the effective number of photons involved in ionization and C a constant. Then the number of ions produced by a laser pulse is given by

$$N_i = Cn_a \int F^N \, dx \, dy \, dz \, dt. \qquad (5.24)$$

It is now assumed that the spatial and temporal variations of F may be separated. We may therefore write

$$F(x, y, z; t) = F_0 \psi(x, y, z)\phi(t). \qquad (5.25)$$

Here F_0 is the value of the flux at some fixed point and time, usually taken as the peak flux on the beam axis in the focal plane.

The total number of photons in the beam, N_p, is given by the integral of F over the focal plane (or any other plane intersecting the entire beam) and over the pulse duration:

$$N_p = F_0 \int \psi(x, y, 0)\phi(t) \, dx \, dy \, dt. \qquad (5.26)$$

Both $\int \psi(x, y, 0) \, dx \, dy = \sigma_1$ and $\int \phi(t) \, dt = \tau_1$ may be determined experimentally, so a value of $F_0 = N_p/\sigma_1\tau_1$ may be obtained from a measurement of N_p.

We may write expression (5.24) as

$$N_i = Cn_a F_0^N \int \psi^N(x, y, z)\phi^N(t) \, dx \, dy \, dz \, dt$$

$$= Cn_a F_0^N V_N \tau_N, \qquad (5.27)$$

where for N-photon ionization the effective volume is

$$V_N = \int \psi^N(x, y, z) \, dx \, dy \, dz$$

and the effective pulse duration is

$$\tau_N = \int \phi^N(t) \, dt.$$

Thus

$$N_i = \frac{Cn_a V_N \tau_N}{(\sigma_1 \tau_1)^N} N_p^N. \tag{5.28}$$

A log–log plot of N_i against N_p yields N. The known distribution functions ψ and ϕ may then be used to find V_N and τ_N and hence, knowing σ_1, τ_1 and n_a we obtain C. The ionization probability is given by

$$v_I = CF_0^N = \frac{N_i}{n_a V_N \tau_N}. \tag{5.29}$$

For xenon, with a ruby laser, Voronov and Delone found that $N = 6.23 \pm 0.14$, with $v_I^{(Xe)} = 10^{5.5 \pm 1.7} \, s^{-1}$ when $F = 10^{30.25 \pm 0.25} \, cm^{-2} \, s^{-1}$ (about 5×10^{11} W cm^{-2}). The ionization energy of a xenon atom is normally 12.13 eV: since the energy of a ruby laser photon is 1.79 eV, it is to be expected that seven photons are needed to ionize the atom. The lower value of N observed indicates that intermediate states are involved.

Voronov et al. (1966a) carried out similar experiments with krypton at low pressures, and found that $N = 6.31 \pm 0.11$. The ionization energy of the krypton atom is 14.0 eV so that eight ruby laser photons might be expected to be needed. They obtained an ionization probability $v_I^{(Kr)} = 10^{6.3 \pm 2.4} \, s^{-1}$ for a photon flux F of $10^{30.8 \pm 0.3} \, cm^{-2} \, s^{-1}$. In other experiments (1966b) these authors compared the ionization probabilities of krypton, xenon and argon, and found that when $F = 10^{30.6 \pm 0.3} \, cm^{-2} \, s^{-1}$, $v_I^{(Xe)}/v_I^{(Kr)} = 10^{0.9 \pm 0.3}$ and $v_I^{(Xe)}/v_I^{(A)} = 10^{2.1 \pm 0.5}$.

Multi-photon ionization measurements were also made for xenon and krypton at λ 1.06 μm by Bystrova et al. (1967, see also Barhudarova et al. (1968)) using a neodymium laser and at λ 0.53 μm by Delone & Delone (1968) with a frequency-doubled neodymium laser. Measurements for xenon at λ 1.06 μm were also made by Louis-Jacquet (1970), who used a proportional counter to detect very small numbers of electrons at pressures as high as 760 Torr.

Peressini (1966) used a 1 GW ruby laser to produce ionization in argon and xenon at low pressures, and collected the electrons with two electrodes. Allowing for secondary ionization, which appeared to be due mainly to the energy (estimated at 60 eV) with which electrons left the focal region, it was found that at a pressure of 0.3 Torr the number of electrons liberated by the laser pulse reached a saturation value when the electric field of the laser was 7×10^7 V cm^{-1} for argon, and 5.5×10^7 V cm^{-1} for xenon. Under these conditions nearly all the atoms in the focal region were presumably ionized: electron–ion inverse Bremsstrahlung would then cause very rapid heating of the electrons, and would explain the large energies observed. This saturation effect was also observed by Chin et al. (1969) in mercury vapour and in xenon. It has been discussed by Chin & Isenor (1970) and by Arutyunyan et al. (1970).

Saturation was also noted by Agostini et al. (1970a, see also 1968, 1970b, c and 1971b) who investigated multi-photon ionization of all the rare gases at λ 1.06 and 0.53 μm. Their results are shown in Figures 5.3(a) and (b). The values of N and N_0 (the number of photons required for direct photoionization) are shown for both wavelengths in Table 5.1 together with the values of the electric field needed to

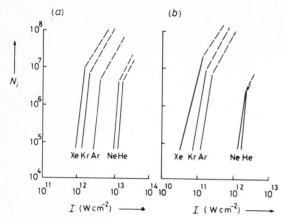

5.3 Number of ions collected, as function of irradiance at focus: (a) $\lambda = 1\cdot06$ μm, (b) $\lambda = 0\cdot53$ μm. (After Agostini *et al.* (1970a))

produce an ionization probability of 10^7 s^{-1}. The substantial differences between values of N and N$_0$ demonstrate the importance of intermediate bound states in the ionization process.

The role of bound states was again shown clearly by the experiments of Baravian *et al.* (1970, 1971) on multi-photon ionization of neon by a ruby laser. The wavelength of a ruby laser is temperature dependent, varying by about $0\cdot06$ Å per degree. At a temperature of about 230 K a resonance occurs between twelve laser photons

TABLE 5.1: *Multi-photon ionization parameters for the rare gases* $\lambda = 1\cdot06$ μm

Atom	N_0	N	Electric field (V cm^{-1}) giving $v_I = 10^7$ s^{-1}
He	21	$18\pm0\cdot3$	$(1\pm0\cdot25)\,10^8$
Ne	19	$13\cdot7\pm0\cdot3$	$(9\pm2\cdot2)\,10^7$
A	14	$10\cdot3\pm0\cdot3$	$(5\pm1\cdot2)\,10^7$
Kr	12	$9\pm0\cdot3$	$(3\cdot5\pm0\cdot9)\,10^7$
Xe	11	$8\cdot7\pm0\cdot3$	$(3\pm0\cdot7)\,10^7$

$\lambda = 0\cdot53$ μm

Atom	N_0	N	Electric field (V cm^{-1}) giving $v_I = 10^7$ s^{-1}
He	11	$9\cdot2\pm0\cdot3$	$(3\cdot7\pm0\cdot8)\,10^7$
Ne	10	$7\cdot3\pm0\cdot3$	$(3\cdot5\pm0\cdot7)\,10^7$
A	7	$5\cdot7\pm0\cdot3$	$(1\cdot15\pm0\cdot2)\,10^7$
Kr	6	$5\cdot5\pm0\cdot6$	$(9\cdot2\pm1\cdot8)\,10^6$
Xe	6	$4\cdot1\pm0\cdot3$	$(7\cdot5\pm1\cdot5)\,10^6$

After Agostini *et al.* (1971*b*).

and the $(11p_{3/2})$ excited state of neon, 21·4294 eV above the ground state. By varying the temperature of the ruby laser rod, Baravian *et al.* tuned the output wavelength over 2 Å through the resonance. They found that the number of electrons generated by the laser pulse rose to ten times the off-resonance value. Their results were in good agreement with the theory of Gold & Bebb (1966).

A similar resonance with an intermediate state was observed by Delone & Delone (1969, see also Delone *et al.* (1972)) in the ionization of potassium by narrow-band $(\Delta\lambda \sim 4 \text{ Å})$ neodymium laser light. By tuning the wavelength the laser light could be brought into resonance with the three-photon transition to the bound $(4f)$ level. Figure 5.4(*a*) shows the dependence of the number of ions collected on the irradiance at two different wavelengths. The induced resonance due to the Stark shift of the $(4f)$ level at high irradiance may be seen in Figure 5.4(*b*). The slope N of these log–log

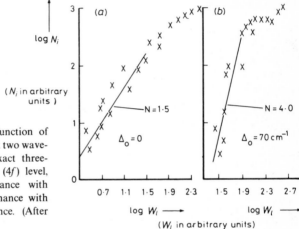

5.4 Number of ions collected, as function of irradiance for neodymium laser light at two wavelengths in potassium vapour: (*a*) exact three-photon resonance with undisplaced $(4f)$ level, (*b*) 70 cm^{-1} deviation from resonance with undisplaced $(4f)$ level, showing resonance with Stark–shifted level at high irradiance. (After Delone *et al.* (1972))

plots was measured at low irradiances and is plotted as a function of the deviation from resonance with the unperturbed $(4f)$ level in Figure 5.5. The results are in good agreement with perturbation theory (Zon *et al.* (1971), Davydkin *et al.* (1971)).

Ionization in caesium vapour was studied by Fox *et al.* (1971) and by David *et al.* (1968) using ruby lasers, and by Hall (1966) at the second harmonics of both ruby and neodymium lasers and also at λ 5750 Å (using Stokes radiation from ruby-pumped deuterium). In their experiments Fox *et al.* found that circularly-polarized light was twice as effective as linearly-polarized light in producing ionization. Held *et al.*

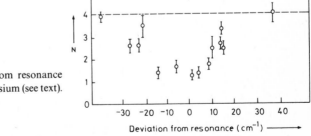

5.5 N as function of deviation from resonance with unperturbed $(4f)$ level of potassium (see text). (After Delone *et al.* (1972))

(1971a, b; 1972b) measured the ionization produced in both caesium and potassium by the fundamental and second harmonic of a neodymium laser. Evans & Thonemann (1972b) measured the intensity dependence of the three-photon ionization of caesium by a ruby laser, and found evidence for two-photon resonance with the (9D) states at an irradiance of 1.4×10^{10} W cm^{-2}. Held et al. (1973) studied the wavelength dependence of the four-photon ionization of caesium using a tuned neodymium laser at constant irradiance. Their results are shown in Figure 5.6 and indicate clearly the three-photon resonance at λ 10 589 Å with the (6f) intermediate level. The results near the (6f) resonance agree well with calculations by Chang & Stehle (1973). In

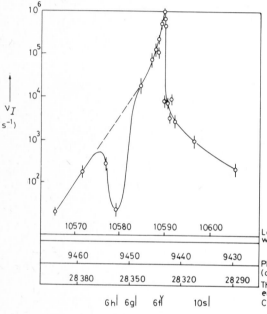

5.6 Variation of ionization probability in caesium with laser wavelength. $I = 1.4 \times 10^8$ W cm^{-2}. (After Held et al. (1973))

another experiment, multi-photon ionization of Cs$_2$ molecules by neodymium laser light was observed by a time-of flight method (Held et al. (1972a)). However, this effect saturates at very low laser irradiances.

Delone et al. (1969) investigated the five-photon ionization of sodium by a neodymium laser. In this case there is no close resonance with an intermediate state and as was expected the value of N was very close to 5 over the range of irradiances used.

The two-stage ionization of rubidium by two mixed laser frequencies has been studied by Ambartsumyan et al. (1971).

Bakos et al. (1972a, b; see also Arslanbekov et al. (1971)) used a tuned neodymium laser to investigate the five-photon ionization of triplet metastable helium, and observed a resonance ascribed to the intermediate $(1 3^3 S)$ state.

The ionization of hydrogen molecules at low pressures by neodymium laser light (λ 1.06 μm) was studied by Berezhetskaya et al. (1970), Agostini et al. (1971a) and Lu Van et al. (1972). Figure 5.7 shows the currents of H$^+$ and H$_2^+$ ions, analysed with

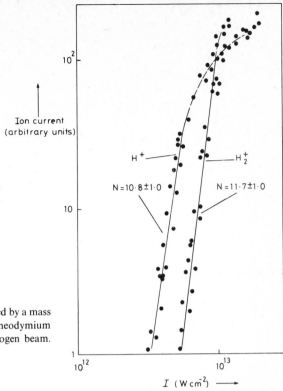

5.7 Currents of H^+ and H_2^+ ions analysed by a mass spectrometer, as functions of irradiance of neodymium laser light focused on a molecular hydrogen beam. (After Lu Van *et al.* (1972))

a magnetic mass spectrometer, as functions of the incident laser irradiance. Eleven- and twelve-photon transitions appear to dominate the formation of H^+ and H_2^+ respectively. Similar experiments at $\lambda\,0.53\,\mu m$ using the second harmonic of neodymium laser light (Mainfray *et al.* (1972), Lu Van *et al.* (1973)) gave the results shown in Figure 5.8.

Lu Van & Mainfray (1972) also investigated the ionization of molecular nitrogen at $\lambda\,0.53\,\mu m$.

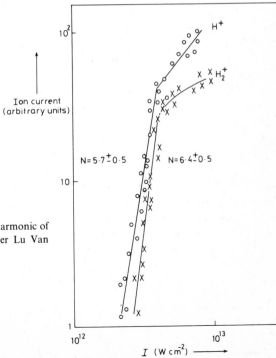

5.8 As for Figure 5.7, using the second harmonic of neodymium laser light at $\lambda\,5300\,\text{Å}$. (After Lu Van *et al.* (1973))

Chin (1971) found that when a ruby laser pulse was focused in iodine vapour only I^+ ions could be detected. In D_2O vapour several ions were formed, among which D_2O^+ predominated, while in CCl_4 vapour the CCl^+ ion was the most abundant.

Chin (1970) also measured the time of flight of ions generated by a ruby laser pulse from the background gas of an ordinary vacuum system at a base pressure of 10^{-6} Torr. He found that at least twenty free electrons were created in the focal volume, and identified many hydrocarbons among the ions, the most abundant of which were C^+, H^+ and C_2^+. The threshold irradiance required to generate twenty electrons was 4×10^{10} W cm^{-2}. Chin concluded that the multi-photon ionization of impurities can play an important role in the initiation of breakdown in gases by laser pulses. Similar experiments were made by Evans & Thonemann (1972a) at a pressure of 10^{-8} Torr with a hydrocarbon partial pressure of 10^{-9} Torr. They detected CH_2^+, $C_2H_n^+$ and $C_4H_n^+$ ions (n being about 6) and others with masses in the range 100 to 120, and noted that the ions and electrons had quite high energies, in the region of 6 eV.

Further evidence concerning the nature of the processes occurring at the initiation of breakdown was obtained at higher pressures by Naiman et al. (1966) by focusing a ruby laser beam in a cloud chamber. The vapours used were methyl alcohol or dimethyl methylphosphonate. When breakdown occurred, a plasma was formed at the focus, giving rise to extensive swirling clouds expanding to 5 cm diameter in a few seconds. At lower laser irradiances, breakdown did not occur, but condensation was nevertheless observed at the focus; this was ascribed to nucleation on charged particles produced by the laser light. Naturally-occurring electrons from other parts of the chamber were swept out by the clearing electric field, and no cosmic ray tracks were observed in the focal volume when ionization by the laser beam occurred. Dust was eliminated by the mode of operation of the cloud chamber. These results confirm that a laser pulse capable of producing some ionization is not necessarily capable of producing the complete avalanche which leads to breakdown.

Chalmeton & Papoular (1967) studied the light emitted by argon and other gases at the focus of a neodymium glass laser beam, at times before and after the very rapid increase in absorption of laser light which characterizes breakdown. They used a photomultiplier to detect light emitted at 90° to the laser beam, and excluded the laser wavelength (λ 1·06 μm) by choosing a photocathode with a long-wavelength cut-off below λ 8000 Å and by interposing a blue-transmitting filter. Repeated observations with a series of different neutral attenuators enabled the light intensity to be measured over a wide range as a function of time without saturating the photomultiplier. A comparison of the light emission in a number of wavelength bands a few hundred Å wide between λ 3100 and 7500 Å indicated that the radiation was predominantly due to an electron–neutral atom Bremsstrahlung continuum. If this was so, then at low ionization levels the light intensity was proportional to the number of free electrons present (the electron temperature appears to have been constant at around 15 000 K, judging from the relative intensity of emission at different wavelengths (Chalmeton & Papoular (1966)). Figure 5·9 shows the total light intensity, in arbitrary units, plotted against time for four pressures of argon. The laser pulse power is also shown, with the time of breakdown at each pressure

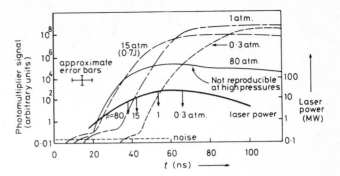

5.9 Light emission from argon at focus of neodymium laser, as function of time. The laser pulse shape is also shown, with times of breakdown at four pressures. (After Chalmeton & Papoular (1967))

indicated by an arrow. According to these authors, it appears that no discontinuity occurs in the luminosity curve at breakdown. The slope of the emission curve does nevertheless suddenly increase: at this moment, the results for A, He, N_2 and O_2 at 760 Torr show that the intensity of light emission is the same within a factor of 2 for each gas. Chalmeton (1969a, b) subsequently made further studies of the pre-breakdown phase using a proportional counter to collect free electrons generated in H_2 and CO_2.

Isenor & Richardson (1971a, b) observed luminescence from molecular gases and vapours irradiated by CO_2 laser light at irradiances well below the breakdown threshold. The luminescence consists mainly of band spectra of dissociation products. It has no distinct pressure or intensity threshold, and has even been observed in gases irradiated by a few hundred watts of continuous CO_2 laser light (Bordé et al. (1966a, b), Karlov et al. (1970)). When a spark was produced in the gas, Isenor & Richardson found that it was preceded by considerable luminosity for about 100 ns, during which time the amount of laser light transmitted through the focal region fell sharply. It appears that molecular excitation processes are important in the development of breakdown at $10.6\ \mu m$.

Veyrie (1968) deduced values of the irradiance at which ionization is initiated in several gases from experimental measurements of the time at which breakdown occurs, as indicated by the onset of perceptible absorption of laser light. For a neodymium glass laser with a pulse duration of 30 ns this 'pre-ionization threshold' was between 1 and $2 \times 10^{10}\ W\ cm^{-2}$ for argon, helium, air and deuterium.

4. DEVELOPMENT OF IONIZATION LEADING TO BREAKDOWN

4.1. CASCADE THEORY OF BREAKDOWN

The starting point of the cascade, or avalanche, theory of breakdown of a gas by an electric field is the assumption that a small number of electrons exist in the volume subjected to the field. Classically, the electrons are accelerated by the field in the intervals between collisions with neutral atoms. The collisions tend to produce an isotropic electron energy distribution. If the field is large enough, some electrons will eventually gain enough energy to cause ionization of gas atoms by collision. Secondary electrons liberated in this way will be accelerated and cause further ionization.

The rate at which an electron gains energy from the electric field may be written as

$$P_e = \frac{e^2 E_e^2}{m_e v_{ea}}, \tag{5.30}$$

where v_{ea} is the collision frequency for momentum transfer between electrons and atoms and E_e is the 'effective field', taking into account the reduced effectiveness of high-frequency fields in giving energy to the electrons (see, for example Macdonald (1966)). The effective field is given by

$$E_e^2 = E^2 \left(\frac{v_{ea}^2}{v_{ea}^2 + \omega^2} \right), \tag{5.31}$$

where ω is the angular frequency of the electric field. The collision frequency is proportional to the gas pressure, so there is a pressure at which the effective field is a maximum, corresponding to $v_{ea} = \omega$. In general v_{ea} is also a function of the electron velocity. However, for helium and hydrogen the collision frequency is effectively independent of velocity, so for these gases we may write

$$v_{ea} = Bp, \tag{5.32}$$

where $B = 2.4 \times 10^9$ Torr^{-1} s^{-1} for helium and 5.9×10^9 for hydrogen.

Meyerand & Haught (1964) pointed out that the classical oscillatory energy of an electron in a microwave field near breakdown is typically 10^{-3} eV while the energy of a single microwave photon is only about 10^{-5} eV; whereas at optical frequencies, although the classical oscillatory energy of an electron is only about 10^{-2} eV the energy of a ruby laser photon is as large as 1.7 eV. They argued that the classical theory which applies for microwaves should be replaced by quantum theory for laser radiation frequencies. The appropriate description of cascade breakdown then involves heating of the electrons through inverse Bremsstrahlung absorption of laser photons in collisions with neutral atoms, followed by ionization by the hot electrons. Quantum-mechanical calculations of this kind were made by Wright (1964). In fact, however, provided the radiation frequency ω is much greater than the collision frequency v_{ea}, and the photon energy $\hbar\omega$ is much less than the ionization energy χ_0, the classical and the single-photon quantum descriptions give very similar results (Zeldovich & Raizer (1964), Ryutov (1964), Browne (1965)). A modification to the inverse Bremsstrahlung theory at high pressures, such that $\omega \lesssim v_{ea}$, has been proposed by Yasojima & Inuishi (1973) to bring classical and quantum treatments into agreement.

In a d.c. or low-frequency a.c. field, electrons may be carried out of the gas to the electrodes before gaining enough energy to cause ionization. In a high-frequency field the distance travelled by an electron before the direction of the field is reversed is usually very small, and electrons are unlikely to be lost by this mechanism. There are, however, competing processes.

If the electric field acts only in a limited region, electrons may be lost by diffusion down the electron density gradient. They may vanish through recombination with positive ions, or attachment to heavy, slow molecules. Electrons may lose energy through inelastic collisions with gas atoms or molecules, resulting in excitation. This is likely to be particularly important in molecular gases with excited states of low energy. Excitation may be the first step towards ionization by subsequent processes, but it may instead be followed by radiative decay from the excited state, in which case the energy is lost (except in the case of resonance radiation, which may be trapped). If the net effect of all these processes is an increase in the number density of electrons with time, breakdown will eventually occur provided the electric field continues for long enough. Assuming that conditions are uniform in space, the net rate of increase of electron density may be written

$$\frac{dn_e}{dt} = (v_i - v_a - v_d)n_e + v_r n_e^2. \tag{5.33}$$

Here $v_i n_e$ is the total rate of ionization, $v_a n_e$ is the rate of attachment, $v_d n_e$ is the net rate of diffusion of electrons out of the region considered and $v_r n_e^2 = v_r n_e n_i$ the rate of recombination, all per unit volume. The coefficients v_i, v_a and v_d are all functions of the gas pressure and v_i depends on the electric field. In the early stages of ionization recombination is relatively unimportant: assuming that v_i, v_a and v_d are independent of time, equation (5.33) may then be integrated to give $n_e(t)$, the value of n_e at time t after starting with an initial electron density n_{e0} at $t = 0$:

$$n_e(t) = n_{e0} \exp\left[(v_i - v_a - v_d)t\right]. \tag{5.34}$$

Thus if $v_i > v_d + v_a$ the electron density will increase exponentially, and if the ionizing field is maintained for long enough breakdown will occur. The breakdown threshold for c.w. radiation, ignoring recombination, is therefore given by the condition

$$v_i = v_d + v_a. \tag{5.35}$$

If only a short pulse of radiation is applied, of duration τ_L, the condition for breakdown in the absence of recombination is that

$$\ln\left(\frac{n_{eb}}{n_{e0}}\right) \leqslant (v_i - v_a - v_d)\tau_L. \tag{5.36}$$

Here n_{eb} is the electron density corresponding to breakdown. For microwaves n_{eb} is generally taken to be the critical density n_{ec} at which the plasma frequency equals the radiation frequency (provided $n_a \geqslant n_{ec}$ as is usually the case). At laser frequencies it is customary to take n_{eb} as the electron density n_{ebl} above which electron–ion inverse Bremsstrahlung becomes more effective in heating the electrons than the electron–neutral atom process. Once this electron density is exceeded, ionization will proceed very rapidly to full ionization, at low pressures, or perhaps to the plasma frequency limit if $n_a \geqslant n_{ec}$, because as the electron temperature rises both n_e and n_i increase and so, therefore, does the rate of heating by electron–ion inverse Bremsstrahlung. Phelps (1966) has shown that a satisfactory approximation to n_{ebl} is $10^{-3}n_a$.

The net rate of energy gain by an electron is equal to $P_e - L_e$, where L_e represents energy losses due to inelastic and elastic collisions. Thus the ionization rate is given by

$$v_i = \frac{P_e - L_e}{\chi}. \tag{5.37}$$

In terms of the laser irradiance I we may write (from (5.30, 5.31))

$$P_e = \frac{e^2 v_{ea} I}{m_e(v_{ea}^2 + \omega^2)\varepsilon_0 c}. \tag{5.38}$$

The breakdown condition (5.36) then becomes

$$I \geqslant \frac{m_e \varepsilon_0 c(v_{ea}^2 + \omega^2)}{e^2 v_{ea}} \left\{ \left[\frac{1}{\tau_L} \ln\left(\frac{n_{eb}}{n_{e0}}\right) + v_a + v_d \right] \chi + L_e \right\}. \tag{5.39}$$

In a given experimental situation, a characteristic diffusion length Λ_d may be defined which incorporates all geometrical effects upon the diffusion loss from the region in which the electric field acts. Then

$$v_d = \frac{D}{\Lambda_d^2}, \tag{5.40}$$

where D, the diffusion coefficient, is a function of the pressure and of the electric field. In the case of a laser beam focused by a lens of small aperture, the important parameter is the minimum diameter of the focused beam. If a large-aperture, short focal length lens is used, the axial variation must be taken into account too. Haught et al. (1966) approximated the focal region as a cylinder of radius

$$r_f = \frac{f(\Delta\alpha)}{2} \tag{5.41}$$

and axial length

$$l_f = (\sqrt{2} - 1)f^2\frac{(\Delta\alpha)}{2r_L}, \tag{5.42}$$

where f is the focal length of the lens, r_L the laser beam radius at half the peak irradiance, as it enters the lens, and $(\Delta\alpha)$ is its divergence in radians. The volume of this cylinder is the 'focal volume' V_f:

$$V_f = \frac{\pi}{8}f^4\frac{(\Delta\alpha)^3}{r_L}(\sqrt{2} - 1). \tag{5.43}$$

For a cylindrical cavity of length l_f and diameter r_f, microwave theory (Macdonald

(1966)) shows that the characteristic diffusion length Λ_d is given by

$$\frac{1}{\Lambda_d^2} = \left(\frac{\pi}{l_f}\right)^2 + \left(\frac{2\cdot4}{r_f}\right)^2. \tag{5.44}$$

For sufficiently large l_f, therefore, $\Lambda_d \simeq r_f/2\cdot4$.

4.2. CALCULATIONS OF GAS BREAKDOWN THRESHOLDS

The theory of cascade ionization was first developed for microwaves (see for example Macdonald (1966)) and has been applied to laser-induced gas breakdown by Minck (1964), Zeldovich & Raizer (1964), Ryutov (1964) and many other authors.

Several detailed calculations of breakdown thresholds have been made for laser pulses of sufficient duration for cascade processes to dominate the development of ionization from the few initial electrons. Analytic semi-empirical expressions were derived by Young & Hercher (1967) and by Morgan et al. (1971) for the high- and low-pressure limits in the region where $\omega \gg v_{ea}$. The breakdown threshold is defined by the attainment of a specified degree of ionization $n_e/n_a = i_B$, which as has already been mentioned is usually taken as 10^{-3} following Phelps (1966). In a monatomic gas at 20°C, $n_a = 3\cdot3 \times 10^{16} \, p_{(\text{Torr})} \, \text{cm}^{-3}$.

At low pressures, where diffusion is the main loss mechanism, Morgan et al. found that for a triangular laser pulse of total duration $2\tau_L$ the peak irradiance at the breakdown threshold is given approximately by

$$\hat{I}_{th} \simeq \frac{n_a}{c_i}\left(\frac{\omega}{v_{ea}}\right)^2\left[\frac{1}{\tau_L}\ln(2\cdot3 i_B V_f n_a) + \frac{2D}{\Lambda_d^2}\right]. \tag{5.45}$$

If $\hat{I}$ is in watts cm^{-2}, c_i is equal to $(2\cdot6 \text{ to } 4\cdot9) \times 10^{23}$ for helium, $(3\cdot8 \text{ to } 7\cdot6) \times 10^{23}$ for neon and $(2\cdot6 \text{ to } 3\cdot8) \times 10^{21}$ for argon.

At high pressures, diffusion may be neglected and in non-attaching gases the main loss mechanism is recombination. In this case (still assuming that $\omega \gg v_{ea}$)

$$\hat{I}_{th} \simeq \frac{n_a}{c_i}\left(\frac{\omega}{v_{ea}}\right)^2\left[\frac{1}{\tau_L}\ln(i_B V_f n_a) + Ri_B n_a \frac{\omega}{v_{ea}}\left(\frac{\pi}{2c_i}i_B\frac{n_a}{\hat{I}_{th}\tau_L}\right)^{\frac{1}{2}}\right], \tag{5.46}$$

where R is the recombination coefficient. It should be noted that under these conditions local variations in the radiation field are important and it is difficult to compare experimental results with theory. In the low-pressure case, on the other hand, electrons diffusing rapidly within the focal volume see an averaged field which is easier to determine.

Chan et al. (1973) included additional terms to allow for the effect of electron energy losses due to inelastic collisions (see Section (5.3)). Afanas'ev et al. (1972) pointed out that in molecular gases inelastic collisions resulting in vibrational excitation may be very important.

With a short laser pulse, it may be that neither recombination nor diffusion is important, the threshold being governed principally by the pulse duration. Here

$$\hat{I}_{th} \simeq \frac{n_a}{c_i}\left(\frac{\omega}{v_{ea}}\right)^2 \frac{1}{\tau_L} \ln\left(i_B V_f n_a\right). \tag{5.47}$$

In this case as the pulse length is varied the pulse energy per unit area, $\hat{I}_{th}\tau_L$, remains constant at threshold.

Since v_{ea} varies as n_a and D varies as n_a^{-1} the threshold irradiance varies with the gas pressure p as p^{-2} in the diffusion-dominated low-pressure limit, as $p^{-\frac{4}{3}}$ for the high-pressure limit and as p^{-1} in the pulse-duration controlled case. At very high pressures we may no longer assume that $\omega \gg v_{ea}$. We may then expect the breakdown threshold to reach a minimum not far from the pressure at which $v_{ea} = \omega$, when E_e/E is greatest (expression (5.31)), and then to rise with increasing pressure.

Detailed calculations of the growth of ionization in several gases were made by Phelps (1966). He used a free–free absorption coefficient, due to Holstein (1965), for electrons of given energy absorbing photons of frequency ω in collisions with neutral atoms. This coefficient involves the momentum transfer cross section, which is known experimentally for helium and argon (Frost & Phelps (1964): see also Druyvesteyn & Penning (1940)) and for nitrogen (Engelhardt et al. (1964)). Assuming that the electrons lose energy by elastic and inelastic collisions and by stimulated emission of Bremsstrahlung, the Boltzmann equation was used to obtain the resulting electron energy distribution functions for these three gases. The rates of exciting and ionizing collisions were then calculated as functions of E/ω: the former are shown in Figure 5.10 for both ruby and microwave frequencies. Assuming a cylindrically

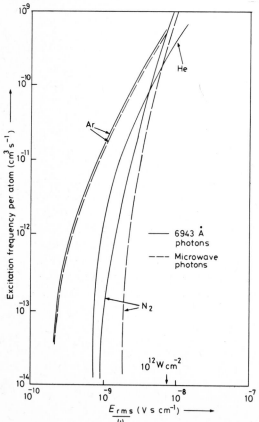

5.10   Frequency of exciting collisions as function of E/ω for microwave and ruby laser photons. (After Phelps (1966))

symmetrical laser light distribution $I(r) = I_0 \exp(-r^2/r_f^2)$, with characteristic radius r_f, and allowing for diffusion losses (taking into account ambipolar diffusion), it was then possible to calculate the variation of the electron density as a function of time. A single initial electron was assumed to be present in the focal volume. The development of ionization in argon, assuming that all excited atoms are immediately photoionized (as was earlier suggested by Minck (1964): see also Barynin & Khoklov (1966)), is shown in Figure 5.11 for three different peak electric fields. Other results are compared with experiment in Section (5.1.1) below.

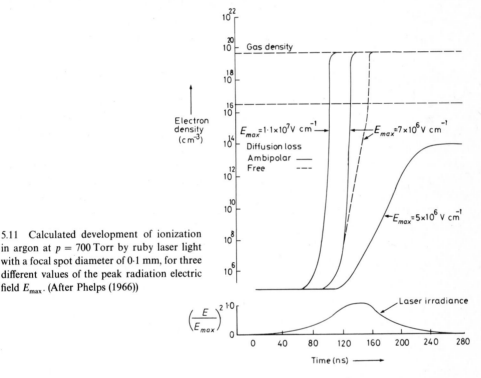

5.11 Calculated development of ionization in argon at $p = 700$ Torr by ruby laser light with a focal spot diameter of 0·1 mm, for three different values of the peak radiation electric field E_{max}. (After Phelps (1966))

A very detailed study of breakdown in air has been made by Kroll & Watson (1972) for a wide range of frequencies. Nielsen et al. (1971) have calculated the breakdown threshold of deuterium by a ruby laser.

As we have seen, if the laser pulse length is decreased the threshold irradiance for cascade ionization rises. Recombination and diffusion become less important. With very short pulses and at low pressures, multi-photon ionization may become more important than the development of the cascade in producing breakdown. This situation was discussed by Bunkin & Prokhorov (1967) and by Ireland & Grey Morgan (1973), who concluded that at low pressures the breakdown threshold, for a very short pulse of duration τ_L, varies as $(\tau_L)^{-1/N}$, N being the number of photons involved in ionization of a gas atom. At high pressures easily-ionized impurity atoms may control the breakdown threshold.

If a free electron gains much more energy from the radiation field than the ionization energy of an atom, the cross-section for ionization or excitation decreases with increasing electron energy. Afanas'ev et al. (1968; 1969a, c) discussed this effect and concluded that at very high irradiances the development of the cascade will be slowed down. There is thus an optimum irradiance for cascade breakdown. The situation is complicated, however, at these high irradiances by the onset of non-linearity in the absorption of energy from the field, and by the occurrence of direct multi-photon ionization.

5. MEASUREMENTS OF BREAKDOWN THRESHOLDS

In general, the breakdown threshold is defined as the lowest irradiance (or the corresponding electric field) at the focal spot at which significant absorption of energy from the laser pulse occurs, usually accompanied by a visible spark. With ruby and neodymium lasers the power at which breakdown occurs is sharply defined, but with carbon dioxide lasers the transition from zero probability of breakdown to 100% probability takes place over a significant range of power.

The detailed structure of the focused laser beam in space and time may be very important in determining the beam power required for breakdown, especially for short pulses at high pressures when there is little diffusion.

When a laser beam is focused by a lens the distribution of irradiance in the focal spot is determined by the mode structure in the laser oscillator, by the effects of any amplifiers and apertures in the system, and by the parameters of the lens. If the laser oscillator is confined to TEM_{00n} modes (often known as 'single transverse mode operation') the distribution of irradiance I across the almost parallel output beam, of power P, will be approximately Gaussian:

$$I(r) = \frac{P}{\pi r_L^2} \exp\left[-\left(\frac{r}{r_L}\right)^2\right]. \tag{5.48}$$

Here r_L is the radius at which the irradiance falls to $1/e$ times its central value. (The radius at half the central irradiance is $(\ln 2)^{\frac{1}{2}} r_L = 0.83 r_L$).

If now the beam is focused by a perfect lens of focal length f, the distribution of irradiance near the focus may be described in terms of the dimensionless coordinates

$$\left.\begin{array}{l} u = \left(\frac{2\pi}{\lambda}\right)\left(\frac{r_L}{f}\right)^2 z \\[2ex] v = \left(\frac{2\pi}{\lambda}\right)\left(\frac{r_L}{f}\right) r, \end{array}\right\} \tag{5.49}$$

where z is the distance along the axis from the focal plane. Near $u = 0$, $v = 0$,

$$I(u, v) \simeq I_0\left(\frac{1}{1+u^2}\right) \exp\left[-\frac{v^2}{(1+u^2)}\right], \tag{5.50}$$

where

$$I_0 = \left(\frac{2\pi}{\lambda}\right)^2 \left(\frac{r_L}{f}\right)^2 \frac{P}{\pi}$$

is the peak irradiance, on axis in the focal plane. Along the optic axis ($v = 0$) the irradiance falls to $I_0/2$ when $u = \pm 1$, i.e. when $z = \pm(\lambda/2\pi)(f/r_L)^2$. In the focal plane plane ($u = 0$) the irradiance falls to I_0/e when $r = r_f = (\lambda/2\pi)(f/r_L)$. Almost two-thirds of the beam power is contained within a circle of radius r_f, which may be used as a definition of the radius of the focal spot. We may compare r_f with the radius r_A of the first minimum of the Airy pattern produced at the focus of a perfect lens with an aperture radius r_l, *uniformly* illuminated by parallel light: $r_A = (1\cdot22\lambda/2)(f/r_l)$ In this case just over four-fifths of the beam power is contained within a circle of radius r_A.

Single-element lenses are used in most experiments with focused high-power laser beams, and significant spherical aberration is often present, though it may be minimized by a suitable choice of curvatures (see for example Strong (1958)). The effect of spherical aberration on the distribution of irradiance in the focal region has been calculated by Innes & Bloom (1966), Zeldovich & Pilipetskii (1966), Evans

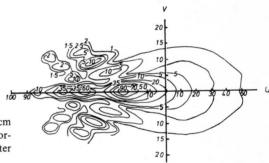

5.12   Distribution of intensity near focus of a 5 cm focal length, $f/5$ simple lens. The intensity is normalized to 100 at the principal maximum. (After Evans & Grey Morgan (1969))

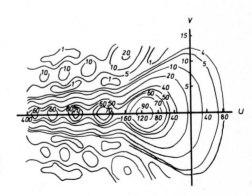

5.13   As for Figure 5.12 but for a 1·8 cm focal length, $f/1·8$ simple lens. (After Evans & Grey Morgan (1969))

& Grey Morgan (1968, 1969) and Aaron *et al.* (1974). Figures 5.12 and 5.13 show lines of constant irradiance, normalized to 100 at the principal maximum, for two different lenses. The beam is incident from the left. In many laser systems the output beam consists of a mixture of several transverse modes and the resulting distribution at the focus is even more complex. The pattern may also change considerably with time during the laser pulse.

The focal spot dimensions used in determining the threshold irradiances (or the corresponding radiation electric fields) quoted in the following sections are often not measured directly but are calculated from the beam divergence and the focal length of the lens on the assumption that aberrations may be neglected. They must therefore be regarded with circumspection.

5.1. OBSERVATIONS WITH Q-SWITCHED RUBY AND NEODYMIUM LASERS (10^{-8} TO 10^{-7} s)

5.1.1. Dependence of breakdown threshold on gas density and dimensions of focal volume

The first measurements of the pressure dependence of the breakdown threshold were made by Minck (1964) in hydrogen, helium, nitrogen and argon, using a ruby laser producing 25 ns pulses. Between 200 and 7.5×10^4 Torr, the threshold laser power fell as the pressure was increased for all the gases. The dimensions of the focal volume were only estimated, so absolute values of the threshold irradiance were not obtained, but the results were consistent with cascade theory.

Meyerand & Haught (1963) observed breakdown with a ruby laser of 30 ns pulse duration in helium and argon at pressures between 1.5×10^3 and 10^5 Torr. They determined the diameter of the focal spot, which was about 200 μm, by measuring the hole produced by the laser pulse in a thin gold foil; this was in agreement with the value calculated from the known beam divergence and the optical parameters of the focusing lens. The results were in qualitative agreement with those of Minck; the absolute value of the electric field at the focus of the laser beam at breakdown was between 10^6 and 10^7 V cm^{-1}. Macdonald (1966) pointed out that the results for helium are in excellent agreement with cascade theory using microwave parameters.

Gill & Dougal (1965), using a ruby laser with a pulse duration of 50 ns, made measurements of breakdown irradiance in high pressure helium, nitrogen and argon between 5000 and 1.5×10^6 Torr. The spot diameters were determined using the metal foil method. Their results are shown in Figure 5.14. Well-defined minima in the breakdown peak electric field occur at about 5×10^5 Torr in helium (4×10^5 V cm^{-1}), at 7.5×10^4 in nitrogen (4.5×10^5 V cm^{-1}) and at 1.5×10^5 Torr in argon (2.5×10^5 V cm^{-1}). These pressures correspond approximately to the condition $\nu_{ea} = \omega$, as is to be expected (see Section (4.2)). Below 10^5 Torr the results are in satisfactory agreement with those of Meyerand & Haught (1963).

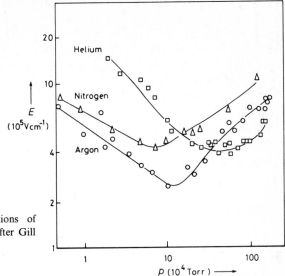

5.14 Breakdown thresholds as functions of pressure, for 50 ns ruby laser pulses. (After Gill & Dougal (1965))

Minck & Rado (1966*a*) made similar measurements in nitrogen, argon, methane, hydrogen, helium and xenon in pressure ranges from 76 to 7.6×10^5 Torr. Though they give no absolute values of the breakdown electric field, their results for nitrogen, argon and helium are qualitatively in agreement with those of Gill & Dougal. The results for methane, hydrogen, helium and xenon are shown in Figure 5.15. The steep dip in the curve for xenon was attributed to the rapid variation in density around the critical point which is at $16.6°C$ and 4.4×10^4 Torr: the upward trend towards higher pressures did not continue in measurements made on liquid xenon.

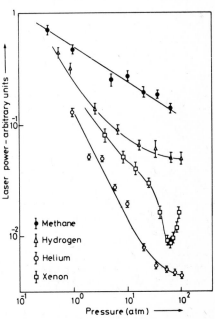

5.15 Breakdown thresholds as functions of pressure for *Q*-switched ruby laser pulses. (After Minck & Rado (1966*a*))

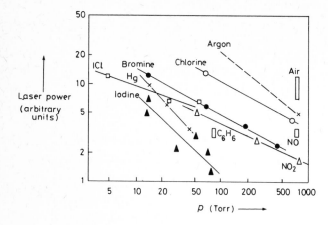

Laser power
(arbitrary
units)

5.16 Breakdown thresholds as functions of pressure for 12 ns ruby laser pulses. (After Holzer *et al.* (1971))

The breakdown of a number of other gases and vapours by a ruby laser was studied by Holzer *et al.* (1971) up to atmospheric pressure. Their results are shown in Figure 5.16. By combining their observations with those of Tomlinson *et al.* (1966) (see Section (5.1.2)), Holzer *et al.* were able to plot the threshold pressure at fixed laser power against ionization energy, and obtained the quite regular relationship shown in Figure 5.17 (only pressures below the $v_{ea} = \omega$ minimum are considered here).

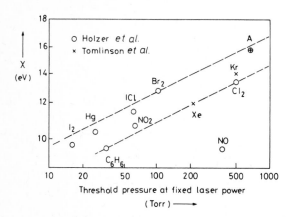

5.17 Threshold pressure at a fixed ruby laser power as function of ionization energy χ. (After Holzer *et al.* (1971))

Morgan *et al.* (1971) found that the threshold irradiance for 40 ns ruby laser pulses in helium varied as p^{-1} over the range from 2×10^3 to 1.6×10^4 Torr, suggesting that neither diffusion nor recombination play an important part in controlling breakdown under these conditions (see Section (4.2)). In argon there was evidence that recombination is important at the higher pressures: the pressure dependence of the breakdown threshold, shown in Figure 5.18, varies from p^{-1} at low pressures to about $p^{-\frac{1}{2}}$ at the highest pressures. In order to fit the results in argon it was necessary to take a value of $3.9 \times 10^9 \, p_{Torr}$ for the average electron–atom collision frequency, suggesting a mean electron energy of less than 5 eV (Macdonald (1966), p. 24).

Guenther & Pendleton (1972) studied the breakdown of deuterium by a ruby laser

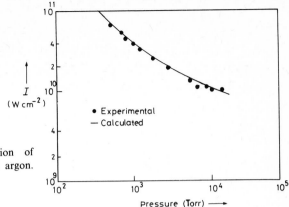

5.18 Breakdown threshold as function of pressure for 40 ns ruby laser pulses in argon. (After Morgan *et al.* (1971))

and found that between 150 and 600 Torr the threshold irradiance varied as $p^{-0.87\pm0.08}$

When breakdown occurs in a dense gas, the attenuation of laser light passing through the focal volume increases abruptly. Tomlinson (1965a) presented a set of oscillograms of the intensity of transmitted laser light taken with a ruby laser focused in argon at 1850 Torr which are shown in Figure 5.19. Breakdown occurs later as the laser pulse peak power is reduced, and is sometimes delayed until as long as 25 ns

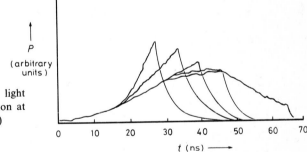

5.19 Oscillograms of ruby laser light transmitted through a focus in argon at 1850 Torr. (After Tomlinson (1965a))

after the peak of the laser pulse has occurred. Very similar results have been obtained with a neodymium laser (Chalmeton (1968)). It is apparent that the growth of ionization from initiation to the breakdown level takes quite a considerable time, as is to be expected on the basis of cascade theory. Tomlinson observed similar effects when laser pulses of constant shape were used to break down gases at different pressures. The field strength and the time of breakdown of several gases by a ruby laser were also measured by Waynant & Ramsey (see Phelps (1966)), whose results for argon are compared with Phelps' calculations in Figure 5.20: it seems that some photo-ionization occurred. For N_2 and air the results also agreed well with theory, assuming that all molecules in states of energy greater than 6·7 eV were immediately photo-ionized.

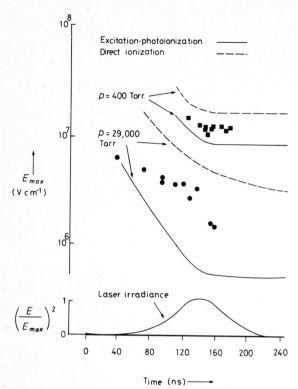

5.20 Comparison of theoretical and experimental breakdown times in argon at $p = 400$ Torr and at 2.9×10^4 Torr, as functions of peak radiation electric field, for ruby laser pulses with the shape shown. (After Phelps (1966))

Experiments with caesium vapour using a ruby laser were first reported by Rizzo & Klewe (1966; see also Klewe & Rizzo (1968)), and subsequently by Okuda et al. (1969). Both sets of breakdown measurements are shown in Figure 5.21. Okuda et al. defined the threshold as the minimum laser irradiance required to produce a detectable current to a probe. Their two plots (1) and (2) refer to small and large area probes respectively. The slope is in each case consistent with a two-photon ionization process. These authors suggest that Rizzo & Klewe's measurements involved plasmas generated at the windows of the caesium cell, and not at the internal focal volume.

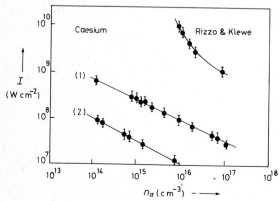

5.21 Breakdown threshold in caesium vapour as function of density, for ruby laser pulses: (1) small probe, (2) large probe (see text). (After Okuda et al. (1969))

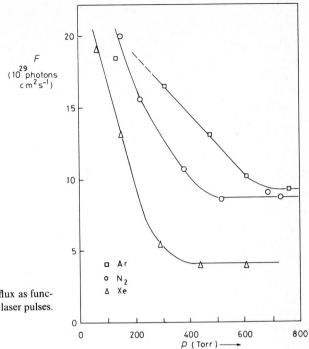

F
(10^{29} photons
$cm^2 s^{-1}$)

5.22   Breakdown threshold photon flux as function of pressure for 35 ns neodymium laser pulses.
(After Bergqvist & Kleman (1966))

□ Ar
○ N$_2$
△ Xe

p (Torr)——→

Breakdown due to neodymium laser radiation at 1·06 μm in the noble gases and
in some molecular gases was studied by Bergqvist & Kleman (1966). Their results
for argon, xenon and nitrogen are shown in Figure 5.22 up to atmospheric pressure.
For xenon, the threshold continued to fall slowly up to 4×10^3 Torr.

Abrikosova & Shcherbina-Samoilova (1968) studied the breakdown of helium in
both the gaseous and the liquid states in a cryostat, arguing that threshold parameters should be substantially unaffected by a change in phase, and by temperature
in the range up to room temperature. In this way breakdown at high densities (up to
the equivalent of 5×10^5 Torr at room temperature) can be investigated without the
need for a high-pressure chamber. They used a ruby laser giving a 20 ns pulse: the
focal spot diameter was estimated to be 32 μm. The threshold irradiances observed
were about twice as high as those given by Gill & Dougal for this range of densities,
but showed the same general behaviour. The discrepancy can rather easily be
explained by uncertainty in focal spot size, though Abrikisova & Bochkova (1969)
suggest that dust particles may contribute to the initial ionization rate in high-
pressure gases. Winterling et al. (1969) observed a marked increase in the threshold
on cooling helium from the lambda point to 1·5 K. Hunklinger & Leiderer (1971)
ascribed this effect to the more rapid settling of solid impurities in superfluid He II.

Geometrical effects may only be ignored if the focal volume is very large, or if the
pressures are so high that diffusion-like losses may be neglected. In fact, geometrical
factors are important over a wide range of conditions, as was first made clear by
Haught et al. (1966). They measured breakdown thresholds in helium, neon, argon

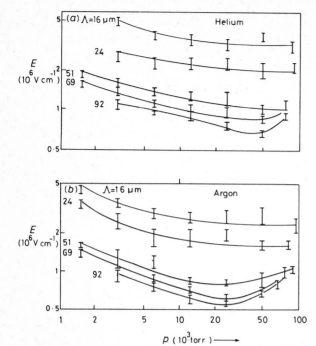

5.23 Breakdown threshold as function of pressure for 50 ns neodymium laser pulses (a) in helium (b) in argon. (After Haught et al. (1966))

and air with both ruby and neodymium lasers, over the range of pressures between 10^3 and 10^5 Torr, using a variety of different focusing lenses so as to vary Λ_d, the characteristic diffusion length. Some of their results with helium and with argon using a neodymium laser of 50 ns pulse duration are shown in Figures 5.23 (a) and (b) respectively. The breakdown threshold varies as $\Lambda_d^{-0.75}$ in all the gases studied. Belland et al. (1971b) observed the same dependence of the threshold field on Λ_d with 1·6 ns and 40 ns neodymium laser pulses in atmospheric air, for values of Λ_d in the region of 10 to 100 μm. Figure 5.24 shows a comparison between their experimental results and the predictions of simple cascade theory allowing only for losses by electron diffusion (see also related calculations by Haynes et al. (1971)). The discrepancies at larger values of Λ_d suggest that diffusion-like losses, perhaps involving

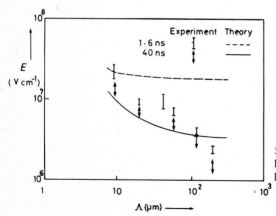

5.24 Measured and calculated breakdown thresholds in air at N.T.P. as functions of Λ_d for neodymium laser pulses (see text). (After Belland et al. (1971b))

inelastic collisions (Chan *et al.* (1973)), are present at these pressures, though Belland *et al.* suggested that the self-focusing observed in related experiments might be involved. Dougal & Gill (1968) found that for helium, with a small value of Λ_d $(2 \times 10^{-4}$ cm), cascade theory with electron diffusion fitted the pressure dependence of the breakdown threshold satisfactorily up to 2000 atm.

Mitsuk *et al.* (1966) made similar measurements in krypton and xenon at pressures up to 580 Torr with a ruby laser of 60 ns pulse duration. The breakdown threshold increased rapidly as the focal length of the focusing lens was reduced below 10 cm, particularly at low pressures. Insufficient data are given for values of Λ_d to be calculated, but they probably lay between 10 and 100 μm.

Smith & Tomlinson (1967) investigated breakdown using a single transverse and axial mode ruby laser which produced single pulses with smooth electric field distributions in both space and time. The breakdown threshold fields measured in argon at 10^4 and 4.8×10^4 Torr were compared with values obtained with multimode lasers by extrapolation from results given by Buscher *et al.* (1965) and Haught *et al.* (1966) and were found to be the same within the experimental uncertainty (a factor of about 2).

Smith & Haught (1966) also investigated the breakdown threshold in mixtures of argon and neon, using a neodymium laser. Their results at a constant total pressure of 5.2×10^4 Torr are shown in Figure 5.25; even one percent of neon reduces the breakdown threshold from 3.2×10^6 to 1.9×10^6 V cm^{-1}, at which it remains unchanged up to the highest neon concentration studied (20%). However, as may be seen from Figure 5.26, this effect was only found at relatively high pressures; below 5×10^3 Torr the addition of 1% neon had no effect on the threshold. The breakdown threshold for pure neon is always higher than that for argon, so some interaction between the different atoms must be responsible for the observed effect. A study of the dependence of the breakdown threshold on the parameter Λ_d at 5.2×10^4 Torr showed that whereas for pure argon the breakdown electric field varied as $\Lambda_d^{-0.96}$, for argon with 10% neon it only varied as $\Lambda_d^{-0.78}$, suggesting that the presence of neon reduced the diffusion-like losses during the development of ionization. Mulchenko &

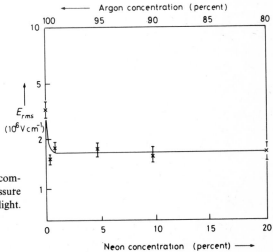

5.25 Breakdown threshold as function of composition of an argon–neon mixture at a pressure of 5.2×10^4 Torr, for neodymium laser light. $\Lambda_d = 16\ \mu$m. (After Smith & Haught (1966))

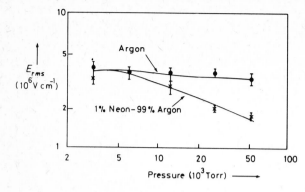

5.26 Breakdown threshold as function of press-ure, for argon and for argon with 1% neon. Neodymium laser, $\Lambda_d = 16\,\mu$m. (After Smith & Haught (1966))

Raizer (1971) repeated these experiments with both ruby and neodymium lasers. Their measurements, which covered the full range of gas compositions from pure argon to pure neon, are shown in Figures 5·27 and 5·28. In the case of neodymium laser light, a small admixture of neon to pure argon reduced the threshold, in agree-ment with the result of Smith & Haught. However, a similar effect was also observed when a small amount of argon was added to pure neon. The magnitude of the effect increased with increasing pressure. With ruby laser light, varying the gas composition caused only a monotonic change in breakdown threshold between the pure gas values, at all pressures from 760 to 7.6×10^4 Torr. The frequency dependence of

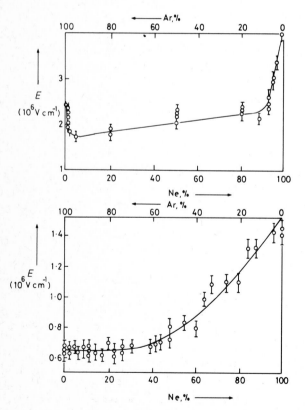

5.27 Breakdown threshold as function of com-position of an argon–neon mixture at a pressure of 80 atm., for neodymium laser light. $\Lambda_d = 17.5\,\mu$m. (After Mulchenko & Raizer (1971))

5.28 Breakdown threshold as function of com-position of an argon–neon mixture at a pressure of 80 atm., for ruby laser light. $\Lambda_d = 26\,\mu$m. (After Mulchenko & Raizer (1971))

these effects seems to rule out an explanation based on variations in diffusion rates. It is probable that some collisional interaction involving an excited atom is responsible.

Chin & Isenor (1967) measured breakdown thresholds with a ruby laser in argon–Freon 12 (CCl_2F_2) mixtures, and Young et al. (1968) subsequently carried out the same experiment in more detail. The results given in the second paper are shown in Figure 5.29. At a total pressure of 250 Torr, the effect of adding increasing proportions of freon to the argon is to reduce the breakdown threshold electric field monotonically,

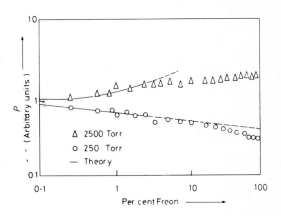

5.29 Breakdown threshold as function of composition of an argon–Freon 12 mixture at two pressures, for ruby laser light. Focal area ~ 10^{-4} cm². Theoretical curves are based on Young & Hercher (1967) (After Young et al. (1968))

the value for pure freon being twice that for pure argon. The low-pressure result is explained by supposing that freon has a rather higher multi-photon ionization rate than argon. The high-pressure result is ascribed to an increased rate of loss of electrons by attachment to the freon molecules.

Breakdown thresholds for neodymium laser light in mixtures of rare gases with mercury vapour have been measured by Mitsuk et al. (1971).

5.1.2. Frequency dependence of breakdown threshold

Akhmanov et al. (1965) measured breakdown thresholds in atmospheric air using neodymium glass laser light at λ 1·06 μm and its harmonic at λ 0·53 μm, and found that the threshold electric field was always higher for the shorter wavelength, as is to be expected if cascade ionization is the dominant cause of breakdown.

Haught et al. (1966) compared thresholds obtained with ruby and with neodymium glass lasers in argon and helium for a constant value of Λ_d in the pressure range from 10^3 to 10^5 Torr. Their results, shown in Figures 5.30(a) and (b), are again consistent with cascade theory.

Tomlinson et al. (1966) measured breakdown thresholds with ruby and neodymium lasers in xenon, krypton, argon, neon, helium, oxygen, nitrogen and air at pressures between 100 and 2500 Torr, and again found higher thresholds for the ruby laser than for the neodymium laser. In all the noble gases, the dependence of the breakdown electric field on pressure was as $p^{-0.5}$ for the neodymium laser and as $p^{-0.35}$ for ruby. This was unexpected, because the focal spot size was smaller for the ruby laser, so

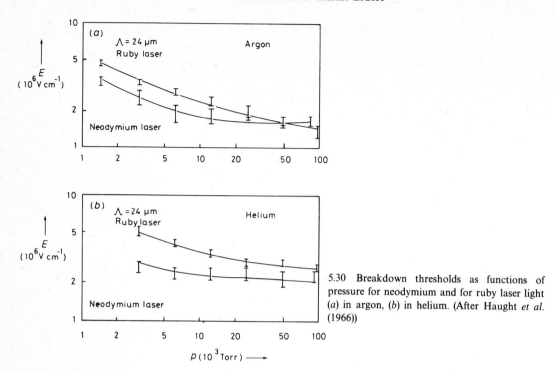

5.30 Breakdown thresholds as functions of pressure for neodymium and for ruby laser light (a) in argon, (b) in helium. (After Haught et al. (1966))

the pressure dependence should have been stronger for the ruby laser if electron diffusion limited the rate of ionization growth. In the molecular gases studied, the pressure dependence was less strong.

Buscher et al. (1965) examined breakdown in xenon and argon by fundamental and second harmonic radiation from both ruby and neodymium lasers, i.e. at wavelengths of 0·69 and 0·35, 1·06 and 0·53 μm. Their results are shown in Figure 5.31. A pronounced maximum in the breakdown threshold is apparent, which the authors estimate to occur at about λ 0·6 μm in argon and about λ 0·9 μm in xenon. The most surprising feature is the low breakdown threshold for short wavelengths. Two lenses of different focal lengths were used, and the focal spot size measured in the experiments varied widely from one wavelength to another.

Similar experiments were carried out by Barthelemy et al. (1968) in atmospheric air, using the second, third and fourth harmonics of a neodymium laser. The results are shown in Figure 5·32. These measurements are more informative than those of Buscher et al. because they were all made with the same silica focusing lens, and because the beam divergences at all four wavelengths were found to be identical (3×10^{-4} rad) and the pulse lengths were quite similar (33 ns at λ 1·06 μm, 28 ns at λ 0·53 μm and 25 ns at λ 0·35 and 0·26 μm). The experiments were repeated by Eremin et al. (1971), whose results were in good agreement with those of Barthelemy et al. The wavelength dependence of the threshold irradiance indicates a maximum between λ 0·53 and 1·06 μm. This is not consistent with simple cascade theory,

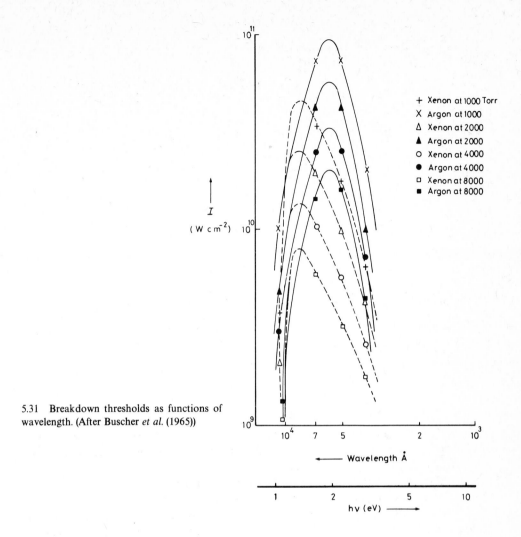

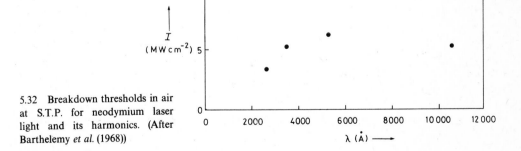

5.31 Breakdown thresholds as functions of wavelength. (After Buscher *et al.* (1965))

Legend for upper figure:

+ Xenon at 1000 Torr
X Argon at 1000
△ Xenon at 2000
▲ Argon at 2000
○ Xenon at 4000
● Argon at 4000
□ Xenon at 8000
■ Argon at 8000

I (W cm^{-2})

Wavelength Å

$h\nu$ (eV)

5.32 Breakdown thresholds in air at S.T.P. for neodymium laser light and its harmonics. (After Barthelemy *et al.* (1968))

I (MW cm^{-2})

λ (Å)

according to which the threshold should vary monotonically as λ^{-2}. Multi-photon processes may be important at short wavelengths.

A more detailed study of the wavelength dependence of the breakdown thresholds in the region of the maximum became possible when tunable dye lasers were developed. Alcock *et al.* (1969c) used two different dyes pumped by a ruby laser to cover the ranges $\lambda\,7200$–$7500\,\text{Å}$ and $\lambda\,8100$–$8600\,\text{Å}$, and obtained pulses of up to $20\,\text{MW}$ peak power of $25\,\text{ns}$ duration. The results of their breakdown measurements on xenon and argon are shown in Figure 5.33. They are in reasonable agreement with those of Buscher *et al.* (1965), and confirm that the location and shape of the peak

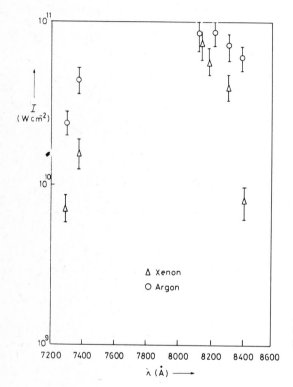

5.33 Wavelength dependence of breakdown thresholds in xenon and argon at a pressure of 8 atm. (After Alcock *et al.* (1969c))

depends on the gas investigated. In air at atmospheric pressure, the breakdown threshold decreased monotonically with increasing wavelength. A dye laser was also used by Morgan *et al.* (1971) to study the wavelength dependence of the threshold in argon. At $10\,640\,\text{Torr}$ a maximum occurred near $\lambda\,8150\,\text{Å}$. However, at lower pressures it was found that the threshold at $\lambda\,8010\,\text{Å}$ was above that at $\lambda\,1{\cdot}06\,\mu\text{m}$ and below that at $\lambda\,6943\,\text{Å}$.

5.2. OBSERVATIONS WITH MODE-LOCKED RUBY AND NEODYMIUM LASERS (10^{-12} TO 10^{-10} s)

Smith & Tomlinson (1967) produced breakdown in air and other gases with a neodymium laser which could be operated without mode-locking, using a Kerr cell Q-switch to give a 40 ns pulse, and could also be Q-switched and mode-locked, using a saturable dye cell (Stetser & De Maria (1966)) to give a train of very brief pulses whose envelope had the same 40 ns duration as the non-mode-locked pulse. The interval between successive spikes was 7 ns; the duration of a spike was less than the resolving time of the detector ($1\cdot5 \times 10^{-10}$ s). The light transmitted through the focus in atmospheric air was detected by a fast photodiode whose output was displayed on an oscilloscope: the incident light signal was detected by another photodiode whose output was delayed by 60 ns and displayed on the same trace. Up to the moment of breakdown the two pulse shapes were identical.

The two oscillograms shown in Figure 5.34 were obtained with pulses of equal total energy. As may be seen, the attenuation of the pulses was similar for the normal Q-switched pulse (a) and for the mode-locked pulse (b), and the time to breakdown was the same for both. The average breakdown threshold electric field in each case was 6×10^6 V cm^{-1} (irradiance $\sim 10^{11}$ W cm^{-2}) although in the mode-locked case the instantaneous field must have reached at least 6×10^7 V cm^{-1} (irradiance $\sim 10^{13}$ W cm^{-2}) and from a comparison with other measurements of the duration of similar mode-locked laser spikes could have been as high as 6×10^8 V cm^{-1} (irradiance $\stackrel{<}{\sim} 10^{15}$ W cm^{-2}). It appeared that breakdown depends only on the average energy input to the focal volume, over a period of at least 10^{-8} seconds. However, when Alcock et al. (1968d) repeated the experiment with a mode-locked laser, they

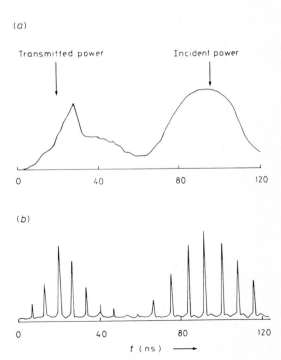

5.34 Incident and transmitted neodymium laser pulses focused in air at S.T.P.: (a) normal Q-switched pulse. (b) mode-locked train of spikes (see text). (After Smith & Tomlinson (1967))

found that on varying the spike separation in the range 4 to 8 ns both the average and the instantaneous peak threshold irradiances remained substantially constant. This suggested that breakdown was due to a single spike: if several spikes contributed to breakdown, varying the interval between spikes would affect the loss processes and hence the threshold irradiance. De Michelis (1970) again repeated the experiment, varying the dimensions of the focal volume. The instantaneous peak threshold irradiance with the mode-locked train was found to be independent of the diffusion length Λ_d, in contrast to the results of Section (5.1.1) for smooth nanosecond pulses. Again, breakdown appears to be due to a single spike.

Alcock & Richardson (1968) produced single mode-locked pulses lasting about 10^{-11} s by cavity dumping a mode-locked neodymium laser. The output pulse was passed through a cell containing a saturable absorber, to reduce the intensity of the background laser light preceding the main pulse, and then through an amplifier to give a final energy of 100 mJ. A 3 cm lens focused the output pulse into air, nitrogen or argon in a pressure vessel. The pressure dependence of the observed breakdown threshold in nitrogen and argon between 500 and 6000 Torr is shown in Figure 5.35(a).

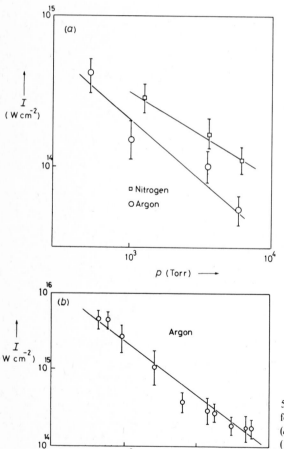

5.35   Breakdown thresholds as functions of pressure for single mode-locked neodymium laser pulses: (a) in nitrogen and argon (after Alcock & Richardson (1968)), (b) in argon (after Dewhurst et al. (1971))

It is very similar to that observed with pulses lasting many nanoseconds (Tomlinson *et al.* (1966)) but the electric field values are higher by about two orders of magnitude, being as high as 3×10^8 V cm^{-1} in air at 760 Torr. Despite the very short duration of the pulse, it seems from the observed pressure dependence that multi-photon ionization cannot be the only process involved in breakdown under the conditions of the experiment. Dewhurst *et al.* (1971) extended the measurements to pressures as low as 40 Torr in argon. As may be seen in Figure 5.35(*b*) they found that the threshold irradiance still varied as $p^{-0.7}$ throughout the range of pressures used. Measurements in helium gave the same pressure dependence. Ireland & Grey Morgan (1974) measured breakdown thresholds in argon and nitrogen using 20 ps pulses of fundamental and second harmonic neodymium laser light (λ 1·06 μm and 0·53 μm). Their results are shown in Figures 5·36(*a*) and (*b*). At λ 1·06 μm the pressure dependence of the threshold is similar to that reported by Alcock & Richardson, though the irradiances are

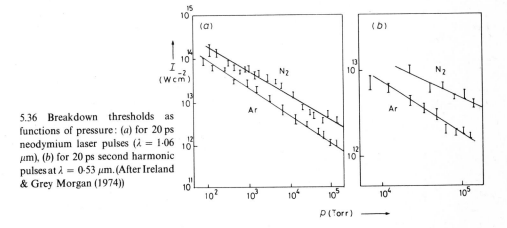

5.36 Breakdown thresholds as functions of pressure: (*a*) for 20 ps neodymium laser pulses ($\lambda = 1·06$ μm), (*b*) for 20 ps second harmonic pulses at $\lambda = 0·53$ μm. (After Ireland & Grey Morgan (1974))

lower. The absence of a transition to a weakly pressure-dependent multi-photon breakdown regime at low pressures is surprising. The presence of a significant amount of background light, spread over several nanoseconds, might have caused cascade breakdown even at low pressures (Ireland & Grey Morgan (1973)), but this explanation was ruled out in Ireland & Grey Morgan's experiments, in which the background irradiance was less than 10^{-4} of the peak irradiance.

In contrast to these observations, Figure 5.37(*a*) shows breakdown thresholds measured by Krasyuk *et al.* (1969, 1970*b*) over a wide range of pressures in nitrogen and in argon and helium for single mode-locked ruby laser pulses lasting 30 to 100 ps. The threshold becomes almost independent of pressure at low pressures, as would be expected on the basis of purely multi-photon or tunnelling ionization. A similar transition was observed in nitrogen by Dewhurst *et al.* (1971) with single mode-locked neodymium laser pulses at a pressure of 1000 Torr (Figure 5.37(*b*)). Krasyuk & Pashinin (1972) also compared the thresholds for breakdown by the fundamental and by the second harmonic of mode-locked ruby laser pulses. When the frequency was doubled the threshold fell by a factor of 20 for argon and 300 for nitrogen,

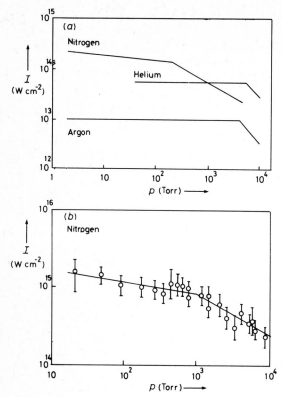

5.37 Breakdown thresholds as functions of pressure : (a) in helium, nitrogen and argon, for single mode-locked ruby laser pulses (after Krasyuk *et al.* (1969, 1970*b*)), (*b*) in nitrogen for single mode-locked neodymium laser pulses (after Dewhurst *et al.* (1971))

consistent with ionization being due to multi-photon processes rather than tunnelling. The threshold was only weakly dependent on pressure even up to 4500 Torr. A reduction in the threshold for air by a factor of 2 or 3 on doubling the radiation frequency of a train of mode-locked neodymium laser pulses was noted by Orlov *et al.* (1971).

The breakdown threshold for air at S.T.P. was measured for several pulse durations, using a neodymium glass·laser with various cavity configurations, by Wang & Davis (1971, see also Alcock & Richardson (1972)). Their measurements agreed with previous work.

5.3. OBSERVATIONS WITH CARBON DIOXIDE LASERS

Breakdown studies using carbon dioxide laser light at λ 10·6 μm are of particular interest, because the wavelength is intermediate between those of near-visible lasers and microwaves. One of the characteristics of carbon dioxide lasers is a relatively high pulse repetition rate, which is generally greater than 1 per second and often as high as 50 or 100 per second, compared to 1 pulse every several minutes for high power ruby or neodymium lasers. Several authors have noted that although break-down may be difficult initially, once it has been achieved (perhaps with some pre-ionization) it is easier to produce with subsequent pulses, presumably because some residual ionization remains in the focal volume (Chan *et al.* (1973)).

Generalov *et al.* (1970*a*) studied the breakdown of high-pressure inert gases using a well-focused 10 KW *Q*-switched laser. The pulse length was 0·3 to 1·5 μs. Their results for xenon are shown in Figure 5.38, and are in qualitative agreement with the cascade breakdown theory.

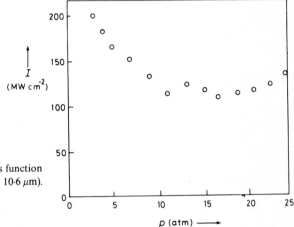

5.38   Breakdown threshold in xenon as function of pressure, for CO_2 laser radiation ($\lambda = 10\cdot6\ \mu$m). (After Generalov *et al.* (1970*a*))

In some early experiments with a carbon dioxide laser, Smith (1970) found it necessary to provide pre-ionization to produce breakdown in argon. Subsequent observations (Smith (1971), Brown & Smith (1973)) showed that for large pre-ionization electron densities the threshold was relatively low and showed no volume dependence. In the absence of pre-ionization, breakdown took on a statistical character and the threshold fell as the focal volume was increased, suggesting that small numbers of impurity particles might be involved.

Breakdown thresholds measured in argon by Franzen (1972) are shown in Figure 5.39. They are in satisfactory agreement with those of Smith (1970). The laser pulse

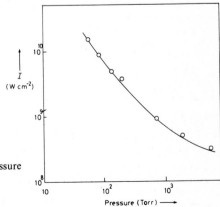

5.39   Breakdown threshold in argon as a function of pressure for CO_2 laser radiation. (After Franzen (1972))

duration was 200 ns and the diameter of the focal spot, using a 5 cm focal length mirror, was 110 μm. Hill *et al.* (1972) determined thresholds for hydrogen, nitrogen, argon and helium using similar laser pulses of about 200 ns duration focused by a 3·8 cm lens to a radius of 100 μm. The threshold was defined as the irradiance which gave a breakdown probability of 0·5. No pre-ionization was used except for hydrogen. The results are shown in Figure 5.40. The measurements for argon show a pressure

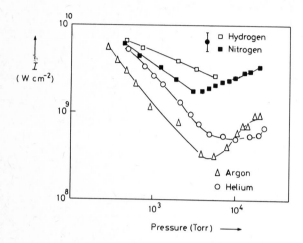

5.40 Breakdown thresholds as functions of pressure, for CO_2 laser radiation. (After Hill *et al.* (1972))

dependence in good agreement with cascade theory, particularly if recombination is taken into account, but the threshold values are rather high. In the molecular gases vibrational excitation may be important. A comparison with Figure 5.14 for ruby laser light shows that the pressures for minimum breakdown irradiance are again approximately determined by the condition $v_{ea} = \omega$.

Chan *et al.* (1973) made an extensive study of breakdown in helium, neon, argon, oxygen and air with 160 ns pulses. In order to explain the results on the basis of cascade theory, additional energy loss terms due to inelastic collisions were introduced empirically; the threshold irradiance is then given by expression (5.39) with

$$
\left.
\begin{aligned}
v_a &= h_a v_{ea} \\[4pt]
v_d &= \frac{D}{\Lambda^2} \\[4pt]
L_e &= \frac{\chi}{\ln 2}\left(\alpha + \frac{\beta}{\Lambda_d^2}\right)v_{ea}.
\end{aligned}
\right\}
\tag{5.51}
$$

In this work (n_{eb}/n_{e0}) was taken to be 10^{13}. Table 5.2 shows the parameters v_{ea}, h_a, α and β used to fit the observations (Λ_d in cm). The diffusion parameter D was taken to equal $\chi/3m_e v_{ea}$, i.e. ambipolar diffusion was neglected. The collision frequencies v_{ea} were taken from other published data, averaged over the electron energy up to the ionization potential, except for argon where, as was noted by Morgan *et al.* (1971) for ruby laser light, a lower value of 3×10^9 s^{-1} Torr^{-1} was found to be necessary, suggesting a mean electron energy of less than 4 eV. Figure 5.41 shows the experimental

TABLE 5.2: Gas breakdown parameters for CO_2 laser light

Gas species	Ionization potential (eV)	v_{ea} (10^9 s^{-1} Torr^{-1})	h_a (10^{-4})	α (10^{-4})	β (10^{-8} cm^{-2})
He	24·5	2·50	0·0	0·0	0·26
Ne	21·6	1·60	0·0	0·0	0·46
Ar	15·8	3·00	0·0	0·1	0·16
O_2	12·1	4·40	7·5	2·6	1·64
Air	15·0	5·20	1·5	3·6	1·17

After Chan *et al.* (1973).

and theoretical results for argon. For air and argon an improved fit was obtained if the values of v_{ea} in the table were halved. Provided the pulse length exceeds 10 ns, the duration is not important: the threshold is correctly given by the irradiance, and not the energy in the pulse.

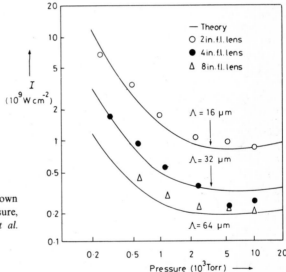

5.41 Measured and calculated breakdown thresholds in argon as functions of pressure, for CO_2 laser radiation. (After Chan *et al.* (1973))

Lencioni (1973) studied the effect of dust on the air breakdown threshold and showed that naturally-occurring dust can lower threshold measurements for spot sizes greater than 100 μm. For highly-filtered clean air the threshold approached an asymptotic value at large spot sizes, in good agreement with the theory of Kroll & Watson (1972). The breakdown of air containing aerosol particles was investigated experimentally by Marquet *et al.* (1972, see also Hull *et al.* (1972)) and discussed by Canavan & Nielsen (1973).

The influence of a transverse d.c. electric field on CO_2 laser induced breakdown

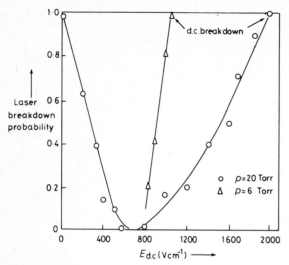

5.42   Effect of a transverse d.c. electric field on the probability of breakdown of air by a CO_2 laser pulse ($I = 4.4 \times 10^{11}$ W cm^{-2}). (After Tulip & Seguin (1973a))

was studied by Tulip & Seguin (1973a). The probability of breakdown in air at an irradiance of 4.4×10^{11} W cm^{-2} is shown as a function of the d.c. field in Figure 5.42. The fall in breakdown probability up to 600 V cm^{-1} at a pressure of 20 Torr was ascribed to the removal of charges from the focal region, and the subsequent rise as the field was increased might have been due to an enhancement of the cascade process.

Tulip & Seguin (1973b) also made a study of the ionization produced by the laser pulse at air pressures below the 19 Torr at which breakdown occurred. The number of ion pairs collected rose linearly from 3×10^9 to 6.5×10^9 per pulse as the pressure was increased from 1 to 18 Torr. This result is difficult to explain on cascade theory: dust may again be involved.

Moody (1973) produced breakdown in argon at pressures between 1 and 7 atm by focusing a pulsed high-power CO_2 laser within the focal volume of a c.w. laser. The effect of the c.w. irradiance of up to 6.2×10^5 cm^{-2} was to raise the breakdown irradiance by up to 38 % at a pressure of 1 atm. The effect was ascribed to a lowering of the collision frequency in the focal volume as a result of heating of neutral atoms. Plasmas initiated by the pulsed laser were sustained by the c.w. laser at pressures in the region 2·3 to 3 atmospheres.

The effect on the breakdown threshold of a spark discharge in the vicinity of the focal region was investigated by Robinson (1973). A marked lowering of the threshold was observed in argon, and smaller effects in nitrogen and helium, but none in carbon dioxide, air or a laser mixture ($1:18:1::CO_2:He:N_2$). Figure 5.43 shows results for argon with two different spacings between the 3 mm spark gap and the focal volume, with a delay of a few microseconds between the spark and the laser pulse. Lowering of the threshold could still be observed as long as a millisecond after the spark. The spark had no effect if a CaF_2 window was placed between it and the focal volume, cutting out radiation below λ 1200 Å.

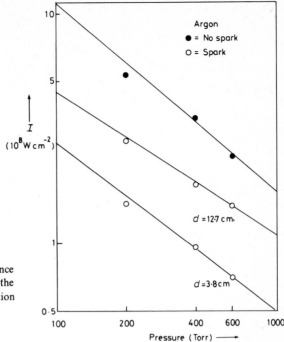

5.43 Effect of a spark discharge at a distance d from the focal region of a CO_2 laser on the breakdown threshold for argon, as a function of pressure. (After Robinson (1973))

Tulip *et al.* (1971) have pointed out that very high irradiances can be obtained if light within a laser cavity is brought to a focus. Air breakdown could be obtained at pressures down to 100 Torr without pre-ionization in a cell placed inside a laser whose normal output consisted of 100 mJ, 200 ns pulses. Berger & Smith (1972) considered the limitations imposed by internal breakdown on the operating pressure of pulsed CO_2 lasers. For beam diameters larger than 0·2 cm the threshold for break-down of a laser mixture $(1:5:1::CO_2:He:N_2)$ by 70 ns pulses was found to be given by $5·0\,p^{-0·7}\,J\,cm^{-2}$, p being the total pressure in atmospheres. Berger & Smith suggested that this probably applied to shorter pulses also, and concluded that efficient operation is limited to the pressure range below 2 to 4 atmospheres.

5.4. EFFECTS OF MAGNETIC FIELDS ON BREAKDOWN THRESHOLDS

Electrons in a magnetic field spiral around the lines of force with the gyro frequency. This can affect breakdown thresholds in two ways: it changes the rate at which electrons gain energy from the electric field, and it reduces the rate of diffusion of electrons perpendicularly to the magnetic field (Lax *et al.* (1950)). The first effect is equivalent (see Macdonald (1966), p. 137) to changing the effective electric field E_e (see Section (4.1)) to

$$E_{ec} = \frac{Ev_{ea}}{2^{\frac{1}{2}}}\left[\frac{1}{v_{ea}^2+(\omega-\omega_{ce})^2}+\frac{1}{v_{ea}^2+(\omega+\omega_{ce})^2}\right]^{\frac{1}{2}}. \tag{5.52}$$

For a given value of the collision frequency v_{ea}, E_{ec} thus reaches a maximum value when the radiation frequency ω equals the electron gyro-frequency ω_{ce}. If v_{ea} is much smaller than ω, in the absence of a magnetic field $E_e \simeq E v_{ea}/\omega$, but when $\omega = |\omega_{ce}|$ the resonance value of E_{ec} is almost equal to $E/2^{\frac{1}{2}}$. This is the low-pressure resonance situation. At very high pressures, i.e. for v_{ea} much greater than both ω and $|\omega_{ce}|$, the effective electric field is scarcely affected by the magnetic field.

The second effect of the magnetic field, the reduction of the rate at which electrons diffuse across the magnetic field, may be described (see Macdonald (1966), p. 141) as increasing the linear dimensions of the interaction region by the factor $(v_{ea}^2 + \omega_{ce}^2)^{\frac{1}{2}}/v_{ea}$ in directions perpendicular to the field. Again the effect is small at high pressures.

Provided the classical theory is valid for laser breakdown, we may predict, therefore, that the effects of a magnetic field will not be large unless $v_{ea} \lesssim \omega, |\omega_{ce}|$, and will be greatest for a given v_{ea} when $\omega_{ce} = \omega$. For neodymium laser light ($\omega = 1.8 \times 10^{15}$ rad s^{-1}), resonance would require a magnetic field as high as 10^8 G. If we take $v_{ea} = 5 \times 10^9 \, p \, s^{-1}$, with p in Torr, then for a magnetic field to produce a substantial change in the effective electric field requires also that $p \lesssim 3 \times 10^5$ Torr.

Early experiments with a ruby laser by Litvak & Edwards (1966) showed that fields up to 100 kG had little influence on the breakdown thresholds in air or argon. Chan et al. (1968) found that a 200 kG field, perpendicular to both the axis of the ruby laser beam and the electric field vector of the laser light, had no effect on the breakdown threshold in air, butane or helium at 760 Torr. On the other hand Vardzigulova et al. (1967), using a neodymium laser, reported that a 210 kG field parallel to the

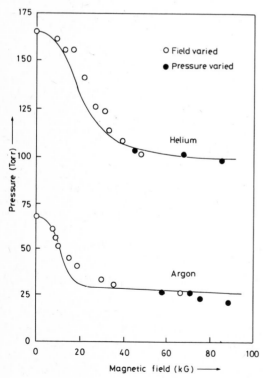

5.44   Breakdown threshold pressures for argon and helium, as functions of magnetic field, for CO_2 laser pulses. Magnetic field direction parallel to laser beam axis. (After Cohn et al. (1972a))

laser axis definitely lowered the breakdown threshold, though to an unspecified extent, at pressures of 30, 160 and 760 Torr in air. In all these experiments the magnetic fields used were over 400 times too small to show resonance effects. So far as diffusion rates are concerned, if the direction of the field is parallel to the axis of the focal volume, assumed to be a fairly long cylinder, the rate of loss of electrons by free diffusion will be reduced by a factor of 2 or more when the magnetic field is switched on provided $v_{ea} \leqslant |\omega_{ce}|$. For a field of 2×10^5 G, if $v_{ea} = 5 \times 10^9 \, p_{Torr} \, s^{-1}$ this reduction is to be expected at pressures below 650 Torr, while a factor of 10 reduction should occur at 65 Torr. The experimental results reported by Chan et al. and Vardzigulova et al. may not be inconsistent, bearing in mind the different directions of the magnetic fields used.

Cohn et al. (1972a) observed marked lowering of the pressure at which breakdown occurred with a carbon dioxide laser in helium and argon when a magnetic field was applied parallel to the laser beam axis. The threshold criterion was that breakdown occurred with 50% of the laser pulses. The results are shown in Figure 5.44, together with curves based on cascade theory taking into account the reduction in the rate of diffusion of electrons across the field. Cascade theory also gave a fairly satisfactory explanation of the results of subsequent experiments (Cohn et al. (1972b)) in which measurements were made of the effect of the magnetic field on the time required for breakdown.

6. SELF-FOCUSING OF LASER BEAMS IN GASES

Non-linear contributions to the susceptibility of a medium cause intensity-dependent variations in refractive index. If the refractive index increases with intensity a Gaussian beam may be self-focused, because the beam produces its own positive lens (Askaryan (1962), Chiao et al. (1964), Talanov (1964)).

Let us consider a non-linear medium in which the refractive index $\mu_{total} = \mu + \mu_2 E^2$, where E is the amplitude of the radiation field. In the absence of diffraction, a plane Gaussian beam of initial radius r and power $P = \pi r^2 \mu^2 \varepsilon_0 c E^2 / 2$ will be brought to a focus in a distance (Kelley (1965), see also Yariv (1967))

$$z_f \simeq r^2 \sqrt{\frac{\pi c \varepsilon_0 \mu^3}{2 \mu_2 P}}. \tag{5.53}$$

In the absence of self-focusing, diffraction would double the beam radius in a distance

$$z_d = \frac{\sqrt{2\pi r^2}}{\lambda}, \tag{5.54}$$

λ being the wavelength of the light. The threshold for self-focusing, when the focusing effect just balances the diffraction spreading of the beam, is found by equating z_f and z_d. Then

$$P = P_{sf} = \frac{c\varepsilon_0 \mu^3 \lambda^2}{4\pi\mu_2} \tag{5.55}$$

in MKS units $[(\mu_2)_{esu} = 9 \times 10^8 \, (\mu_2)_{MKS}]$. It is important to note that it is the beam power, not the irradiance, that determines the threshold. However, at threshold the self-focusing length, allowing for diffraction, is infinite. In a real situation the non-linear medium is of limited extent and geometrical constraints must be considered.

More detailed calculations of the self-focusing process have been made for a Gaussian wavefront (Dyshko et al. (1967), Abramov et al. (1969), Ledenev et al. (1972)), which indicate that at powers well above threshold, several focal points will occur on the beam axis. Bespalov & Talanov (1966) have shown that well above threshold a Gaussian beam is unstable against perturbations due to inhomogeneities and aberrations and will decay into several self-focusing channels. As the focusing proceeds, the irradiance in the beam increases and additional non-linearities may be involved: heating, excitation (including multi-photon excitation) and ionization are to be expected and stimulated scattering may become important (Kelley & Gustafson (1973)). Taking into account also the time-dependence of the beam power, a very complex situation arises in the subsequent development of the channel. The motion of the self-focused spot may give rise to the appearance of filaments; this has been discussed by Lugovoi & Prokhorov (1968).

Self-focusing may be caused by several different mechanisms, which can be divided into two groups. In the first group the non-linear susceptibility is associated with the orientation (optical Kerr effect) or the excitation of the atoms or molecules of the medium. The response is very rapid. In the second group the effect is due to displacement of the medium resulting from electrodynamic or thermal forces, and is therefore relatively slow.

Among the fast processes, the optical Kerr effect has a response time constant of order 10^{-11} s; coefficients have been given for the inert gases (New & Ward (1967)). The effects of excitation were discussed by Askaryan (1966). An excited atom is more easily polarized than the same atom in the ground state, so excitation leads to an increase in the refractive index. Javan & Kelley (1966) have pointed out that at a frequency just above resonance with an absorption transition, the anomalous dispersion causes the refractive index to increase with intensity as a result of saturation of the transition. This too may lead to self-focusing. Key et al. (1970) estimated the cubic non-linear polarizability of an atom when resonances occur at energies near both $\hbar\omega$ and $2\hbar\omega$ above the ground state, and concluded that this could be more important than the linear polarizability. The non-linear polarizability of molecules as a result of electronic distortion in strong fields has been discussed by Brewer & Lee (1968). It should be noted that excitation is often accompanied by ionization: the presence of free electrons, which causes a decrease in refractive index, can more than compensate the self-focusing effect of excited states, leading instead to an expansion of the beam.

The radial gradient of the electric field in a Gaussian beam produces electro-dynamic, or 'ponderomotive', forces on the medium which may cause self-focusing both in a neutral gas and in a fully-ionized plasma. The time-averaged force per unit volume on a medium of dielectric constant ε and relative permeability μ' is given (Landau & Lifshitz (1960)) by

$$\mathbf{F}_m = \frac{(\varepsilon-1)\varepsilon_0}{2}\mathbf{V}(\overline{E^2}) + \frac{(\mu'-1)\mu_0}{2}\mathbf{V}(\overline{H^2}) + \frac{(\varepsilon\mu'-1)}{c^2}\frac{\partial}{\partial t}(\overline{\mathbf{E}\times\mathbf{H}}), \qquad (5.56)$$

ignoring collisional effects. For high-frequency fields $\mu' = 1$ so the second term is zero. The third term is only important while the field is being established. In a neutral gas $\varepsilon > 1$ and the first term produces electrostriction—a compression into regions of higher electric field. In a fully-ionized gas $\varepsilon \simeq 1-(\omega_{pe}^2/\omega^2) < 1$ and so the first term now represents a force expelling plasma from the higher field regions. However, in both cases the refractive index will increase towards the centre of a Gaussian beam, and self-focusing is possible in principle. This effect has been discussed by Hora (1969a) and by Palmer (1971, 1972) who has examined the close connection between the onset of self-focusing and the occurrence of stimulated scattering processes in a plasma. The time required for ponderomotive forces to produce self-focusing is of the order of the beam diameter divided by the sound speed.

Heating due to absorption of light in a Gaussian beam may give rise to a number of thermal effects. If $(\partial\varepsilon/\partial T)_\rho > 0$, ρ being the mass density of the medium, a rapid transient self-focusing effect may occur (Petrishchev & Talanov (1971)). Thereafter, pressure and temperature gradients will affect the density distribution. Thermal effects will accompany electrodynamic ponderomotive effects and will tend to reduce the density near the beam axis: in a neutral gas they may be expected to tend to cancel the self-focusing effect of electrostriction, whereas in a fully-ionized plasma they may reinforce the electrodynamic effect.

Basov et al. (1967a, 1968a) found that when a 1 GW, 15 ns neodymium laser pulse was focused by a long focal length lens (2·5 m) in air, several hundred separate sparks were produced over a distance of more than 2 m. Self-focusing was suggested as a possible explanation for the effect. Korobkin et al. (1967) observed that breakdown in air due to a 100 MW, 15 ns ruby laser pulse focused by a 50 mm lens occurred at several discrete points and suggested that this too might be due to self-focusing. Evidence for the occurrence of self-focusing associated with breakdown in air at S.T.P. was found by Korobkin & Alcock (1968) using a 3 MW Q-switched ruby laser operated in a single longitudinal and single transverse mode with a pulse duration of 8 ns. Photographs of the focal volume in ruby laser radiation scattered at 90° to the direction of the axis of the focused beam showed bright filaments or points at breakdown. The filaments were less than 5 μm in diameter and ran parallel to the beam axis. Several were observed when the laser pulse was focused with a 10 cm focal length lens giving a focal spot diameter of 50 μm, but only one was produced by a 3·5 cm focal length lens. Time-resolved streak photography in laser light using an image-converter camera indicated that the filaments recorded without time resolution were formed by the motion of very bright breakdown points towards the laser, as had been suggested previously in connection with breakdown in solids by Lugovoi

& Prokhorov (1968). Streak photographs were also taken in the light of the plasma formed at breakdown, which suggested that the small breakdown point was very hot. The angular distribution of forward-scattered laser light was studied, after blocking the directly transmitted beam. The scattered light was found to be concentrated into a cone of 30° included angle. It was coherent, it occurred at the moment of breakdown and it contained a third of the laser power, suggesting that it was not in fact scattered, but diffracted from a narrow channel. Spectroscopic studies of this light (Alcock *et al.* (1969b)) showed that it was broadened and displaced towards longer wavelengths by about 0·12 Å : a very similar effect is observed when self-focusing occurs in liquids. These observations are consistent with the occurrence of self-focusing leading to the formation of channels with diameters of about 1·7 μm, in which the power density reaches $4·4 \times 10^{13}$ W cm^{-2}, corresponding to an E field of $1·3 \times 10^8$ V cm^{-1}. (The beam diameter might have been even smaller since it was derived on the assumption that the light was diffracted from a plane circular aperture, whereas in fact it would probably be more gradually defocused (Brewer *et al.* (1968))). This would be sufficient to cause full ionization in a fraction of a nanosecond, leading locally to strong absorption and rapid heating of the plasma. Direct multi-photon ionization would be the most important ionizing process; diffusion losses would be rapid from so narrow a channel and would hinder the development of an avalanche.

Complex filamentary structures were observed by Tomlinson (1969) in several gases in the region of the focal volume of a ruby laser. The filaments were shown to grow towards the lens during the 30 ns laser pulse. The breakdown thresholds were in agreement with cascade theory and did not appear to be influenced by self-focusing effects, and Tomlinson concluded that if self-focusing was responsible for the filaments it must have developed after the ionization processes began. A similar conclusion was reached by Key *et al.* (1970), who made schlieren observations of self-focused filaments produced in argon by a 6 ns neodymium laser pulse, and also by Alcock *et al.* (1970) and Richardson & Alcock (1971a, b; see also Alcock (1972)) from a study of breakdown in several gases by a single mode ruby laser, using photographic, spectroscopic and interferometric methods. These experiments are discussed further in Chapter 6. Ionization probably occurs by a stepwise process involving excited atoms, so that at some stage during the avalanche process leading to breakdown a partially-ionized gas may exist with sufficient excited atoms to overcome the defocusing effect of the free electrons. Key *et al.* (1970) found it necessary to invoke the cubic non-linear polarizability of excited atoms to account for their results on this basis. Alternatively the ponderomotive effect proposed by Hora (1969a) might have been responsible for self-focusing in these experiments, though if this were the case self-focusing should also have been observed in Thomson scattering experiments where no anomalies were in fact reported (Alcock *et al.* (1970)).

Schlieren observations of self-focused filaments were also described by Belland *et al.* (1971b) with a multimode neodymium laser producing 1·6 ns pulses of 1·5 GW. A single filament appeared initially, accompanied later by two more forming an arrow. This fork-like effect was noted earlier by Savchenko & Stepanov (1968) and by Alcock *et al.* (1970) at powers well above the breakdown threshold. About 80 % of the incident light was scattered into a cone outside the laser beam itself, and 10 %

was found within the solid angle of the laser beam. Only about 10% was therefore absorbed in the breakdown plasma.

Evidence for self-focusing of mode-locked laser pulses in air was obtained by Alcock *et al.* (1969a), who produced breakdown with trains of pulses from a ruby laser and also from a neodymium laser. Individual pulses produced very small separate breakdown regions: most of the forward scattered light was contained in a cone of 15° to 20° included angle, suggesting filament diameters of a few microns. The spectrum of the forward scattered neodymium laser light was broadened to 200 Å, compared to the 75 Å breadth of the incident light, and slightly displaced to longer wavelengths.

Bunkin *et al.* (1971) found that self-focusing occurred in air, nitrogen and argon with a single mode-locked ruby laser pulse of 20 to 100 ps duration. With a 15 cm lens many breakdown points were observed over a distance of several centimetres along the beam axis near the focus. In air and nitrogen at S.T.P., when the laser pulse power was about 1.5×10^9 W, strong scattering of laser light occurred without being accompanied by breakdown. A two-stage self-focusing process was proposed, the first being due to the non-linear polarizability of molecules and the second due to the presence of excited molecules.

Self-focusing of CO_2 laser light in BCl_3 and in SF_6 was first reported by Karlov *et al.* (1973). The effect was observed in BCl_3 at pressures above 0.1 atm with 5 to 10 kW pulses lasting 20 μs. The value of μ_2 derived from the experiments was $(4.6 \pm 0.7) \times 10^{-7} p_{atm}$ cgs esu. In SF_6 the effect was considerably greater and μ_2 was estimated to be about $5 \times 10^{-5} p_{atm}$ cgs esu. In each case the self-focusing was ascribed to saturation of the anomalous dispersion in resonant absorption. The laser radiation produced dissociation of the gas, which was followed by recombination accompanied by visible luminescence: this displayed the shape of the laser beam clearly in BCl_3.

6

Plasmas Formed by Light in Gases

1. INTRODUCTION

In the previous chapter we considered the conditions under which breakdown may be produced in gases by laser light. The ionized region is relatively strongly absorbing from the moment when breakdown occurs; the remainder of the laser pulse causes rapid heating, thus generating a high temperature plasma. Absorption mechanisms have already been discussed in Chapter 2; we shall now consider the dynamics of these plasmas.

If the intensity of the focused laser pulse increases after breakdown, it is to be expected that the region of ionization will extend away from the focus, towards the focusing lens, until the peak of the laser pulse has passed. This is simply the result of the breakdown condition being fulfilled over a larger volume. The effect is often described as a 'breakdown wave'. Experimentally, asymmetric plasma boundary velocities of the order of 10^7 cm^{-1} have been observed, the velocity being greatest towards the lens. However, a breakdown wave is not the only mechanism which might cause such an effect. Rapid expansion of the heated gas can produce a shock wave, which resembles a detonation wave if the shock front absorbs the laser light strongly. The plasma will tend to be distorted from a sphere because more energy per unit mass will be absorbed on the side nearest the focusing lens. Also, the plasma generated by the laser pulse emits intense light over a broad spectrum, mainly in the ultra-violet. Some of this radiation may produce ionization on being absorbed in the cooler surrounding gas, which will then absorb incoming laser light more strongly. Self-focusing effects must also be considered.

The time scale for these events is that of the laser pulse itself, and is thus very short. After the pulse has ended, the plasma will continue to expand, though more slowly, and will cool over a period of many microseconds.

We shall discuss the theory of these processes and then review experimental results.

2. BREAKDOWN WAVE THEORY

In this section we follow the theory given by Raizer (1965a). Consider a Gaussian laser beam of convergence angle 2α, brought to a focus of radius r_f. We suppose that breakdown, corresponding to the attainment of an electron density n_{eb}, occurs at the focus at time $t = t_b$, and subsequently occurs further and further away from the focus, along the axis of the beam towards the focusing lens, as the laser pulse power continues to rise. Figure 6.1 shows the geometrical situation. It is assumed that the

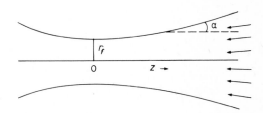

6.1   Geometry of focal region. (After Raizer (1965a))

laser pulse irradiance on axis at the focus, $I_0(t)$, would vary with time as shown in Figure 6.2 in the absence of any absorption.

We suppose that breakdown is the result of avalanche ionization, and we neglect all loss mechanisms such as diffusion. Thus we may write for the electron density

$$\frac{dn_e}{dt} = v_i n_e, \tag{6.1}$$

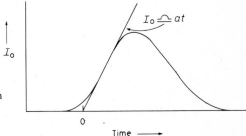

6.2   Time variation of irradiance on laser beam axis in focal plane. (After Raizer (1965a))

so that starting from an initial value n_{e0} at $t = 0$,

$$n_e(t) = n_{e0} \exp \left[\int_0^t v_i \, dt \right]. \tag{6.2}$$

The ionization rate v_i is taken to be proportional to the laser light irradiance I, so that

$$v_i = bI, \tag{6.3}$$

b being a constant. On the axis of the laser beam

$$I(z, t) \simeq I_0(t) \frac{r_f^2}{r^2}, \tag{6.4}$$

r being the beam radius at an axial distance z from the focus.

The time t_{bz} at which breakdown will occur at distance z from the focus is given by

$$n_e(z, t_{bz}) = n_{eb},$$

i.e.

$$n_{e0} \exp \left[b \int_0^{t_{bz}} I(z, t) \, dt \right] = n_{eb}$$

or

$$\frac{br_f^2}{r^2} \int_0^{t_{bz}} I_0(t) \, dt = \ln \left(\frac{n_{eb}}{n_{e0}} \right). \tag{6.5}$$

When breakdown occurs at the focus, we know that

$$b \int_0^{t_b} I_0(t) \, dt = \ln \left(\frac{n_{eb}}{n_{e0}} \right).$$

We approximate the beam radius at distance z from the focus by

$$r = r_f + z \tan \alpha,$$

and obtain

$$\frac{\int_0^{t_{bz}} I_0(t) \, dt}{\int_0^{t_b} I_0(t) \, dt} = \frac{r^2}{r_f^2} = \left(1 + \frac{z \tan \alpha}{r_f} \right)^2. \tag{6.6}$$

To simplify the calculation, the variation of the laser light irradiance with time is next approximated by a straight line (see Figure 6.2), the time now being measured from the time of zero power. Then $I_0(t) = (\text{const.})t$, say, so

$$z = \frac{r_f}{\tan \alpha} \left[\left(\frac{t_{bz}}{t_b} \right) - 1 \right]. \tag{6.7}$$

The velocity with which the breakdown region moves towards the laser is then given by

$$U_b = \frac{dz}{dt_{bz}} = \frac{r_f}{t_b \tan \alpha}. \tag{6.8}$$

As is to be expected, high velocities are associated with a rapid laser pulse rise time and a slowly converging beam. If for example $r_f = 10^{-2}$ cm, $t_b = 5$ ns and $\tan \alpha = 0.1$, $U_b = 2 \times 10^7$ cm s^{-1}. However, if $r_f = 3 \times 10^{-3}$ cm, $t_b = 20$ ns and $\tan \alpha = 0.7$, $U_b = 2 \times 10^5$ cm s^{-1}, which is so low that the breakdown wave will probably be replaced by a faster process.

It is important to remember that the angle of convergence of a real Gaussian beam becomes smaller in the immediate vicinity of the focus. The initial velocity of the breakdown region will therefore be higher than that calculated above, to which it will however soon fall. Real focused laser beams will also have more complex intensity distributions, especially if spherical aberration is present (see Chapter 5, Section (5)), so irregular development of the plasma is to be expected. The self-focusing effects discussed in Section (6) of Chapter 5 may also be important.

If the resulting plasma is sufficiently transparent to laser light, the breakdown wave can propagate in both directions along the laser beam axis from the focal plane (Canto et al. (1968)).

3. DETONATION WAVE THEORY

As the plasma formed by the laser is heated it expands, and if the process is sufficiently rapid it will generate a shock wave travelling outwards into the undisturbed gas. The shock wave heats the gas as it passes through it, which may cause ionization and make the gas strongly absorbing to laser light, so that in the vicinity of the laser beam axis more energy is fed from the laser pulse into the shock wave. As was pointed out by Ramsden & Savic (1964) (see also Raizer (1965a) and Champetier (1965)), the process is analogous to the propagation of a detonation wave in an explosive material, the reaction energy of the explosive being replaced in this case by the absorbed laser pulse energy. There is a very important difference, however: the energy gained per particle from a chemical explosive is approximately constant, while the energy gained from the laser pulse depends on the velocity of the wave, the power of the pulse, and the geometry of the focused beam.

We shall summarize the classical theory of detonation waves, and then consider the effects to be expected when the shock wave is supported by laser radiation.

The medium into which the shock front moves is supposed to be an ideal gas, at rest in the laboratory coordinate system. The shock front is taken to be plane, and of such an extent that lateral effects may be neglected. The velocity of propagation in the laboratory frame is U. It is convenient to define the velocities of the gas on each

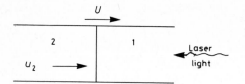

Velocities in laboratory coordinates

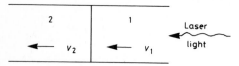

Velocities relative to the shock front

6.3   Laser-driven detonation wave (see text)

side of the front relative to that of the front. Thus in the undisturbed region (1) (see Figure 6.3) $v_1 = -U$, while behind the shock front in region (2), where the velocity of the gas in the laboratory frame is u_2, $v_2 = u_2 - U$.

To define the condition of the gas we must know the pressure p, temperature T, and density ρ. In the case of an ideal gas the equation of state is

$$p = (\gamma - 1)c_V T \rho, \tag{6.9}$$

where the adiabatic ratio $\gamma = c_p/c_V$, c_p and c_V being the specific heats per unit mass at constant pressure and constant volume respectively, which we shall assume for the moment to be constant.

Conservation of mass, momentum and energy give the following three relationships between conditions on each side of the shock front:

$$\rho_1 v_1 = \rho_2 v_2, \tag{6.10}$$

$$p_1 + \rho_1 v_1^2 = p_2 + \rho_2 v_2^2 \tag{6.11}$$

and

$$\tfrac{1}{2}v_1^2 + c_V T_1 + \frac{p_1}{\rho_1} + Q = \tfrac{1}{2}v_2^2 + c_V T_2 + \frac{p_2}{\rho_2}. \tag{6.12}$$

Here Q is the energy supplied to unit mass of gas as it passes through the front.

We wish to find the velocity of the shock front $U = -v_1$ and the condition of the gas behind it in region 2, given the condition of the gas in region 1 and the value of Q. We have so far only four equations (6.9, 6.10, 6.11 and 6.12) to determine five quantities. From (6.10) and (6.11) we may write

$$v_1^2 = \left(\frac{p_2 - p_1}{\rho_2 - \rho_1}\right)\frac{\rho_2}{\rho_1} \tag{6.13}$$

and

$$v_2^2 = \left(\frac{\rho_1}{\rho_2}\right)^2 v_1^2. \tag{6.14}$$

Making use of the energy equation (6.12) and the equation of state with (6.13) and (6.14) we find

$$\frac{p_2}{p_1} = \frac{\left(\frac{\gamma+1}{\gamma-1}\right) - \frac{\rho_1}{p_2} + 2Q\frac{\rho_1}{p_1}}{\left(\frac{\gamma+1}{\gamma-1}\right)\frac{\rho_1}{\rho_2} - 1}. \tag{6.15}$$

This is known as the Hugoniot equation for a detonation, and reduces to the usual Hugoniot equation for a normal shock if $Q \to 0$.

If we define

$$\left.\begin{aligned} \mathscr{P} &= \frac{p_2}{p_1}, \\ \mathscr{V} &= \frac{\rho_1}{\rho_2}, \end{aligned}\right\} \tag{6.16}$$

we may plot the Hugoniot equation on a $\mathscr{P}, \mathscr{V}$ diagram. Figure 6.4 shows the Hugoniot curve for a normal shock $(Q = 0)$, AH, and also the isentropic adiabat AI given by $\mathscr{P} = \mathscr{V}^{-\gamma}$. The tangents of the two curves coincide at the point $(1, 1)$ which represents the conditions in the undisturbed gas. The maximum density ratio which can occur across a shock front is found from expression (6.15) when

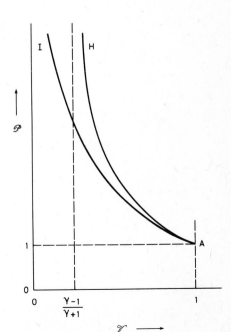

6.4 $\mathscr{P}, \mathscr{V}$ diagram showing isentrope AI and shock Hugoniot curve AH

$\mathscr{P} = p_2/p_1 \to \infty$: here

$$\frac{\rho_2}{\rho_1} = \frac{\gamma+1}{\gamma-1} \quad \left(\tfrac{p_2}{p_1} \to \infty\right). \tag{6.17}$$

Thus the Hugoniot curve approaches $\mathscr{V} = (\gamma-1)/(\gamma+1)$ asymptotically as $\mathscr{P}$ increases. In contrast, the isentrope approaches $\mathscr{V} = 0$ as $\mathscr{P}$ increases. In this chapter we need consider only values of $\mathscr{P}$ greater than unity (we shall examine the Hugoniot curve for $\mathscr{P} \leqslant 1$ in Chapter 7).

Figure 6.5 shows the Hugoniot curve for a detonation with a constant value of Q. The velocity of the shock front in the undisturbed gas, given by (6.13) may be written in terms of $\mathscr{P}$ and $\mathscr{V}$:

$$v_1^2 = \frac{\mathscr{P}-1}{1-\mathscr{V}}\left(\frac{p_1}{\rho_1}\right). \tag{6.18}$$

On the $\mathscr{P}$, $\mathscr{V}$ diagram, therefore, all points corresponding to a given value of v_1 lie on the same straight line, passing through $(1, 1)$, with slope

$$\frac{d\mathscr{P}}{d\mathscr{V}} = -\frac{\rho_1}{p_1}v_1^2.$$

In general, a line such as (AFG), corresponding to an allowed value of the velocity, intersects the detonation Hugoniot curve at two points. However, as was noted by Chapman, there is one line which is tangential to the Hugoniot curve at C, and corresponds to the lowest possible detonation velocity. The point C is known as the Chapman–Jouguet point for the detonation.

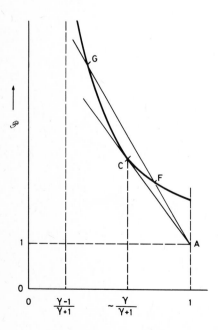

6.5 $\mathscr{P}, \mathscr{V}$ diagram showing detonation Hugoniot for constant Q

From (6.15) and (6.18), writing $(\gamma+1)/(\gamma-1) = A$ and eliminating $\mathscr{P}$,

$$\left[v_1^2 \frac{\rho_1}{p_1}(1-\mathscr{V})+1\right](A\mathscr{V}-1) = A-\mathscr{V}+2Q\frac{\rho_1}{p_1}. \tag{6.19}$$

For fixed values of $2Q\rho_1/p_1 = B$ and of $v_1^2\rho_1/p_1 = C$ we have

$$[C(1-\mathscr{V})+1](A\mathscr{V}-1) = A-\mathscr{V}+B \tag{6.20}$$

which, being a quadratic equation in $\mathscr{V}$, yields in general two values corresponding to the two intersections F and G of the velocity line AFG with the Hugoniot curve FCG. At the Chapman–Jouguet point, $\mathscr{V} = \mathscr{V}_C$ must be single-valued, i.e. from (6.20)

$$\mathscr{V}_C = \frac{(A+1)(C+1)}{2AC} \tag{6.21}$$

or

$$C = \frac{\gamma}{\mathscr{V}_C - \gamma(1-\mathscr{V}_C)}. \tag{6.22}$$

Hence from (6.18) the corresponding value of $\mathscr{P}$ is

$$\mathscr{P}_C = C(1-\mathscr{V}_C)+1 = \frac{\mathscr{V}_C}{\mathscr{V}_C-\gamma(1-\mathscr{V}_C)}. \tag{6.23}$$

Thus from (6.22) and (6.23)

$$C = \gamma\frac{\mathscr{P}_C}{\mathscr{V}_C}. \tag{6.24}$$

If p_2 and ρ_2 are the Chapman–Jouguet-point values of pressure and density, equation (6.24) may be rewritten as

$$v_1^2 = \frac{\gamma p_2 \rho_2}{\rho_1^2}, \tag{6.25}$$

so that

$$v_2^2 = \frac{\gamma p_2}{\rho_2}. \tag{6.26}$$

Thus (as was noted by Jouguet) at the Chapman–Jouguet point the velocity of the gas behind the detonation front, relative to the front, is equal to the sound velocity behind the front $V_{s2} = (\gamma p_2/\rho_2)^{\frac{1}{2}}$. By a rather lengthy argument it can be shown (see for example Shchelkin & Troshin (1965)) that a steady detonation wave can only occur when this condition, known as the Chapman–Jouguet condition, is satisfied. In this case equation (6.26) provides the last of the five equations we need, the others being (6.9), (6.10), (6.11), and (6.12).

Equations (6.13) and (6.14) together with the Chapman–Jouguet condition (6.26) give

$$\frac{1}{\mathscr{V}} = \frac{(1+\gamma)C/\gamma}{1+C}. \tag{6.27}$$

Writing

$$D = \frac{(\gamma + 1)}{\gamma} \frac{Q}{c_V T_1} = \frac{(\gamma + 1)}{\gamma} \frac{\rho_1}{p_1}(\gamma - 1)Q,$$

equations (6.13), (6.15) and (6.27) give

$$v_1^2 = \gamma \frac{p_1}{\rho_1}[(1 + D) \pm (D^2 + 2D)^{\frac{1}{2}}]. \tag{6.28}$$

The positive sign applies to a detonation wave. When the energy delivered to unit mass of the gas in the detonation is large compared to the initial internal energy $c_V T_1$, i.e. when $D \gg 1$, expression (6.28) for the detonation wave velocity reduces to

$$U^2 = v_1^2 \simeq 2\gamma \frac{p_1}{\rho_1} D$$

$$= 2(\gamma^2 - 1)Q. \tag{6.29}$$

$$(D \gg 1)$$

The temperature behind the front may be found for these conditions. From (6.9) we have

$$c_V T_2 = \frac{p_2}{(\gamma - 1)\rho_2}$$

which from (6.26) gives

$$c_V T_2 = \frac{v_2^2}{\gamma(\gamma - 1)}. \tag{6.30}$$

According to (6.29), $C = v_1^2 \rho_1/p_1 \gg 1$ when $D \gg 1$, so by (6.27)

$$\mathscr{V} \simeq \frac{\gamma}{1 + \gamma}. \tag{6.31}$$

$$(C \gg 1)$$

Then from (6.14)

$$\left(\frac{v_2}{v_1}\right)^2 \simeq \left(\frac{\gamma}{1 + \gamma}\right)^2, \tag{6.32}$$

so from (6.29), (6.30) and (6.32)

$$c_V T_2 = \frac{\gamma}{(1 + \gamma)(\gamma^2 - 1)} v_1^2 \tag{6.33}$$

$$= \frac{2\gamma}{1 + \gamma} Q. \tag{6.34}$$

The flow velocity of the gas behind the front, in laboratory coordinates,

$$u_2 = U - v_2 \simeq \frac{U}{1 + \gamma} \quad \text{if} \quad C \gg 1 \tag{6.35}$$

from (6.32). This is the velocity one might expect to measure from Doppler shift observations on line radiation emitted by gas atoms.

Detonation in a chemical explosive is a combustion process initiated by a shock (see for example Courant & Friedrichs (1948)). The shock raises the density, temperature, pressure and entropy of the medium and sets off the exothermic chemical reaction in a thin layer immediately behind the shock front. The reaction energy then causes a fall in pressure and density to the Chapman–Jouguet point, together with a further rise in temperature and entropy. In the case of a laser-driven 'detonation' no chemical reaction is involved. The initial shock front changes the transparent cold gas into a dense, cool, absorbing plasma (the inverse Bremsstrahlung absorption coefficient (expression 2.84) varies approximately as $\rho^2 T^{-\frac{3}{2}}$). Incident laser energy is therefore deposited immediately behind the shock front. The resulting fall in pressure and density to the Chapman–Jouguet point, together with the rise in temperature, produces a hot plasma behind the detonation which is only weakly absorbing. If laser light of irradiance I is incident on the front, where a fraction f of the laser power is absorbed, the energy received by unit mass of the gas, Q, is given* by

$$Q = \frac{If}{U\rho_1}. \qquad (6.36)$$

The Hugoniot curve must now be modified to allow for the velocity dependence of Q (Raizer, (1965a)): it passes through the point $(1, 1)$ on the $\mathscr{P}, \mathscr{V}$ plot vertically (Figure 6.6). The preceding analysis in this section remains valid when Q is a function of U.

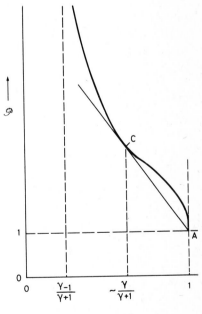

6.6 $\mathscr{P}, \mathscr{V}$ diagram showing detonation Hugoniot for laser-driven detonation. (After Raizer (1965a))

---

* Ramsden & Savic (1964) inadvertently used ρ_2 instead of ρ_1.

From (6.29) and (6.36), then,

$$U = \left[2(\gamma^2 - 1)\frac{If}{\rho_1}\right]^{\frac{1}{3}}. \tag{6.37}$$

This is the steady-state solution for a plane wave in light of constant irradiance.

In a focused laser beam, the irradiance varies with position and decreases on axis as the shock front moves towards the lens. For a Gaussian beam whose semi-angle of convergence is α, the irradiance I_z on the axis at a distance z from the focus is related to the irradiance I_0 on axis at the focus by the expression

$$I_z \simeq I_0 \frac{rf^2}{2z^2(1 - \cos\alpha)}, \tag{6.38}$$

where r_f is the radius of the focal spot. Thus from (6.37)

$$U = \frac{dz}{dt} = \left[\frac{(\gamma^2 - 1)r_f^2 I_0 f}{z^2(1 - \cos\alpha)\rho_1}\right]^{\frac{1}{3}}. \tag{6.39}$$

Assuming that I_0 is constant, it follows that

$$z = \left(\frac{5}{3}\right)^{\frac{3}{5}} t^{\frac{3}{5}} \left[\frac{(\gamma^2 - 1)r_f^2 I_0 f}{(1 - \cos\alpha)\rho_1}\right]^{\frac{1}{5}}. \tag{6.40}$$

In this simple form, therefore, the theory predicts that the front of the plasma will move towards the lens, its position varying as $t^{\frac{3}{5}}$. The assumption of constant laser power during the period of the expansion is not unreasonable if breakdown occurs, and measurements are made, near the peak of the laser pulse, since z varies only as $I_0^{\frac{1}{5}}$. Daiber & Thompson (1967) examined the case of a laser pulse with a Gaussian shape in time. The laser pulse reaches its maximum power after an interval t_{bm} following initial breakdown. The distance travelled by the front after breakdown up to the moment when the laser pulse has reached its peak is given by equation (6.40) with the factor $t^{\frac{3}{5}}$ replaced by t_{bm}^j where

$$j = \frac{3}{5} \frac{(\beta/3)^{\frac{1}{2}} t_{bm}}{(\pi/4)^{\frac{1}{2}} \operatorname{erf}\left[(\beta/3)^{\frac{1}{2}} t_{bm}\right]}. \tag{6.41}$$

Here β is a measure of the pulse width, which is inversely proportional to the square of the pulse duration. A larger exponent is thus predicted if breakdown occurs well before the peak of the pulse.

The effect of a finite beam diameter is to produce a plasma which can expand laterally. The lateral boundary velocity V_{bl} is given approximately by the speed of sound in the plasma. From expression (6.32), therefore, the ratio of the lateral expansion velocity to the axial velocity U is

$$\frac{V_{bl}}{U} \simeq \frac{\gamma}{1 + \gamma} \tag{6.42}$$

and thus lies between $\frac{2}{3}$ and $\frac{1}{2}$. With a finite beam diameter the shock front thickness may become important too. Raizer (1965a) concluded that in order to apply the plane wave theory given above to a beam of radius r with a detonation front thickness Δz, the effective laser irradiance If should be replaced by If', where

$$f' = \frac{f}{(1 + \Delta z/r)}. \tag{6.43}$$

If therefore the absorption length of the plasma for the laser light (which defines the front thickness) is comparable with or greater than the beam radius, significant slowing down of the detonation wave is to be expected.

3.1. EFFECTS OF IONIZATION

The theory given above assumed constant values of the specific heats of the medium. In reality, a transition takes place from a cold neutral gas to a strongly-ionized plasma. It is necessary to allow for some of the laser energy absorbed by the detonation wave going into ionization (and dissociation in the case of molecular gases) and excitation, and for an increase in the number of particles per unit mass which must be heated as electrons are released. The subject of strong shocks in gases has been reviewed by Gross (1971).

In general, shocks in room temperature gases begin to produce some ionization above about Mach 20 (the Mach number is the shock speed in units of the sound speed $(\gamma p_1/\rho_1)^{\frac{1}{2}}$). The effects of dissociation and ionization in hydrogen are seen in Figures 6.7 and 6.8. Dissociation begins to be important at speeds as low as Mach 5. The density ratio across the shock reaches values as high as 24, though eventually it

Right 6.7 Density ratio across shock waves in hydrogen as function of shock Mach number. (After Gross (1971))

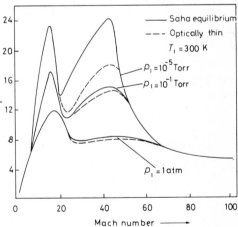

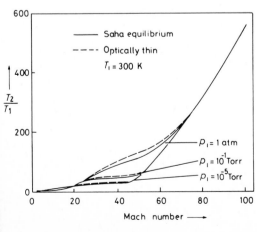

Left 6.8 Temperature ratio across shock waves in hydrogen as function of shock Mach number. (After Gross (1971))

falls towards the limiting value expected for a monatomic gas $((\gamma+1)/(\gamma-1) = 4$ when $\gamma = \frac{5}{3})$ when the hydrogen is hot and fully ionized.

A detailed calculation was made for helium by Key (1969) on the assumption of thermal equilibrium between the populations of neutral, singly- and doubly-ionized atoms, using the Saha equation (equation (2.57)). Figure 6.9 shows the equilibrium proportions of singly- and doubly-ionized atoms as functions of temperature, for a

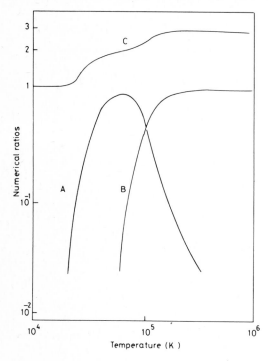

6.9   Populations in thermal equilibrium in helium at a density of 1.72×10^{20} nuclei cm^{-3}, showing proportions of atoms that are (A) singly ionized and (B) doubly ionized: (C) shows ratio (total number of particles/number of nuclei). (After Key (1969))

total density of nuclei of 1.72×10^{20} cm^{-3}. This value was taken as the Chapman–Jouguet-point density for an initial cold helium pressure of 4 atmospheres, and assumes a γ of $\frac{5}{3}$. The number of electrons per nucleus is also shown. Figure 6.10 shows the detonation wave velocity as a function of the absorbed irradiance, while Figure 6.11 shows the temperature behind the wave as a function of the wave velocity.

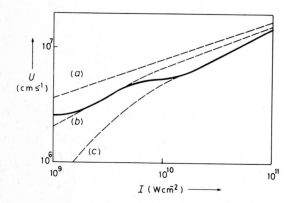

6.10   Velocity of laser-driven detonation wave in helium at a pressure of 4 atm, taking ionization into account, as function of absorbed irradiance (see text). (After Key (1969))

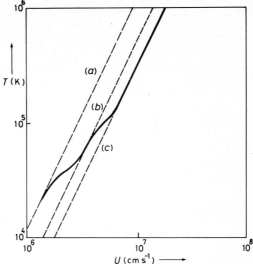

6.11   Temperature behind laser-driven detonation wave in helium initially at a pressure of 4 atm., as function of detonation wave velocity (see text). (After Key (1969))

The full curves represent complete solutions: the dotted curves show approximate solutions assuming (a) no ionization, (b) all atoms singly ionized, (c) all atoms doubly ionized behind the detonation wave.

Zeldovich & Raizer (1966) have made use of an effective adiabatic exponent γ^*. When c_V is no longer a constant the equation of state (6.9) may be written

$$\gamma - 1 = \frac{p}{W_s \rho}, \qquad (6.44)$$

where W_s is the specific internal energy. The effective adiabatic exponent γ^* is the fixed value of γ giving the best fit over the range of p and ρ considered.

It is convenient to approximate the internal energy over the same range of p and ρ by an expression of the form

$$W_s = \frac{aT^x}{\rho^y}. \qquad (6.45)$$

In this case γ^*, x and y are not independent. From the Maxwell thermodynamic relations and the second law (see for example Wilson (1966))

$$\left[\frac{\partial(W_s)}{\partial(1/\rho)} \right]_T = T \left[\frac{\partial p}{\partial T} \right]_{1/\rho} - p \qquad (6.46)$$

so

$$\gamma^* - 1 = \frac{y}{x-1}. \qquad (6.47)$$

Extensive theoretical calculations have been made for air on the assumption of thermal equilibrium. For temperatures between 10^4 and $2\cdot 5 \times 10^5$ K and densities between $10^{-3}\rho_A$ and $10\rho_A(\rho_A = 1.29 \times 10^{-3}$ g cm^{-3} is the density of air at 760 Torr and 273 K), Zeldovich & Raizer give the following expression for W_s:

$$W_s = 8.3\left(\frac{T}{10^4}\right)^{1.5}\left(\frac{\rho_A}{\rho}\right)^{0.12} \text{ eV/molecule}$$

$$= 27T^{1.5}\left(\frac{\rho_A}{\rho}\right)^{0.12} \text{ J kg}^{-1},$$

(6.48)

where T is in K. The corresponding value for γ^* is 1.24. Between 2.5 and 5×10^5 K, Raizer (1965a) gives an expression which omits the weak density dependence:

$$W_s = 4.6 \times 10^9\left(\frac{T}{5 \times 10^5}\right)^{\frac{3}{2}} \text{ J kg}^{-1}.$$

(6.49)

From 5×10^5 to 10^6 K, Raizer gives

$$W_s = 4.5 \times 10^9\left(\frac{T}{5 \times 10^5}\right)^{\frac{7}{4}} \text{ J kg}^{-1},$$

(6.50)

with $\gamma^* = 1.33$.

The relation between the temperature in the plasma and the velocity of the boundary, U, is of considerable practical interest. When c_V is a function of temperature expression (6.33) becomes

$$W_s = \frac{\gamma^*}{(1+\gamma^*)(\gamma^{*2}-1)}U^2.$$

(6.51)

Thus in air in the region between 10^4 and 2.5×10^5 K expressions (6.48) and (6.51) give

$$T \simeq \left[\frac{\gamma^*}{(1+\gamma^*)(\gamma^{*2}-1)}\frac{1}{2.7 \times 10^5}\right]U^{\frac{4}{3}}$$

$$= 2.4 \times 10^{-4}U^{\frac{4}{3}} \text{ K},$$

(6.52)

where U is in cm s^{-1}.

Similarly between 5×10^5 and 10^6 K

$$T \simeq 6.6 \times 10^{-3}U^{\frac{8}{7}} \text{ K}.$$

(6.53)

In terms of the lateral expansion velocity $V_{bl} = [\gamma^*/(\gamma^*+1)]U$, the corresponding expressions are (Komissarova et al. (1970))

$$T \simeq 5.4 \times 10^{-4}V_{bl}^{\frac{4}{3}} \qquad (10^4 \leqslant T \leqslant 2.5 \times 10^5 \text{ K})$$

(6.54)

and (Korobkin et al. (1967))

$$T \simeq 1.25 \times 10^{-2}V_{bl}^{\frac{8}{7}} \qquad (5 \times 10^5 \leqslant T \leqslant 10^6 \text{ K}).$$

(6.55)

These expressions are plotted in Figure 6.12.

Ionization processes in laser-heated plasmas have been discussed by Afanasev et al. (1969b) for conditions in which Saha's equation is not valid.

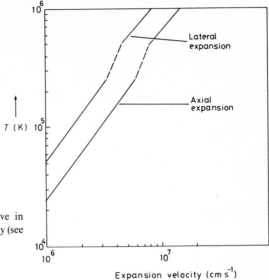

6.12   Temperature behind detonation wave in air at S.T.P. as function of expansion velocity (see text)

4. PLASMA-RADIATION-PROPAGATED WAVE THEORY

Once a plasma has been formed, it emits radiation whose wavelength distribution depends on the temperature and density. In the case of a plasma formed by a laser in a gas originally at atmospheric pressure and room temperature, heated to about 10^5 K, most of the energy emitted is in the far ultraviolet, and has an absorption length of several millimetres in the hot plasma, but only a fraction of a millimetre in the surrounding cold gas (see for example Zeldovich & Raizer (1966) Chapter 9). Thus a layer of gas outside the plasma, although transparent to laser light, is heated by the plasma radiation, and on reaching a sufficiently high temperature will be ionized to such an extent that it will become strongly absorbing for the laser light. The layer will then be further heated very rapidly until its temperature becomes so high that it is transparent. By this time a new layer of plasma nearer the laser will have become strongly absorbing, so the boundary of the plasma will move towards the laser.

   A simple model may be used to illustrate the mechanism of this radiation-propagated ionization wave. We are concerned now with radiation from the plasma, which is emitted in all directions, and it facilitates calculations to consider either an infinite plane wave or a spherical wave. We shall use the latter model and quote results for the plane wave. We suppose the plasma to be formed immediately the gas temperature T_g reaches a critical value $T_{g_{crit}}$ (which will depend on the nature of the gas).

The total power in the spherically-converging laser beam is P_0 and the plasma temperature, which is supposed to be uniform over the whole plasma, is $T_p = T_e = T_i$. We consider two cases:

(a)   when the plasma is optically thick for all wavelengths of interest, and emits as a black body; and

(b)   when the plasma, though absorbing laser radiation strongly, is optically thin for the bulk of the Bremsstrahlung radiation it emits.

4.1. OPTICALLY THICK PLASMA

Ignoring radiation losses compared to the energy supplied by the laser, if the radius of the plasma is r at time t, and $Z = 1$ so that $n_e = n_i$,

$$3n_e \kappa_B T_p \tfrac{4}{3}\pi r^3 = P_0 t, \tag{6.56}$$

n_e being the electron density in the plasma. The rate of emission of radiation by a black body at T_p, per unit area, is (see (2.24))

$$\mathcal{M}_B = \sigma_s T_p^4, \tag{6.57}$$

σ_S being the Stefan–Boltzmann constant. The total power radiated from the plasma is therefore $4\pi r^2 \mathcal{M}_B$. The plasma radiation power received per unit area of a spherical shell of cool gas of radius r' (where $r' > r$) is

$$\frac{r^2}{r'^2}\mathcal{M}_B \exp\left[-(r'-r)/\lambda_a\right],$$

where λ_a is the average absorption length of the plasma radiation in the cool gas, and is assumed to be constant. The rate at which energy is absorbed per unit volume at a point at radius r' is therefore given by

$$P_A(r') = \frac{r^2}{r'^2}\frac{\mathcal{M}_B}{\lambda_a}\exp\left[-(r'-r)/\lambda_a\right]. \tag{6.58}$$

If the plasma boundary reaches radius r' at time t', the total energy received by a volume element at r' up to that time must be sufficient to raise the temperature of the gas to $T_{g\,crit}$, i.e.

$$\int_0^{t'} P_A(r')\,dt = \tfrac{3}{2}n_m\kappa_B T_{g\,crit}, \tag{6.59}$$

where n_m is the number density of gas molecules (assuming that the gas is only slightly ionized at $T_{g\,crit}$), or from (6.56), (6.57) and (6.58)

$$\int_0^{t'} \frac{\sigma_s}{\lambda_a}\left(\frac{P_0}{4\pi n_e \kappa_B}\right)^4 \frac{t^4}{r^{10} r'^2}\exp\left[-(r'-r)/\lambda_a\right]\,dt = \tfrac{3}{2}n_m\kappa_B T_{g\,crit}. \tag{6.60}$$

Introducing the velocity of the plasma boundary, $V_b = dr/dt$, allows a change of variable to r, so that

$$\int_0^{r'} \frac{\sigma_S}{\lambda_a}\left(\frac{P_0}{4\pi n_e \kappa_B}\right)^4 \frac{t^4}{r^{10}r'^2 V_b} \exp\left[-(r'-r)/\lambda_a\right]dr = \tfrac{3}{2}n_m \kappa_B T_{g\text{crit}}. \tag{6.61}$$

Provided $t^4/r^{10}V_b$ varies as r^n, where $n \geqslant 0$, the integration may be carried out approximately for $r' \gg \lambda_a$ to give

$$\sigma_S\left(\frac{P_0}{4\pi n_e \kappa_B}\right)^4 \frac{t'^4}{V_b r'^{12}} = \tfrac{3}{2}n_m \kappa_B T_{g\text{crit}}. \tag{6.62}$$

The absorption length λ_a has disappeared on integration.

For generality we now replace t' and r' by t and r, and recalling that $V_b = dr/dt$ we may write

$$t^4 = Xr^{12}\frac{dr}{dt}, \tag{6.63}$$

where

$$X = \left(\frac{4\pi n_e \kappa_B}{P_0}\right)^4\left(\frac{3n_m \kappa_B T_{g\text{crit}}}{2\sigma_S}\right). \tag{6.64}$$

Thus

$$t^5 = \tfrac{5}{13}Xr^{13}, \tag{6.65}$$

so that $r \propto t^{5/13}$ and

$$
\begin{aligned}
V_b &= \left[\frac{1}{X}\left(\frac{5}{13r^2}\right)^4\right]^{\frac{1}{5}} \\
&= \left(\frac{5P_0}{52\pi r^2 n_e \kappa_B}\right)^{\frac{4}{5}}\left(\frac{2\sigma_S}{3n_m \kappa_B T_{g\text{crit}}}\right)^{\frac{1}{5}}.
\end{aligned}
\tag{6.66}
$$

We note from (6.63) that $t^4/(r^{10}V_b)$ varies as r^2, in agreement with the assumption made in obtaining (6.62).

The plasma temperature

$$
\begin{aligned}
T_p &= \frac{P_0 t}{4\pi r^3 n_e \kappa_B} \\
&= \left(\frac{15 n_m T_{g\text{crit}} P_0}{104\pi n_e \sigma_S r^2}\right)^{\frac{1}{5}}.
\end{aligned}
\tag{6.67}
$$

A similar analysis for the one-dimensional case with a constant laser irradiance I yields

$$V_b = \left(\frac{I}{3n_e \kappa_B}\right)^{\frac{2}{3}}\left(\frac{2\sigma_S}{3n_m \kappa_B T_{g\text{crit}}}\right)^{\frac{1}{3}}. \tag{6.68}$$

and

$$T_p = \left(\frac{n_m T_{g_{crit}} I}{2n_e \sigma_S}\right)^{\frac{1}{4}},$$

(6.69)

which are almost exactly the same as (6.66) and (6.67) when $I \sim P_0/4\pi r^2$.

4.2. OPTICALLY THIN PLASMA

We again ignore radiation losses compared to the energy supplied by the laser. The plasma, though optically thin for most of its own emitted radiation, is opaque to the laser light, so equation (6.56) still holds. However, the total radiated power is now $4\pi r^3 P_\beta/3$ where P_β is the rate of Bremsstrahlung emission per unit volume. Now

$$P_\beta = \xi T_p^{\frac{1}{2}},$$

(6.70)

where ξ is given by (2.72), so from (6.56)

$$P_\beta = \xi\left(\frac{P_0 t}{4\pi r^3 n_e \kappa_B}\right)^{\frac{1}{2}}.$$

(6.71)

Thus in place of (6.60) we have

$$\int_0^{t'} \frac{\xi}{3\lambda_a}\left(\frac{P_0}{4\pi n_e \kappa_B}\right)^{\frac{1}{2}} \frac{t^{\frac{1}{2}} r^{\frac{3}{2}}}{r'^2} \exp\left[-\left(\frac{r'-r}{\lambda_a}\right)\right] dt = \frac{3}{2} n_m \kappa_B T_{g_{crit}}.$$

(6.72)

Provided that $t^{\frac{1}{2}} r^{\frac{3}{2}}/V_b$ varies as r^n, where $n \geqslant 0$, the integration may again be carried out approximately for $r' \gg \lambda_a$ giving

$$\frac{\xi}{3}\left(\frac{P_0}{4\pi n_e \kappa_B}\right)^{\frac{1}{2}} \frac{t'^{\frac{1}{2}}}{V_b r'^{\frac{1}{2}}} = \frac{3}{2} n_m \kappa_B T_{g_{crit}}.$$

(6.73)

Replacing t' and r' by t and r,

$$t^{\frac{1}{2}} = X' r^{\frac{1}{2}} \frac{dr}{dt},$$

(6.74)

where

$$X' = \left(\frac{4\pi n_e \kappa_B}{P_0}\right)^{\frac{1}{2}}\left(\frac{9n_m \kappa_B T_{g_{crit}}}{2\xi}\right).$$

(The condition for integration is therefore satisfied, with $n = 2$). Thus the velocity of the plasma boundary

$$V_b = \frac{1}{X'^{\frac{2}{3}}}$$

$$= \left(\frac{P_0}{4\pi n_e \kappa_B}\right)^{\frac{1}{3}}\left(\frac{2\xi}{9n_m \kappa_B T_{g_{crit}}}\right)^{\frac{2}{3}},$$

(6.75)

and is therefore independent of the radius and hence also of time.

The plasma temperature

$$T_p = \frac{P_0 t}{4\pi r^3 n_e \kappa_B}$$

$$= \left(\frac{9 n_m T_{g\text{crit}} P_0}{8\pi n_e \xi r^3}\right)^{\frac{2}{3}}.$$

(6.76)

In the one-dimensional case with constant laser irradiance, I,

$$V_b = \left(\frac{I}{n_e \kappa_B}\right)^{\frac{1}{3}} \left(\frac{\xi z}{3 n_m \kappa_B T_{g\text{crit}}}\right)^{\frac{2}{3}}$$

(6.77)

and

$$T_p = \frac{I}{n_e \kappa_B} \frac{1}{V_b}$$

$$= \left(\frac{3 n_m T_{g\text{crit}} I}{n_e \xi z}\right)^{\frac{2}{3}}.$$

(6.78)

Again, if we set $I \sim P_0/4\pi r^2$ and $r = z$, expressions (6.77) and (6.78) are almost the same as (6.75) and (6.76) respectively.

The results obtained above are of only heuristic value. Gas dynamic effects and the energy involved in the process of ionization have been completely neglected, so that equation (6.56) is unrealistic. This is particularly important for the black body case. However, a more detailed analysis using a similar approach to the problem should be useful. Raizer (1965a) carried out computations for a semi-infinite cylindrical plasma created in air of normal density, and concluded that in the region $5 \times 10^5 < T_p < 10^6$ K the velocity of the plasma boundary varied approximately as $I^{\frac{1}{3}}$.

The processes described in Sections (2), (3) and (4) above may all play a part in the development of a plasma created by a laser pulse in a gas. As was noted in Section (2), the breakdown wave has a very high velocity near the focus of a Gaussian beam and probably dominates the initial stages of expansion. Subsequently a detonation wave is likely to develop. The radiation wave may be important at very high laser irradiances when very hot plasmas are generated, if the propagation velocity exceeds that of the detonation wave. No shock wave is produced under these conditions and, in contrast to the detonation wave which heats the ions preferentially, a radiation wave supplies energy initially to the electrons in the plasma.

5. BEHAVIOUR OF PLASMA AFTER LASER PULSE HAS CEASED

When energy is no longer being supplied to the plasma, none of the models described so far apply. The plasma still expands, however, and we may expect the boundary to behave in a similar manner to a blast wave.

The approximate theory of spherical blast waves given by Taylor (1950a) and Sedov (1959) is valid for strong shocks with large Mach numbers. A more accurate treatment was given by Sakurai (1953), who also considered plane and cylindrical blast waves. Although by the end of the laser pulse the plasma may have developed a marked asymmetry, during the subsequent blast wave phase it fairly rapidly becomes almost spherical. The early development of the shape of the blast wave from a laser-heated plasma has been calculated in detail by Panarella & Savic (1968).

If energy W is supplied instantaneously at a point in a gas of density ρ_0 at time $t = 0$, Taylor & Sedov found that the radius r_s of the resulting shock wave, ignoring effects of ionization and excitation, is given by

$$r_s = t^{\frac{2}{5}} \left(\frac{W}{\rho_0} \right)^{\frac{1}{5}} Y(\gamma), \tag{6.79}$$

where $Y(\gamma)$ is a numerical factor of order unity which depends only on γ. A detailed discussion of the development of the wave is given by Zeldovich & Raizer (1966). For $\gamma = 1.23$, they give $Y(\gamma) = 0.93$. The effect of ionization has been treated by Gross (1964).

In the case of a cylindrical blast wave the radius varies at $t^{\frac{1}{2}}$ and as $W^{\frac{1}{4}}$. The effects of ionization on cylindrical waves in helium plasmas were investigated by Ahmad et al. (1969a).

Ahlborn (1972) and Ahlborn & Strachan (1973) considered the possibility of a shock wave being driven by radiation emitted by the hot plasma after the laser pulse has ceased. On the assumption of black body radiation from a spherical plasma the radius is predicted to vary as $t^{\frac{3}{5}}$. It has not been established that such a mechanism will apply to laser-heated plasmas.

6. COMPUTER CALCULATIONS

Detailed computer calculations of the development of a plasma formed in deuterium by a spherically-convergent 100 J laser pulse of peak power 20 GW have been described by Kidder (1968). Figures 6.13(a), (b) and (c) for deuterium initially at S.T.P. show the electron temperature, the pressure and ρ_0/ρ respectively as functions of radius at different times. Calculation for deuterium at 56 times the S.T.P. density $(10^{-2} \text{ gm cm}^{-3})$ are shown in Figure 6.14. As a means of obtaining very high

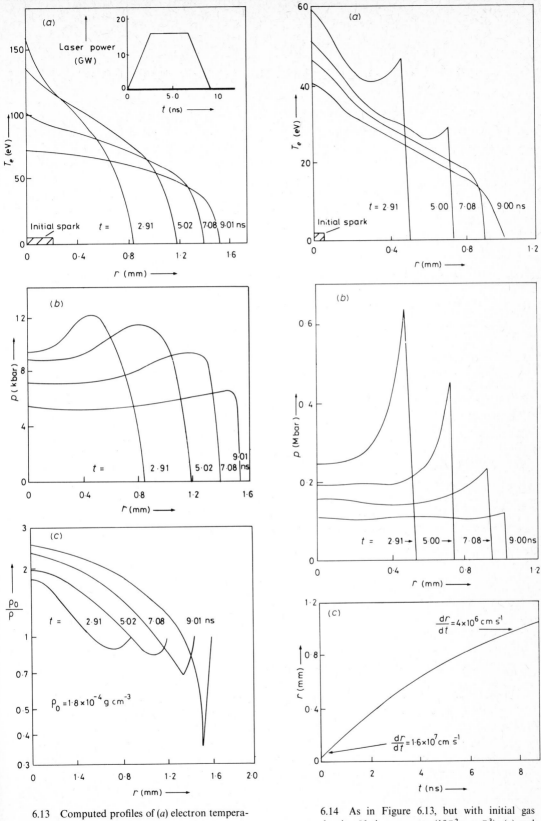

6.13 Computed profiles of (a) electron temperature, (b) pressure and (c) (ρ_0/ρ) in plasma produced in deuterium initially at S.T.P. by the laser pulse shown inset in (a). Spherical symmetry is assumed. (After Kidder (1968))

6.14 As in Figure 6.13, but with initial gas density 56 times greater (10^{-2} g cm^{-3}): (a) and (b) show profiles of the electron temperature and the pressure: (c) shows time dependence of plasma radius. (After Kidder (1968))

temperature plasmas this approach is not very promising. Scaling relations indicate that the temperature increases only as the cube root of the laser power. At very high powers the calculation is probably unrealistic because breakdown would probably occur at several points and the available energy would be distributed over more particles.

7. HEATING OF MAGNETICALLY-CONFINED PLASMAS BY LASER BEAMS

A number of authors have discussed the heating of relatively large volumes of plasma by laser light, a topic which we have already mentioned in Chapter 4, Section (4). Interest has been centred on longer wavelengths, where the absorption coefficient is greater, and in particular on the 10·6 μm wavelength of the exceptionally efficient CO_2 laser. In addition to the references given in Chapter 4, one-dimensional heating has been investigated theoretically by Steinhauer & Ahlstrom (1971), while Engelhardt et al. (1971), Martineau (1971), Martineau & Pepin (1971) and Fuchs (1973) have considered the heating of θ-pinch plasmas. Shatas et al. (1972) have discussed laser heating of a dense plasma focus. Jassby (1972) has considered the heating of a dense arc plasma. The use of very high magnetic fields to reduce the thermal conductivity of the heated plasma has been discussed by Cohn & Lax (1972).

8. DIAGNOSTIC TECHNIQUES FOR PLASMAS FORMED IN GASES

Laser-heated plasmas are characterized by their high temperature, high density, small size and very short lifetime. Although the afterglow may persist for several microseconds, the highest temperatures are attained during the laser pulse, which is usually much shorter. The short time scale of events and the small dimensions of the plasma give rise to considerable experimental difficulties, and severely limit the choice of diagnostic techniques. A further source of frustration is the low repetition rate attainable with very high power lasers. Nevertheless, several effective experimental methods have been developed for studying laser-heated plasmas. Before discussing the results it may be helpful to describe the more important techniques briefly. Thomson scattering has been discussed in detail in Chapter 3.

8.1. HIGH-SPEED PHOTOGRAPHY

The earliest studies of laser-generated plasmas were made by Ramsden & Davies (1964) using an image-converter camera in the streak mode. The image of the plasma was focused on to a narrow slit, and the slit was then imaged on to the photocathode of the image converter. In the image-converter tube the electrons emitted from the photocathode are accelerated and focused to form an image of the slit on a phosphor screen, and this image is electronically swept across the screen at right angles to the slit. Recent developments have improved the time resolution available with image converter tubes to a few picoseconds (Bradley *et al.* (1971)).

Framing image-converter cameras, in which a sequence of two-dimensional images of the plasma are formed with short exposures at short intervals, have also been extensively used. Exposure times as short as 300 ps have been achieved.

An optical shutter using the optical Kerr effect (Duguay & Hansen (1969)) allows two-dimensional images to be recorded with exposure times of only 5 ps. Richardson & Sala (1973) obtained a sequence of such photographs of laser-generated plasmas by placing an optical Kerr effect shutter in front of a relatively slow streaking image-converter camera.

8.2. SHADOW AND SCHLIEREN PHOTOGRAPHY

In contrast to the methods described in the previous section, both shadow and schlieren photography (see for example Weinberg (1963); Ascoli-Bertoli (1970)) record the effect of the refracting plasma on the wave front of light passing through it from an external source. Time resolution is normally achieved by using single short pulses of light, though other methods are available. One problem is the synchronization of the laser producing the plasma and the photographic light source.

The first reported shadow photographs of laser-produced plasmas in gases were obtained by Malyshev *et al.* (1966a), who solved the problem of synchronization very simply by using as the photographic light source a part of the ruby laser pulse producing the plasma, delayed by reflections between widely separated mirrors. The arrangement is shown in Figure 6.15. The beam splitter (S) which divides the incoming laser beam produces multiple reflections, and all but one of the reflected beams are eliminated by the aperture (A). A red filter (F) protects the photographic plate (P) from white light. The time resolution was not very precise, because the laser pulse lasted 50 ns, but measurable shadowgrams were obtained for times centred on delays of 1, 31, 64, and 110 ns relative to the time of arrival of the main pulse at the focus by suitable adjustments of the mirrors.

Evtushenko *et al.* (1966a) synchronized two ruby lasers by Q-switching them from the same rotating prism, and used one laser for creating the plasma and the other for photography. The arrangement is shown in Figure 6.16. P_1 is the rotating prism and P_2, which deflects the beam from the ruby R_2 into P_1, is above the beam from R_1 to P_1. The relative timing of the two laser pulses was adjusted by changing the angle of one of the lasers: one minute of arc corresponded to 95 ns.

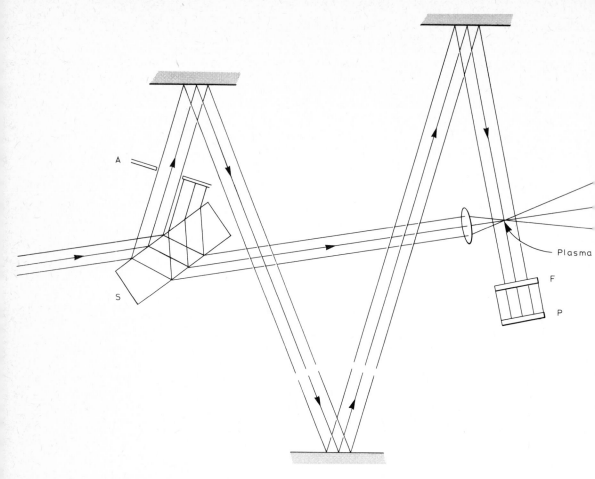

6.15 Arrangement for shadow photography of laser-generated plasma (see text). (After Malyshev *et al.* (1966a))

Korobkin *et al.* (1967) refined the single-laser method of Malyshev *et al.* in three ways. First, the laser light which passed through the plasma region was used for photography. This reduced the exposure time, because as soon as breakdown occurred the transmitted light was cut off by the plasma. Second, a saturable dye filter was placed in the beam transmitted through the plasma, giving a much steeper rising edge to the pulse. In this way, with a 30 ns laser pulse photographs were obtained with an exposure time of only 5 or 6 ns. Third, the light after being made uniform and parallel with a diffusing screen and a system of lenses, and being passed through the plasma, was brought to a focus at a small diaphragm before expanding on to the photographic plate. This greatly reduced the intensity at the plate of light emitted by the plasma.

A very simple optical system for obtaining a shadowgram was devised by Askaryan *et al.* (1969). A circular hole of 3 mm diameter was drilled in the centre of the lens used to focus the ruby laser beam which generated the plasma. In this way the central part of the 15 mm diameter laser beam remained parallel, and after passing the focal region was recorded photographically. By photometry, knowing the time variation of the

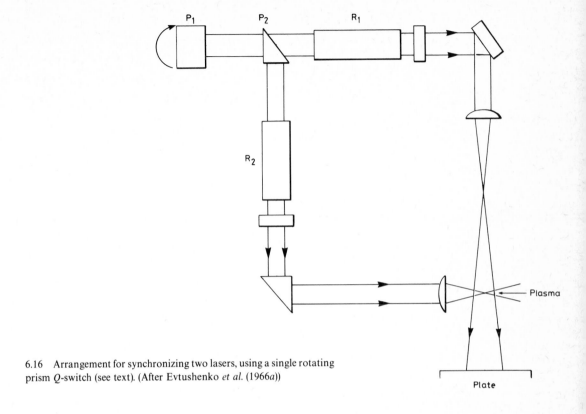

6.16   Arrangement for synchronizing two lasers, using a single rotating prism Q-switch (see text). (After Evtushenko *et al.* (1966a))

laser pulse intensity, it was possible to make an estimate of the transverse expansion velocity of the plasma.

It is possible to use sources other than lasers if one is studying the relatively long-duration blast wave phase of the plasma. In this case time resolution of the shadow image is conveniently achieved using an image-converter camera (Büchl *et al.* (1968): Hohla *et al.* (1969)).

An elegant method for studying the rapidly varying early stage of plasma formation was developed by Alcock *et al.* (1968a, b; see also Hamal *et al.* (1969)). Two lasers were used, a ruby laser Q-switched by a Pockels cell for producing the plasma, and a mode-locked neodymium glass laser, from which second harmonic pulses were obtained, for photography. Figure 6.17 shows the experimental arrangement. The lasers were synchronized by using a spark gap (SG), broken down by light from the mode-locked laser, to switch the Pockels cell (PC). The mode-locked laser produced a train of pulses at λ 1·06 μm, 5·5 ns apart with a duration of several picoseconds. The ADP crystal converted the λ 1·06 μm light into green light at λ 5 300 Å and a copper sulphate cell (C) absorbed the rest of the infrared. After passing through the plasma the parallel beam was focused either on to the schlieren knife edge (K) or on to a round 1 mm diameter obstacle which permitted observation of the entire shock front. A magnified image of the plasma was recorded on the film (F).

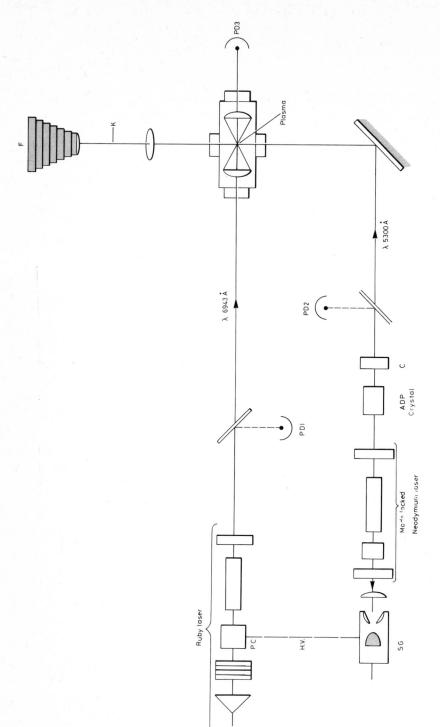

6.17 Arrangement for plasma generation and schlieren photography using a *Q*-switched laser with frequency doubling. (After Hamal *et al.* (1969))

8.3. INTERFEROMETRIC AND HOLOGRAPHIC STUDIES

Interferometric techniques (see for example Françon (1966)) using ruby laser light sources are very suitable for studies of laser-produced plasmas. Good time resolution is attainable. The narrow bandwidth of the source permits good discrimination against the self-luminosity of the plasma, and the long coherence length of the light facilities optical adjustment.

The technique usually makes use of a Mach–Zehnder interferometer, the plasma being generated in one arm. In a simple arrangement used by Alcock et al. (1966a) a small fraction of the light from the ruby laser used to produce the plasma is split off and fed into the interferometer. An optical 'delay line' consisting of two parallel mirrors enables interferograms to be obtained up to 100 ns or so after breakdown of the gas. This system is rather limited in its usefulness, and Alcock et al. went on to develop a better method using two ruby lasers. The arrangement is shown in Figure 6.18. The Pockels cell Q-switch of the second laser was triggered by a pulse initiated

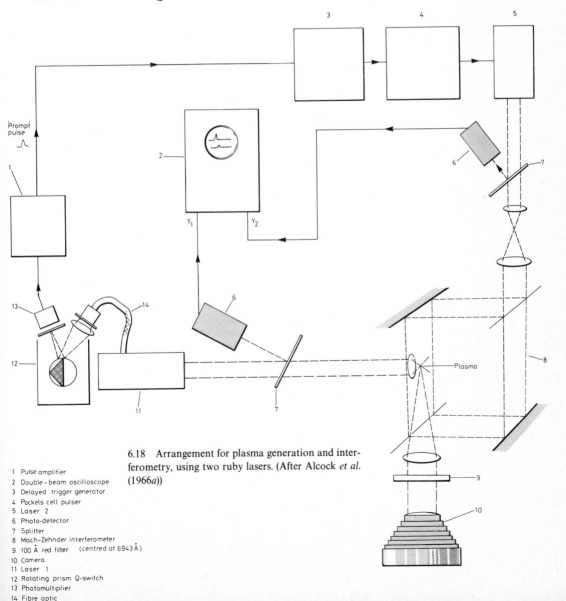

6.18   Arrangement for plasma generation and interferometry, using two ruby lasers. (After Alcock et al. (1966a))

1 Pulse amplifier
2 Double-beam oscilloscope
3 Delayed trigger generator
4 Pockels cell pulser
5 Laser 2
6 Photo-detector
7 Splitter
8 Mach–Zehnder interferometer
9 100 Å red filter   (centred at 6943 Å)
10 Camera
11 Laser 1
12 Rotating prism Q-switch
13 Photomultiplier
14 Fibre optic

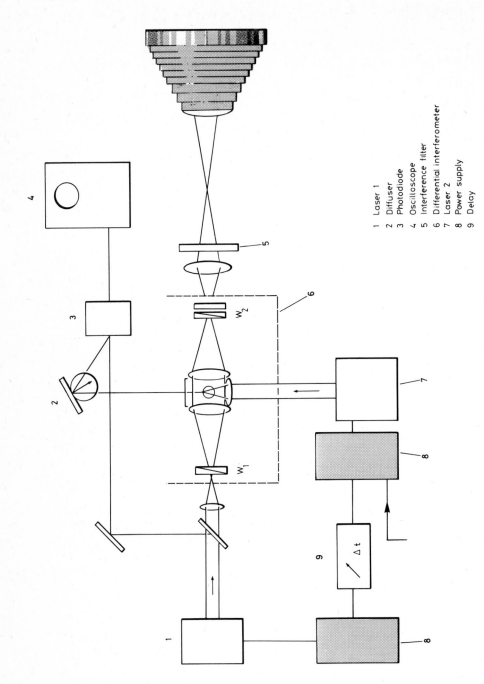

6.19 Arrangement for plasma generation and interferometry, using a differential polarizing interferometer. W_1, W_2 are Wollaston prisms. (After Hugenschmidt (1970))

1 Laser 1
2 Diffuser
3 Photodiode
4 Oscilloscope
5 Interference filter
6 Differential interferometer
7 Laser 2
8 Power supply
9 Delay

by light reflected from the rotating mirror Q-switch of the first, and passed through a variable delay. The pulse duration of both lasers was 30 ns; a Kerr cell was sometimes used to reduce the interferogram exposure time to 6 ns.

A differential polarizing interferometer has been used for laser-generated plasma studies by Hugenschmidt (1970). The arrangement is shown in Figure 6.19. The two wavefronts of orthogonal polarizations are sufficiently separated on passing through the system to produce two separate images on recombination, and only one of the two elements of wavefront forming each image is perturbed by the plasma. Such an arrangement is relatively easy to set up.

Another important improvement in the technique was to make simultaneous measurements at two wavelengths. This allows us to separate the contributions of free electrons and neutral atoms to the refractivity. Alcock & Ramsden (1966) first carried out two-wavelength interferometry on laser heater plasmas using the fundamental and the second harmonic wavelengths of a ruby laser. Their apparatus was basically the same as that of Figure 6.18 except that a 4 cm ADP crystal was placed in the beam of the ruby laser, serving to generate ultraviolet light at $\lambda\,3472\,\text{Å}$ as the inteferometer light source, and a beam splitter made from Corning CS7–37 glass divided the red and ultraviolet beams after they emerged from the interferometer, so that separate images were obtained in the two wavelengths.

The shadow photographs described in the previous section may be regarded as single beam holograms of the plasma. For a general description of holographic techniques, the reader is referred to Jahoda (1971). Ostrovskaya & Ostrovskii (1966), Zaidel et al. (1966) and Zaidel & Ostrovskii (1968) used the triple delayed beam system shown in Figure 6.20 to produce three shadow images with relative delays of 40 ns, making use of the light transmitted through the focal region from the ruby laser. The distance of the plate from the plasma was 50 cm. These holograms were then reconstructed in helium–neon laser light at $\lambda\,6328\,\text{Å}$. The real image appeared at a distance $50 \times 6943/6328$ cm from the hologram. A lens, placed at twice its focal length beyond the real image plane, produced a unit magnification image on a film. A small obstacle midway between the lens and the film removed zero-order images and much of the light from the virtual image.

Better holograms are obtained using the two-beam technique. Kakos et al. (1966) used the apparatus of Figure 6.20 but with an additional wedged glass plate with reflecting and semi-reflecting areas which divided each of the three image-forming beams into two, separated by $1.5°$ and coinciding at the photographic plate. Only one of each pair of beams traversed the plasma, the other providing a reference wavefront. With this apparatus interferograms were also obtained, using the two-exposure method (Gabor et al. (1965), Heflinger et al. (1966)). One exposure was made when a plasma was present, the other when it was absent (by reducing the laser output power). A wedge in the secondary beams produced uniformly spaced reference fringes, which were made easier to measure by placing lenses in each beam so as to give divergent beams at the plate.

This technique was later developed to provide a sequence of five interferograms at different times (Komissarova et al. (1968)). On the assumption of cylindrical symmetry, an Abel inversion (see for example Bockasten (1961)) of the results gave the radial

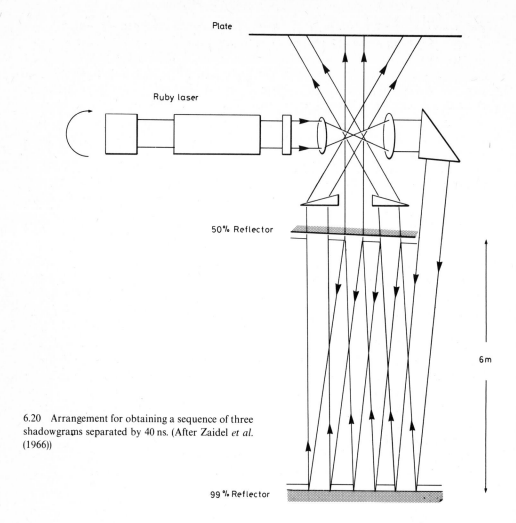

6.20   Arrangement for obtaining a sequence of three shadowgrams separated by 40 ns. (After Zaidel *et al.* (1966))

distribution of the refractive index perpendicular to the laser beam axis. A camera giving a sequence of 10 holograms with 1 ns exposures separated by 2 ns intervals was described by Buges & Terneaud (1970).

A further development was the use of two wavelengths simultaneously for holography, which was demonstrated by Komissarova *et al.* (1969, 1970) using the fundamental and the second harmonic of a ruby laser.

8.4. SPECTROSCOPIC OBSERVATIONS OF LIGHT EMITTED BY PLASMA

The simplest way to investigate the spectrum of light from a plasma is to image the plasma on to the slit of a spectrograph and record the spectrum photographically.

This method integrates the light emitted by the plasma over its entire duration, and in common with more sophisticated spectroscopic methods it also integrates along the line of sight. In laser-heated plasmas conditions change rapidly, and there are large gradients of temperatures and densities, so such averages may not be very informative. This is particularly true in the case of hot hydrogen plasmas, in which line emission occurs only during the relatively cool early and late phases of plasma development, and the interesting hot region emits only a continuum whose spectrum varies rapidly. In the other gases with several electrons, or in hydrogen with a high-Z impurity, it is possible to obtain information about the state of the hottest region of the plasma by studying the emission lines from the most highly ionized atoms observed. Line intensities and relative intensities have been used to determine the electron temperature, while the Stark and Doppler effects influencing line profiles may often be separated to give values of electron densities and ion velocities and temperatures (see Griem (1964)).

Time-resolved spectroscopy is possible with nanosecond, or even picosecond (Bradley & Sibbett (1973)) resolution, but is much easier with microsecond resolution. It has been used mainly for studying the relatively slowly varying afterglow phase of the plasma. One method is to pass the plasma light through a monochromator, place a photomultiplier behind the exit slit and record the output signal on an oscilloscope. This gives the variation in time of light intensity entering the monochromator at a particular wavelength. In order to obtain the profile of a spectral line, it is necessary to make repeated observations at slightly different wavelength settings of the monochromator and to ensure that the plasmas analysed are closely reproducible from shot to shot. This method was used by Breton et al. (1965), Durand & Veyrie (1966), Litvak & Edwards (1966) and Ahmad et al. (1969a). Ahmad et al. also used interference filters and photomultipliers to obtain the ratio of the intensities of a line and of the adjacent continuum.

Rotating-mirror (Evtushenko et al. (1966b)) and image-converter (Ahmad et al. (1969a)) streak-cameras have been used to obtain time-resolved records of the spectrum in the output plane of a spectrograph.

8.5. X-RAY INTENSITY MEASUREMENTS

The emission spectrum of a plasma was discussed in Chapter 2, where it was shown that the Bremsstrahlung and recombination radiation from a fully-ionized plasma is emitted mainly in the far ultraviolet and soft X-ray wavelength region. From a study of this short-wavelength radiation it is possible to deduce the temperature of the plasma. When plasmas are produced in gases by laser pulses, if the ambient gas pressure is high the radiation may be strongly absorbed before it reaches the detector. Thus in air the method is limited to fairly low pressures, usually below 1 atm.

The first application of this method to laser-generated plasmas in gases was by Mandelstam et al. (1965), who used Geiger–Müller counters behind thin aluminium or beryllium windows to measure the X-ray flux in the transmitted wavelength ranges from plasmas in air. The counter time constant was long, about 10^{-5} s, so no time resolution was possible. From a knowledge of the transmission of the foils, the geometry of the experiment and the absorption of the air, the electron temperature could be estimated.

Instead of measuring the absolute value of the X-ray flux, Alcock *et al.* (1966*b*) measured the relative flux transmitted through two foils of different thicknesses. This method was originally developed by Jahoda *et al.* (1960) for θ-pinch plasmas. Alcock *et al.* used plastic scintillators and fast photomultipliers, which gave a time resolution of 20 ns. Beryllium foil windows, 50 and 125 μm thick, covered the two scintillators. Similar measurements were made by several other authors. Calculations necessary for the interpretation of the measurements in terms of the electron temperature are given for aluminium and beryllium foils by Elton & Roth (1967), for aluminium, beryllium and polyethylene by Puell *et al.* (1970*a*), for beryllium by Guenther & Pendleton (1972) and for aluminium and gold by Robouch & Rager (1973), who have considered in detail the use of silicon detectors for temperatures above 1 keV.

The application of the method to laser-produced plasmas has been discussed by Ahmad & Key (1972), who concluded that absolute measurements are more accurate than two-foil ratio measurements at electron temperatures up to about 50 eV. This is because quite a strong absorption is necessary in one foil for the ratio method, resulting in a small signal-to-noise ratio.

8.6. MICROWAVE MEASUREMENTS

Microwave techniques are useful for studying the later phases in the development of plasmas generated by laser pulses, when the electron density is low and the self-luminosity is weak. They do not provide much information about the initial behaviour of the plasma, apart from a measure of its size.

Microwave propagation experiments on plasmas produced in gases by laser pulses were first reported by Askaryan *et al.* (1965*a*, 1966), who made cut-off measurements with 8 mm waves to determine the size of the ionized volume in which the electron density exceeded the critical value.

Microwave interferometry in a Fabry–Perot cavity at a wavelength of 2 mm was used by Bize *et al.* (1967) to study the later stages of development of plasmas produced in air. The perturbation caused by a laser-generated plasma in a microwave cavity was studied by Hall (1969*a*). A self-tuning oscillator and frequency measuring device allowed the measurement of the resonance frequency shift of the cavity due to the presence of the plasma. This shift is proportional to the plasma volume. Also, the change in Q of the cavity was measured, which gave a value for the surface conductivity of the plasma.

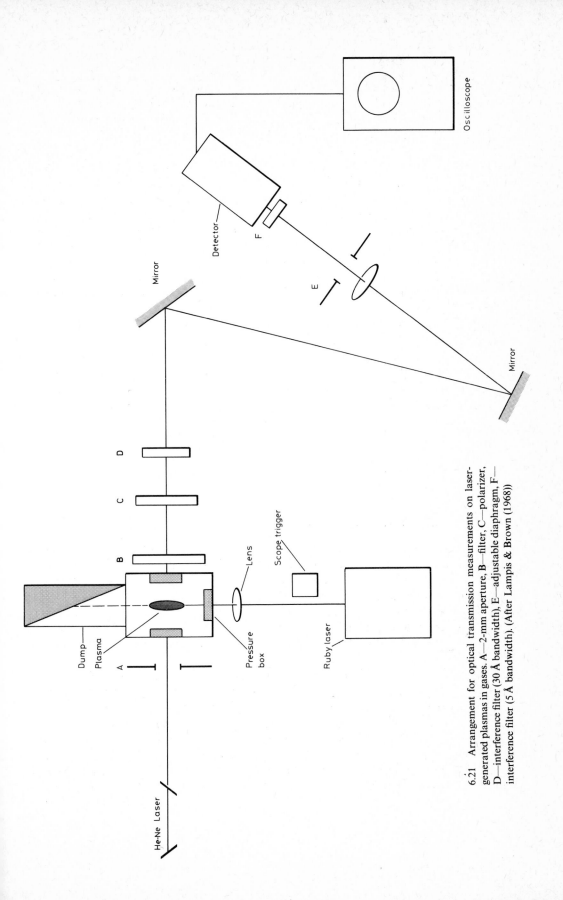

6.21 Arrangement for optical transmission measurements on laser-generated plasmas in gases. A—2-mm aperture, B—filter, C—polarizer, D—interference filter (30 Å bandwidth), E—adjustable diaphragm, F—interference filter (5 Å bandwidth). (After Lampis & Brown (1968))

8.7. OPTICAL TRANSMISSION MEASUREMENTS

The absorption of the incident laser light by the plasma it generates has been measured by several authors.* The effect of the plasma on a narrow probing beam of steady intensity has received little attention, yet as was shown by Lampis & Brown (1968) this very simple diagnostic technique gives a great deal of information about the later stages of development of the plasma. Lampis & Brown used a c.w. (continuous wave) helium–neon gas laser beam of 2 mm diameter as the probe, and studied plasmas generated in helium at 10 atm by a Q-switched ruby laser. The arrangement is shown in Figure 6.21. By measuring the absorption, the increase in divergence of the probing laser beam, and the diameter of the plasma as functions of time, it was possible to deduce the electron density and temperature in the plasma.

9. STUDIES OF PLASMAS FORMED BY RUBY AND NEODYMIUM LASER PULSES

In this section we shall describe the results of experiments in which plasmas were produced by ruby or neodymium laser pulses, mostly in air but also in other gases, over an extensive range of pressures. Wide ranges of laser pulse powers and durations have been employed. We shall consider first experiments using laser pulses of more than 1 ns duration, mostly from Q-switched lasers, and then turn to the effects of ultra-short mode-locked pulses. Within each group we shall discuss results roughly in the order of increasing laser power.

9.1. PULSES OF DURATION GREATER THAN 10^{-9} s

9.1.1. Laser power up to ~ 10 MW

In 1964, using an image-converter camera, Ramsden & Davies (1964) obtained streak photographs of plasmas produced in atmospheric air by a 20 ns, 0·3 J ruby laser pulse focused by a lens of 8 mm focal length. The plasma boundary nearest the laser was found to move towards the laser with an initial velocity of about 10^7 cm s^{-1}. Ramsden & Savic (1964) extended the measurements and took framing-mode photographs of the plasma. During the laser pulse the displacement of the boundary towards the lens varied as $t^{0·65}$, while after the pulse had ended the dependence was as $t^{0·4}$. The total energy absorbed by the gas was 0·2 J. The results indicate that the plasma boundary behaved as a radiation-supported detonation wave during the laser pulse, and a blast wave afterwards. Both streak and framing camera photographs also showed that a part of the plasma furthest away from the laser tended to move in the

---

* See for example Berry *et al.* (1964), Tomlinson (1965*b*), Fecan *et al.* (1966), Sun *et al.* (1967).

opposite direction. With an estimated electron density of about 10^{20} cm^{-3} the plasma was probably not completely opaque to laser light, so some energy could be transmitted to the rear of the shock wave. Ramsden & Davies recorded the time-integrated spectrum of light emitted at 90° to the laser beam axis by the plasma in air and found that it consisted of an intense continuum, together with a few NII lines and a sharp line close to the laser wavelength but displaced by up to 3 Å to the short wavelength side. This line was ascribed to laser light scattered from the plasma. The displacement, interpreted as a Doppler shift, was in good agreement with the observed initial velocity of the luminous plasma boundary towards the laser of 10^7 cm s^{-1} obtained from streak photographs. Occasionally a weak line was seen in the scattered spectrum, shifted to longer wavelengths, which was consistent with the observation of a weaker luminous front moving away from the laser.

When plasmas formed in air were photographed through a narrow-band interference filter at λ 6943 Å the scattering region, viewed at 90° to the laser beam axis, appeared as a small central spot 0·1 mm in diameter and 0·5 mm long. The volume of the self-luminous plasma was much greater, being 1 mm in diameter and 1·5 mm long. The intensity of the laser light at 90° in air could be explained in terms of Thomson scattering from a plasma with $n_e = 5 \times 10^{19}$ cm^{-3}. In plasmas formed in a helium jet the corresponding value of n_e was 5×10^{17} cm^{-3}, which was in good agreement with the value deduced from Stark broadening observations on the He I line at λ 5876 Å. The line width of the scattered light was found to be less than the resolving limit of the spectrograph (0·4 Å), which suggested that $\alpha > 1$ if Thomson scattering was responsible (see Chapter 3) and set an upper limit of 10 eV on the ion temperature. Mandelstam et al. (1965) have pointed out that this is not consistent with the observed expansion velocity according to detonation wave theory. As we shall see later, there is evidence that other scattering processes were more important.

The time-integrated quantity of ruby laser light emerging in different directions from plasmas produced in air was measured by Young et al. (1966). The Q-switched, single-frequency laser had a peak power of 5 MW and a pulse duration of 30 ns. As may be seen in Figure 6.22, the emergent laser light energy varied as $\cos^2 \phi$ in the plane, normal to the laser beam axis, containing the polarization vector of the incident light, the polarization vector being at $\phi = 90°$. In planes containing the laser beam axis, the emergent laser light was strongly peaked in the forward direction, with a small peak in the backward direction. The self-luminosity of the plasma was found to be emitted uniformly in all directions. The spectral intensity of the continuous radiation was greatest at about 4000 Å, corresponding to a black body temperature of about 10^4 K. Visual characteristics of the spark in air were also described by these authors; the risk of retinal burns makes repetition of such observations undesirable without very careful precautions.

Further observations were made on plasmas produced in air and other gases (at unspecified pressures) with a similar ruby laser by Savchenko & Stepanov (1968), who obtained time-integrated photographs of the focal region, in laser light, from two directions simultaneously, perpendicular to the laser beam axis and at 180° to each other. They found that the two photographs of a single event differed widely in the intensity, location and size of the luminous regions, which had a beadlike appearance

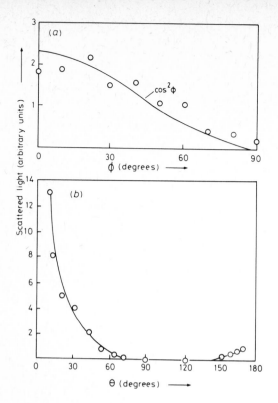

6.22 Angular distribution of laser light scattered from plasma (a) in plane normal to laser beam axis (polarization vector **E** is at $\phi = 90°$), (b) in plane containing laser beam axis ($\theta = 0°$ for forward scattering). (After Young *et al.* (1966))

observed also by Korobkin *et al.* (1967). The total amounts of laser light scattered in the two directions were unequal, and depended on the direction of rotation of the Q-switching prism and the distribution of the laser pulse energy across the beam. The direction of the electric field of the laser light was vertical, the viewing directions horizontal. By taking photographs through a plate of iceland spar, images were obtained in light of two orthogonal polarizations. It was found that the laser light emerging horizontally from the focal region was mainly polarized vertically, i.e. parallel to the direction of polarization of the incident laser light. This is to be expected whether the light is reflected or Rayleigh scattered. However, similar photographs of light emerging vertically showed that most of this light was polarized parallel to the beam axis. Scattered light should have very low intensity in this direction, so this observation suggests that much of the laser light coming from the breakdown region is reflected by the plasma.

Savchenko & Stepanov found similar effects in hydrogen, nitrogen, oxygen, carbon dioxide, chlorine and methane. The inert gases behaved differently. In xenon and krypton a diffuse luminous volume was observed with no structure. The photographs obtained through the red filter were almost the same as those through a violet one, and might indeed have been due to the plasma self-luminosity; the light was completely unpolarized in each case. In argon, helium and neon a very faint structure was discernible through the red filter. In all gases the plasma as viewed in its own self-luminosity was diffuse and showed no structure. These results suggest that the structure which appears in photographs taken in laser light may sometimes be due not to the development of separate breakdown regions, but to reflections of the coherent light from irregularities that develop on the surface of the plasma.

236

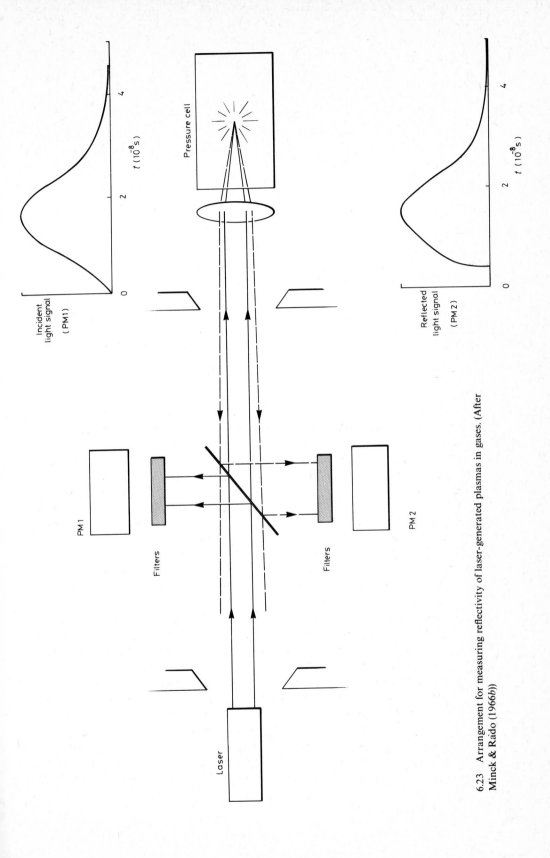

6.23 Arrangement for measuring reflectivity of laser-generated plasmas in gases. (After Minck & Rádo (1966*b*))

Minck & Rado (1966b) investigated the laser light returned towards the laser from plasmas in argon and in nitrogen. Their method is shown in Figure 6.23. The ruby laser, Q-switched by a rotating prism, produced pulses of up to 10 MW peak power of 20 ns duration. The reflectivity of the plasma, which attained a fairly constant value after breakdown, is plotted in Figure 6.24 for argon as a function of pressure, at various laser powers. As may be seen, an abrupt increase in reflectivity occurred above a certain pressure. The low-pressure background reflectivities are ascribed to

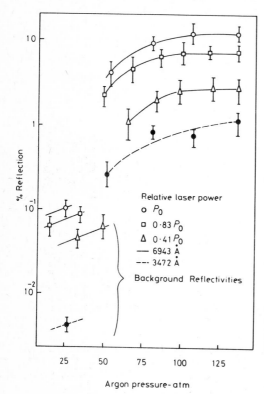

6.24　Observed pressure dependence of reflectivity of laser-generated plasmas in argon for ruby laser light ($\lambda = 6943$ Å) and for its second harmonic ($\lambda = 3472$ Å). (After Minck & Rado (1966b))

reflections from the equipment. When second harmonic light was added to the incident laser beam by introducing an ADP crystal into its path, it was found that the reflectivity for the second harmonic behaved very similarly to that for the fundamental, reaching 0·01 at high pressures. The wavelength of the reflected light was measured using an interferometer, and found to be displaced from that of the incident light by no more than 10^{-2} cm^{-1} to the blue. Interpreted as a Doppler shift, this implies a velocity of the reflector of the order of 10^{4} cm s^{-1} or less. Minck & Rado concluded that, at a sufficiently high initial gas pressure, the laser pulse creates an almost stationary region which is highly reflecting because the electron density is so great that the plasma frequency equals the laser frequency. If only single ionization occurs, and the number density of atoms in the ionized region remains equal to that of the initial cold gas, this situation can occur at 85 atmospheres in argon.

Clearly, it can occur at lower filling pressures if multiple ionization is produced by the laser pulse.

Interferometric studies were made by Alcock *et al.* (1966a) of plasmas produced in air by 7 MW, 30 ns ruby laser pulses focused by an 8 mm lens. Some of the results are shown in Figure 6.25. The behaviour of the general shape of the plasma confirms the photographic results already described above. A sharp positive fringe shift at the boundary shows the presence of a shock front. In the centre of the plasma, a large negative shift is seen during the early stages of development, which is as much as four fringes in the interferogram recorded at 35 ns. This corresponds to an electron density of about 3×10^{19} cm^{-3}. Larger displacements were observed at earlier times, and the maximum electron density must have exceeded 5×10^{19} cm^{-3}, corresponding to full ionization. In the blast wave phase the mass density profile in the shock front

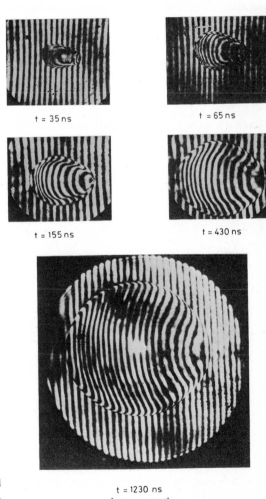

t = 35 ns

t = 65 ns

t = 155 ns

t = 430 ns

6.25 Interferograms of plasma produced in air at S.T.P. (After Alcock *et al.* (1966a), by courtesy of A. J. Alcock)

t = 1230 ns

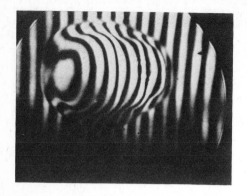

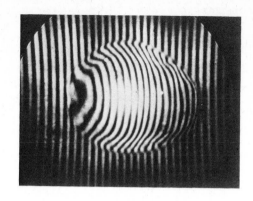

λ = 6943 Å

λ = 3471·5 Å

Time Delay = 170 ns

6.26   Simultaneous interferograms at λ 6943 Å and λ 3471·5 Å of plasma produced in air at S.T.P. (After Alcock & Ramsden (1966), by courtesy of A. J. Alcock)

was analysed and found to agree well with Sakurai's theory (Sakurai (1953)), as did the radius of the shock wave.

Alcock & Ramsden (1966) later obtained simultaneous interferograms at two wavelengths (the fundamental and the second harmonic of the ruby laser wavelength). Figure 6.26 shows a pair of simultaneous interferograms of a plasma in air at S.T.P., while Figure 6.27 shows how the average electron density was found to vary with time.

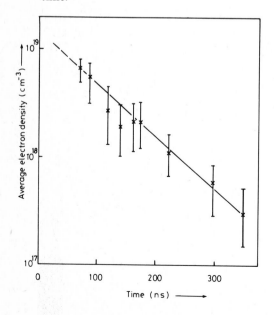

6.27   Electron density as function of time in plasma produced in air at S.T.P. (After Alcock & Ramsden (1966))

Microwave studies of laser-generated plasmas were made by Askaryan *et al.* (1965*a*, 1966), who used 8 mm microwaves ($n_{ec} \sim 10^{13}$ cm^{-3}). Cut-off measurements showed that the ionized volume created by a laser pulse in air grew to a diameter of the order of 1 cm, and persisted for hundreds of microseconds. Evidence was obtained for the occurrence of a fast photo-ionization halo with $n_e \gtrsim 10^{13}$ cm^{-3} outside the front bounding the main plasma. The halo lasted many microseconds in argon and hydrogen, but was much shorter in duration in oxygen, probably as the result of attachment. Bize *et al.* (1967) obtained similar results with 2 mm microwaves; the plasma electron density rapidly rose to the cut-off value of $2 \cdot 5 \times 10^{14}$ cm^{-3} and stayed above it for 150 μs.

A detailed investigation was carried out by Hall (1969*a*) of the perturbation caused by laser-generated plasmas in a resonant microwave cavity ($\lambda = 13 \cdot 6$ cm, $2 \cdot 8 \times 10^9$ Hz). In air at pressures only slightly above atmospheric, the results were in good agreement with blast wave theory. At higher pressures, the method used suggested that after an initial expansion the effective radius of the plasma contracted. However, what was actually measured was the radius of the shell at which the electron temperature was about 4200 K and the electron density was about 10^{15} cm^{-3}. At higher pressures the lower conductivity of the plasma resulting from a reduced Mach number led to the effective plasma radius being smaller than the true radius. A strong photo-ionization halo surrounding the plasma can cause a greater perturbation of the microwave cavity than the main plasma itself. For this reason the method is restricted to gases in which electrons attach rapidly to neutral molecules, such as air, CO_2 and O_2.

Helium plasmas produced at a pressure of 600 Torr by a small neodymium laser, giving an irradiance at the focus of 5×10^{10} W cm^{-2}, were investigated spectroscopically by Durand & Veyrie (1966). The profile of the He II line at $\lambda\, 4686$ Å was determined, from which, using the theory of Griem, Kolb and Shen (see for example Griem (1964)) the electron density could be calculated. The results are shown in Figure 6.28; the time scale zero is at the peak of the 30 ns laser pulse. From consideration of the equilibrium populations of neutral atoms and singly- and doubly-ionized atoms, which are manifested in the relative intensities of He I and He II lines, an

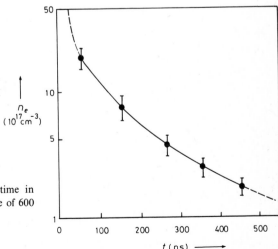

6.28 Electron density as function of time in plasma produced in helium at a pressure of 600 Torr. (After Durand & Veyrie (1966))

estimate of 3.6×10^4 K was obtained for the average temperature of the plasma 40 ns after the laser pulse maximum.

A very careful and detailed study of the decay of plasmas produced in hydrogen by a 0·2 J, Q-switched ruby laser was carried out by Litvak & Edwards (1966). Measurements of the Stark-broadened profile of H_α were used to find the electron density, making use of Griem's tables (Griem (1964)). H_β was not used because H_γ was so broad that these lines overlapped. The electron temperature was found, following Griem, from the ratio of the intensity of H_α to that of the adjacent continuum. The measurements were made as a function of time over the range of pressures from 1 to 70 atmospheres, using a monochromator and photomultiplier and varying the pass band from shot to shot. The cooling effects of radiation and expansion were discussed, taking recombination into account. The maximum temperature and electron density occurred at about the time of the laser pulse, and for all pressures were about 10^5 K and 10^{19} cm^{-3}. It should be remembered, however, that the light was probably not emitted from the hot central region of the plasma.

Time-resolved line profiles were obtained from the afterglow of laser-generated plasmas in helium and helium–hydrogen mixtures at pressures of 1 to 10 atm by Evtushenko et al. (1966b). A rotating-mirror streak camera was used to record the spectrum in the output focal plane of a spectrograph. From measurements of the breadths of H_α and He I λ 5876 Å the early maximum electron density was calculated to be about 5×10^{18} to 10^{19} cm^{-3}, which agreed reasonably well with the value deduced from the time-integrated line breadth of He II λ 4686 Å.

Ahmad et al. (1969a, b) studied the development of plasmas produced in helium by a ruby laser of 10 MW peak power and 30 ns pulse duration, focused by a lens of 2·7 cm focal length. By varying the helium pressure from 1 to 4 atmospheres, the optical thickness of the plasma for laser light could be varied. Image-converter streak photographs were obtained in plasma light and also in scattered laser light (Figure 6.29), which show very clearly how at the higher pressure the expansion of the plasma is only towards the laser, because most of the incident laser light is absorbed

6.29   Streak photographs of plasmas produced in helium : (a), (b) in plasma light ; (c), (d) in scattered light. In (a) and (c), at high pressure, the plasma is optically thick. In (b) and (d), at low pressure, it is optically thin. (After Ahmad et al. (1969b), by courtesy of M. H. Key)

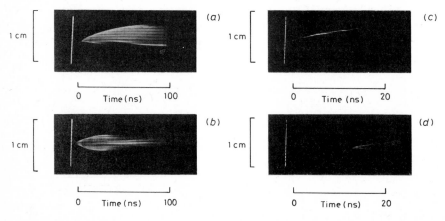

in the boundary nearest the lens, whereas at the lower pressure the plasma develops symmetrically, both sides being more or less equally heated. From the velocity of the boundary, making use of the detailed calculations of Key (1969), the temperature and density of the plasma were deduced, and hence the maximum electron density in the shock front. The He I line at λ 5876 Å was used to give the electron density as a function of time in plasmas produced in helium at a pressure of 4 atm (Ahmad et al. (1969a)). The spectrum recorded by an image-converter streak camera showed that for the first 100 ns the line was strongly self-reversed, due to the high population of 2^3p metastable excited atoms in the shock front surrounding the plasma. Measurements of the profile at later times were made with a monochromator and photomultiplier. The electron density of the emitting region is shown as a function of time in Figure 6.30. The electron temperature was found from the ratio between the

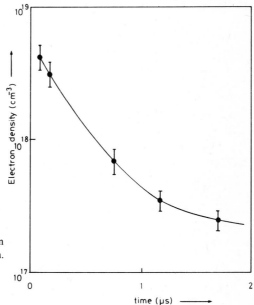

6.30   Electron density as function of time in plasma generated in helium at a pressure of 4 atm. (After Ahmad et al. (1969a))

intensities of the line and of the adjacent continuum, using a monochromator and photomultiplier, which involved making two measurements from successive shots, and also from a single shot by means of two interference filters and photomultipliers. The results are shown in Figure 6.31.

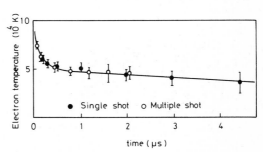

6.31   Electron temperature as function of time in plasma generated in helium at a pressure of 4 atm. (After Ahmad et al. (1969a))

Ahmad & Key (1969) also obtained streak photographs of more complex plasma structures produced by focusing a laser pulse on to a shock front generated in helium by another laser pulse one microsecond earlier.

Ahmad *et al.* (1969*b*) considered the possibility that the laser light emerging from the focal volume was reflected from the plasma boundary, rather than scattered by the plasma electrons. They calculated the reflectivity of the plasma as a function of the electron density and the angle of incidence by finding the refractive index, ignoring the complex term, and applying Fresnel's equations. Results for light with the direction of polarization lying in the plane of reflection, corresponding to the experimental situation, are shown in Figure 6.32 for a central ray deviated through 90° (angle of incidence 45°) and for the outermost rays entering an $f/8$ lens. Figures

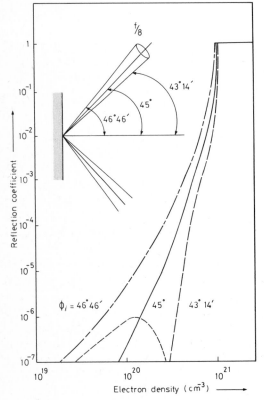

6.32   Calculated reflectivity of plasma for ruby laser light with direction of polarization lying in plane of reflection, for three angles of incidence ϕ_i. (After Ahmad *et al.* (1969*b*))

6.33(*a*), (*b*) show how such reflections might arise in the optically thick and optically thin cases respectively. The estimated experimental values of the apparent reflectivity were $1 \cdot 2 \times 10^{-6}$ at 1 atmosphere and $2 \cdot 2 \times 10^{-3}$ at 4 atmospheres, in reasonable agreement with the values expected from the plasma parameters. The very rapid variation with pressure of the intensity of laser light emerging from the plasma is not consistent with either Thomson or Rayleigh scattering. Furthermore, the scattered

(a)

(b)

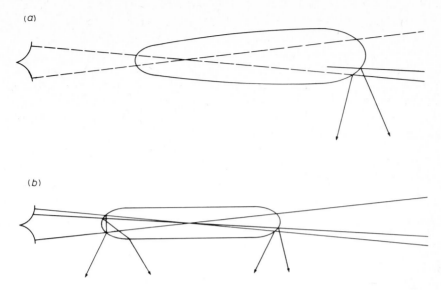

6.33   Sketches showing how reflections may arise from (a) optically thick and (b) optically thin plasmas. (After Ahmad *et al.* (1969*b*))

light intensity would be a minimum in the direction of observation used in the experiments, and even in the orthogonal direction would be very much lower than is actually observed. Thus the emergent laser light is almost certainly reflected from the plasma, in agreement with the conclusions of Savchenko & Stepanov (1968).

We have already noted in Chapter 5 the occurrence of self-focused filaments when breakdown occurs in gases. Tomlinson (1969) investigated plasmas produced by a 5 MW ruby laser pulse of 30 ns duration focused by a 3 cm focal length lens to a 30 μm diameter spot in several gases. In air at S.T.P., time-integrated photographs taken in laser light at 90° to the laser beam axis showed the narrow central filament observed by Korobkin & Alcock (1968), and streak photographs confirmed that the scattering region was only 30 to 60 μm long at any instant and moved from the focal point towards the lens, coinciding with the boundary of the luminous plasma. The observed thickness is similar to the absorption length expected (see Figure 2.6) for ruby laser photons. The scattered power at 90° was measured, and found to be larger by several orders of magnitude than could be produced by Thomson scattering. This result was obtained over a wide range of experimental conditions in air and in nitrogen. The experiments were repeated in the inert gases helium, argon, krypton and xenon at atmospheric pressure: the scattered power was lower than that observed in air by three or four orders of magnitude in all these gases. A close study of the time-integrated photographs obtained in laser light showed that the filaments observed were not continuous but consisted of closely spaced, discrete scattering centres. The centres were extremely small, their dimensions being less than the 14 μm resolution of the camera. In argon, the spacing between the centres was much greater than in air. The scattering centres ceased to be visible when two or three new centres had appeared

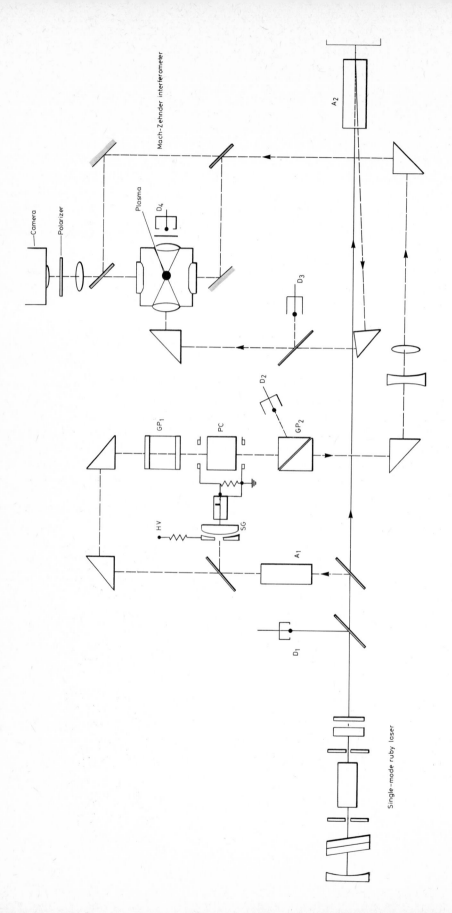

6.34 Arrangement for plasma generation and short-exposure (700 ps) interferometry. A—laser amplifier, SG—spark gap, PC—Pockels cell, GP—Glan prism, D—photodiode. (After Richardson & Alcock (1971b))

nearer the laser, suggesting that a single centre was not optically thick, but that three of them absorbed strongly. This suggests that each centre had a thickness of about 10 μm. In xenon, no filaments or scattering centres were observed.

Tomlinson concluded that the laser radiation collected from the plasma generated in air was due to Fresnel reflections from small regions of high density in the plasma. He pointed out that the microstructure was strongly dependent on the gas in which the plasma was produced, and did not appear to be correlated with the focusing optics, thus ruling out any explanation based on local enhancement of the radiation field by aberrations of the focusing lens (Evans & Grey Morgan (1969)), and indicating that self-focusing occurred in the plasma after breakdown. Alcock *et al.* (1970, see also Alcock (1972)) carried out similar experiments using a single-mode ruby laser and reached the same conclusion. The molecular gases nitrogen, freon and methane were characterized by almost continuous filaments, while in all the noble gases the scattering points were widely separated. Wilson (1970) found that the average distance between breakdown centres in air decreased as the pressure was increased from 200 to 760 Torr. At pressures approaching 760 Torr the formation of continuous filaments began to be noticeable. No scattering centres were observed in argon.

In order to study self-focused filaments in gases interferometrically, spatial resolution of less than 10 μm is necessary. Also, the development of the filament is rapid and time resolution of 1 ns is desirable. Richardson & Alcock (1971a, b) succeeded in observing filaments produced in a number of gases by a single-mode ruby laser, using a Mach–Zehnder interferometer. The laser beam was focused to a spot whose diameter at half maximum intensity was 190 μm. A small fraction of the laser beam was extracted by a beam splitter, and a fast electro-optic shutter selected a 700 ps pulse for the interferometer (Figure 6.34). The fringes were recorded photographically with a magnification of × 70. Interferograms taken in argon at 1 atm at different times are shown in Figure 6.35. The filament has a minimum diameter of not more

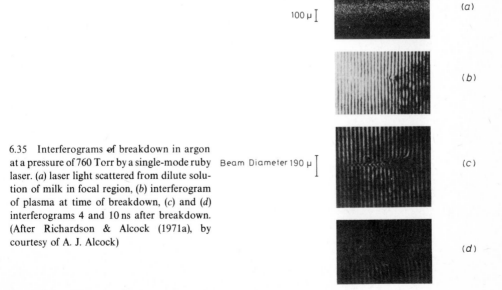

100 μ

Beam Diameter 190 μ

(a)

(b)

(c)

(d)

6.35 Interferograms of breakdown in argon at a pressure of 760 Torr by a single-mode ruby laser. (a) laser light scattered from dilute solution of milk in focal region, (b) interferogram of plasma at time of breakdown, (c) and (d) interferograms 4 and 10 ns after breakdown. (After Richardson & Alcock (1971a), by courtesy of A. J. Alcock)

than 13 μm and has a lower refractive index than the ambient gas, indicating that it is strongly ionized. It appears at the time of breakdown, and grows to a length of 300 μm. A larger region of plasma then develops at the end of the filament further from the laser. The filament continues to grow towards the laser at a velocity of about 4×10^7 cm s^{-1}, and additional larger regions of plasma appear at intervals of 0·2 to 0·4 mm. As they appear, the filament gradually decays on the side further from the laser. When one of these larger regions is formed, a narrow central core about 100 μm long occurs with an electron density much higher than that of the main plasma. In molecular gases, filaments were also observed, but the larger plasma formed at the end of the filament away from the laser grew steadily towards the laser, in contrast to the discontinuous development seen in argon.

Schlieren observations by Key *et al.* (1970) provided further direct evidence for the occurrence of self-focused filaments. A plasma was produced in argon at 760 Torr by a 2 MW neodymium glass laser pulse of 6 ns duration. A fraction of the laser light was split off and allowed to cause breakdown in a second gas cell at the common focus of two lenses. Breakdown occurred on the leading edge, so the transmitted pulse was thus shortened to 1 ns. It was then frequency doubled, and further shortened to 0·5 ns, by being passed through an ADP crystal. This short green pulse was used for photography, passing through the region of the main argon plasma 2 ns after the leading edge of the main laser pulse, and at right angles to it. The schlieren photograph showed a single filament of less than 10 μm diameter, with relatively large plasma blobs at intervals along the filament, in which very small plasma blobs were also seen. To within about a nanosecond, therefore, self-focusing and breakdown are simultaneous.

Finally, in this sub-section, we may note that plasmas have also been produced with very long laser pulses. Vanyukov *et al.* (1966) used a non-Q-switched neodymium laser with a pulse duration of about a millisecond and an output energy of upwards of 1000 J. In this mode, the laser output consists of a rapid, more or less random succession of spikes. The plasma in air at S.T.P. was photographed: it extended for 20 to 30 mm from the focus towards the 10 cm lens.

9.1.2. Laser power between ~ 10 and ~ 100 MW

Electron density distributions in plasmas produced by 20 to 30 MW Q-switched ruby laser pulses have been determined holographically (Komissarova *et al.* (1968), Iznatov *et al.* (1971)) and interferometrically (Hugenschmidt (1970)). In air at 1 atm (Komissarova *et al.* (1968)), in hydrogen at 1·75 atm (Iznatov *et al.* (1971)) and in xenon at 1·4 atm (Hugenschmidt (1970)), rather flat electron density distributions with a central maximum of about 10^{19} cm^{-3} were observed shortly after the laser pulse ended. In high-pressure hydrogen, however, large electron density gradients were observed. Figure 6.36 shows some of these results, while Figure 6.37 shows the shapes of the incident and transmitted pulses for various pressures of hydrogen.

When holograms of plasmas produced in air at atmospheric pressure by similar laser pulses were reconstructed (Ostrovskaya & Ostrovskii (1966), Zaidel *et al.* (1966)) it was noticed that the laser beam was focused to a narrow bright line at a

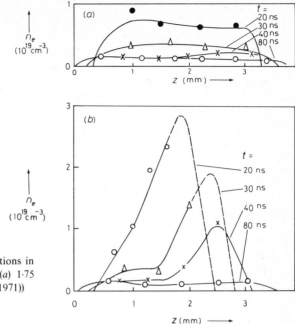

6.36 Axial electron density distributions in hydrogen plasmas. Initial pressure (a) 1·75 atm., (b) 11 atm. (After Iznatov et al. (1971))

certain distance from the plane of the real image. This was ascribed to the defocusing effect of the plasma, which may be regarded in its early stages of development as a negative cylindrical lens. From the focal length and the known dimensions of the plasma the refractive index may be calculated, neglecting aberrations, and hence the electron density (ignoring the contributions of neutral atoms). The average value of n_e over the first 120 ns after breakdown, obtained from twenty holograms, was 2.4×10^{19} cm^{-3}.

Two-wavelength holography (Komissarova et al. (1969, 1970)) using ruby laser light and its second harmonic showed a marked reduction in heavy-particle density in the central region of plasma, associated with the formation of the shock wave. With this technique it was possible to study the structure of the plasma at earlier times. The distributions of electron density and ion temperature deduced from holographic

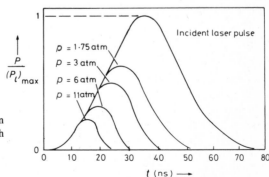

6.37 Transmitted ruby laser pulse shapes in hydrogen at four pressures compared with incident pulse. (Iznatov et al. (1971))

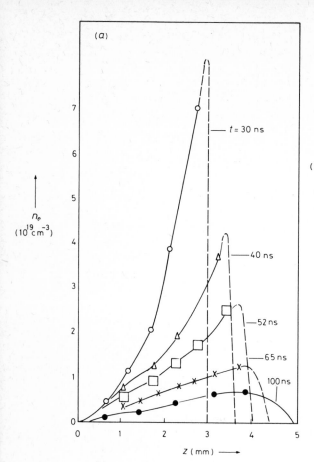

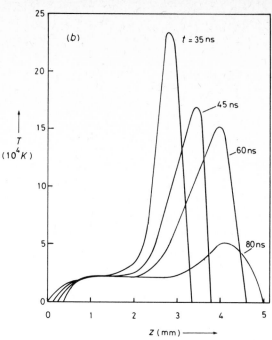

6.38 Axial (a) electron density and (b) ion temperature distributions in plasma formed by a ruby laser pulse in air at S.T.P. Laser light is incident from the right. (After Komissarova *et al.* (1970))

interferograms of the plasma produced by a 0·7 J, 30 ns ruby laser pulse are shown in Figures 6.38(a) and (b) respectively. Time is measured from the moment of breakdown, which occurred about 15 ns before the peak of the laser pulse. It is apparent that a hot region about 0·5 mm thick occurs at the boundary moving towards the laser. The ion temperature approaches $2·5 \times 10^5$ K shortly after the time of the laser pulse maximum.

In early spectroscopic studies of plasmas produced in air by 30 MW ruby laser pulses Mandelstam *et al.* (1964, see also Pashinin *et al.* (1965)) noted, in addition to a strong continuum, lines of N I, N II, O II and also H_α. Many of the lines were multiplets whose components, separated by several ångstroms, were unresolved, indicating a high electron density. From the half-intensity width of the N II line λ 3995 Å, which was measured to be 10 Å, an electron density of about 2×10^{18} cm^{-3} was deduced. From the relative intensities of the N II lines λ 5179 and 5045 Å, the electron temperature was found to be in the region of $(3 \text{ to } 6) \times 10^4$ K.

Breton *et al.* (1965) determined the time dependence of the continuous spectrum emitted from plasmas produced in nitrogen at pressures between 1 and 25 atm by a 30 MW neodymium laser. The intensity decreased with increasing gas pressure. The maximum emission occurred at wavelengths below λ 4000 Å.

Plasmas produced in atmospheric air by 50 MW Q-switched ruby laser pulses focused by lenses of 25 to 75 mm focal length were studied by Mandelstam *et al.*

(1965). X-ray emission studies using a Geiger counter (see Section (8.5)) indicated an electron temperature of about 60 eV, the uncertainty being less than 15 eV. Spectroscopic measurements on the laser light scattered from the plasma confirmed the general pattern found by Ramsden & Davies (1964). Light scattered through 180° (backwards towards the laser) was examined; the shift was double that observed at 90°, confirming its Doppler origin. The velocity derived from the shift varied with the irradiance I roughly as $I^{\frac{1}{3}}$: at maximum laser power it was about 10^7 cm s^{-1}. Detonation wave theory predicts that with an irradiance of 10^{11} W cm^{-2} the axial velocity of expansion will be $\sim 10^7$ cm s^{-1} and the temperature $\sim 7 \times 10^5$ K, in satisfactory agreement with experiment (the irradiance was actually 2×10^{11} W cm^{-2} but allowance has to be made for the energy of lateral expansion).

The breadth of the scattered laser line was only about 1 Å, comparable to that observed by Ramsden & Davies (<0.4 Å). Interpreted as the breadth of a Thomson scattering ion component ($\alpha \sim 25$, see Chapter 3) this gave an ion temperature much lower than the observed electron temperature. The electron–ion equilibration time was short ($\lesssim 10^{-10}$ s) under these conditions so the electron and ion temperatures should have been almost the same. An alternative explanation for the observed line breadth is that the light is simply reflected from the moving shock front, whose motion varies in the non-uniform laser beam.

Malyshev *et al.* (1972) used the Thomson scattering method to study the central region of plasmas produced in atmospheric air by laser pulses of 0·6 to 1·2 J, at a time 11 μs after the laser pulse. Under the conditions of observation the value of the parameter α was about 3.3, but the electron density was not so large that reflections were a problem. For a scattering angle of 90° the electron lines were about 60 Å away from the ruby laser wavelength, giving an electron density of about 10^{17} cm^{-3}. The relative intensities of the electron lines and the central Rayleigh-scattered peak were also measured, and combined with the electron density gave electron temperatures of $(1·2$ to $1·4) \times 10^4$ K.

The afterglow of plasmas generated by 50 MW (2·5 J, 50 ns) ruby laser pulses in helium at 10 atm was investigated by Lampis & Brown (1968) using a narrow helium–neon laser beam as a probe. Some results are shown in Figures 6.39 and 6.40. Detailed

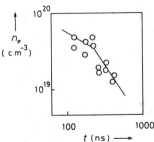

Left 6.39   Electron density as function of time after breakdown, in plasma formed by a ruby laser pulse in helium at a pressure of 10 atm. (After Lampis & Brown (1968))

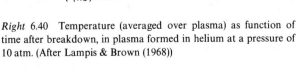

Right 6.40   Temperature (averaged over plasma) as function of time after breakdown, in plasma formed in helium at a pressure of 10 atm. (After Lampis & Brown (1968))

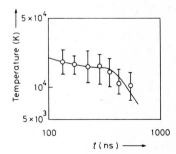

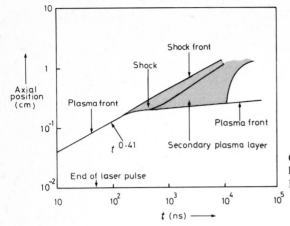

6.41 Axial expansion of plasma formed by a ruby laser pulse in helium at a pressure of 10 atm. (After Lampis & Brown (1968))

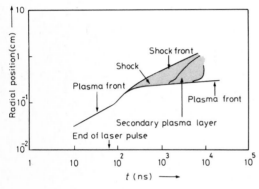

6.42 Radial expansion of plasma formed in helium at a pressure of 10 atm. (After Lampis & Brown (1968))

information was obtained about the structure of the plasma as may be seen from Figures 6.41 and 6.42 which show the axial expansion of the plasma towards the laser, and the radial expansion. The initial expansion of the plasma after the laser pulse ends is at a velocity proportional to $t^{0.41}$ in good agreement with blast wave theory. Later, a secondary plasma layer appears to develop outside the main plasma, in agreement with the microwave observations of Askaryan et al. (1965a, 1966) already described. After about 30 μs the shock waves are reflected back by the walls of the pressure chamber and a complex pattern develops.

Plasmas produced in high-pressure helium (up to 20 atm) by 100 MW laser pulses were used by Baravian et al. (1972) to investigate recombination processes. The variations of the average electron temperature and density with time, as determined spectroscopically, are shown in Figures 6.43(a) and (b).

Detailed studies have been made of plasmas produced in air by 100 MW ruby laser pulses focused by a 50 mm lens. Korobkin et al. (1967) measured the velocity of the luminous front towards the laser, using an image-converter camera, and compared this value with the velocity deduced from the Doppler shift of scattered light.

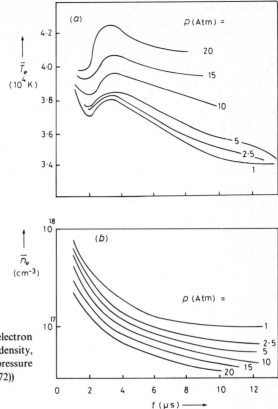

6.43   Time dependence of (a) average electron temperature and (b) average electron density, determined spectroscopically, in high-pressure helium plasmas. (After Baravian *et al.* (1972))

The initial velocity of the luminous region was about 4.5×10^7 cm s^{-1}, whereas the Doppler shift gave only 1×10^7 cm s^{-1}. In streak photographs taken in scattered laser light, the plasma boundary appeared to move discontinuously towards the laser along the axis over a distance of several millimetres, an effect noted with higher laser powers by Daiber & Thomson (1967). A time-integrated photograph of the spark taken in laser light suggested that the air was breaking down at several points along the lens axis, the separation between these points being of the order of 0·3 to 0·5 mm. The streak photographs were interpreted as being due to the successive development of breakdown regions, the boundary of the plasma produced at each point moving smoothly towards the lens over a distance of about 0·3 mm, and then vanishing, presumably as the laser pulse energy was absorbed in a new region nearer the lens. The movement of the boundary of a single breakdown region could be adequately described by detonation wave theory and explained the Doppler shift measurements. By rotating the direction of the viewing slit through 90° about the horizontal axis it was possible to measure the transverse velocity of the plasma boundary, which was about 5×10^6 cm s^{-1}. Taking this to be equal to the acoustic velocity, the ion temperature was found to be about 5×10^5 K.

0   5 m m

6.44   Schlieren photographs of plasmas formed in gases. The images produced by a train of mode-locked pulses are super-imposed. (Reproduced by courtesy of A. J. Alcock)

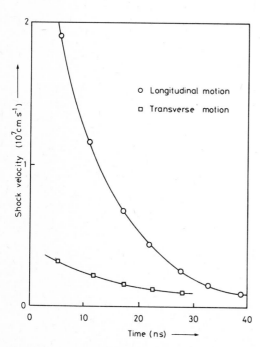

6.45   Longitudinal and transverse plasma boundary velocities in air at S.T.P. as functions of time. (After Alcock *et al.* (1968*a*))

Plasmas produced under very similar conditions were studied by Alcock *et al.* (1968*a*, *b*; see also Hamal *et al.* (1969)) who obtained very clear schlieren photographs using a train of mode-locked pulses as the schlieren light source (see Section (7.2)). Figure 6.44 shows the superimposed images of the plasma boundary, from which the velocities along and perpendicular to the ruby laser beam axis could be deduced. Some results are shown in Figure 6.45. The shock front thickness was found to be not more than 0·02 mm. The schlieren patterns observed changed in detail if a different ruby was used in the laser. In further work (Alcock *et al.* (1968*c*)) this technique was used to study the development of plasmas in neon, nitrogen, argon and air over a wide range of pressures (60 to 1520 Torr) with laser pulses having peak powers from 35 to 300 MW and durations from 14–24 ns focused by a lens of 3·5 cm length. The initial velocity of motion of the plasma boundary was found to vary as the square root of the initial gas pressure—a result which is difficult to explain on the basis of a detonation wave. Streak photographs taken in laser light gave velocities in good agreement with the schlieren results. However, streak photographs in plasma light showed that the self-luminous region extended beyond the edge of the highly-ionized region, so that excitation and probably ionization occurs ahead of the main plasma. Figure 6.46 compares axial velocities obtained by all three methods in air at atmospheric pressure, for a 50 MW, 30 ns ruby laser pulse. From the lateral expansion velocity of the plasma the maximum ion temperature for a 100 MW laser pulse was found to be about $4·5 \times 10^5$ K, in agreement with the observation of Korobkin *et al.* (1967).

Measurements for argon at 760 Torr, with 150 MW laser pulses, were analysed in detail. Up to the time of peak laser power, the positions of the boundaries of the disturbances recorded by all three methods vary together approximately as $t^{0·65}$: afterwards the position of the plasma boundary, as defined by schlieren and by laser light scattering, varies as $t^{0·21}$, which is slower than a blast wave, while the luminous front moves as $t^{0·4}$, as a blast wave should. It appears that the blast wave is detached from the main plasma near the time of peak laser power. Similar results were obtained over a wide range of pressures and laser powers, though the initial slope varied somewhat.

Schlieren measurements by Büchl *et al.* (1968) on blast waves produced by 40 MW ruby laser pulses showed good agreement with Sakurai's theory at Mach numbers

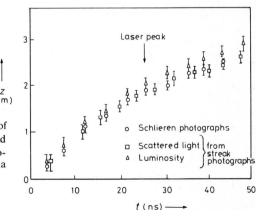

6.46 Variation with time of axial positions of plasma boundaries in air at S.T.P., as recorded by schlieren photography, and by streak photography in scattered laser light and in plasma luminosity. (After Alcock *et al.* (1968*b*))

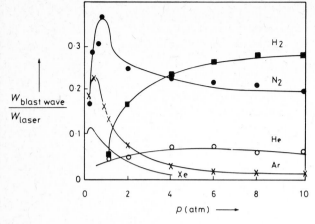

6.47 Fraction of incident laser energy observed in blast waves produced in gases, as function of pressure. (After Büchl *et al.* (1968))

exceeding 2 or 3 over a wide range of pressures, from 200 Torr to 10 atm, in hydrogen and nitrogen, and in helium, argon and xenon. The proportion of the incident laser energy found in the blast wave is plotted as a function of pressure in Figure 6.47.

9.1.3. Laser power between ~100 MW and ~1 GW

Electron temperatures in plasmas formed by ruby laser pulses with powers up to a few hundred megawatts were determined from X-ray measurements by Alcock *et al.* (1966*b*) using the two-foil ratio method and by Ahmad & Key (1972) using the absolute intensity method (see Section (8.5)). Alcock *et al.* found that the results were not very reproducible from shot to shot; in air at 400 Torr, electron temperatures between 40 and 180 eV were obtained with 300 to 400 MW laser pulses, while in neon at around 600 Torr the temperature varied between 77 and 92 eV. The temperatures could be correlated, on the basis of a detonation wave, with the velocity of the luminous front obtained from streak photographs. Ahmad & Key found that using absolute measurements of the X-ray intensity, reasonably consistent values of the electron temperature could be obtained. Figure 6.48 shows the behaviour of the temperature as a function of the laser power (focused by a 3 cm lens) for atmospheric

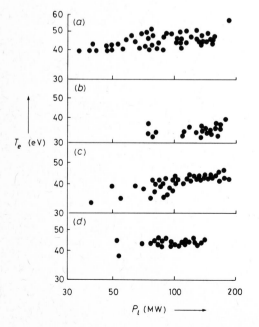

6.48 Electron temperatures from X-ray intensity measurements as functions of laser power, for plasmas in (*a*) air at 760 Torr, (*b*) helium at 3040 Torr, (*c*) helium at 1520 Torr and (*d*) helium at 760 Torr. Error bars are shown on the right. (After Ahmad & Key (1972))

air and for helium at three different pressures. The duration of the high-temperature phase was typically 10 ns in air, and as long as 40 ns in helium. The electron temperatures in air are somewhat higher than the ion temperatures of less than 3×10^5 K obtained with ~ 300 MW laser pulses by Ambartsumyan *et al.* (1965) from measurements of the transverse expansion velocity. However, Ambartsumyan *et al.* used a longer focal length lens (5 cm) and noted that the breakdown was very non-uniform.

Neusser *et al.* (1971) investigated plasmas produced by focusing a ruby laser into a narrow jet of nitrogen expanding from 40 atm into a vacuum. The electron temperature was deduced from X-ray measurements using several different pairs of foils; Figure 6.49 shows how T_e varied with the incident laser power in an almost rectangular, 9 ns pulse.

Plasmas produced by 5 J, 200 MW ruby laser pulses in helium at 360 Torr were studied by Braerman *et al.* (1969). The Stark-broadened profiles of the lines He II

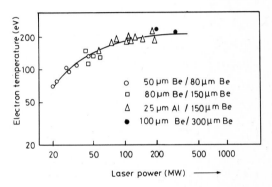

6.49 Electron temperature from X-ray intensity measurements as function of laser power, for plasma produced in a jet of nitrogen at a pressure of 40 atm. (After Neusser *et al.* (1971))

λ 4686 Å, He I λ 5876, 4471 and 3889 Å were measured to give values of the electron density. For the electron temperature in the early stages when the plasma was very hot, no He I lines were observable, so the line-to-continuum intensity ratio of He II λ 4686 was used. During this stage, $7 < T_e < 50$ eV. Between 200 ns and 1 μs the ratio of the intensities of He II λ 4686 Å and He I λ 5876 Å was a convenient indicator, being strongly dependent on T_e in the range $3 < T_e < 7$ eV. After 1 μs, the line-to-continuum intensity ratio for He I λ 5876 was used. The conditions under which the measurements may be reliably interpreted are discussed by these authors. Their results for the variation of the electron temperature and density with time are shown in Figures 6.50 (a) and (b). Again the maximum electron density was near 10^{19} cm^{-3}. The maximum temperature, in the region of 50 eV, is quite close to that derived by Ahmad & Key from X-ray measurements in helium plasmas at 760 Torr.

Daiber & Thompson (1967) obtained both streak and framing camera photographs of plasmas produced in air and other gases by a 10 J ruby laser pulse of 25 ns duration. In air at atmospheric pressure, the position of the plasma boundary as it moved towards the focusing lens varied with time as $t^{0.83}$ during the laser pulse and then, during the blast wave phase, as $t^{0.21}$. Under other conditions of gas pressure, focal length, pulse energy, etc. the time dependence in the initial phase varied between $t^{0.5}$ and $t^{0.9}$, but in the blast wave phase all experiments, including schlieren photo-

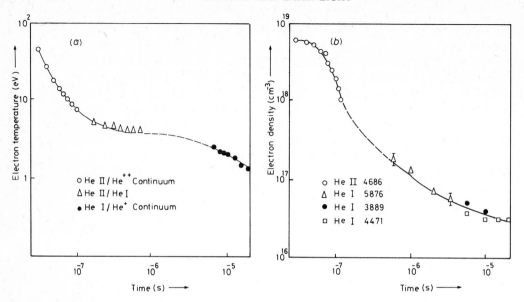

6.50 (a) Electron temperature and (b) electron density as functions of time in plasmas produced in helium at a pressure of 360 Torr. (After Braerman et al. (1969))

graphs, gave very similar results, close to $t^{0.2}$, in contrast to the theoretical prediction of $t^{0.4}$. Basov et al. (1972b) have suggested in another context that radiative energy losses might result in a slower expansion. At high pressures the plasma boundary was sometimes found to make discontinuous jumps towards the lens during the heating pulse. Pumping out and refilling the gas chamber between shots was found to reduce the probability of this effect. Streak photographs of plasmas produced in air at the somewhat lower pressure of 330 Torr (Daiber & Thompson (1970)) by similar laser pulses showed the same effect of discontinuous breakdown: the velocity of the individual breakdown fronts was consistent with detonation wave theory, in agreement with the observations of Korobkin et al. (1967).

Daiber & Thompson (1970) also used an X-ray technique to measure electron temperatures of 40 to 160 eV for plasmas produced in air at 330 Torr by ruby laser pulses of 300 to 600 MW of 10 to 25 ns duration. With three detector channels with different foils, two simultaneous measurements of temperature could be made. These agreed within 10%, and indicated that the thermal Bremsstrahlung continuum did indeed dominate the spectrum from 5 to 15 Å. The X-ray emission was found to occur only while laser energy was being strongly absorbed by the plasma. Spatially-resolved measurements showed that the electron temperatures obtained from one part of the plasma agreed with those obtained from the plasma as a whole. For short laser pulses of 10 to 12 ns duration, the temperature fitted well on the basis of detonation wave theory with the initial axial velocity of the plasma measured by streak photography, confirming the observation of Alcock et al. (1966b), but for longer pulses the observed velocity was higher than was predicted by theory for the measured temperature.

Daiber & Winans (1968) studied the visible and near ultraviolet spectrum emitted from the plasmas generated in nitrogen by ruby laser pulses focused by lenses of 6 or 10 cm focal length. In addition to lines of N I and NII, the λ 4867 Å line of the N III spectrum appeared when the laser peak power was raised to 700 MW. No band spectrum was observed. These authors also studied the spectrum from plasmas in argon, and found broad lines of A I and A II, together with a strong continuum. The continuum came from a small central region of the plasma, very much smaller than the region emitting A I lines but only slightly smaller than the region emitting A II. The A I lines at λ 7948 and 8015 Å were found to be 2·8 Å wide.

Measurements on plasmas generated in deuterium by ruby laser pulses of a few hundred megawatts lasting 4 nanoseconds were described by Guenther & Pendleton (1972). Only a small fraction (20%) of the laser pulse energy was absorbed at a deuterium pressure of 600 Torr, and even less at lower pressures. Streak photographs indicated discontinuous growth of the plasma with initial velocities of 4×10^7 cm s^{-1} towards the laser and 10^6 cm s^{-1} away from it at 600 Torr. The maximum velocity fell to 2×10^7 cm s^{-1} at 200 Torr. The markedly asymmetric growth in plasmas which were only weakly absorbing is not understood. Holographic interferometry yielded electron densities between 2×10^{18} and 10^{19} cm^{-3}. Two-foil X-ray measurements gave electron temperatures of up to about 45 eV, which did not vary much with the laser power, or with the gas pressure between 200 and 500 Torr. The expansion of the plasma after the laser pulse agreed well with Sakurai's blast wave theory (see Section (5)). The Taylor–Sedov theory, valid only for strong shocks, was satisfactory in the early stages of the blast wave phase while the shock velocity was at least Mach 4. About 15% of the incident laser pulse energy was contained in the blastwave. The observed absorption coefficient (0·6 cm^{-1} at 600 Torr) of the plasma for the laser light agreed well with the predicted value for the inverse Bremsstrahlung process at the observed electron density and temperature, and gave a satisfactory energy balance.

A particularly interesting experiment was described by Mead (1970, see also Presby (1971), and Mead (1971)) in which twelve spherically-symmetrically placed ruby laser beams obtained from the same oscillator were focused on a point in a target chamber. The optical paths were made equal to ensure simultaneous arrival of the laser energy in all twelve channels. A total energy of up to 2 J was available in a pulse of duration 10 ns. The target chamber was filled with deuterium, hydrogen or helium, and a multipulse schlieren system with 24 ns separation between exposures was used to study the development of the plasma after the laser pulse ended.

Problems were encountered at high powers when multiple breakdown along individual laser beams upset the spherical symmetry of the plasma. There was also a tendency for the plasma to grow preferentially towards the focusing lenses. The symmetry of the plasma was improved by focusing each beam 2 mm beyond the common centre. Figure 6.51 shows measurements of the velocity of the plasma boundary as a function of its radial position for four plasmas with different absorbed energies. Also shown are the results of computer predictions, and lines with a slope of $-1·5$ corresponding to spherical blast wave theory. The observed slopes, varying

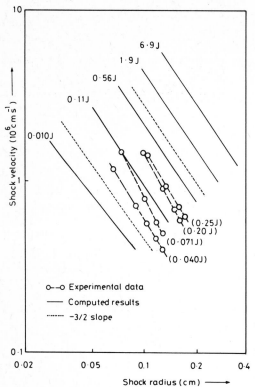

6.51 Radial variation of measured and calculated shock velocities for spherical plasmas in deuterium initially at a pressure of 1 atm. (After Mead (1970))

between -1.6 and -1.9, are in reasonable agreement with Taylor's theory and with the computations. Between 4 and 13% (with an average of 9%) of the laser pulse energy was absorbed by the plasma. Absorption only takes place after breakdown, which occurred close to the middle of the laser pulse, so the proportion of post-breakdown energy absorbed was nearer 20%.

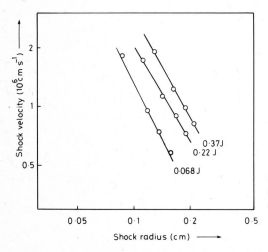

6.52 Measured shock velocities for spherical plasmas in hydrogen initially at a pressure of 1 atm (see text). (After Mead (1970))

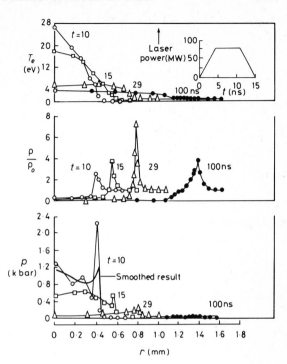

6.53 Computer calculations of electron temperature, mass density and pressure distributions, for spherical deuterium plasma heated by laser pulse shown (inset). Initial deuterium pressure is 1 atm. (After Mead (1970))

Similar results for hydrogen plasmas are shown in Figure 6.52: the velocities for the same energy and radius are close to $\sqrt{2}$ times those measured in deuterium, as is predicted by blast wave theory. Difficulties were experienced in obtaining satisfactory results with helium, but the available data were again in reasonable agreement with theory.

The computer code also provided predictions of other plasma parameters. Figure 6.53 shows the predicted electron temperature, mass density and pressure distributions at four different times after the start of the laser pulse. The plasma is essentially a hot, low-density bubble with a strong shock at the periphery, where the density is several times that of the undisturbed gas.

9.1.4. Laser power above 1 GW

A number of experiments have been reported in which plasmas were produced by gigawatt neodymium glass laser pulses focused in deuterium and in air. An extensive series of photographs of plasmas produced by gigawatt laser pulses in deuterium at several pressures was described by Veyrie (1968). These showed clearly how the development of the plasma in time is limited quite closely to the conical regions containing the laser beam. The neodymium glass laser produced pulses of 30 joules in 30 ns. As the gas pressure was increased from 300 Torr to 2 280 Torr the velocity in the cone furthest from the focusing lens became smaller and smaller as a result of the increasing absorption coefficient. A dark central region could be seen, probably corresponding to the region where the deuterium was fully ionized.

Bobin *et al.* (1968*a*) used a neodymium glass laser producing 10 joule pulses of 5 ns duration. Streak photographs showed three stages in the development of the plasma in deuterium. In the first stage the plasma propagates at a velocity consistent with the radiation-supported detonation wave model. After a few nanoseconds, however, the speed of propagation suddenly increases by a factor as great as 10, and can be as high as 8×10^7 cm s^{-1} for a time not longer than the duration of the laser pulse. During this second stage the luminous front is less sharply defined than in the first stage, and there is some evidence that light is not as strongly absorbed as it is in plasma produced by a longer laser pulse of equal total energy. Similar effects were observed in air. In the third stage, the plasma expands more slowly, behaving like an orthodox blast wave. Figure 6.54 shows measured velocities in deuterium at various initial pressures, compared with the maximum velocity attained using a 30 ns pulse of the same energy, and Figure 6.55 shows the second-stage velocity in air.

The second-stage velocity is so high that the volume of plasma created is not given sufficient energy to reach a very high temperature. At an initial pressure of 1190 Torr in deuterium (about 10^{20} atoms cm^{-3}), the plasma only absorbed about a half of the laser power, corresponding to a linear absorption coefficient of about 4 cm^{-1}. Assuming that the absorption is due to electron–ion inverse Bremsstrahlung, the formula for the absorption coefficient together with Saha's equation lead to values of only 2×10^4 K for the temperature and 5×10^{18} cm^{-3} for the electron density.

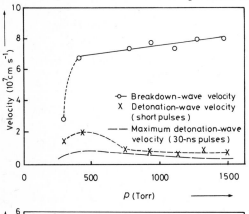

6.54   Velocities derived from streak photographs of plasmas formed in deuterium by 5 ns laser pulses, compared with maximum velocities observed using 30 ns pulses of the same energy. (After Bobin *et al.* (1968*a*))

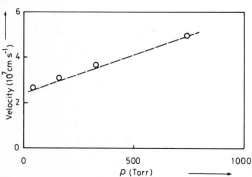

6.55   Second-stage velocities derived from streak photographs of plasmas formed in air by 5 ns laser pulses. (After Bobin *et al.* (1968*a*))

The radiation from such a plasma cannot sustain a radiation breakdown wave, so it is probable that the fast stage is due to an avalanche breakdown wave (see Section (2)). It is difficult to understand why the plasma should not become more highly ionized and consequently much hotter as the laser pulse continues after breakdown has occurred. Bobin et al. suggest that the avalanche process stops when the electron density becomes sufficiently high for the electron–electron thermalization time to be shorter than the reciprocal of the effective electron–atom collision frequency, v_{ea}. The thermalization time for electrons at a temperature T_e K is given (Spitzer (1962)) by

$$\tau_{ee} = \frac{0 \cdot 266 \, T_e^{\frac{3}{2}}}{n_e \ln \Lambda} \, \text{s}, \tag{6.80}$$

where $\ln \Lambda$ is about 9 and n_e is in cm^{-3}. For deuterium at 1190 Torr, v_{ea} is about $10^{13} \, \text{s}^{-1}$. If $T_e = 2 \times 10^4$ K, $v_{ea}\tau_{ee} = 1$ when $n_e \sim 10^{18} \, \text{cm}^{-3}$. Under these conditions non-linear effects may be important; the Bremsstrahlung absorption coefficient will fall with increasing irradiance above about $6 \times 10^{12} \, \text{W cm}^{-2}$, when $\eta \simeq 1$ (see Chapter 2, Section (8.1.3)). In any case, it is apparent that the occurrence of a breakdown wave leads to a large volume of cool plasma, and is therefore undesirable when a very high temperature plasma is required.

As has already been noted in Chapter 5, Section (5), the effects of aberrations in the lens focusing the laser beam can be quite important. Theoretical calculations of flux distributions in the focal region suggest that several separate breakdown regions may occur along the axis as the laser pulse power increases. This was demonstrated experimentally, for plasmas produced in deuterium using several different focusing lenses, in a set of streak photographs taken in plasma light by Veyrie et al. (1967).

Two-foil X-ray measurements were made by Vanyukov et al. (1968) on plasmas in air and in air–deuterium mixtures at 70 to 160 Torr, generated by 6 GW laser pulses of 20 ns duration. Electron temperatures between 200 and 400 eV were measured using combinations of beryllium, nickel and aluminium foils.

9.2. MODE-LOCKED PULSES

Because of the very short duration of a mode-locked ruby or neodymium pulse (typically 10^{-10} to 10^{-11} s) no measurements have yet been reported of the development of the plasma during the pulse. The breakdown of the gas has been discussed in Chapter 5; at the present time we can only add some information about the decay of the plasma after the pulse has ended.

Streak photographs were taken by Alcock & Richardson (1968) of plasmas produced in air, nitrogen and argon using a neodymium laser emitting a single pulse lasting only a few picoseconds, which was focused by a lens of 2 cm focal length. They showed a central bright core with a more or less constant diameter of 0·1 mm, which often appeared to be hollow. The core was surrounded by a luminous shell which expanded symmetrically in all directions at 2×10^8 cm s^{-1} for 1 ns and then shrank at 2×10^7 cm s^{-1}. Surrounding the shell was a weakly-luminous region which initially expanded at at least 4×10^8 cm s^{-1}, the fastest rate measurable with the streak camera, and decayed in a few nanoseconds. It was suggested that this outer region was due to photo-excitation or ionization by radiation from the core of the plasma. Similar results were obtained in all three gases. When the laser produced two pulses separated by 8 ns, the second pulse caused the central core to expand asymmetrically towards the laser, and its diameter increased by about 0·2 mm faster than could be measured. A new luminous shell and weak outer region developed.

Kaitmazov et al. (1968) obtained photographs of the plasma generated in air at S.T.P. by a train of mode-locked neodymium laser pulses separated by the laser cavity transit time, which could be varied between 4 and 33 nanoseconds. A discontinuous sequence of small plasmas was observed whose spacing depended on the energy in each spike and on the time interval between spikes. For time intervals greater than 15 ns the spacing of the breakdown points was independent of the spike interval, perhaps being limited by the volume capable of being ionized by the spike energy. For a spike of 0·1 J, the maximum volume ionized was about 0·5 mm^3. Figure 6.56 shows a time-integrated photograph of the plasma due to a similar laser, obtained by Andrews (1970). Photographs taken by Kaitmazov et al. in laser light scattered through 90° showed a series of small bright scattering points along the laser beam axis. Similar observations by Alcock et al. (1969a), which set an upper limit of 5 μm to the diameter of some of the scattering points, have already been mentioned in Chapter 5 as evidence for the occurrence of self-focusing.

Streak photographs of plasmas produced by trains of mode-locked pulses, obtained by Alcock et al. (1968d), showed that the arrival of a new pulse moved the luminous plasma boundary towards the laser by about 2 mm at an initial speed of at least

6.56   Time-integrated photograph of plasma formed in air at S.T.P. by train of mode-locked neodymium glass laser spikes. (Reproduced by courtesy of A. J. Andrews)

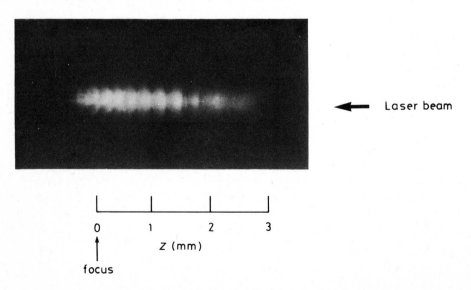

0   1   2   3

Z (mm)

focus

2×10^7 cm s^{-1}, after which the boundary moved relatively slowly until the next pulse arrived. Similar results with rather higher velocities were reported by Gorbenko *et al.* (1969); in these experiments the boundary of the region of initial breakdown was observed to move towards the laser at about 3×10^8 cm s^{-1}. Such a high velocity is not likely to result from a gas-dynamic process, and might again be due to photo-ionization by the hot plasma.

High-speed framing camera observations have been made by Richardson & Sala (1973) with exposure times of 5 ps, using an optical Kerr effect shutter and a swept image-converter camera. The mode-locked pulses forming the plasma also opened the Kerr effect shutter. By adjusting the lengths of the optical paths, photographs were obtained of the plasmas at the times when laser pulses arrived there. Very bright regions were observed at the plasma boundary nearest the laser in photographs taken up to 100 ns after breakdown. These are unlikely to have been due to directly reflected laser light since the S 11 photocathode of the image-converter was insensitive to the neodymium laser wavelength.

Marked departures from equilibrium were reported by Gudzenko *et al.* (1969) in plasmas generated by a train of mode-locked spikes in sodium vapour with $n = 10^{16}$ cm^{-3}. The $4d$–$3p$, $5d$–$3p$ and $6d$–$3p$ transitions appeared for a few nano-seconds after each spike following the second or third in a mode locked train, and were all much brighter than the usual yellow $3p$–$3s$ doublet. The effect was not observed in the emission from plasmas generated by 30 ns Q-switched pulses.

10. PLASMAS HEATED BY CARBON DIOXIDE LASERS

10.1. PLASMAS FORMED BY PULSED LASERS IN COLD GASES

Plasmas produced in air and other gases by 5 μs CO_2 laser pulses with an initial peak power of 1 MW were studied by Gravel *et al.* (1971). When the laser pulse was focused by a 12·5 cm focal length mirror in air at 200 Torr, the initial movement of the plasma boundary towards the mirror at 3×10^6 cm s^{-1} slowed down after about 1 μs, as may be seen from the streak photograph (Figure 6.57), and the region of maximum luminosity moved back towards the focus. A filament then appeared which expanded in the same direction as the incident radiation at a velocity of 10^6 cm s^{-1}. Similar results were obtained with other gases. With long laser pulses lasting several microseconds, significant density changes can occur as a result of expansion transverse to the beam axis. After the initial expansion of the plasma, radiation is not absorbed strongly in the plasma boundary nearest the mirror but penetrates progressively into the bulk of the plasma as the density falls. Eventually the laser beam passes through the plasma and is absorbed strongly in the boundary between the plasma and the neutral gas furthest away from the mirror.

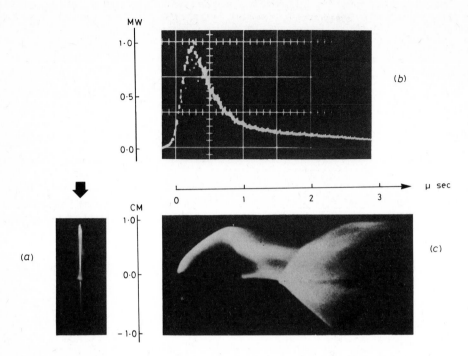

6.57 Plasma produced in air at a pressure of 200 Torr by CO_2 laser pulse (a) time-integrated photograph of plasma (laser light propagated downwards), (b) laser pulse stage, (c) streak photograph of plasma on same time scale as (b), and same displacement scale as (a). (After Gravel et al. (1971), by courtesy of A. J. Alcock)

A similar study was made by Tomlinson (1971) of plasmas generated in argon at 2500 Torr by a pulsed CO_2 laser of 200 kW peak power. The laser pulse power reached a peak after 200 ns with a half-width of about 300 ns, and then continued at about 50 kW for several microseconds. Streak photographs showed that the luminous plasma moved bodily towards the laser from the focus with an initial velocity of about 6×10^5 cm s^{-1}, then slowed down after 400–500 ns. The velocity of the boundary nearest the laser fell to the sound velocity (about 3×10^4 cm s^{-1}) whilst the boundary furthest from the laser travelled rapidly back to the focus and then continued to move away from the laser at the sound speed, so that a uniform expansion took place for several microseconds. No filamentary structure was reported in the plasma. Tomlinson found that the initial motion of the luminous plasma was consistent with detonation wave theory, and suggested that the wave was attenuated by the loss of energy to lateral expansion as the absorption length for laser light in the plasma increased (Raizer (1965a); see expression 6.43 above). The maximum temperature in the plasma was estimated from the initial expansion velocity to be not more than 4 eV.

The development of a filament of plasma travelling in the same direction as the laser light in air at 760 Torr was confirmed by Richardson & Alcock (1971c) using

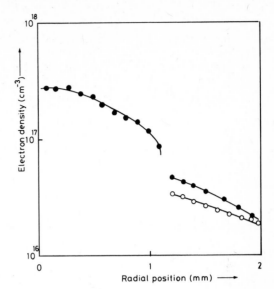

6.58 Electron density profile transverse to laser beam axis, in plasma formed in helium at 0·75 atm. by a CO_2 laser, 5 μs after laser pulse. Densities were obtained from Stark broadening: up to 1 mm, He II λ 4686 Å; above 1 mm, He I λ 6678 Å (dots) and He I λ 4713 Å (circles). (After George *et al.* (1971))

two-wavelength interferometry. The electron density in the main body of the plasma 1·7 μs after breakdown was only about 5×10^{17} cm^{-3} and was quite uniform, so the plasma transmitted most of the laser light, which was absorbed near the far boundary and initiated a new plasma expanding in the direction of propagation of the laser beam.

The high repetition rate of carbon dioxide lasers (typically 10–100 pulses per second) makes spectroscopic observations of the plasmas they produce particularly convenient. Very detailed studies have been made of the emission spectrum from plasmas produced in helium at 0·75 atm using 1 to 2 MW pulses (George *et al.* (1971): Ya'akobi *et al.* (1972)). The plasma density distribution, transverse to the laser beam axis at a time 5 μs after the end of the 500 ns laser pulse, was calculated from Stark broadening measurements and is shown in Figure 6.58. The presence of a low-density halo is of particular interest. The positions of the boundaries of regions emitting He I λ 5876 Å and He II λ 4686 Å were measured as functions of time (Figure

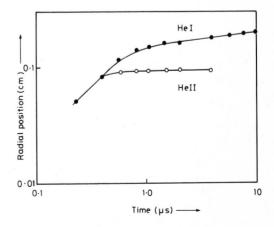

6.59 Positions of boundaries of regions emitting HeI (λ 5876 Å) and He II (λ 4686 Å) lines as functions of time. Plasma was generated by a CO_2 laser in helium at 0·75 atm. (After George *et al.* (1971))

6.59). Between 200 and 400 ns the two boundaries coincide and their position varies as $t^{0.7}$. Some care is needed in interpreting these results.

Plasmas were produced in hydrogen using 10 MW laser pulses by Offenberger & Burnett (1972). Streak photographs showed a similar general pattern to that observed with lower laser powers. At low pressures ($p < 100$ Torr) the initial velocity of the boundary towards the laser agreed well with detonation wave theory. At higher pressures the initial expansion velocity was rather higher than detonation wave theory predicted. This was ascribed to the occurrence of a breakdown wave during the fast-rising leading edge of the laser pulse. The results are compared with theory in Figure 6.60. The subsequent decay of the detonation wave and the appearance of a luminous front moving away from the laser occurred after several hundred nano-seconds when the laser power was still as great as 1 MW. An interpretation in terms of Raizer's criterion for energy to be expended in transverse expansion was not entirely satisfactory, particularly at low pressures. For hydrogen at 50 Torr the temperatures involved when the detonation wave decayed were in the region of 1·26 to 1·0 eV, and here a comparison with theory is difficult because the effective γ for hydrogen is varying rapidly, as may be seen from Figures 6.7 and 6.8. Measurements of the H_α emission line profile and the ratio of the line intensity to that of the adjacent con-tinuum for a gas pressure of 75 Torr gave an electron density of about 5×10^{17} cm^{-3} and an electron temperature of about 6 eV, at a time 600 ns after the start of the laser pulse. Some of the incident laser light was found to be reflected back into the f/7·5 focusing system at pressures above 100 Torr. The maximum proportion of the incident power reflected was 2%: this occurred at a pressure of 150 Torr.

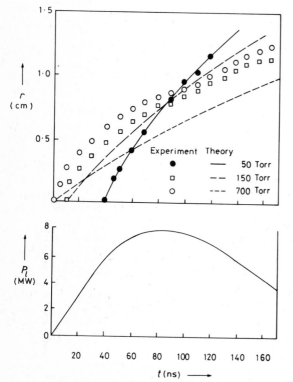

Experiment	Theory	
●	——	50 Torr
□	—— —	150 Torr
○	----	700 Torr

6.60 Axial position of luminous boundary of plasma produced in hydrogen, as function of time. Plasma was generated by a CO_2 laser pulse of shape shown. Theoretical curves assume a laser-driven detonation wave. (After Offenberger & Burnett (1972))

Large irradiances can be obtained by arranging for light within a laser cavity to pass through a focus. Plasmas were generated in this way in air by Tulip *et al.* (1971) and by Karlov *et al.* (1972). Karlov *et al.* noted self-modulation of the laser just below the breakdown threshold, leading to a spiky output lasting three times as long as the normal 10 μs laser pulse. This effect was ascribed to self-defocusing resulting from the development of the electron cascade at the focus to a density of (5 to 10) $\times 10^{-16}$ cm^{-3}, sufficient to defocus the beam but not high enough to cause substantial absorption.

Plasmas formed by carbon dioxide laser light falling on solid targets in gases are described in Chapter 8, Section (7.2).

10.2. HEATING OF EXISTING PLASMAS BY LASER RADIATION

A 0·2 J, 180 ns carbon dioxide laser pulse was focused into a helium θ-pinch plasma by Engelhardt *et al.* (1972). Spectroscopic measurements indicated that the laser pulse probably produced a slight increase in the electron temperature, which was in the region of 3·5 eV. The electron density was rather low (about 2×10^{17} cm^{-3}) compared to the critical density for λ 10·6 μm so the absorption coefficient was very small.

Heating of an argon arc with an even lower electron density (2×10^{16} cm^{-3}) was reported by Jassby & Marhic (1972) with a laser giving a 250 ns, 0·8 J pulse (followed by a low power tail). The electron temperature, deduced from relative line intensities in the A II spectrum, was in the region of 1·5 eV and was found to rise by 0·7 eV when the laser pulse was applied.

Chapelle *et al.* (1973) have described experiments in which c.w. laser light (500 W, focused to 10^5 W cm^{-2}) was passed through a glow discharge in helium at 0·2 Torr ($n_e \sim 10^{11}$ to 10^{12} cm^{-3}). The population of excited states was found to drop in the presence of the laser light, perhaps as a result of photo-ionization: the electron temperature did not rise.

10.3 PLASMAS SUSTAINED BY C.W. LASERS

If breakdown is produced in a gas by a high-power laser pulse, it is possible to sustain a plasma indefinitely with a very much lower power c.w. laser. The theory has been given by Raizer (1970*a, b*) and by Steverding (1972). Generalov *et al.* (1970*b*) obtained a steady plasma in xenon at pressures between 3 and 4 atm with a 150 W carbon dioxide laser focused by a mirror of 2·5 cm focal length. The initial breakdown was accomplished by a pulsed high-power CO$_2$ laser. The plasma, about 1 mm in length, lay between the focal plane and the laser, within the focal cone, and was quite stable. It emitted very intense white light. Subsequently Generalov *et al.* (1971) investigated plasmas produced in this way in argon and xenon by a 300 W laser. The pressure dependence of the minimum power needed to sustain the continuous plasma is shown in Figure 6.61 for a horizontal laser beam. For a given laser power, there is a

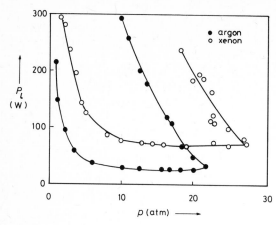

6.61 Minimum CO_2 laser power necessary to sustain a plasma in a horizontal laser beam, as function of gas pressure. (After Generalov *et al.* (1971))

maximum pressure above which the plasma cannot be maintained. This is probably the result of strong convective effects and perhaps the buoyancy of the hot, low-density plasma. No high-pressure limit was observed when the beam travelled vertically upwards to the focus. By adding 1 % of hydrogen to argon spectroscopic estimates of the electron density were obtained from H_β line profiles (this was not possible in xenon, where Xe I lines interfered). Between 4 and 16 atm the electron density was about 5×10^{17} cm^{-3}. Assuming thermal equilibrium, Saha's equation gave values of T_e of $(1 \cdot 5$ to $1 \cdot 3) \times 10^4$ K. At 2 atm, T_e rose to $2 \cdot 3 \times 10^4$ K.

Franzen (1972, 1973) used a pulsed CO_2 laser beam, collinear with the c.w. beam, to produce initial breakdown, and was able to sustain a plasma in argon at 4·4 atm with 150 W of c.w. laser power in the TEM_{11} mode, focused by a 5 cm focal length mirror. An intense axial core was observed in the luminous cone. The irradiance was 2×10^6 W cm^{-2}. In argon at 2.4 atm 300 watts were required, in good agreement with Generalov's results. Franzen also measured c.w. laser thresholds for maintaining plasmas in Kr and Xe.

A steady plasma was sustained in atmospheric air by Smith & Fowler (1973) with 2 kW of c.w. laser light. Breakdown was initiated by an electrical discharge. The plasma was stationary when a lens of not more than 14 cm focal length was used to focus the laser light: when a mirror with a 25 cm or longer focal length was used the plasma ran up the beam towards the laser with an initial speed of about 10 m s^{-1}, slowed, stopped for about 30 ms and vanished. The steady plasma was observed to have an intense core a few millimetres thick and about 1 cm long. About 60 % of the laser light was absorbed. The absorption coefficient had a maximum value of 0·6 cm^{-1}, from which an electron temperature of $1 \cdot 4 \times 10^4$ K was deduced. A luminous column of incandescent gas was seen extending for tens of cm above the plasma, in slow convective motion. These steady plasmas are of considerable practical interest (for example, as light sources) and are likely to receive much attention in future.

11. MAGNETIC EFFECTS

The effects of a steady external magnetic field on the development of laser-heated plasmas in gases have been investigated by several authors. Askaryan *et al.* (1965*b*) produced plasmas in a field of 10 kG and recorded the signal induced in a coil a few cm in diameter surrounding the plasma region. The generation of the plasma was accompanied by an expulsion of magnetic field, which did not return for several microseconds. Subsequent measurements (Askaryan *et al.* (1967*a*)) showed that the magnetic moment of the plasma was proportional to the laser energy absorbed. Further experiments on the diamagnetism of these plasmas were also reported (Askaryan *et al.* (1968, 1970)).

Vardzigulova *et al.* (1967) and Askaryan *et al.* (1972) studied plasmas generated in air in a strong magnetic field of $2 \cdot 1 \times 10^5$ gauss parallel to the laser beam axis. They found that the total light emission detected by a photocell increased by about 50% when the magnetic field was applied. Chan *et al.* (1968) investigated the effect of a 200 kG field, perpendicular to the laser beam axis, on plasmas in air, butane and helium at S.T.P. The duration of the plasma luminosity was increased somewhat by the presence of the field, particularly in helium. Streak photographs showed that in helium the rate of expansion was significantly reduced by the field, as may be seen in Figure 6.62, but the effect was small in air and imperceptible in butane. The difference in behaviour of these gases was ascribed to the much larger electron collision cross sections in butane and air as compared to helium, in which the mean free path under these conditions is roughly equal to the cyclotron radius.

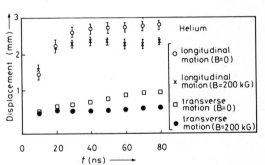

6.62 Positions of luminous boundaries of plasmas produced in helium at S.T.P. with and without a magnetic field perpendicular to the laser beam axis, as functions of time. The line of sight of the streak camera was parallel to **B**. (After Chan *et al.* (1968))

Chan *et al.* (1971*b*) also made measurements of the magnetic moments of plasmas in a steady field of up to 8 kG, and found the pressure dependence of the probe coil signals in propane, air, hydrogen and helium. The signal amplitude was always less than a tenth of that which would be expected in the absence of collisional diffusion of the plasma across the magnetic field, a result consistent with observations by Savchenko & Stepanov (1966).

Kaitmazov *et al.* (1971) pointed out that when a plasma in a magnetic field has expanded until the kinetic and magnetic pressures are in equilibrium, it will continue to expand significantly by diffusion across the field unless the penetration depth during the remainder of the plasma lifetime is small compared to its radius. The penetration depth depends on the conductivity, and hence on the electron temperature, of the plasma. Thus if diffusion is to be limited at equilibrium between kinetic and magnetic pressures the plasma temperature must still be high. Strong magnetic fields are therefore necessary. Kaitmazov *et al.* found that magnetic fields of 4×10^5 G produced significant effects on the geometry of plasmas generated in air at atmospheric pressure by 2 to 3 J Q-switched neodymium laser pulses or trains of mode-locked pulses. The expansion of the plasma was mainly in the direction of the magnetic field.

Fay & Jassby (1972) reported that when an argon arc plasma was heated by a pulsed carbon dioxide laser an axial magnetic field (up to 10 kG) reduced the radial heat losses significantly.

Laser-generated plasmas were found by Korobkin & Serov (1966) to have their own 'spontaneous' magnetic fields, which were ascribed to asymmetric motion of the shock front towards the laser. These authors found it necessary to surround the plasma with black paper to prevent spurious signals being generated in the search coils by the photo effect. They also took the precaution of collecting the signal from the search coil via a transformer with an earthed centre-tapped primary winding. The magnetic field was only detected during the time when the plasma was being heated by the laser pulse.

7

Theory of Plasmas Formed by Light Incident on Solids

1. INTRODUCTION

As we have seen in the previous chapter, the plasma formed by focusing laser light into a chamber containing a gas occupies a substantial volume, which increases if the laser pulse power or duration is increased, so that the extra energy is shared between more particles. In principle a small solid target in vacuum, which contains a fixed number of atoms, should be more readily heated to very high temperatures. However, the interaction between the light and the target now takes place initially in a thin surface layer with very steep density gradients, which may not absorb light efficiently.

The processes involved in the initial stages of generating a plasma from a solid target are complex. It is necessary to consider in turn the heating of the target while in the solid state, and the subsequent melting, vaporization and ionization. With high irradiances ionization can occur very rapidly and, since the energy required to ionize an atom is small compared to the energy needed to heat it to temperatures of some millions of degrees K, in many calculations the initial stages of plasma production from a solid target are ignored and a cold, fully-ionized plasma is assumed to exist with ion and electron densities equal to the solid atom density.

The dynamics of plasmas formed from solid targets are governed by the geometry of the target and of the laser beam, and of course by the time-dependence of the laser pulse power. Although early work was concerned with spherical targets we shall find it convenient to begin by considering plane targets. The introduction of spherical symmetry involves important changes in the flow equations which, as we shall see, may afford a solution to the problem of attaining the very high densities and temperatures needed to produce useful yields of thermonuclear energy.

Finally, we shall discuss some magnetic field effects associated with these plasmas.

2. INITIAL STAGES OF PLASMA FORMATION FROM SOLID TARGETS IN VACUUM

When a laser pulse is incident on an opaque solid target two extreme situations may be identified. At very low irradiances the light causes only a rise in temperature by conduction below the surface with no change of phase. If the irradiance is very great, multi-photon ionization takes place at the surface during a few cycles of the electric field. Between these extremes there is a wide variety of intermediate situations in which changes of phase, the pressure due to vaporization, thermionic emission and shock wave generation may be important. Not only the time dependence of the irradiance of the laser pulse but also the many thermal, optical and mechanical properties of the target material and their temperature and pressure dependence must be taken into consideration in any complete theoretical treatment of the interaction.

2.1. THERMAL CONDUCTIVITY CALCULATIONS

When light falls on an opaque material, some of it is reflected and the rest is normally absorbed by electrons in the conduction band, which are thus raised to higher energy states. Energy is then transferred from these electrons by collisions with other electrons and with phonons, so that the solid tends to reach an equilibrium at a higher temperature. In order that a single instantaneous temperature may be defined at every point, the light must not cause a substantial change in the internal energy of the region where absorption occurs during the relaxation times involved, which are of the order of 10^{-12} to 10^{-13} seconds. Provided this condition is satisfied it is possible to discuss classically the effects of thermal conductivity on the temperature distribution in the solid.

In the simplest situation at low irradiances, the target temperature never reaches the melting point, even at the surface, and reasonably accurate calculations may be performed using single values for the thermal conductivity and other parameters in the usual diffusion equation.

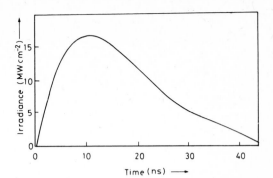

7.1.   Laser pulse shape assumed in calculating Figure 7.2. (After Ready (1965))

Ready (1965) carried out detailed calculations of the temperature rise due to the absorption of a light pulse of the form shown in Figure 7.1, using a one-dimensional model. Figure 7.2 shows the results for a semi-infinite copper target. As may be seen, the depth at which the temperature reaches a tenth of the surface value is only a few microns even after 40 ns. The rate of cooling is rapid, the surface temperature falling

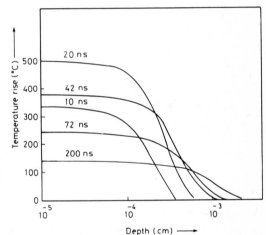

7.2   Temperature distributions in a semi-infinite copper target heated by a light pulse of shape shown in Figure 7.1. (After Ready (1965))

to a quarter of the maximum value in 200 ns. Such rapid changes in temperature are accompanied by sudden expansion processes which can generate shock waves in the target. These effects have been discussed by Steverding (1970).

A one-dimensional model of thermal diffusion is valid only provided the diameter of the laser beam at the surface is much greater than the penetration depth during the pulse. For most of the laser pulses used in plasma production this condition is obeyed. The more complicated heat transfer calculations necessary for situations encountered in welding and cutting applications using microsecond or millisecond pulses (see for example Ready (1971)) are outside the scope of this book.

More accurate calculations must take into account the temperature dependence of the optical properties of the target surface. This was done for metals on the basis of the Drude–Zener free-electron model by Libenson *et al.* (1968), who showed that the surface temperature begins to increase exponentially at constant flux density.

2.2. EFFECTS OF MELTING AND VAPORIZATION

The next stage in heating a solid target begins when the target temperature is raised to the melting point. In many applications, such as welding, this transition is of course very important. Regarded as a step towards the production of plasma, however, the transient liquid phase is not of great significance. The latent heat of fusion is small compared with the latent heat of vaporization or the ionization energy. Also, within the short duration of a high-power laser pulse the molten target material will not be displaced significantly. The most important consideration may well be the change in the optical characteristics of the surface. Bonch-Bruevich *et al.* (1968) measured the time dependence of the reflectivities of several metal surfaces heated by laser pulses, and deduced the relationship shown schematically in Figure 7.3 between the surface temperature and the absorptance of silver. The melting point occurs at the discontinuity in the diagram. It is apparent from these results that a marked increase in the rate of heating by light of constant irradiance must occur when surface melting takes place.

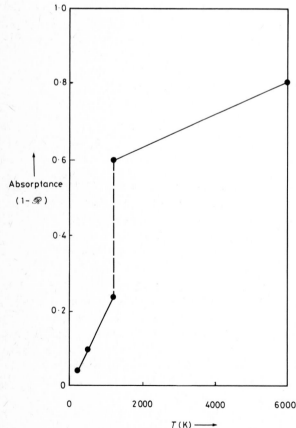

7.3 Temperature dependence of absorptance of a silver surface, showing discontinuity near melting point. (After Bonch-Bruevich *et al.* (1968))

When the boiling point of the target is reached more complex effects arise. If the irradiance is not too high, the rate of evaporation will not be very great, and the tenuous vapour layer evolved from the molten or solid surface will not absorb light strongly. This situation was investigated by Anisimov *et al.* (1966). At higher ir-radiances the initial rate of evaporation will be greater, and the resulting vapour density may become sufficiently large for the bulk of the incident flux to be absorbed in the region occupied by the vapour phase.

When considering vaporization it is important to remember that the boiling point of a material is pressure-dependent. The radiation pressure due to a focused laser beam absorbed or reflected by a surface can be very great; for example, a pulse of 100 MW peak power absorbed in an area of 10^{-4} cm^2 exerts a peak pressure of 300 atmospheres. Askaryan & Moroz (1962; see also Hughes (1964)) pointed out that, when evaporation is taking place, even greater pressures arise due to the recoil on the surface from departing particles. If I is the effective irradiance, v the final velocity of the vapour flow and W_v the energy of vaporization and acceleration per unit mass, the recoil pressure is given approximately by

$$p \simeq \frac{Iv}{W_v}.$$
(7.1)

This pressure, known as the *ablation pressure*, is much greater than the radiation pressure—by a factor of 10^4 or more. The boiling point (and indeed many other parameters) of the target material under these extreme conditions may only be es-timated at present. Optical properties may be strongly affected in some materials by transitions from metallic to non-metallic behaviour, or *vice versa* (see for example Mott & Davis (1971), Chapter 5).

A detailed discussion of the dynamics of the vaporization process has been given by Krokhin (1971).

2.3. THERMAL IONIZATION

The temperature in the hottest region of the vapour generated by the laser pulse will rise as long as the local rate of absorption of energy from the beam is sufficiently great to overcome cooling by expansion and conduction (at this stage radiative cooling may be neglected). Eventually, the vapour may reach a sufficiently high temperature for a significant number of atoms to be ionized by collisions (the possibility of multi-photon ionization is discussed below). If the vapour is still sufficiently dense for thermodynamic equilibrium to hold, we may use Saha's equation to relate the den-sities n_e, n_i and n_a of electrons, ions and neutral atoms at a temperature T:

$$\frac{n_e n_i}{n_a} = \frac{2\mathscr{U}_i}{\mathscr{U}_a}\left(\frac{2\pi m_e \kappa_B T}{h^2}\right)^{\frac{3}{2}} \exp\left(\frac{-\chi}{\kappa_B T}\right).$$
(7.2)

Here $\mathscr{U}_i$, $\mathscr{U}_a$ are the partition functions for the singly-ionized and neutral atoms respectively (the partition function is the average value of the degeneracy g over all states, weighted according to their populations: $\mathcal{U}_i/\mathcal{U}_a$ nearly always lies between $\frac{1}{6}$ and 2), and χ is the ionization energy of the neutral atom. In the early stages of ionization, only singly-ionized atoms are present, so $n_e = n_i$.

The strong absorption occurring in a fully-ionized gas is due to free–free transitions of electrons in collision with positive ions (inverse Bremsstrahlung). At the very low electron densities and high neutral atom densities occurring in a slightly-ionized vapour, electrons are more likely to absorb photons during free–free transitions in collisions with neutral atoms, which though less effective are more frequent (see Chapter 2, Section (8.1.4)). The presence of even a small proportion of free electrons causes a marked increase in the absorption coefficient of a gas, and hence an increase in the rate of heating, leading to a greater degree of ionization, and so on. Electron–ion inverse Bremsstrahlung absorption soon becomes the dominant heating process. On the assumption that this stage has been reached, let us now calculate the rate of heating of a plasma. From Chapter 2, expression (2.83), the net absorption coefficient for inverse Bremsstrahlung at frequency ω may be written as

$$a_\omega = \frac{n_e^2 e^6 \bar{g}}{6\varepsilon_0^3 ch\omega^3 m_e^2}\left(\frac{m_e}{6\pi\kappa_B T}\right)^{\frac{1}{2}}\left[1 - \exp\left(-\frac{\hbar\omega}{\kappa_B T}\right)\right]. \tag{7.3}$$

Here we have assumed that the dispersion term is close to unity (because the electron density is well below the critical value), that no non-linear effects need be considered, and that the positive ions are all singly charged. Stimulated emission has been allowed for.

We next insert into (7.3) the value of n_e^2 found from Saha's equation (7.2) with the result that

$$a_\omega = \frac{n_a e^6 \bar{g}}{12\sqrt{3}\pi^2 \varepsilon_0^3 ch^4 \omega^3}\frac{\mathcal{U}_i}{\mathcal{U}_a}\kappa_B T \exp\left(-\frac{\chi}{\kappa_B T}\right)\left[1 - \exp\left(-\frac{\hbar\omega}{\kappa_B T}\right)\right]. \tag{7.4}$$

Suppose now the laser beam irradiance is I watts cm^{-2}. The power absorbed per unit area of the beam in a depth l of plasma will be $I[1 - \exp(-a_\omega l)]$: provided $l \ll 1/a_\omega$ it will be approximately $Ia_\omega l$. The mean absorption rate per unit volume in this region is therefore Ia_ω watts cm^{-3}. Provided the degree of ionization is not too large, the total number of particles per unit volume will not greatly exceed n_a. Thus in thermal equilibrium the mean power absorbed per particle, P, is Ia_ω/n_a. As may be seen from (7.4), P is independent of density and increases as the temperature rises, as long as the degree of ionization remains low. This result was obtained by Archbold et al. (1964). Figure 7.4 shows the quantity $(a_\omega/n_a)\mathcal{U}_a/\mathcal{U}_i\bar{g}$ cm^2 plotted as a function of temperature for four values of the ionization energy χ. The rate of temperature rise due to this process is given by $Ia_\omega/n_a\kappa_B$ K s^{-1}, κ_B being Boltzmann's constant in joules K^{-1}. As an example, consider a vapour with an ionization energy of 8 eV at a temperature of 6000 K. Taking $\mathcal{U}_a/\mathcal{U}_i\bar{g} \simeq 1$, a_ω/n_a is about 10^{-23} cm^{-2}, so in this case the rate of temperature rise in K s^{-1} is numerically roughly equal to the irradiance in watts cm^{-2}.

As the temperature rises, the rate of heating increases initially. However, as full ionization is approached the curves of Figure 7.4 are no longer valid: the power

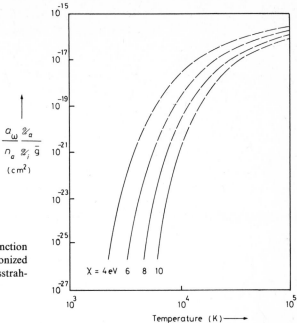

$$\frac{a_\omega}{n_a} \frac{\mathcal{Z}_a}{\mathcal{Z}_i} \frac{1}{g}$$

(cm^2)

7.4 Temperature dependence of function determining rate of heating of slightly-ionized plasma by electron–ion inverse Bremsstrahlung. (After Archbold et al. (1964))

X = 4eV   6   8   10

Temperature (K) ⟶

absorbed per particle begins to fall with increasing temperature and becomes density dependent.

The initial ionization of a small spherical hydrogen target was considered by Martineau & Tonon (1969a), who took into account the effects of three-body recombination, which is the dominant recombination process at high densities.

2.4. MULTI-PHOTON IONIZATION OF SOLIDS

The theory of multi-photon ionization in gases has been discussed in Chapter 5. The very high particle densities occurring in solids, and in vapour layers generated by intense light, will cause strong Stark broadening, thereby favouring quasi-resonant processes as well as depressing the ionization energy. Calculations have been made of the multi-photon photocurrent from metals, on the Sommerfeld model, by Silin (1970). Some experimental observations with a gold target have been reported by Farkas et al. (1972).

At very high irradiances tunnelling theory (Chapter 5, Section (2)) shows that ionization can be extremely rapid. With neodymium or ruby laser light ionization occurs within one cycle if $I \gtrsim 10^{15}$ W cm^{-2}.

The stimulated Raman effect induced by laser light in a dense medium can produce anti-Stokes photons of higher frequency. A sequence of similar processes can generate high-order anti-Stokes photons of considerably greater energy than that of the original laser photons. Mennicke (1971a) showed experimentally that ruby laser light at λ 6943 Å focused to an irradiance of about 10^{12} W cm^{-2} in solid hydrogen, deuterium or nitrogen produces anti-Stokes lines up to at least the seventh order in hydrogen, the tenth in deuterium and the twelfth in nitrogen, all these lying in the region of λ 2300 Å. Excitation followed by ionization is to be expected with photons of this wavelength: if even higher order photons are produced, direct single-photon ionization may occur.

3. REFLECTIVITY OF A PLASMA BOUNDARY

Light of frequency ω approaching a uniform semi-infinite plasma in which the electron density exceeds the critical value

$$n_{ec} = \omega^2 \frac{m_e \varepsilon_0}{e^2} \tag{7.5}$$

will be reflected. For neodymium laser light ($\lambda = 1.06 \,\mu$m), $n_{ec} \simeq 10^{21}$ cm^{-3}, while for carbon dioxide laser light ($\lambda = 10.6 \,\mu$m) $n_{ec} \simeq 10^{19}$ cm^{-3}. Assuming that the plasma frequency ω_{pe} is much greater than ω and that ω is much greater than the effective collision frequency ν^*_{ei} (see expression 2.85) the reflection coefficient is given (Kidder (1971)) by

$$\mathscr{R} = \left[1 + \frac{2\nu^*_{ei}}{\omega_{pe}} \right]^{-1}, \tag{7.6}$$

while the penetration depth is given by

$$d = \frac{c}{2\omega_{pe}}. \tag{7.7}$$

Figure 7.5 shows how the absorptances of fully-ionized deuterium, lithium deuteride, carbon and aluminium, for light at the neodymium laser wavelength, vary with

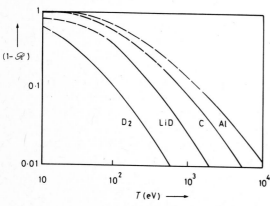

7.5   Absorptances at λ 1.06 μm of surfaces of fully-ionized materials, at their normal solid densities, in vacuum, as functions of temperature. (After Kidder (1971))

temperature, assuming that the materials remain at their normal solid densities.

The boundary of such an 'overdense' plasma in a vacuum is not in reality discontinuous, and incident light will pass through layers of lower electron density, where it will be to some extent absorbed, both before and after being reflected. We shall calculate the overall reflectivity of the plasma for light of low irradiance such that nonlinear effects may be ignored, following Dawson *et al.* (1968). The problem has also been discussed by Mulser (1970).

Let us consider a plane wave of irradiance I incident perpendicularly in the z direction on the plane boundary of an overdense, fully-ionized plasma. The radiation will propagate into the plasma (whose density we shall assume to increase monotonically from zero), being absorbed as it goes, until at some depth z_c it reaches the layer of critical density. Here it will be reflected, and on its way outwards will again suffer absorption, finally emerging with irradiance I_r. We may define the effective overall reflectivity of the plasma boundary, $\mathcal{R}$, as I_r/I. We may write, for radiation of frequency ω,

$$\mathcal{R}_\omega = \left(\frac{I_r}{I}\right)_\omega = \exp\left[-2\int_{-\infty}^{z_c} a_\omega(z)\,dz\right]$$

$$= \exp[-2\mathcal{D}], \tag{7.8}$$

$\mathcal{D}$ being the optical depth of the plasma up to the critical density layer.

To get an estimate of the value of $\mathcal{R}_\omega$ we suppose that the plasma has a uniform temperature, and that the electron density varies linearly with z, being zero when $z = 0$. Then from expressions (2.42) and (1.41)

$$a_\omega(z) = \mathcal{L}\left[\frac{n_e(z)}{\omega}\right]^2\left[1 - \left(\frac{\omega_{pe}(z)}{\omega}\right)^2\right]^{-\frac{1}{2}}, \tag{7.9}$$

where $\mathcal{L}$ is a constant. Now

$$\left[\frac{\omega_{pe}(z)}{\omega}\right]^2 = \frac{n_e(z)}{n_{ec}} = \frac{z}{z_c}, \tag{7.10}$$

so

$$\mathcal{D} = \int_0^{z_c} a_\omega(z)\,dz = \mathcal{L}\left(\frac{n_{ec}}{\omega}\right)^2\int_0^{z_c}\left(\frac{z}{z_c}\right)^2\left(1 - \frac{z}{z_c}\right)^{-\frac{1}{2}}\,dz$$

$$= \tfrac{16}{15}\mathcal{L}\left(\frac{n_{ec}}{\omega}\right)^2 z_c. \tag{7.11}$$

Clearly, the effective reflectivity depends very strongly on the plasma electron density profile assumed: if the density initially rises rapidly and then slowly approaches a maximum just equal to n_{ec}, the absorption will be significantly greater than in the case of a linear density variation over the same distance.

The frequency of the incident light is also very important. This may be illustrated by considering radiation from two lasers, of frequencies ω_1 and ω_2 entering uniform-temperature plasmas with a constant electron density gradient. The effective reflectivities are

and

$$\left.\begin{aligned}\mathscr{R}_{\omega_1} &= \exp\left[-\tfrac{32}{15}\mathscr{L}\left(\frac{n_{ec_1}}{\omega_1}\right)^2 z_{c_1}\right]\\[2mm]\mathscr{R}_{\omega_2} &= \exp\left[-\tfrac{32}{15}\mathscr{L}\left(\frac{n_{ec_2}}{\omega_2}\right)^2 z_{c_2}\right].\end{aligned}\right\}$$ (7.12)

Since the density gradient is constant,

$$\frac{z_{c_1}}{z_{c_2}} = \left(\frac{\omega_1}{\omega_2}\right)^2$$ (7.13)

so

$$\frac{\ln(\mathscr{R}_{\omega_1})}{\ln(\mathscr{R}_{\omega_2})} = \left(\frac{\omega_1}{\omega_2}\right)^4 = \left(\frac{\lambda_2}{\lambda_1}\right)^4.$$ (7.14)

Thus if, for example, we compare the reflectivities of a plasma boundary (with a linearly increasing electron density) for light from a carbon dioxide laser ($\lambda = 10\cdot6\ \mu m$) and from a neodymium glass laser ($\lambda = 1\cdot06\ \mu m$) we find that

$$\mathscr{R}_{Nd} = (\mathscr{R}_{CO_2})^{10^4},$$ (7.15)

so a plasma which is almost perfectly reflecting for CO_2 laser radiation, with $\mathscr{R}_{CO_2} = 0\cdot999$, will reflect scarcely any neodymium laser light, the corresponding value of $\mathscr{R}_{Nd}$ being about 10^{-4}. From these considerations it seems that short-wavelength lasers are preferable for plasma heating.

Shearer (1971) discussed the reflectivity of the boundary for light falling on it obliquely, at an angle of incidence ϕ_i to the normal, and found that

$$\frac{I_r(\phi_i)}{I} = \exp(-2\mathscr{D}\cos^5\phi_i).$$ (7.16)

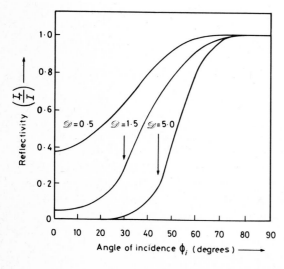

7.6 Reflectivity of plasma boundary as function of angle of incidence for three values of $\mathscr{D}$, the optical depth at normal incidence up to the critical density layer. (After Shearer (1971))

Figure 7.6 shows the reflectivity as a function of ϕ_i for three values of $\mathscr{D}$. The laser beam is refracted away from the immediate vicinity of the critical density layer where most of the absorption takes place, and despite the longer path length in the plasma the reflectivity increases with the angle of incidence. In these calculations optical resonance absorption, discussed in Section (10.2) of Chapter 2, was not taken into account. This effect reduces the reflectivity, as was shown by Freidberg *et al.* (1972) who considered a plasma with a linear density gradient rising from $n_e = 0$ at $z = 0$ to $n_e = n_c$ at $z = z_c$ where $\omega_{pe} = \omega_0$. Figure 7.7 shows the power absorption coefficient as a function of ϕ_i, $k_0 = 2\pi/\lambda_0$ and z_c. Collisional effects were included by Mueller (1973) using the normal value of v_{ei} at low irradiances and the non-linear

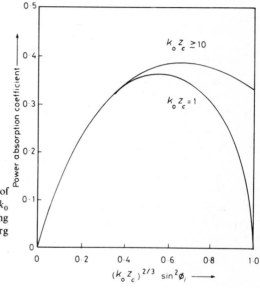

7.7 Power absorption coefficient as function of angle of incidence ϕ_i, vacuum wave number k_0 and depth of critical density surface, z_c, allowing for optical resonance absorption. (After Freidberg *et al.* (1972))

expression corresponding to (2.90) for $\eta > 1$. The results of the computation are shown in Figure 7.8. Here z_c is in units of the incident laser wavelength λ_0. Both neodymium and CO_2 lasers are considered. The results in (a), (b), (d) and (e) are for low irradiances ($\eta \ll 1$). The absorption becomes very strong over a narrow angular range for the high irradiances of (c) and (f); however, for irradiances three times greater than these values the peak absorptances fall to about 0·5.

At high irradiances the other non-linear effects discussed in Chapter 2 become important, and the critical density region of the plasma may become strongly absorbing. Other reflection mechanisms may also occur at high irradiances. Shearer & Duderstadt (1973) calculated the fraction of the incident energy absorbed by deuterium initially at solid density, for laser pulses over a wide range of wavelengths, allowing for the parametric ion-acoustic instability. The computer programme took into account the expansion of the hot plasma formed at the target surface, which will be discussed in later sections of this chapter. The results for 4 ns and 100 ps pulses, shown

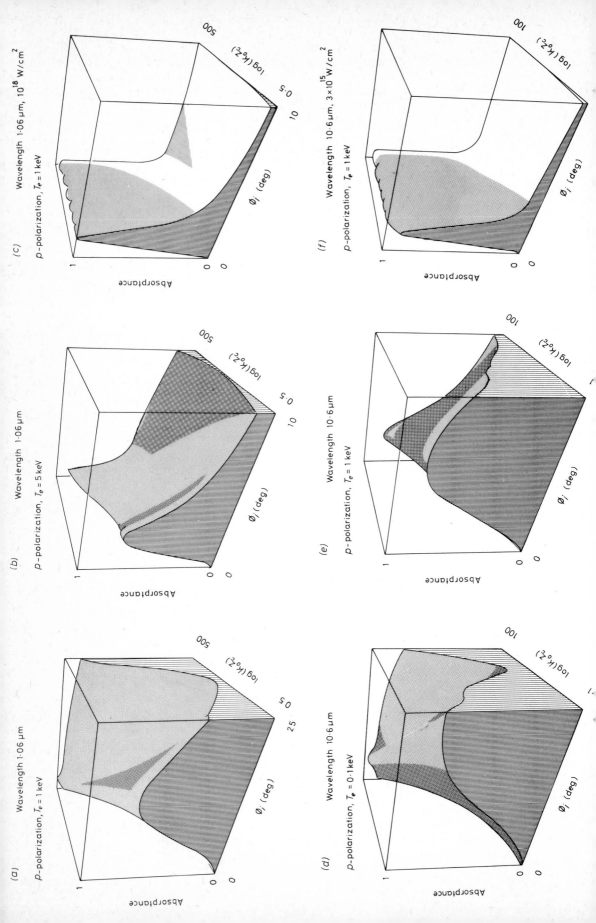

(c) Wavelength 1·06 μm, 10^{18} W/cm^2
p-polarization, $T_e = 1$ keV
log (k_0^2/z)
500
0·5
1·0
ϕ_i (deg)
Absorptance
0
1

(b) Wavelength 1·06 μm
p-polarization, $T_e = 5$ keV
log (k_0^2/z)
500
0·5
1·0
ϕ_i (deg)
Absorptance
0
1

(a) Wavelength 1·06 μm
p-polarization, $T_e = 1$ keV
log (k_0^2/z)
500
0·5
2·5
ϕ_i (deg)
Absorptance
0
1

(f) Wavelength 10·6 μm, 3×10^{15} W/cm^2
p-polarization, $T_e = 1$ keV
log (k_0^2/z)
100
ϕ_i (deg)
Absorptance
0
1

(e) Wavelength 10·6 μm
p-polarization, $T_e = 1$ keV
log (k_0^2/z)
100
ϕ_i (deg)
Absorptance
0
1

(d) Wavelength 10·6 μm
p-polarization, $T_e = 0·1$ keV
log (k_0^2/z)
100
ϕ_i (deg)
Absorptance
0
1

On facing page   7.8   Absorptance as function of angle of incidence ϕ_i, vacuum wave number k_0 and depth of critical density surface, z_c, for Nd and CO_2 lasers, allowing for optical resonance, collisional effects and non-linear inverse Bremsstrahlung. (After Mueller (1973))

in Figures 7.9(a) and (b) respectively, indicate that short-wavelength laser light is more readily absorbed, particularly when delivered in short pulses.

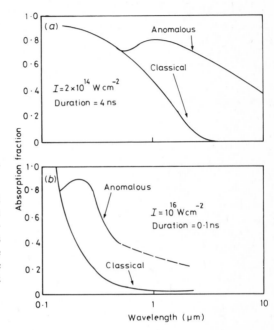

7.9   Calculations of fraction of incident light energy absorbed when plasma is formed from a solid deuterium target, as function of wavelength, for (a) 4 ns, 2×10^{14} W cm^{-2} pulses and (b) 100ps, 10^{16} W cm^{-2} pulses. The 'anomalous' curves allow for the parametric ion-acoustic instability. (After Shearer & Duderstadt (1973))

4. PLASMAS FORMED FROM SEMI-INFINITE SOLID TARGETS

We now investigate the development of the plasma formed by a laser pulse incident on the surface of a massive solid target. As the laser irradiance increases the nature of the interaction changes. Several possible regimes may be distinguished.

(1)   At low irradiances ($I \lesssim 10^8$ W cm^{-2}) the rate of evaporation is relatively small and initially the tenuous vapour is only very slightly absorbing. For short pulses the distributions of density and laser irradiance along the normal to the target surface are as sketched in Figure 7.10(a). The theory has been discussed by Basov *et al.* (1966). For pulses lasting several microseconds Vilenskaya & Nemchinov (1969) predict that

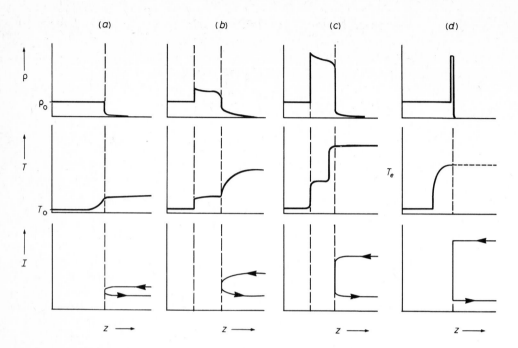

7.10 Sketches showing profiles of density, temperature and incident and reflected light irradiance for solid targets (see text).

the effects of increased absorption in an outer layer of evaporated material, condensed as a result of adiabatic cooling, can lead to complicated pressure oscillations in the vapour. The effects of thermal conductivity in the vapour under these conditions have been discussed by Volosevich & Levanov (1970).

(2) As the irradiance is increased the vapour will become ionized and optically moderately thick, and will tend to shield the target surface. However, if the fraction of the laser flux reaching the surface is thus reduced the rate at which the target material is evaporated is also reduced, so as expansion proceeds the optical density falls, allowing a larger proportion of laser light to penetrate to the surface and generate a denser vapour. The balance between these effects leads to a 'self-regulating' regime, first discussed by Krokhin (1964, 1965). A shock wave may be created in the solid by the ablation pressure, and a dense layer of cool material develops near the surface, which increases in thickness as time proceeds (Figure 7.10(b)).

(3) At even higher irradiances the plasma is generated and heated to a very high temperature in a thin absorbing layer at the surface of the dense region and has only a small optical thickness, so the self-regulating process no longer occurs (Figure 7.10(c)). The absorbing layer may be treated as a laser-heated deflagration front moving into the target behind the shock front (Kidder (1968), Fauquignon & Floux (1970)).

(4) At very high irradiances non-linear thermal conduction by electrons becomes important. If the rise time of the laser pulse is sufficiently short a region of high temperature extends into the target before any significant flow of target material can take place (Zeldovich & Raizer (1966) (Figure 7.10(d)).

The first regime does not produce very hot plasmas and will not be considered further: we shall analyse the last three regimes in some detail.

4.1. SELF-REGULATING PLASMA EVOLVED FROM SHOCK-COMPRESSED SURFACE LAYER

Here we shall follow the analysis given by Caruso & Gratton (1968a). Other treatments have been given by Afanas'ev et al. (1966a, b), Caruso et al. (1966), Afanas'ev & Krokhin (1967) and Krokhin (1971).

Figure 7.11 shows schematically the model adopted. Assuming that we are concerned with processes taking place near the focus of the laser beam, within the Gaussian waist, a one-dimensional treatment is appropriate for short pulses (Nemchinov (1967)). The solid target has an initial density ρ_0, and the velocity of the shock

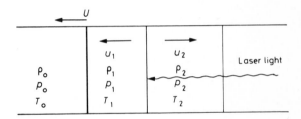

7.11 Self-regulating plasma produced from laser-heated solid (see text)

wave propagating within it, away from the light source, is U in the laboratory frame of reference. The strongly absorbing layer, a partially-ionized plasma in which $n_e \simeq n_{ec}$, has a density ρ_1. The plasma, which is assumed to be fully ionized, is described by its mean density ρ_2, its temperature T_2 (we assume $T_e = T_i$) and its mean velocity of expansion from the surface of the dense layer u_2 (towards the light source, so u_2 is negative). It is assumed that $\rho_2 \ll \rho_0$; that the plasma frequency is always less than the frequency of the incident light; and that the energy needed to ionize unit mass of the target, W_I, is small compared to the total energy per unit mass supplied to it. The mean velocity of expansion of the plasma is taken to be equal to the isothermal sound velocity, i.e.

$$u_2^2 = (n_e + n_i)\frac{\kappa_B T_2}{\rho_2}$$

$$= \frac{(1+Z)\kappa_B T_2}{m_i}, \tag{7.17}$$

m_i being the ion mass. A single species of ion of charge equal to Z proton charges has been assumed; Green (1970) has shown that this does not lead to serious errors.

The irradiance I, which is assumed not to vary with time, is absorbed in a characteristic length $1/a_\omega$. Reflection is ignored. From expression (2.84) of Chapter 2 we find that we may write

$$a_\omega = \mathcal{B}\rho_2^2|u_2|^{-3} \qquad (7.18)$$

where $\mathcal{B}$ is a constant.

Dimensional arguments are now used to relate the quantities ρ_2 and u_2 to the dimensionally independent parameters $\mathcal{B}, I$ and the time t. Any dimensionless quantity must be regarded as time-independent, because no dimensionless quantity can be formed from $\mathcal{B}, I$ and t. Thus the ratio of the thickness of the plasma l_2 to the absorption length $1/a_\omega$ must be a constant, i.e.

$$l_2 a_\omega = c_0. \qquad (7.19)$$

Also it is found that

$$u_2 = c_1(\mathcal{B}t)^{\frac{1}{8}}I^{\frac{1}{4}} \qquad (7.20)$$

and

$$\rho_2 = c_2(\mathcal{B}t)^{-\frac{3}{8}}I^{\frac{1}{4}}, \qquad (7.21)$$

where c_1 and c_2 are constants to be determined.

For the total mass of solid material converted into plasma per unit area of surface up to time t we have

$$M' = \int_0^t \rho_2|u_2|\,dt = |c_1|c_2\mathcal{B}^{-\frac{1}{4}}t^{\frac{3}{4}}I^{\frac{1}{2}}. \qquad (7.22)$$

Also the pressure in the plasma

$$p_2 = \rho_2 u_2^2 = c_1^2 c_2(\mathcal{B}t)^{-\frac{1}{8}}I^{\frac{3}{4}}. \qquad (7.23)$$

It is possible to relate c_1 and c_2 to c_0. At time t,

$$l_2 = \int_0^t |u_2|\,dt = \tfrac{8}{9}|u_2|_t t. \qquad (7.24)$$

Thus

$$c_0 = l_2 a_\omega = \tfrac{8}{9}|u_2|_t a_\omega$$
$$= \tfrac{8}{9}\left(\frac{c_2}{c_1}\right)^2. \qquad (7.25)$$

If the energy delivered to the solid is ignored compared to that delivered to the plasma, the power balance equation is approximately

$$I = \frac{d}{dt}(\rho_2 u_2^2 l_2)$$

$$= u_2^2 \frac{d}{dt}(\rho_2 l_2) + \rho_2 l_2 \frac{d}{dt}(u_2^2) \tag{7.26}$$

$$= \tfrac{8}{9} c_1^3 c_2 I,$$

so

$$c_1^3 c_2 = \tfrac{9}{8}. \tag{7.27}$$

Then from (7.25) and (7.27)

$$\left. \begin{array}{l} c_1^8 = \dfrac{9}{8c_0} \\[3mm] c_2^8 = \left(\dfrac{9}{8}\right)^5 c_0^3 \end{array} \right\} \tag{7.28}$$

On physical grounds, c_0 is unlikely to be very far from unity. If $c_0 \ll 1$ the plasma will be transparent, and the laser energy will be largely available for producing more evaporation, causing the plasma density, and hence the absorption coefficient, to increase. If $c_0 \gg 1$ the plasma will be almost opaque and little further evaporation can occur, but continued heating and expansion of the plasma will cause the absorption coefficient to decrease. We have therefore a self-regulating system. In view of the large powers of c_1 and c_2 in (7.28) it is reasonable to take $c_1 \simeq c_2 \simeq 1$.

Let us now consider the shock wave generated in the solid. As in Section (3) of Chapter 6, we write down the equations of conservation of mass and momentum to relate conditions on each side of the shock front:

$$-\rho_0 U = \rho_1(u_1 - U) \tag{7.29}$$

$$\rho_0 U^2 = \rho_1(u_1 - U)^2 + p_1. \tag{7.30}$$

The initial pressure in the unperturbed solid p_0 has been ignored compared to the pressure p_1 driving the shock wave. Also, $p_1 = p_2$ depends so weakly on the time (expression (7.23)) that it may be regarded as a constant.

Solving (7.29) and (7.30) we obtain

$$u_1^2 = \left(1 - \frac{\rho_0}{\rho_1}\right)\frac{p_1}{\rho_0} \tag{7.31}$$

and

$$U^2 = \left(\frac{\rho_1}{\rho_1 - \rho_0}\right)\frac{p_1}{\rho_0}. \tag{7.32}$$

The velocity u_1 of the dense surface layer is due almost entirely to the pressure exerted by the expanding plasma. The velocity V_a with which the surface retreats into the shock-compressed solid as the result of ablation alone is given by

$$V_a = \frac{M'}{\rho t} = \frac{1}{\rho_0}(\mathscr{B}t)^{-\frac{1}{4}}I^{\frac{1}{2}} = \frac{\rho_2}{\rho_1}u_2, \tag{7.33}$$

so

$$\frac{V_a^2}{u_1^2} = \frac{\rho_0\rho_2}{\rho_1^2}\frac{U}{u_1} \ll 1. \tag{7.34}$$

None of the laser light actually penetrates the dense surface layer, so the power delivered to it per unit area is just $p_1|u_1| \simeq I|u_1/u_2|$ from (7.20) and (7.23). Thus, making use of (7.31), we find that the fraction of the laser pulse power reaching the dense layer is less than $(\rho_2/\rho_0)^{\frac{1}{2}}$ which is normally very small. Given a sufficiently large irradiance, the resulting shock wave can nevertheless produce strong ionization in the solid, provided

$$p_1|u_1| > \rho_0 U W_I. \tag{7.35}$$

A necessary, though not sufficient, condition for this to be true is that $I > \rho_0 u_2 W_I$, i.e. $I > (\mathscr{B}t)^{\frac{1}{2}}(\rho_0 W_I)^{\frac{3}{2}}$. Caruso & Gratton estimated that this model is valid for pulses lasting a few nanoseconds with $10^9 \leqslant I \leqslant 10^{14}\,\mathrm{W\,cm^{-2}}$ for hydrogen isotope targets. In this range the radiation pressure is small compared to p_1.

Detailed computer calculations based on a similar model were reported by Kidder (1968), who found good agreement with the predictions of expression (7.23) for the pressure induced in deuterium by light fluxes of 10^{13} to $10^{14}\,\mathrm{W\,cm^{-2}}$. Mulser & Witkowski (1968), using another very similar model, also carried out numerical calculations on the plasma produced from a block of solid hydrogen by ruby laser light of irradiance $10^{12}\,\mathrm{W\,cm^{-2}}$. Their results are given in Figure 7.12, which shows how the temperature distribution and the laser irradiance in the plasma vary with time, and Figure 7.13 which shows the depth of ablation ($\int_0^t V_a\,dt$) as a function of time. More detailed calculations were made by Mulser et al. (see Büchl et al. (1971), Mulser et al. (1973)) for plasmas produced from solid deuterium by neodymium laser light.

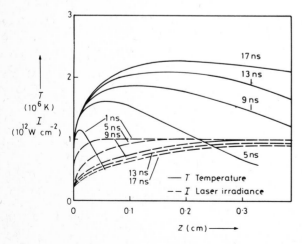

7.12   Laser irradiance and plasma temperature as functions of distances from surface of a solid hydrogen target. (After Mulser & Witkowski (1968))

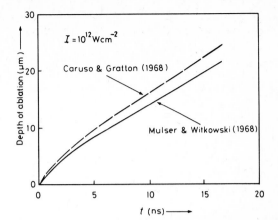

7.13 Depth of ablation from solid hydrogen target as function of time. (After Mulser & Witkowski (1968))

Figure 7.14 shows the results for an irradiance of 5×10^{12} W cm^{-2}. Mulser *et al.* found that thermal conductivity effects become important when the irradiance exceeds 10^{13} W cm^{-2}. Their numerical calculations for $I = 10^{15}$ W cm^{-2} with solid deuterium are shown in Figure 7.15. Similar calculations for lower ruby laser irradiances ($\sim 10^{11}$ W cm^{-2}), which took into account the energies of vaporization and ionization, were made for aluminium targets by Goldman *et al.* (1974).

7.14 Calculations of heating and expansion of plasma formed from a plane solid deuterium target by neodymium laser light of irradiance $I_0 = 5 \times 10^{12}$ W cm^{-2}. The 'mass coordinate' is $(1/\rho_0) \int \rho \, dz$: the corresponding spatial coordinate z is shown below. (After Mulser *et al.* (1973))

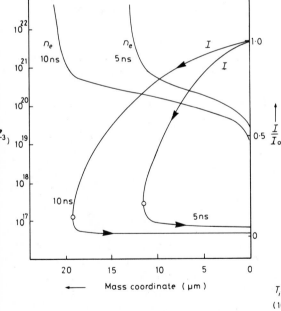

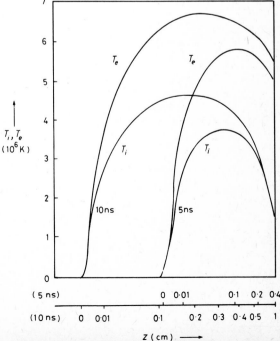

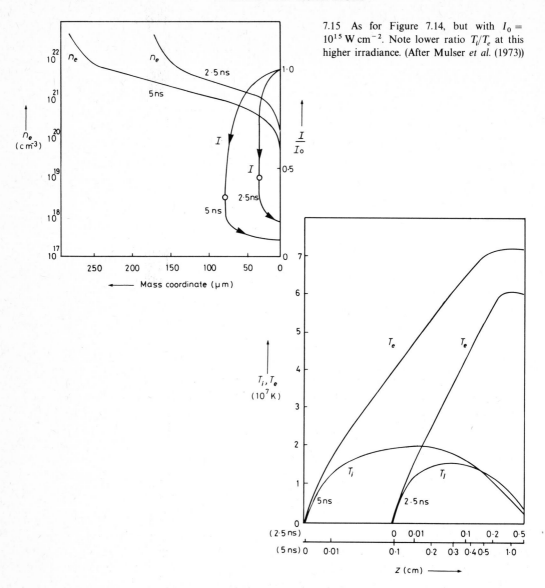

7.15 As for Figure 7.14, but with $I_0 = 10^{15}\,\mathrm{W\,cm^{-2}}$. Note lower ratio T_i/T_e at this higher irradiance. (After Mulser *et al.* (1973))

For a laser pulse lasting several nanoseconds focused to a spot of diameter much less than a millimetre, the one-dimensional model is not appropriate. Lateral expansion was first taken into account by Nemchinov (1967). Puell (1970) considered a model in which the three separate regions are distinguished: the undisturbed target (I); a region of one-dimensional flow at high density, near the surface, in which most of the heating takes place (II); and a region of three-dimensional, approximately adiabatic, expansion (III) (Figure 7.16). The boundary between regions II and III is arbitrarily taken to be at a distance from the target surface which is equal to the focal spot radius r_f. Puell obtained a steady state solution and found that for a laser

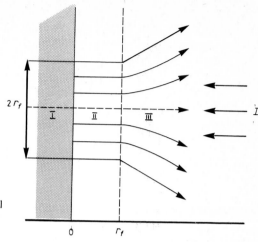

7.16 Sketch of interaction zone, taking lateral expansion into account. (After Puell (1970))

irradiance I the maximum value of the electron temperature, which occurs at the boundary between regions II and III, is given by

$$\kappa_B T_{e_{max}} = \zeta^{-\frac{2}{3}} \left(\frac{3m_i \psi r_f}{50} \right)^{\frac{2}{9}} I^{\frac{4}{9}}. \tag{7.36}$$

Here ζ is a factor which depends on the ratio T_i/T_e and varies from 1 when $T_i = 0$ to $(Z+1)/Z$ when $T_i = T_e$. The total number of particles detached by the laser pulse, N, is given by

$$N = \pi r_f^2 \zeta^{-\frac{1}{3}} \left(\frac{2}{3m_i \psi r_f} \right)^{\frac{2}{3}} \int_{-\infty}^{\infty} \left(\frac{I}{5} \right)^{\frac{2}{3}} dt \tag{7.37}$$

and the expansion energy of an ion is

$$W_i = 5Z\zeta \kappa_B T_{e_{max}}. \tag{7.38}$$

The parameter ψ is equal to $2\cdot 5 \times 10^{-55} (\omega_{ruby}/\omega_0)$ c.g.s. units, ω_0 being the frequency of the laser light and ω_{ruby} that of a ruby laser.

Caruso & Gratton (1968a) considered the case of a target of limited radius r_t in a laser beam focal spot of much larger radius and obtained a result similar in form to (7.36) with r_f replaced by r_t. For the rate of production of plasma, their result corresponded to the differential of expression (7.37) with respect to time, r_f being replaced by r, the instantaneous radius of the dense phase.

4.2. DEFLAGRATION PRECEDED BY A SHOCK WAVE

Let us now assume that the laser irradiance is sufficiently great, and the pulse sufficiently short, for the plasma produced to be very hot and thus transparent to laser light. At constant irradiance the absorbing surface of the dense region moves into the target as a laser-heated deflagration wave. This regime has been investigated in one dimension by Fauquignon & Floux (1970), and also by Bobin (1971) on the basis of Fraser's (1958) theory of radiation fronts.

The process may be understood if we return to the shock wave theory of Chapter 6, Section (3) and consider the region of the $\mathcal{P}$–$\mathcal{V}$ diagram below $\mathcal{P} = 1$ (Figure 7.17). In a deflagration wave the energy supplied to the wave produces an increase in

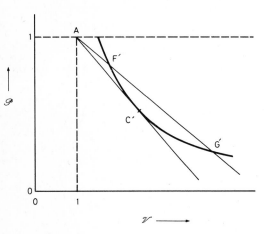

7.17 $\mathcal{P}$–$\mathcal{V}$ diagram for a deflagration wave

volume and a decrease in pressure, in contrast to a detonation wave. The Chapman–Jouguet point C′ now determines the *maximum* allowable velocity for the deflagration wave in the solid: a steady deflagration can propagate at any lower velocity. The greater the energy input per unit mass, the lower is the velocity: F′C′G′ is moved further to the right.

It may be shown (see for example Courant & Friedrichs, (1948, p. 224)) that a deflagration wave is preceded by a shock wave. Thus the velocity of the deflagration wave in the laboratory coordinate system, in which the solid target is initially at rest, can exceed the Chapman–Jouguet velocity, since the deflagration propagates in a medium already moving in the same direction. Figure 7.18 shows the situation.

7.18   Deflagration wave preceded by a shock wave (see text)

U_S and U_D are the shock and deflagration velocities, u_1 is the velocity of the material behind the shock wave and u_2 the velocity of the plasma near the deflagration front, all in laboratory coordinates.

We now have two sets of equations for mass, momentum and energy. For the shock wave,

$$
\left.
\begin{aligned}
\rho_0 U_S &= \rho_1(U_S - u_1) \\
p_0 + \rho_0 U_S^2 &= p_1 + \rho_1'(U_S - u_1)^2 \\
\tfrac{1}{2} U_S^2 + c_V T_0 + \frac{p_0}{\rho_0} &= \tfrac{1}{2}(U_S - u_1)^2 + c_V T_1 + \frac{p_1}{\rho_1}
\end{aligned}
\right\}
\tag{7.39}
$$

while for the deflagration,

$$
\left.
\begin{aligned}
\rho_1(U_D - u_1) &= \rho_2(U_D + u_2) \\
p_1 + \rho_1(U_D - u_1)^2 &= p_2 + \rho_2(U_D + u_2)^2 \\
\tfrac{1}{2}(U_D - u_1)^2 + c_V T_1 + \frac{p_1}{\rho_1} + Q &= \tfrac{1}{2}(U_D + u_2)^2 + c_V' T_2 + \frac{p_2}{\rho_2}
\end{aligned}
\right\}
\tag{7.40}
$$

Here c_V, c_V' are the specific heats at constant volume of the solid and of the plasma respectively, and Q is the energy supplied to unit mass of gas as it passes through the deflagration front. We neglect the ionization energy. From the shock wave equations,

$$
U_S^2 = \left(\frac{p_1 - p_0}{\rho_1 - \rho_0}\right)\frac{\rho_1}{\rho_0},
\tag{7.41}
$$

$$
(U_S - u_1)^2 = \left(\frac{\rho_0}{\rho_1}\right)^2 U_S^2 = \frac{\rho_0}{\rho_1}\left(\frac{p_1 - p_0}{\rho_1 - \rho_0}\right)
\tag{7.42}
$$

and

$$
\frac{p_1}{p_0} = \frac{A - (\rho_0/\rho_1)}{A(\rho_0/\rho_1) - 1}.
\tag{7.43}
$$

Here $A = (\gamma + 1)/(\gamma - 1)$, γ being the adiabatic ratio for the solid. Similarly for the deflagration wave

$$
(U_D - u_1)^2 = \left(\frac{p_1 - p_2}{\rho_1 - \rho_2}\right)\frac{\rho_2}{\rho_1},
\tag{7.44}
$$

$$
(U_D + u_2)^2 = \left(\frac{\rho_1}{\rho_2}\right)^2 (U_D - u_1)^2 = \frac{\rho_1}{\rho_2}\left(\frac{p_1 - p_2}{\rho_1 - \rho_2}\right)
\tag{7.45}
$$

and

$$
\frac{p_2}{p_1} = \frac{A - (\rho_1/\rho_2) + 2Q(\rho_1/p_1)}{A'(\rho_1/\rho_2) - 1},
\tag{7.46}
$$

where $A' = (\gamma' + 1)/(\gamma' - 1)$, γ' being the adiabatic ratio for the plasma. The velocity $(U_D - u_1)$ corresponds to the ablation velocity V_a in Caruso & Gratton's model.

We may suppose that the plasma density ρ_2 immediately behind the deflagration front is known; it is the maximum density into which the laser light can propagate, i.e. it is the density of a fully-ionized plasma whose electron number density equals the critical value for the wavelength of the incident light. Thus

$$\rho_2 = \frac{m_i}{Z}\frac{m_e \varepsilon_0}{e^2}\omega_0^2.$$ (7.47)

For solid targets, $\rho_2 \ll \rho_1$.

The velocity of the plasma behind the deflagration front, relative to the front, v_2, cannot, according to the Chapman–Jouguet condition, exceed

$$v_{2\text{CJ}} = \left(\frac{\gamma' p_2}{\rho_2}\right)^{\frac{1}{2}}.$$ (7.48)

(The plasma is, of course, freely expanding, and can be expected to attain a limiting velocity of expansion

$$v_{\exp} = \frac{2}{\gamma'-1}\left(\frac{\gamma' p_2}{\rho_2}\right)^{\frac{1}{2}},$$ (7.49)

where p_2, ρ_2 are the values just behind the front.) We shall assume that

$$u_2 \simeq v_{2\text{CJ}} = \left(\frac{\gamma' p_2}{\rho_2}\right)^{\frac{1}{2}} \gg U_D, u_1.$$ (7.50)

From (7.45), then,

$$u_2 \simeq \frac{\rho_1}{\rho_2}(U_D - u_1),$$ (7.51)

so

$$(U_D - u_1)^2 \simeq \gamma'\frac{p_2}{\rho_2}\left(\frac{\rho_2}{\rho_1}\right)^2.$$ (7.52)

Also from (7.45), since $\rho_1 \gg \rho_2$,

$$u_2^2 \simeq \frac{\rho_1}{\rho_2}\left(\frac{p_1 - p_2}{\rho_1}\right),$$ (7.53)

so from (7.50) we find we may write approximately

$$p_1 - p_2 = \gamma' p_2$$

or

$$p_1 = (1+\gamma')p_2.$$ (7.54)

Next from (7.46)

$$\begin{aligned}\frac{p_2}{p_1} &= \frac{1}{1+\gamma'} \\ &= \frac{A(\rho_2/\rho_1)-1+2Q(\rho_2/\rho_1)}{A'-(\rho_2/\rho_1)} \\ &\simeq \frac{(2Q\rho_2/p_1)-1}{A'}\end{aligned} \tag{7.55}$$

for small values of ρ_2/ρ_1. Thus, approximately,

$$p_1 = \frac{(1+\gamma')2Q\rho_2}{\gamma' A'}, \tag{7.56}$$

$$p_2 = \frac{2Q\rho_2}{\gamma' A'} \tag{7.57}$$

and

$$u_2^2 = \frac{2Q}{A'}. \tag{7.58}$$

From (7.52),

$$(U_D - u_1)^2 \simeq \frac{2Q\rho_2^2}{A'\rho_1^2}. \tag{7.59}$$

For a strong shock, we may suppose that

$$\rho_1 \simeq A\rho_0, \tag{7.60}$$

so

$$(U_D - u_1)^2 \simeq \frac{2Q\rho_2^2}{A^2 A'\rho_0^2}. \tag{7.61}$$

The initial pressure in the solid, p_0, may be neglected, so from (7.45)

$$\begin{aligned}U_S^2 &\simeq \frac{p_1}{\rho_1-\rho_0}\frac{\rho_1}{\rho_0} = \frac{p_1}{\rho_0}\frac{A}{A-1} \\ &= \frac{2(\gamma'+1)Q\rho_2 A}{(A-1)(\gamma' A')\rho_0}.\end{aligned} \tag{7.62}$$

From (7.42), (7.60) and (7.62), then,

$$(U_S - u_1)^2 = \frac{2(\gamma'+1)Q\rho_2}{A(A-1)\gamma' A'\rho_0}, \tag{7.63}$$

so

$$u_1 = \left(\frac{2Q\rho_2}{\gamma' A'\rho_0}\right)^{\frac{1}{2}} \left\{ \left[\frac{A(\gamma'+1)}{A-1}\right]^{\frac{1}{2}} - \left[\frac{\gamma'+1}{A(A-1)}\right]^{\frac{1}{2}} \right\}. \tag{7.64}$$

Then from (7.61)

$$U_D = \left(\frac{2Q\rho_2}{\gamma' A'\rho_0}\right)^{\frac{1}{2}} \left\{ \left[\frac{\gamma'\rho_2}{A^2\rho_0}\right]^{\frac{1}{2}} + \left[\frac{A(\gamma'+1)}{A-1}\right]^{\frac{1}{2}} - \left[\frac{\gamma'+1}{A(A-1)}\right]^{\frac{1}{2}} \right\}. \tag{7.65}$$

But $\rho_2 \ll \rho_0$, and A is of order unity, so

$$U_D \simeq \left[\frac{2Q\rho_2(\gamma'+1)(A-1)}{\gamma' A' A\rho_0}\right]^{\frac{1}{2}} \tag{7.66}$$

$$= \left(\frac{A-1}{A}\right) U_S. \tag{7.67}$$

We have now determined all the velocities in terms of Q, ρ_2 and ρ_0. However, Q, the energy absorbed per unit mass on passing through the deflagration front, itself depends on the mass flow rate through the front:

$$Q = \frac{I}{(U_D - u_1)\rho_1}, \tag{7.68}$$

where I is the effective irradiance of laser light after allowing for reflection losses. Thus

$$Q = I\left(\frac{A'}{2Q\rho_2^2}\right)^{\frac{1}{2}},$$

i.e.

$$Q = I^{\frac{2}{3}}\left(\frac{A'}{2\rho_2^2}\right)^{\frac{1}{3}}. \tag{7.69}$$

We find, therefore, that

$$U_S^2 = (\gamma+1)\left(1 - \frac{1}{\gamma'}\right)\frac{\rho_2}{\rho_0}\left[\left(\frac{\gamma'+1}{\gamma'-1}\right)\frac{1}{2\rho_2^2}\right]^{\frac{1}{3}} I^{\frac{2}{3}}, \tag{7.70}$$

$$U_D^2 = \frac{4}{(\gamma+1)}\left(1 - \frac{1}{\gamma'}\right)\frac{\rho_2}{\rho_0}\left[\left(\frac{\gamma'+1}{\gamma'-1}\right)\frac{1}{2\rho_2^2}\right]^{\frac{1}{3}} I^{\frac{2}{3}} \tag{7.71}$$

and

$$u_2^2 = \left[\frac{2(\gamma'-1)I}{\rho_2(\gamma'+1)}\right]^{\frac{2}{3}}. \tag{7.72}$$

Thus the plasma ion temperature

$$T_i = \frac{u_2^2 \mathscr{A}}{\gamma' R}$$

$$= \frac{\mathscr{A}}{R\gamma'} \left[\frac{2(\gamma' - 1)I}{\rho_2(\gamma' + 1)} \right]^{\frac{2}{3}},$$

(7.73)

where $\mathscr{A}$ is the atomic mass number of the ions and R is the gas constant per mole, assuming a fully-dissociated plasma. For a solid deuterium target, $\mathscr{A} = 2$, $\gamma' = \frac{5}{3}$, $\gamma = \frac{7}{5}$, $\rho_0 = 0.17$ g cm^{-3}, and if neodymium laser light ($\lambda = 1.06 \ \mu$m) is used to heat the deuterium $\rho_2 = 3.4 \times 10^{-3}$ g cm^{-3}. R, the gas constant, is 8.3×10^7 erg deg^{-1} mole^{-1}. Then

$$T_i = 4.0 \times 10^{-7} I^{\frac{2}{3}} \text{ K}$$

$$u_2 = 5.3 \ I^{\frac{1}{3}} \text{ cm s}^{-1}$$

$$U_S = 1.0 \ I^{\frac{1}{3}} \text{ cm s}^{-1}$$

$$U_D = 0.83 \ I^{\frac{1}{3}} \text{ cm s}^{-1}.$$

(7.74)

Here I is in ergs cm^{-2} s^{-1}. These values depend on the laser wavelength λ_0 through $\rho_2 \propto n_{ec} \propto \lambda_0^{-2}$, so T_i varies as $\lambda_0^{\frac{4}{3}}$ while U_S and U_D vary as $\lambda_0^{-\frac{2}{3}}$. The same dependence of the temperature on the laser irradiance was deduced for large fluxes by Puell (1970).

The structure of the deflagration front was investigated analytically by Bobin (1971). Viscous effects were found to be unimportant compared to the effects of thermal conductivity. The temperature and density profiles across the front were obtained. For neodymium laser light the characteristic thickness of the front Δz was found to be given by

$$\Delta z \simeq 4 \times 10^{-16} T_{e_{max}}^2 \text{ cm}$$

(7.75)

for a solid deuterium or deuterium–tritium target. The time required for the front to develop in deuterium is about $2.5 \times 10^{-20} T_{e \ max}^{\frac{3}{2}}$ s, with $T_{e_{max}}$ in K, and for nanosecond pulses is generally short compared to the pulse duration.

The rate of neutron emission from the heated region, of volume V, in deuterium is given (see Chapter 4, Section (2)) by

$$RV = \text{const.} \frac{n_i^2}{2} \langle \sigma v \rangle V.$$

(7.76)

Since $V \propto \Delta z$ and $\langle \sigma v \rangle \propto T_i^{5.5}$ at temperatures from 0.5 to 1 keV,

$$RV \propto T_i^{7.5} \quad _{(T_i \lesssim 1 \text{ keV})}.$$

(7.77)

But from (7.74) $T_i \propto I^{\frac{2}{3}}$, so

$$RV \propto I^5 \quad _{(T_i \lesssim 1 \text{ keV})}.$$

(7.78)

For a fixed pulse duration τ_L, therefore, the total neutron yield is given by

$$RV\tau_L \propto W_L^5,$$
$$(\tau_L = \text{const.}, T_i \lesssim 1 \text{ keV}) \qquad (7.79)$$

W_L being the laser pulse energy. This result was given by Bobin (1972). At constant pulse energy, in the region of validity of this model,

$$RV\tau_L \propto \tau_L^{-4}$$
$$(W_L = \text{const.}, T_i \lesssim 1 \text{ keV}) \qquad (7.80)$$

Bobin *et al.* (1972) also used the model to calculate the nuclear energy yield from a solid D–T target heated by neodymium laser light.

Deflagration wave theory was extended by Colombant & Tonon (1973) to deal with plasmas formed from high-Z target materials, where significant proportions of the incident energy go into multiple ionization and also into Bremsstrahlung, recombination and line radiation. Detailed calculations were made for carbon, iron and uranium.

4.3. THERMAL WAVE FOLLOWED BY A RAREFACTION WAVE

We turn now to the effects of laser pulses of even higher irradiance, such as are generated by mode-locked ruby and neodymium lasers. The duration of such a pulse is measured in picoseconds, so even though the energy in the pulse may be only a few joules the irradiances attainable in the focused beam can be extremely large. A pulse of 10 J lasting 10^{-11} s, focused into an area of 10^{-4} cm^2, implies an electric field exceeding 10^9 V cm^{-1}.

When the leading edge of such a terawatt pulse is focused on the surface of a solid target, all the surface atoms in the focal spot are ionized in less than a cycle of the laser light. Indeed, as was remarked by Bunkin & Kazakov (1970a), the focused pulses can *only* interact with a plasma. The surface might then be expected to become almost perfectly reflecting (Bunkin (1969)), since the density of surface electrons will be roughly that of the solid atoms, which exceeds the critical density for ruby or neodymium laser light ($n_{ec} \sim 10^{21}$ cm^{-3}). However, it is in fact observed experimentally that much of the laser pulse energy is absorbed by the target. It is assumed that the energy is delivered to the surface electrons through one of the non-linear processes described in Chapter 2.

The duration of the pulse is generally shorter than, or comparable with, the electron–ion momentum transfer time τ_{ei} of a hot plasma: $\tau_{ei} \simeq 10^{-15} T_e^{\frac{3}{2}}$ s, where T_e is in eV (Spitzer (1962), p. 135). However, the pulse is usually longer than the electron–electron thermalization time. We may therefore speak meaningfully about the electron temperature at the end of the pulse. When the electrons are heated extremely rapidly marked departures from ionization equilibrium must be expected. This is not important for hydrogen isotope plasmas, which are fully ionized relatively easily, but should be taken into account for heavy atom plasmas. The theory has been given by Afanas'ev *et al.* (1970).

The energy of the hot electrons produced by the pulse is transmitted by collisions to electrons in the interior of the target. The mean energies of the surface electrons will greatly exceed the first ionization energies of the atoms of the target, and so we may treat the solid initially as a cold, fully-ionized plasma so far as its thermal conductivity is concerned. The electron thermal conductivity is temperature dependent, varying as $T_e^{\frac{5}{2}}$, so conduction by this process, sometimes known as an 'electron thermal wave', is non-linear, and gives rise to the temperature distribution sketched in Figure 7.19.

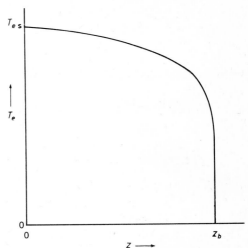

7.19 Sketch of temperature distribution created by non-linear electron thermal conductivity

Zeldovich & Raizer (1966), Caruso & Gratton (1969) and Babuel-Peyrissac *et al.* (1969) considered the heat transfer problem for one-dimensional heat flow – usually a satisfactory approximation for the short time scales involved. The atoms of the target are assumed to be singly ionized and of mass $\mathcal{A} m_p$. The heat flux in the plasma due to the electron thermal conductivity is (Spitzer (1962), p. 144)

$$H = -aT_e^{\frac{5}{2}}\left(\frac{\partial T_e}{\partial z}\right)$$

(7.81)

where $a \simeq 2 \times 10^{-6}$ in c.g.s. K units. The temperature distribution shown in Figure 7.19 is given by

$$T_e(z, t) = T_{es}(t)\left[1 - \frac{z^2}{z_b^2(t)}\right]^{\frac{2}{5}},$$

(7.82)

T_{es} being the temperature at the surface, and may be regarded as almost constant except near the boundary at z_b.

If a laser beam of constant irradiance I is fully absorbed at the surface, Zeldovich & Raizer (1966) have shown that

$$z_b = \text{const.}\, \rho^{\frac{5}{9}} I^{\frac{5}{9}} t^{\frac{7}{9}}$$

(7.83)

and

$$T_{es} = \text{const.}\, \rho^{-\frac{2}{9}} I^{\frac{4}{9}} t^{\frac{2}{9}}.$$

(7.84)

For a very short laser pulse which delivers energy W'ergs per cm^2, essentially instantaneously, Caruso & Gratton (1969) find that

$$z_b \simeq 1 \cdot 3 \times 10^{11} n_e^{-\frac{3}{5}} W'^{\frac{3}{5}} t^{\frac{2}{5}} \text{ cm} \tag{7.85}$$

and

$$T_{es} \simeq 4 \cdot 8 \times 10^4 n_e^{-\frac{2}{5}} W'^{\frac{4}{5}} t^{-\frac{2}{5}} \text{ K,} \tag{7.86}$$

where n_e is in cm^{-3}. The average value of the temperature up to the boundary,

$$\overline{T}_e = 0 \cdot 80 T_{es}. \tag{7.87}$$

At the same time the electrons begin to transfer energy to the ions by collisions. Also the heated region begins to expand—a rarefaction wave moves into the target from the surface. If the electron temperature is sufficiently high, the electron thermal wave propagates more rapidly at first than the rarefaction wave, but slows down and is eventually overtaken. As soon as this 'hydrodynamic separation' occurs, rapid cooling ensues. The rarefaction wave will catch up with the inner boundary of the heated layer at a time t_r such that

$$z_b(t_r) \simeq \left[\frac{3\kappa_B \overline{T}_e(t_r)}{\mathcal{A} m_p} \right]^{\frac{1}{2}} t_r. \tag{7.88}$$

This comes from Caruso & Gratton 1969. For times $\gg T_{ee}$ Replace 3 with $\frac{5}{3}$

Then

$$t_r \simeq 8 \times 10^6 \mathcal{A}^{\frac{3}{4}} W'^{\frac{1}{4}} n_e^{-1} \text{ s,}$$

$$z_b(t_r) \simeq 4 \cdot 5 \times 10^{12} \mathcal{A}^{\frac{5}{8}} W'^{\frac{3}{8}} n_e^{-1} \text{ cm} \tag{7.89}$$

and

$$\overline{T}_e(t_r) \simeq 10^3 \mathcal{A}^{-\frac{1}{4}} W'^{\frac{1}{2}} \text{ K.}$$

Here W' is in erg cm^{-2}, n_e in cm^{-3}.

For a steady laser irradiance I, incident on a solid deuterium target, the corresponding conditions for separation have been found by Babuel-Peyrissac *et al.* (1969) to be

$$t_r = 1 \cdot 2 \times 10^{12} I n_e^{-2} \text{ s}$$

$$z_b(t_r) = 0 \cdot 97 \times 10^{20} I^{\frac{4}{3}} n_e^{-\frac{7}{3}} \text{ cm} \tag{7.90}$$

$$\overline{T}_e(t_r) = 3 \times 10^7 I^{\frac{2}{3}} n_e^{-\frac{2}{3}} \text{ K.}$$

Here I is in ergs cm^{-2} s^{-1} and n_e in cm^{-3}.

In this discussion the relatively slow heating of the ions by the electrons has been ignored. As the plasma expands, more and more of the total energy will be given to the ions, but at the same time it will be converted from thermal to radially-directed kinetic energy.

The very full numerical computing code used by Kidder (1968) for plasmas produced by nanosecond pulses was applied to picosecond pulses by Shearer & Barnes (1970). This 'two-fluid' code allows for hydrodynamic effects with different electron and ion temperatures, and for the effects of electron thermal conductivity. Several different initial density profiles at the deuterium target surface were put into the programme, with different reflection coefficients, to simulate the effects of precursor light leaks from the laser before the main pulse arrived (an important practical problem). The pulse durations were varied between 6 and 750 ps. The spatial distributions of the electron and ion temperatures were calculated at several times and from the known cross-sections for D–D nuclear reactions the neutron yield was obtained. The results are shown in Figure 7.20. It was found that the number of neutrons emitted should

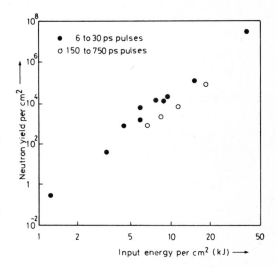

7.20 Calculated neutron yield per unit area of solid deuterium target, as function of laser energy absorbed per unit area. (After Shearer & Barnes (1970))

depend mainly on the amount of energy absorbed from the laser, and not on the details of the density profile. Under some conditions, however, the density gradient allowed the formation of a shock wave, and ion temperatures as high as 10 keV arose in the outer region: this gave only a small additional contribution to the neutron yield. The total neutron yield predicted is also rather insensitive to the duration of the pulse, as may be seen in the diagram.

Figure 7.21 shows the calculated electron and ion temperatures and the electron

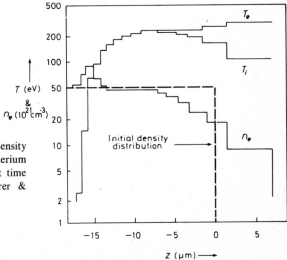

7.21 Calculated temperature and density profiles in plasma from a thick deuterium target heated by a 6 ps laser pulse, at time 31.6 ps after the pulse. (After Shearer & Barnes (1970))

density profile at the time when the ions have attained their maximum temperature. The laser energy absorbed per unit area was about 92 J mm^{-2}. The presence of a region of high density at the temperature front indicates that a shock wave has formed, preceding the rarefaction which is moving into the target from the surface. The average velocity of the boundary towards the laser up to this time is about 2×10^7 cm s^{-1}, which is not very different from the mean thermal velocity of the ions.

Zakharov et al. (1970a) estimated the maximum ion temperature attained in these plasmas by including the effects of electron–ion collisions, making use of average values of the electron and ion temperatures in the heated layer, $\bar{T}_e$ and $\bar{T}_i$, and assuming that the maximum temperature is reached before any gas dynamic process becomes important. The value of $(\bar{T}_i)_{max}$ varies as $W'^{\frac{1}{4}}$. The thickness of the heated layer when $\bar{T}_i = (\bar{T}_i)_{max}$ is given by

$$l_m = \frac{2}{3} \frac{W'}{n_e (\bar{T}_i)_{max}(1 + 1/\bar{Z})}. \tag{7.91}$$

The time during which the ion temperature remains more or less constant near $(\bar{T}_i)_{max}$ is determined by the onset of the expansion process, and is given by

$$t_m = 1\cdot13 \left[\frac{m_i}{3(\bar{T}_i)_{max}(1 + \bar{Z})} \right]^{\frac{1}{2}} l_m. \tag{7.92}$$

These results, together with the nuclear reaction cross sections, enabled Zakharov et al. to calculate the neutron yield to be expected from plasmas generated from solid LiD and D$_2$ targets by pulses of 6 ps duration. Their results are shown in Figure 7.22, and may be seen to be in good agreement with those of Shearer & Barnes (1970) (Figure 7.20). Somewhat higher ion temperatures and neutron yields are predicted from plasmas in which some heavier atoms are added to the deuterium or lithium deuteride (Zakharov et al. (1971a)).

Zakharov et al. (1971b) also calculated the neutron yield expected from solid

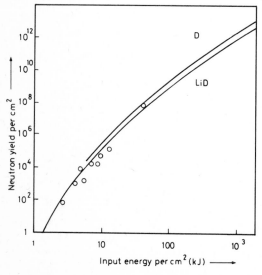

7.22   Calculated neutron yield as function of input energy, per unit area of heated surface, for LiD and D$_2$ targets heated by 6 ps laser pulses. Circles show values calculated by Shearer & Barnes (1970). (After Zakharov et al. (1970a))

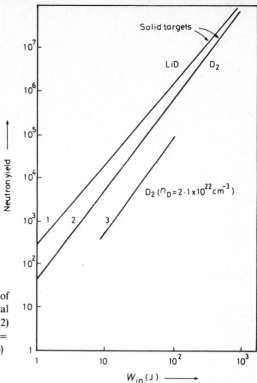

7.23 Calculated neutron yield as function of input energy assuming a hemispherical thermal wave, for LiD and D_2 targets: (1) solid LiD, (2) solid D_2 ($n_D = 5 \times 10^{22}$ cm^{-3}), (3) D_2 ($n^D = 2\cdot1 \times 10^{22}$ cm^{-3}). (After Zakharov et al. (1971b))

deuterium and lithium deuteride targets, assuming that the laser energy is deposited instantaneously at a point on the target surface so that a hemispherical region is heated by conduction. The results are shown in Figure 7.23; the total neutron yield is approximately proportional to the square of the laser pulse energy, in contrast to the prediction of deflagration wave theory (expression (7.79)). Figure 7.24 shows how

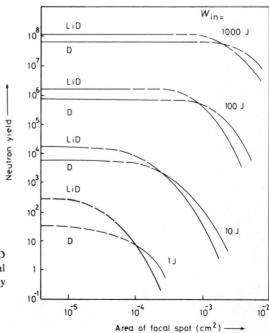

7.24 Calculated neutron yield, using LiD and D_2 targets, as function of area of focal spot, for four values of the absorbed energy W_{in}. (After Zakharov et al. (1971b))

the neutron yield varies with the area of the focal spot receiving the laser energy at the target surface.

The effects of very short pulses have also been discussed analytically by Anisimov (1970) and by Plis *et al.* (1972).

4.4. THE LIMIT OF THERMAL CONDUCTIVITY

The simple concept of thermal conductivity may cease to be valid at very high irradiances. Salzmann (1972) has pointed out that the heat transport across a plane is no longer determined solely by the local temperature gradient if the gradient changes significantly within a distance equal to λ_{ee}, the mean free path for electron–electron collisions. Thus the Fourier theory of heat conduction is invalid if

$$\frac{dT_e}{dx} \gtrsim \frac{T_e}{\lambda_{ee}}. \tag{7.93}$$

Also, an electron temperature gradient is accompanied by an electric field

$$E \sim \frac{\kappa_B}{e} \frac{dT_e}{dx} \tag{7.94}$$

and Bickerton (1973) has noted that if

$$\lambda_{ee} e E \gtrsim \kappa_B T_e, \tag{7.95}$$

which is equivalent to (7.93), some of the electrons will be accelerated to such high energies that their mean free path for subsequent collisions will be very long: they are said to have 'run away', or 'decoupled'.

The heat flux H arises from an asymmetry in the electron velocity distribution, and Bickerton remarked that it cannot exceed the product of the random flux of electrons in one direction across the plane and their mean thermal energy. Thus as an upper limit we have

$$H_{max} = n_e \left(\frac{\kappa_B T_e}{2\pi m_e}\right)^{\frac{1}{2}} \left(\frac{3\kappa_B T_e}{2}\right), \tag{7.96}$$

on the understanding that the electron velocity distribution may not in fact be Maxwellian. Very approximately the same value of H_{max} is obtained if the usual expression (7.81) for the electron thermal conductivity is taken with the limiting temperature gradient $dT_e/dx = T_e/\lambda_{ee}$.

Such severely distorted electron velocity distributions may well allow instabilities to grow. Bickerton concluded that if $T_e \gg T_i$ (as is usually the case in laser heated plasmas) the ion acoustic instability could develop, given sufficient time, limiting the energy flux to a value of about $(m_e/m_i)^{\frac{1}{2}} H_{max}$.

In a plasma with a steep density gradient intense laser light is most strongly absorbed in a layer where the electron density is close to the critical value, determined by the laser wavelength. Thus the maximum laser irradiance which can be assimilated by thermal conduction in the boundary of a dense plasma depends on the wavelength

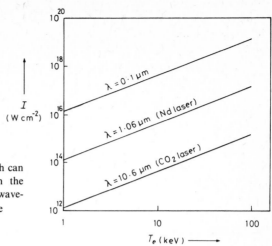

7.25 Upper limits of laser irradiance which can be assimilated by thermal conduction in the boundary of a dense plasma, for three wavelengths, as functions of electron temperature

and on the electron temperature. The upper limit derived from expression (7.96) is shown for three wavelengths in Figure 7.25: the ion acoustic instability might reduce these values substantially.

5. PLASMAS FORMED FROM THIN SOLID FILMS

When the target material is of finite thickness l the results of sections (4.1) and (4.2) will apply up to a time

$$t_s = \frac{l}{U},$$ (7.97)

when the shock wave generated at the front surface (nearest the laser) will reach the other surface. The ablation of the front surface to form plasma will reduce the thickness of the dense layer to zero in a time

$$t_B = \frac{l\rho_0}{V_a\rho_1}.$$ (7.98)

According to Caruso & Gratton (1968a), in the self-regulating regime of Section (4.1)

$$\frac{t_B}{t_s} = \left(\frac{\rho_0}{\rho_2}\frac{U}{u_1}\right)^{\frac{1}{2}} \gg 1,$$ (7.99)

while the deflagration wave theory of Section (4.2) (Fauquignon & Floux (1970)) gives

$$\frac{t_B}{t_s} = \left[\frac{(\gamma'+1)(\gamma+1)}{2\gamma'}\frac{\rho_0}{\rho_2}\right]^{\frac{1}{4}} \gg 1. \qquad (7.100)$$

If the momentum transferred to the target by radiation pressure may be ignored, the total momentum of the entire system, solid, dense phase and plasma together, must be zero. This fact is very useful in analysing the development of the plasma between t_s and t_B. Up to t_s the rate of change of momentum of the plasma towards the laser is balanced by the recoil pressure, which generates the shock wave in the solid. After t_s the momentum of the plasma must be balanced by the momentum of the entire dense phase, which in Caruso & Gratton's model is characterized by average quantities. In reality, further waves must propagate in the dense region, and a very complicated situation arises. Caruso & Gratton simplified the problem by supposing that the equation of motion is just

$$[\rho_0 l - M'(t)]\frac{d\bar{u}_1}{dt} = -u_2\frac{dM'}{dt}, \qquad (7.101)$$

where $\bar{u}_1$ is the average velocity of the dense phase in the laboratory frame of reference. A good approximate solution may be found by supposing u_2 (the expansion velocity of the plasma from the surface of the dense phase, towards the laser) to be independent of time so that

$$\bar{u}_1 = u_2 \ln\left[1 - \frac{M'(t)}{\rho_0 l}\right]. \qquad (7.102)$$

The laboratory-frame velocity of the element of plasma emitted at time t is

$$u_{pt} = \bar{u}_1 + u_2$$
$$= u_2\left[1 + \ln\left(1 - \frac{M'(t)}{\rho_0 l}\right)\right]. \qquad (7.103)$$

The first layer of plasma is thus emitted towards the laser with velocity close to u_2. As more and more of the target is converted into plasma, i.e. as M' increases, u_{pt} falls, reaching zero when $M' = 0.63\rho_0 l$ and then increasing in the direction away from the laser. An approximately symmetrical plasma is generated, provided the duration of the laser pulse is not less than t_B.

Mulser & Witkowski (1969) and Mulser (1970) computed the behaviour of plasmas produced from a 50 μm film of solid hydrogen by light of irradiance 10^{12} W cm^{-2}. In this calculations they took into account the compressibility of solid hydrogen (the pressure generated is of order 10^5 atm) and the effects of heat conduction, and in their second paper the Saha equation was used to determine the degree of ionization. Figure 7.26 shows the density profiles initially and after 0.5, 1.5 and 2 ns. Velocity and temperature profiles are also shown. The velocity of the shock wave is 2.7×10^6 cm s^{-1}. Figure 7.27 shows the depth of ablation for three values of the laser flux: a comparison of the two curves for $I = 10^{13}$ W cm^{-2} shows that heat conduction is

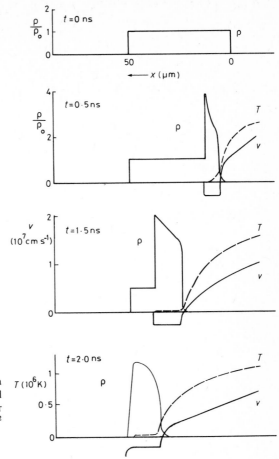

7.26   Calculations of dynamics of plasma formation from a 50 μm film of solid hydrogen. Scales are common to all four diagrams. Laser irradiance is 10^{12} W cm^{-2} from $t = 0$. (After Mulser (1970))

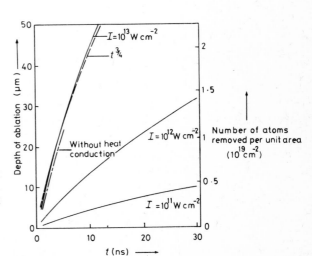

7.27   Depth of ablation of solid hydrogen target calculated as function of time for three constant laser irradiances. The number of atoms removed is shown on the right. (After Mulser (1970)) (1970))

beginning to have a slight influence at this irradiance. On integrating expression (7.33) for the ablation velocity with respect to time we find that the depth of ablation according to the Caruso & Gratton theory should vary as $I^{\frac{1}{2}}t^{\frac{3}{4}}$. As may be seen in Figure 7.27, this relation is closely obeyed. The limitations of one-dimensional theory for small focal spot diameters have been discussed qualitatively by Sigel (1970).

Zakharov *et al.* (1970*b*) considered the case of a thin foil in which all the electrons are heated to a temperature T_{e0} by a short, very intense laser pulse before any expansion or energy transfer to the ions can take place. The ions are subsequently heated by collisions and cooled by expansion. The maximum ion temperature achieved was calculated as a function of T_{e0}. The results are shown in Figure 7.28 for a lithium

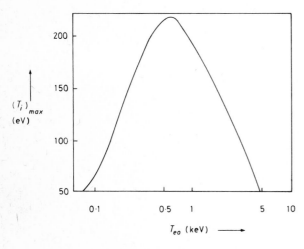

7.28   Maximum attainable ion temperature in LiD plasma slab initially 10^{-2} cm thick with $n_e = 10^{21}$ cm^{-3}, after electrons are heated instantaneously to T_{e0} (After Zakharov *et al.* (1970*b*))

deuteride plasma 10^{-2} cm thick with $n_e = 10^{21}$ cm^{-3}. The maximum ion temperature attained, however high the initial electron temperature, is about 215 eV, and this could only be increased by increasing the initial thickness of the plasma.

6. SPHERICAL PLASMAS WITH UNIFORM TEMPERATURE

Calculations of laser heated plasma dynamics were first presented in 1963 by Basov & Krokhin (1964), who considered the heating of a small spherical target. Similar results were obtained by Dawson (1964), whose treatment we follow here, and by Engelhardt (1963). In these early calculations a simple model was used. The target is supposed initially to be a small, spherical, cool but fully-ionized plasma of radius r_0 and mass M at the critical electron density n_{ec}. The temperature T and the particle densities n_e, n_i within the plasma are assumed uniform, and the laser energy is supposed to be delivered to the plasma at a constant rate P_0. Radiation losses are neglected. The work done by the plasma pressure p in expanding the plasma is taken to be equal to the increase in kinetic energy of radial motion of the plasma. Then if the radius of the plasma boundary is r_b,

$$4\pi r_b^2 \frac{dr_b}{dt} p = \tfrac{1}{2}\bar{M}\frac{d}{dt}\left(\frac{dr_b}{dt}\right)^2. \tag{7.104}$$

Here $\bar{M}$ is a modified value for the total plasma mass, which allows for the fact that there must be a radial gradient of radial velocity. Since the plasma density is assumed to be uniform, if we take the local radial velocity at any instant to be proportional to the distance from the centre we find that $\bar{M} = 3M/5$. For the pressure we have

$$p = (n_e + n_i)\kappa_B T. \tag{7.105}$$

The mass of an electron may be neglected compared to that of an ion, so

$$n_i = \frac{3M}{4\pi r_b^3 m_i}. \tag{7.106}$$

If the ions have an average charge of Z proton charges, $n_e = Zn_i$ and the total number of particles in the plasma

$$N \simeq \frac{M(1+Z)}{m_i}. \tag{7.107}$$

Some of the energy supplied by the laser is converted into energy of radial expansion: the rest raises the temperature of the plasma. Thus

$$P_0 = 4\pi r_b^2 \frac{dr_b}{dt} p + \tfrac{3}{2}N\kappa_B \frac{dT}{dt}. \tag{7.108}$$

Setting $T = 0$ and $dr_b/dt = 0$ when $t = 0$, these equations may be integrated with the results that

$$r_b = \left(r_0^2 + \frac{10}{9}\frac{P_0 t^3}{M}\right)^{\frac{1}{2}} \tag{7.109}$$

and

$$\kappa_B T = \frac{P_0 t}{3N}\left(\frac{2r_0^2 + \tfrac{5}{9}(P_0 t^3/M)}{r_0^2 + \tfrac{10}{9}(P_0 t^3/M)}\right). \tag{7.110}$$

For small values of t, $(t \ll (r_0^2 M/P_0)^{\frac{1}{2}})$, we may write

$$\left.\begin{aligned} r_b &\simeq r_0 \\[4pt] \kappa_B T &\simeq \frac{2}{3}\frac{P_0 t}{N} \end{aligned}\right\} \quad (t \to 0) \tag{7.111}$$

while for large values of t

$$
\left.
\begin{aligned}
r_b &\simeq \left(\frac{10}{9}\frac{P_0 t^3}{M}\right)^{\frac{1}{2}} \\
\kappa_B T &\simeq \frac{P_0 t}{6N}
\end{aligned}
\right\} \quad (t \to \infty)
$$

(7.112)

so for long laser pulses the energy converted into thermal energy of the plasma, $3N\kappa_B T/2$, is only equal to a quarter of the total energy, $P_0 t$, supplied by the laser pulse. The remainder is in the form of radially-directed kinetic energy.

When the laser pulse ends, or when the plasma becomes transparent, no more energy is given to the plasma. If we suppose that heating stops suddenly at time t_h when $T = T_h$ and $r_b = r_h$, the subsequent expansion may be calculated, with the result that

$$
\left.
\begin{aligned}
T &= T_h\left(\frac{r_h}{r_b}\right)^2, \\
\left(\frac{dr_b}{dt}\right)^2 &= \left(\frac{dr_b}{dt}\right)^2_{t_h} + \frac{5N}{M}\kappa_B T_h\left(1 - \frac{r_h^2}{r_b^2}\right)
\end{aligned}
\right\} \quad (t \geqslant t_h).
$$

(7.113)

The results of some numerical calculations on this model are shown in Figure 7.29.

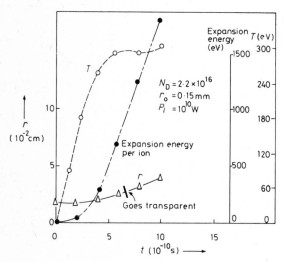

$N_D = 2\cdot2\times10^{16}$
$r_0 = 0\cdot15\,mm$
$P_l = 10^{10}\,W$

Expansion energy per ion

Goes transparent

7.29 Calculations of heating and expansion of a small, fully-ionized, spherical deuterium plasma initially at critical electron density, using a ruby laser. (After Dawson (1964))

The validity of the model depends upon the dimensions of the target and upon the frequency, duration and power of the laser pulse. It is necessary for the product of the diameter of the plasma and the linear absorption coefficient of the plasma at the laser frequency to remain not less than about unity throughout the interaction, otherwise the plasma will be transparent to the laser pulse. For neodymium laser light focused in a plasma with $n_e = 10^{21}$ cm^{-3}, at a temperature of 10^6 K, the absorption coefficient a_ω (see Chapter 2, Section (8.1.2)) is several hundred cm^{-1}, so plasmas as small as $100 \, \mu m$ in diameter will absorb strongly. Haught & Polk (1966) improved the model used above, by taking into account the effects of radial pressure gradients on the spherically-symmetrical plasma expansion and allowing for the time dependence of the laser pulse power, while retaining the assumption of a uniform (though time-dependent) temperature throughout the plasma. They also allowed for the fact that with a very small target the expanding plasma intercepts an increasing area of the laser beam cross-section as it expands. (This effect was also considered by Askaryan & Tarasova (1971).) Fader (1968) developed the theory further, using a hydrodynamic model. The equations were solved numerically, and also by separation of the variables, yielding solutions which were in good agreement. The density and pressure distributions were found to be very nearly Gaussian, except for deviations imposed by the conditions at the plasma boundary. The radial velocity distribution was linear. Haught & Polk (1970) later developed a simpler similarity model in which the density and velocity profiles were assumed to take the asymptotic forms found from the hydrodynamic calculations. They used the earlier integrated form of the momentum and energy equations, but took into account the finite electron–ion collision time and found both the electron and the ion temperatures. The results of some of their numerical calculations are shown in Figure 7.30(a) and (b). In both cases the target is taken to be a speck of lithium hydride of $10 \, \mu m$ radius, and the half-intensity radius of the Gaussian laser beam is also $10 \, \mu m$. In Figure 7.30(a), the laser pulse has a peak power of 1 GW and a rise time, equal to the half-intensity

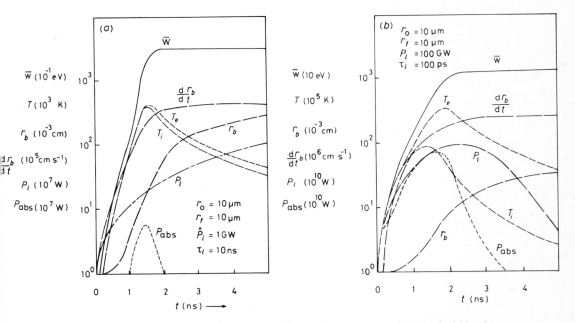

7.30   Calculations of heating and expansion of plasmas formed from small lithium hydride spheres by ruby laser pulses (a) with peak power 1 GW and (b) with peak power 100 GW. (After Haught & Polk (1970))

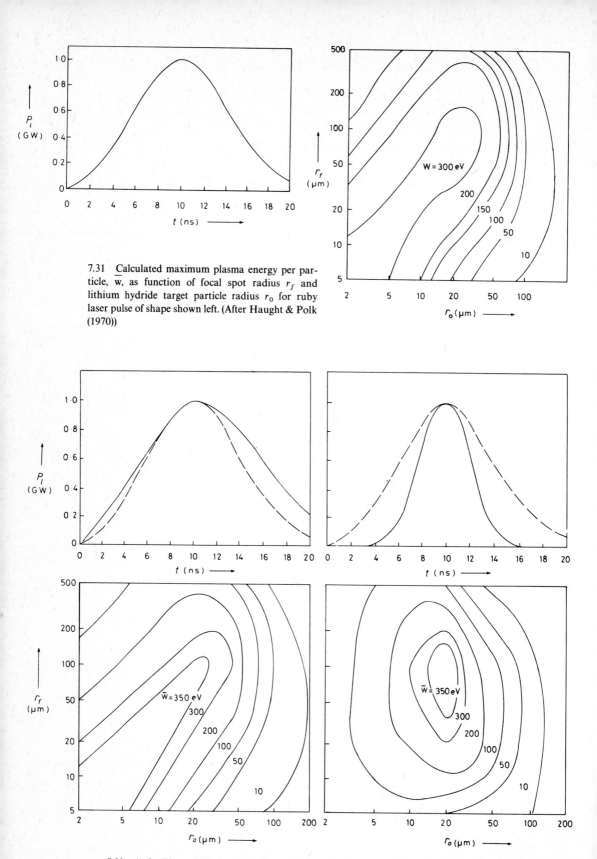

7.31 Calculated maximum plasma energy per particle, $\overline{w}$, as function of focal spot radius r_f and lithium hydride target particle radius r_0 for ruby laser pulse of shape shown left. (After Haught & Polk (1970))

7.32 As for Figure 7.31, showing effects of varying laser pulse shape. (After Haught & Polk (1970))

duration, of 17 ns; while in Figure 7.30(b) the pulse of 100 GW peak power has a rise time and half width of only 0·1 ns. It is apparent that most of the energy in the longer pulse is wasted: the plasma becomes transparent long before the pulse peak power is reached. The energy in the shorter pulse is much more effectively used. With long pulses, therefore, the peak power is irrelevant for this target geometry and beam diameter: what matters is the initial rate at which the power increases. Passing an exponentially growing pulse through additional amplifiers will have no effect on the shape of its leading edge or (in this case) on the plasma it will generate.

Figures 7.31 and 7.32 show how the calculated maximum plasma energy per particle is affected by varying the focal spot size and target dimensions for a pulse of 1 GW peak power with three different time profiles.

Bonnier & Martineau (1972, 1973) and Bonnier et al. (1971) have calculated the time dependence of the radius, temperature and density in spherical plasmas generated by inverse Bremsstrahlung absorption of carbon dioxide laser light, assuming equal, uniform electron and ion temperatures.

If the electrons in a plasma are heated instantaneously to a temperature T_{e0}, the ion temperature will subsequently rise initially but will then fall as the result of expansion. Zakharov et al. (1970b) calculated the maximum attainable ion temperature for a lithium deuteride sphere with $n_{e0}r_0 = 10^{19} \text{ cm}^{-2}$ as a function of T_{e0}, assuming uniform temperatures throughout the plasma. Their results are shown in Figure 7.33 which may be compared with the plane plasma results of Figure 7.28.

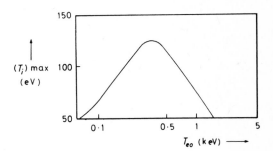

7.33   Maximum attainable ion temperature in a LiD plasma sphere initially with $n_e r_0 = 10^{19} \text{ cm}^{-2}$, after electrons are heated instantaneously to T_{e0}. (After Zakharov et al. (1970b))

The more rapid fall in density in the spherical case results in a lower ion temperature for a given initial thickness and electron temperature.

For this model to be valid it is necessary for the thermal conductivity to be sufficiently high for the thermal diffusion time over the radius of the plasma to be short compared to the laser pulse duration. The conductivity is due to the electrons, and the solution of the one-dimensional thermal diffusion equation yields a characteristic diffusion length Λ_T for a time t:

$$\Lambda_T^2 \simeq 4 \times 10^9 \frac{T^{\frac{5}{2}}}{n_e} t \text{ cm}^2 \qquad (7.114)$$

where T is in K and n_e in cm^{-3}. During a 10 ns pulse, therefore, if $n_e = 10^{21}$, heat will diffuse a characteristic distance of the order of 1 mm at 10^7 K, but only 3 μm at 10^5 K. During a 10 ps pulse heat will only diffuse 35 μm even at 10^7 K and will only travel 0·1 μm at 10^5 K. In general, radiative transfer of energy within the plasma is not effective because the plasma is transparent to most of the radiation it emits. Thus even in small plasmas, of diameter perhaps 100 μm, the assumption of uniform temperature throughout the volume will only hold at high temperatures in pulses lasting a nanosecond or more.

Calculations have been made, using this and similar models, of the dynamics of deuterium-tritium spheres heated by laser pulses (Johnson & Hall (1971), Somon (1972)). Estimates of the minimum laser energy required for 'breakeven', when an equal amount of energy is released from nuclear reactions (see Chapter 4) are in the region of 4×10^7 J (Babykhin & Starykh (1971)) to 10^9 J (Chu (1972)) per pulse. Tamping might reduce these energy requirements somewhat, but they appear to be beyond the capabilities of foreseeable laser systems. However, as we shall see later in this chapter, if we drop the assumption of uniform temperature and explore other possibilities, we find that breakeven may be achieved with very much smaller laser energies.

6.1. HEATING LIMITS IN UNDERDENSE PLASMA

At very high irradiances it is necessary to take into account the non-linear behaviour of the absorption coefficient of the plasma. In an underdense plasma ($n_e < n_{ec}$) the Bremsstrahlung absorption coefficient decreases above the optimum irradiance, as was discussed in Chapter 2, Section (8.1.3), and in the absence of other non-linear absorption mechanisms this limits the rate of heating of the plasma. Hughes & Nicholson-Florence (1968) pointed out, using a simple model, that there is then a limit to the temperature which can be attained with a given initial plasma radius, no matter how much laser power and energy are available. The model used is similar to that of Basov & Krokhin and of Dawson discussed at the beginning of this section. The initial plasma is supposed to be small, cold and fully ionized, with the critical electron density. The temperature and density are assumed to be uniform throughout the plasma during the heating period. However, the rate P at which energy is supplied to the plasma is no longer constant. In order to set an upper limit to the rate of heating it is supposed that the optimum irradiance (expression (2.91)) is maintained over the whole plasma at all times. Thus the total rate of energy input to the plasma is

$$P \simeq \mathcal{Y} \frac{N^2}{r_b^3 T^{\frac{1}{2}}} \qquad (7.115)$$

where

$$\mathcal{Y} = \frac{Z^3 e^4}{8\pi^2 (2\pi m_e \kappa_B)^{\frac{1}{2}} (Z+1)^2 \varepsilon_0^2} \ln\left(\frac{4\kappa_B T}{\gamma \hbar \omega}\right), \qquad (7.116)$$

assuming a single ion species of charge Z. The logarithmic term varies only slowly with temperature, and for neodymium laser light interacting with a plasma in the temperature range $10^6 \lesssim T \lesssim 10^8$ K may be taken to be about 8. Thus $\mathscr{Y}$ is taken to be a constant. Equations (7.105), (7.106), (7.107), (7.108) (with P_0 replaced by P) and (7.115) give

$$\mathscr{Y}Nt = \kappa_B r_b^3 T^{\frac{3}{2}}. \tag{7.117}$$

From (7.104) we have also

$$\frac{3N\kappa_B}{\overline{M}}\left(\frac{\mathscr{Y}N}{\kappa_B}\right)^{\frac{2}{3}} t^{\frac{2}{3}} = r_b^3 \frac{d^2 r_b}{dt^2} \tag{7.118}$$

(the results given by Hughes and Nicholson-Florence do not satisfy this equation).

The solution to equation (7.118) will be of the form $r_b = r_b(r_0, t)$, where r_0 is the initial radius. This may then be used in (7.117) to find $T = T(r_0, t)$. The temperature will pass through a maximum, since $P \rightarrow 0$ as $t \rightarrow \infty$. For a given target material, the maximum temperature attained will be determined solely by the initial radius of the plasma.

6.2. DECAY OF UNIFORM-TEMPERATURE PLASMA

The expansion of the plasma after heating has ceased is of considerable interest to experimentalists, because it is possible to relate the expansion velocity (which is quite easily measured) to the temperature of the plasma before it expanded. The adiabatic expansion of a simple spherical plasma consisting of electrons and only one species of ion has already been calculated on the uniform density model used by Dawson (1964). After heating ceases, the plasma boundary will soon approach an asymptotic velocity given (see expression 7.113) by

$$V_{b\infty}^2 = \frac{5N}{M}\kappa_B T_h \tag{7.119}$$

where N is the total number of plasma particles, M their total mass and T_h the maximum plasma temperature, attained when heating stopped. The radial velocity within the plasma is assumed to be proportional to the distance from the centre.

In Haught & Polk's first modification (1966) of Dawson's model, a radial density gradient was introduced so that

$$n_i(r) = n_i(0)\left(1 - \frac{r}{r_b}\right), \tag{7.120}$$

where r_b is the boundary radius and $n_i(0)$ is the central ion density: $n_i(0) = 3N_i/\pi r_b^3$. If we suppose that all the energy given to the plasma was thermalized at a uniform temperature $T_h = T_i = T_e$ before expansion began, the asymptotic velocity of expansion of the boundary is now given by

$$V_{b\infty}^2 = \tfrac{15}{2}(Z+1)\frac{\kappa_B T_h}{m_i}.$$

(7.121)

The same functional dependence, with different values of the constant, is obtained
with the other models of the expansion process (see also Mirels & Mullen (1963)).
A similar result was obtained by Opower & Press (1966) from consideration of
electrostatic forces on plasma particles.

If the electrons and ions have different temperatures, expression (7.121) becomes

$$V_{b\infty}^2 = \frac{15}{2m_i}(Z\kappa_B T_e + \kappa_B T_i).$$

(7.122)

Demtröder & Jantz (1970) pointed out that in many situations, and particularly
with long laser pulses, the processes of heating and expansion cannot legitimately
be separated. The initial temperature may not be as high as might be inferred from
the expansion energy. Electrons receive energy from the laser light, thermalize it
by electron–electron collisions, and transfer it to ion kinetic energy of expansion.
This cycle may be repeated many times within the laser pulse duration. Demtröder
& Jantz proposed a modified form of expression (7.122):

$$\tfrac{1}{2}m_i V_{b\infty}^2 = \text{const.}\,(nZ\kappa_B T_e + \kappa_B T_i).$$

(7.123)

Here n is a measure of the number of times an electron gains and loses energy to the
ions.

When several species of ions are present in the plasma, the situation becomes
rather complex. Some authors have used expression (7.123) with different values of
Z and m_i to obtain asymptotic expansion velocities for different species of ions from
the same plasma. Mulser (1971) has argued that this is incorrect. There is no doubt
that recombination processes must be considered, as a result of which ionization and
excitation energy may be returned to the plasma as kinetic energy. General discussions
have been given by Afanas'ev & Rozanov (1972) and by Rumsby & Paul (1974).
As was noted by Demtröder & Jantz, when an ion has reached its asymptotic expansion
velocity its translational energy will not be much affected by subsequent recombination
processes. Ions entering an analyser with a certain energy may have been produced
by recombination of more highly charged ions of the same energy. Recombination is
more probable at higher densities: if a radial density gradient exists in the plasma,
the ions which happen to be nearest the boundary are most likely to travel to the
detector without recombination. Because they are nearest the boundary they will
reach the detector first. Thus the most highly ionized atoms will have the highest
recorded expansion velocities.

The effects of both three-body and radiative recombination processes on the
decay of lithium hydride plasmas were discussed by Mattioli (1971), who concluded
that if the plasma expands at a velocity greater than 10^7 cm s^{-1}, recombination is
not energetically important and ionization will be 'frozen in' as a result of the rapidly
decreasing density. The same result was obtained for highly-ionized carbon ions by
Boland et al. (1968) and by Basov et al. (1969a). The effects of other collisional processes have been considered by Allen (1972).

7. SHOCK WAVES IN SPHERICAL PLASMAS

The uniform-temperature theory of the previous section is really more appropriate to the later stages in the development of the plasma. If the irradiance at the surface of a uniformly-irradiated spherical solid target is large enough the ablation pressure will generate a shock wave similar to that discussed for plane geometry in Section (5). However, in spherical geometry the converging inward motion of the material behind the shock front leads to a further adiabatic compression. The shock wave converges to the centre, where it is reflected and travels outwards. The reflected wave propagates through inward-moving compressed material and again increases the density. Theoretical treatments by Guderley (1942) and others are discussed by Zeldovich & Raizer (1966, Chapter XII). Whereas the density ratio across a strong plane shock front is limited to $(\gamma + 1)/(\gamma - 1)$, in the case of a reflected spherical shock with $\gamma = \frac{5}{3}$ the overall density ratio is about 33 (Brueckner (1973)) and when $\gamma = \frac{7}{5}$ it is about 138 (Zeldovich & Raizer (1966)).

Kidder (1968) has computed the behaviour of a 62 μg deuterium sphere of density 5.4×10^{-3} g cm^{-3} (initial radius about 1.4 mm) supplied with 10 kJ of laser energy in 5 ns. Some results are shown in Figures 7.34 and 7.35. The curve showing the position of the light front gives the radius at which the incident light power per steradian has been decreased by absorption by a factor 1/e. Calculations of the number of neutrons produced by 5 ns laser pulses under these and other conditions are shown in Table 7.1. The amount of thermonuclear energy released per neutron is only about

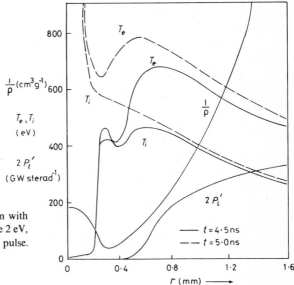

7.34   Heating of a sphere of deuterium with initial density 5.4 mg cm^{-3}, temperature 2 eV, and radius 1.4 mm, by 10 kJ, 5 ns light pulse. (After Kidder (1968))

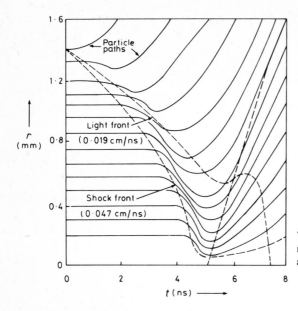

7.35   Particle, shock and light front trajectories on radius–time plot for deuterium sphere: conditions as for Figure 7.34. (After Kidder (1968))

10^{-12} J, so a 10^5 J laser pulse releases only 0·3 J. Most of the thermonuclear reactions take place in the hot dense plasma occurring in the reflected shock wave. Extensive calculations for a variety of pulse shapes have been described by Goldman (1973, see also Lubin *et al.* (1972)).

TABLE 7.1: *Calculated neutron yields from spherical deuterium targets of mass M and density ρ_0, supplied with W joules of laser energy in 5 ns*

W (J)	M (μg)	$1/\rho_0$ (cm^3/g)	Number of neutrons
10^2	2·7	370	5×10^3
10^3	12	370	1×10^7
10^4	62	185	2×10^9
10^5	250	90	3×10^{11}

After Kidder (1968).

In principle, it should be possible to arrange for a second shock wave to travel inwards in the compressed medium behind the first shock front, with a higher velocity adjusted so that both shocks arrive at the centre almost simultaneously. Indeed, a sequence of shock waves of increasing strength could be arranged to arrive in rapid succession at the centre. To produce a high final density it is important that the shocks should not overtake each other before reaching the centre, because this would result in a lower density and a higher temperature. Brueckner (1973) has described very detailed calculations of the development of a plasma from a solid DT sphere of 1 mm diameter by three successive shock waves. The shocks are produced by rapid changes in the spherically-symmetrically applied laser power, which starts at 6.3×10^{11} watts, rises to 6.3×10^{12} watts after 5·47 ns and then to 4×10^{14} watts between 7·21 and 7·42 ns. A total of 60·1 kJ is absorbed (anomalous absorption is allowed for). The density distributions at several times in the final stages of the implosion are shown in Figure 7.36. The maximum density, averaged over the central tenth of the radius, is

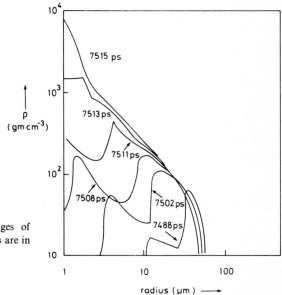

7.36 Density distributions in final stages of implosion of a DT sphere (see text). Times are in picoseconds. (After Brueckner (1973))

1380 g cm^{-3}, corresponding to more than 7000 times the initial density. The central density remains above 100 g cm^{-3} for less than 2×10^{-11} seconds, and the calculated nuclear energy yield is about 500 kJ, or more than eight times the laser energy absorbed. This is a very much more encouraging result than that obtained for a single shock. A number of questions arise in connection with the role of thermal conductivity and the equation of state of very dense plasma, which we shall consider in the next section.

8. ISENTROPIC COMPRESSION

As was noted in Chapter 6, Section (3), the shock Hugoniot is initially tangential to the isentropic curve in $(\mathscr{P}, \mathscr{V})$ space. Thus a sequence of a very large number of very weak shocks is almost equivalent to an isentropic compression (Figure 7.37).

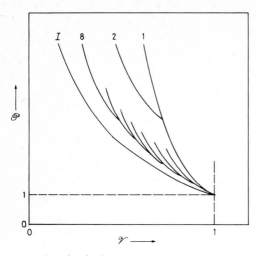

7.37   Sketch showing how a sequence of weak shock waves produces near-isentropic compression (*I*)

A spherical target can be compressed by ablation pressure, and by varying the laser power the compression can be kept approximately isentropic, thus minimizing the energy required to achieve a given density. The temperature T of the core of radius r and density ρ will then vary as $\rho^{\gamma-1}$:

$$T \propto \rho^{\gamma-1} \propto r^{-3(\gamma-1)}. \tag{7.124}$$

The average radial velocity $\bar{v}$ of the compressed material should be close to, but must not exceed the sound speed. Thus

$$\bar{v} \propto T^{\frac{1}{2}} \propto r^{-\frac{3}{2}(\gamma-1)}. \tag{7.125}$$

If r tends to zero when $t = \tau_0$, we find on integrating (7.125) that

$$\bar{v} \propto \left(1 - \frac{t}{\tau_0}\right)^{-3(\gamma-1)/(3\gamma-1)}. \tag{7.126}$$

The pressure p is proportional to T and ρ, so

$$p \propto \rho\bar{v}^2. \tag{7.127}$$

If we suppose that the fraction of the laser power P_L expended on compressing the core is constant (the remainder of P_L, assuming total absorption, will heat up the ablating plasma), we have

$$P_L \propto (\text{pressure} \times \text{area} \times \text{velocity}), \text{ i.e.}$$

$$P_L \propto \bar{v}^{(9\gamma-7)/[3(\gamma-1)]} \tag{7.128}$$

$$\propto \left(1 - \frac{t}{\tau_0}\right)^{-(9\gamma-7)/(3\gamma-1)}.$$

Thus for $\gamma = \frac{5}{3}$, the optimum time variation of laser power is given by

$$P_L \propto \left(1 - \frac{t}{\tau_0}\right)^{-2}. \tag{7.129}$$

This very simple derivation of the optimum pulse shape was given by Clarke *et al.* (1973): Nuckolls *et al.* (1972) find that

$$P_L \propto \left(1 - \frac{t}{\tau_0}\right)^{-3\gamma/(\gamma+1)}$$

$$= \left(1 - \frac{t}{\tau_0}\right)^{-15/8} \qquad (7.130)$$

for $\gamma = \frac{5}{3}$.

The laser power absorbed by the plasma must therefore be increased very rapidly towards the end of the time during which compression takes place. The simple model used here breaks down as $t \to \tau_0$, and no analytical treatment of the final stages has yet been published.

8.1. DEGENERACY

At high densities, as a consequence of the exclusion principle, the Fermi energy of an electron gas becomes important. It is given by

$$w_F = \frac{h^2}{8m_e}\left(\frac{3}{\pi}n_e\right)^{\frac{2}{3}} \qquad (7.131)$$

and is shown plotted as a function of n_e in Figure 7.38. If the temperature $\kappa_B T_e$ is less than the Fermi energy the electrons are said to be *degenerate*. The pressure of the degenerate electron gas is

$$p_F = \frac{2}{5}n_e w_F\left[1 + \frac{5\pi^2}{12}\left(\frac{\kappa_B T_e}{w_F}\right)^2 - \frac{\pi^4}{16}\left(\frac{\kappa_B T_e}{w_F}\right)^4 + \dots\right]. \qquad (7.132)$$

For degenerate, non-relativistic electrons ($n_e \lesssim 10^{30}$ cm^{-3}) the isentropic exponent $\gamma = \frac{5}{3}$: for extremely relativistic electrons $\gamma = \frac{4}{3}$ (see e.g. Landau & Lifshitz (1959), Chapter 5).

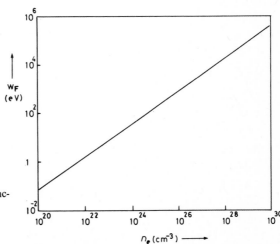

7.38 Fermi energy of electron gas as function of density

The minimum pressure in a gas with an electron density n_e occurs when $\kappa_B T_e \ll \mathrm{w_F}$:

$$(p_F)_{\min} = \tfrac{2}{5} n_e \mathrm{w_F}. \tag{7.133}$$

Thus in order to achieve a plasma electron density of 5×10^{26} cm^{-3}, a minimum pressure of almost 10^{12} atmospheres must be applied. Even higher pressures will be necessary if the electron temperature rises above the Fermi energy during compression.

8.2. COMPUTER CALCULATIONS OF SPHERICAL IMPLOSIONS

Computers have been used to obtain very detailed analyses of the development of the plasma produced in spherical geometry by a laser pulse of a given shape. Nuckolls et al. (1972) published the results of calculations for a drop of liquid deuterium–tritium mixture of radius 0·4 mm (mass 5×10^{-5} g), heated by a spherically symmetric 60 kJ pulse of laser light at a wavelength of 1 μm. The one-dimensional Lagrangian computer code took many physical processes into account, including nuclear reactions and the heating of the plasmas by the charged reaction products (see Chapter 4). Initially a weak laser pulse is applied to produce an absorbing region out to a radius of 1 mm. The main laser pulse is then applied, beginning with a power of 10^{11} W and rising over 20 ns, slowly at first but increasingly rapidly, to a maximum of 10^{15} W. The time dependence of the laser power is shown in Figure 7.39 together with curves showing the behaviour of other parameters.

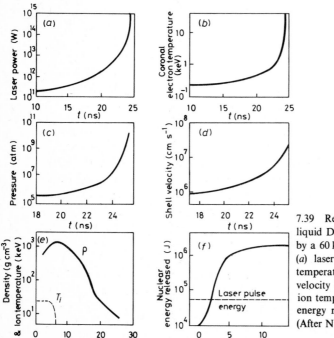

7.39   Results of computations for a spherical liquid DT target of radius 0·4 mm, imploded by a 60 kJ pulse of light of wavelength 1 μm. (a) laser pulse shape, (b) coronal electron temperature, (c) ablation pressure, (d) inward velocity of compressed shell, (e) density and ion temperature in central region (f) nuclear energy released (on a displaced time scale). (After Nuckolls et al. (1972))

When the main pulse is switched on a shock wave travels into the sphere at about 10^6 cm s^{-1}. The shock is weak in the outer region but increases in strength as the result of convergence and heats the central core of the target significantly. The region behind the initial shock is compressed, essentially isentropically, into a dense imploding shell which travels inwards at a speed rising to 3×10^7 cm s^{-1} (Figure 7.39(d)). The ablation pressure during this period is rising from 10^5 to 10^{11} atm: the electrons in the shell are Fermi-degenerate. The internal pressures now become greater than the ablation pressure and the shell slows down, being compressed isentropically to densities exceeding 10^3 g cm^{-3}. The shock-heated central core is somewhat less dense but considerably hotter, the ion and electron temperatures exceeding 10 keV. Nuclear reactions yield 1·8 MJ of fusion energy over a period of about 10^{-11} s, at the end of which the central plasma is expanding rapidly.

The energy gain, defined as the ratio of nuclear energy released to laser energy supplied, was calculated by Nuckolls *et al.* for four different laser pulse energies and

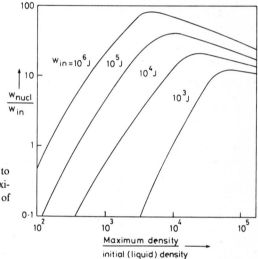

7.40 Ratio of nuclear energy released (w_{nucl}) to laser energy supplied (w_{in}) as function of maximum compression attained, for four values of w_{in}. (After Nuckolls *et al.* (1972))

for several pulse shapes giving different peak compression ratios. The results (Figure 7.40) show that excessive compression is not efficient.

Calculations using a similar code for spherical shells and for spheres were described by Clarke *et al.* (1973). Figure 7.41 shows the behaviour of the density and temperature of the plasma formed from a 7·5 μg solid DT shell, with an inner radius r of 0·54 mm and a thickness equal to $r/56$. The laser pulse energy, at a wavelength of 10·6 μm, was 5·3 kJ, all of which was assumed to be absorbed either in the corona or by some non-linear process at the critical density layer. The power rises from about 6×10^8 W to 4×10^{13} W, varying as $[1-(t/\tau_0)]^{-15/8}$ with $\tau_0 = 30$ ns. The inner 0·75 μg core is compressed to a density of 6×10^3 g cm^{-3}. The electron temperature reaches 10 keV in the absorbing corona, while the average ion temperature in the fully-compressed core is about 8 keV, although it reaches 30 keV briefly. The nuclear energy yield is 58 kJ.

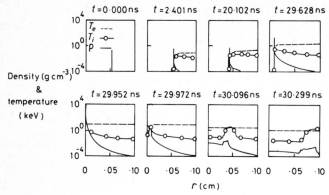

7.41   Density and temperature distributions calculated for a 7·5 μg DT shell imploded by 5·3 kJ of CO_2 laser radiation. (After Clarke et al. (1973))

The results of several computations for different targets are shown in Figure 7.42. The choice of the initial laser power has a considerable effect on the nuclear energy yield, as may be seen from Figure 7.42(a) for several different shells. The optimum initial power varies with the mass M as $M^{1·50}$ for spheres and as $M^{1·38}$ for shells. Figure 7.42(b) for 60 μg spheres shows that the energy gain has a sharp threshold as the input energy is increased. The optimum input energy is about 0·7 kJ/μg.

The computed variation with density of the mean ion temperature in the 6 μg core of a 60 μg sphere is shown in Figure 7.42(d). In this example the laser power was initially $1·3 \times 10^{10}$ W and varied as $[1-(t/\tau_0)]^{-2}$ with $\tau_0 = 20$ ns. The ion temperature reaches a maximum of 95 keV.

Further computations for uniform DT spheres, together with some results for LiDT spheres, were described in detail by Fraley et al. (1974a).

In another paper Fraley et al. (1974b) discussed several different spherically-symmetric target configurations. Hollow spheres of high-Z material containing DT fuel have several advantages over spherical fuel pellets. The high-Z shell can act as a container for the fuel at room temperature. It shields the fuel from fast electrons (see Section (8.4)). It allows the use of longer laser pulses. Even better are high-Z shells with solidified fuel deposited as a thin layer on the inner surface. More complex multi-layer constructions may also be advantageous.

8.3. SYMMETRY AND STABILITY

The very high densities and temperatures predicted by these calculations are strongly dependent on the spherical symmetry of the implosion of the dense core. In any real system it will not be possible to arrange for uniform irradiation of the target from all directions. However, the laser light is absorbed in the corona outside or near the critical density surface, whence energy is transported by conduction to the much denser ablating surface (Figure 7.43). The thickness of the conduction region is considerable, and Henderson & Morse (1974) have shown that even if the laser

7.42 Some results of computations for shell and spherical targets of solid DT. (a) Ratio (output energy (w_{out})/input energy (w_{in})) as function of initial laser power P_{to} for shells of initial inner radius 5.4×10^{-2} cm:

	$M(\mu g)$	$w_{in}(kJ)$		$M(\mu g)$	$w_{in}(kJ)$
(1)	3	2.2	(5)	26	18.2
(2)	5	3.5	(6)	60	43
(3)	7.5	5.3	(7)	250	178
(4)	10.8	7.5			

P_t varies with time as $P_{to}[1-(t/\tau_0)]^{-1.875}$, with $\tau_0 = 30$ ns. (b) Output energy as function of input energy for a sphere with $M = 60$ μg, showing optimum input energy $(w_{in})_{opt}$. (c) Optimum initial laser power $(P_{to})_{opt}$ as function of mass for shells as in (a); and also for spheres with P_t varying as $(1-(t/\tau_0))^{-2}$ with $\tau_0 = 20$ ns. (d) Variation with density of the mean ion temperature in the 6 μg core of a 60 μg sphere. (e) Optimum ratio w_{out}/w_{in} as function of optimum values of P_{to} for spheres and for shells. (f) Optimum ratio w_{out}/w_{in} as function of initial inner radius R of 7.5 μg shells, for two values of τ_0. (After Clarke et al. (1973))

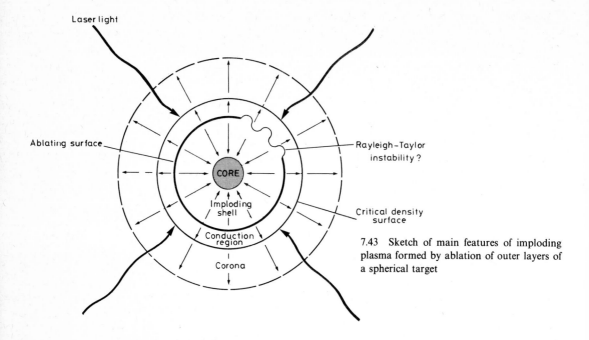

7.43 Sketch of main features of imploding plasma formed by ablation of outer layers of a spherical target

irradiation is non-uniform conduction tends to make the pressure at the ablation surface much more symmetric than might be expected from the laser energy distribution. This smoothing effect improves with increasing radius of the critical density surface, which is an argument in favour of using long-wavelength laser light.

The ablating surface separates two media of different densities, the inward acceleration being applied from the less dense to the denser medium. Under these conditions Rayleigh–Taylor instability (Taylor (1950b); see Chandrasekhar (1961)) can occur, and any perturbation of the spherical surface will grow. At a plane interface, the growth rate is given by

$$\gamma_\lambda = \left(\frac{2\pi a}{\lambda}\right)^{\frac{1}{2}} \tag{7.134}$$

where a is the acceleration and λ the wavelength of the perturbation. The acceleration due to laser ablation might be of order 10^{16} cm s^{-2}, and the longest wavelengths about 10^{-2} cm, giving growth rates of at least 10^9 s^{-1}. Nuckolls $et\ al.$ (1972) suggested that the amplitude of the instability will be restricted by ablative 'fire polishing': any salient high-density region will be nearer to the critical density surface which acts as the heat source and will thus be more rapidly ablated. The effectiveness of this stabilizing process depends on the thickness of the conduction region between the critical density layer and the ablating surface. Shiau $et\ al.$ (1974) concluded from their analysis for homogeneous DT spheres that the ablating surface is not stabilized against small departures from spherical symmetry. However, Henderson $et\ al.$ (1974), using a similar perturbation technique, found that for conditions relevant to laser fusion experiments the surface is positively stable. Neither analysis took into account the possible effects of spontaneously-generated magnetic fields, which would reduce the thermal conductivity.

In the case of a hollow dense shell, containing less-dense fuel, Rayleigh–Taylor instability may occur at the inner surface of the shell towards the end of the implosion when the surface is decelerating, and at the outer surface of the shell during the earlier stages when the shell is accelerating (as in the case of the homogeneous sphere). Fraley et al. (1974b) found that to ensure acceptably low growth rates for the instability modes, the radius of the shell must not exceed five times its thickness.

8.4. EFFECTS OF FAST ELECTRONS

The ablating plasma consists of two regions. In the outer corona, with electron density $n_e \lesssim n_{ec}$, laser energy is absorbed by the plasma electrons directly. Between the critical density surface and the ablating surface is the conduction region (this region really extends into the corona). If the electrons in the corona are heated much more rapidly than they can be cooled by the electrons in the conduction region their energy will rise and their temperature will 'decouple' or 'run away' from that of the conduction region, as discussed in Section (4.4). Laser energy will then be expended mainly in producing a very hot, low-density plasma, and not in ablation and compression. These effects have been considered by Kidder & Zink (1972).

The rate of change of the electron temperature T_c in the conduction region is given by

$$\frac{dT_c}{dt} = \frac{T_H - T_c}{\tau_{ee}} \tag{7.135}$$

where T_H is the electron temperature of the hot corona and τ_{ee} is the equilibration time. The coupling between the electrons in the two regions is indicated by $1/\tau_{ee}$, which varies with the mean electron temperature $\bar{T}_e = (T_H + T_c)/2$ as sketched in Figure 7.44. As we would expect from Section (4.4), the coupling, which depends on the electron density and is limited by n_{ec}, is wavelength dependent. Figures 7.45(a) and (b) show the computed electron and ion temperature distributions in a plasma formed from a 50 μg deuterium sphere heated by 1 ns, 10^5 J laser pulses with Gaussian

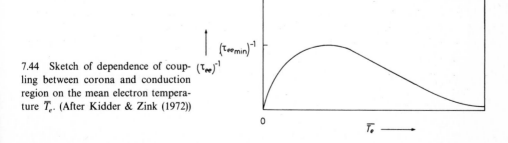

7.44 Sketch of dependence of coupling between corona and conduction region on the mean electron temperature $\bar{T}_e$. (After Kidder & Zink (1972))

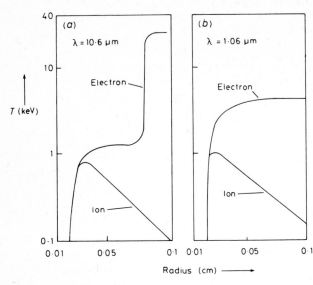

7.45 Radial distribution of temperature in plasmas produced from $50\,\mu g$ deuterium pellets (a) by a CO_2 laser and (b) by a neodymium glass laser. Pulse energy 10^5 J, duration 1 ns. (After Kidder & Zink (1972))

time profiles at wavelengths of $10.6\,\mu m$ and $1.0\,\mu m$. It was assumed that all light penetrating to the critical density layer was absorbed there by non-linear effects. The decoupling at $10.6\,\mu m$ is apparent.

At very high irradiances absorption as the result of collective effects near the critical density surface becomes important, especially at long wavelengths. Strongly non-Maxwellian electron velocity distributions may be expected, with a low-density, high-energy tail containing much of the energy. At the laser irradiances required for nuclear fusion, electrons with energies of several hundred keV may be expected, with mean free paths considerably larger than the dimensions of the plasma. Nuckolls *et al.* (1972) found that these 'suprathermal' electrons can preheat the imploding shell and make compression more difficult. Their effects were discussed in detail by Morse & Nielson (1973).

High-energy electrons will be more rapidly slowed down if high-Z atoms are present in the plasma. As Morse and Nielson pointed out, even if the atoms are not completely stripped of electrons they appear to an energetic electron to have their full nuclear charge. The addition of heavy atoms to the outer layers of a target sphere would help to prevent decoupling and preheating of the imploding shell, and might also give improved ablation momentum transfer. On the other hand the presence of high-Z atoms would mean that the degeneracy pressure would be greater for a given ion density.

9. MAGNETIC EFFECTS

9.1. PLASMA IN AN EXTERNALLY APPLIED FIELD

If a high-conductivity plasma is created in a magnetic field, it can only expand freely along the magnetic lines of force, and in the absence of diffusion must overcome the magnetic pressure in order to expand orthogonally to them. To produce a field-free cavity of volume V in a field B requires energy W given by

$$W = \frac{B^2}{2\mu_0} V. \tag{7.136}$$

When B is expressed in kG, V in cm^3 and W in J, $W = 4 \times 10^{-3} B^2 V$. The maximum volume to which the plasma can expand is found by setting W equal to the plasma energy (Dawson (1964)). The density of the plasma formed by a laser pulse from a small finite target falls more slowly in a magnetic field, and the onset of transparency, which often limits the duration of the heating period, is delayed. As the plasma expands in the field, currents are generated in the surface which result in ohmic heating. In this way some of the expansion energy is converted into thermal energy. The interface between the plasma and the region of greater field strength is unstable against corrugations, which are known as 'interchange' instabilities and are another form of Rayleigh–Taylor instability. The resulting turbulence may again convert some of the expansion energy into energy of random motion, but it also destroys containment. The growth rates of these instabilities do not appear to have been investigated under the conditions which arise in laser-heated plasmas.

The thermal conductivity transverse to a strong magnetic field varies as $T_e^{-\frac{1}{2}} B^{-2}$ (Spitzer (1962)) so if a thin-film target is placed perpendicularly to the magnetic lines of force at the focus of a laser beam, the plasma formed in the focal area will tend to be isolated thermally from the rest of the target material (Shkuropat & Shneerson (1967)).

The expansion of a plasma in a magnetic field has been investigated theoretically by several authors. The simple model of a cylindrical plasma which expands freely, parallel to its axis, along the magnetic field lines but not at all across them was discussed by Shkuropat & Shneerson (1967), using the one-dimensional equivalent of the model of Section (6.1). For a constant effective laser power it is found that after a long time a half of the incident laser energy is converted into thermal energy, in contrast to a quarter for the case of three-dimensional expansion. As compared to a spherical target of the same mass in the absence of a magnetic field, the onset of transparency is markedly delayed by a strong magnetic field provided the initial length of the plasma is much greater than its diameter. Related calculations were made by Nguyen & Parbhakar (1973).

The behaviour of a plasma of uniform temperature and density expanding into a magnetic field after the laser pulse has ended was investigated by Bhadra (1968), who for simplicity assumed not only a spherically symmetric uniform plasma but also a spherically symmetric magnetic field. This configuration, though physically unrealizable, shows a number of interesting general features. If the plasma has zero resistivity, it bounces periodically off the magnetic field, the maximum radius of the plasma boundary being

$$(r_b)_{max} \simeq \left(\frac{\mu_0 N \kappa_B T_h}{8\pi B^2} \right)^{\frac{1}{3}}$$

(7.137)

and the period

$$\tau_b \simeq \frac{(r_b)_{max}}{V_{b\infty}}.$$

(7.138)

Here N is the total number of particles, T_h the temperature at the end of the laser pulse and $V_{b\infty}$ the asymptotic expansion velocity. If the resistivity is finite and not too great, the plasma diffuses outwards across the field, and the bounce radii increase with time. A sufficiently resistive plasma simply diffuses steadily outwards without bouncing.

Bernstein & Fader (1968) considered the case of a spherical, resistive, collision-dominated plasma, which was assumed to retain its spherical shape as it expands in a uniform magnetic field. Viscous and electrostatic forces were neglected. Numerical results were given for a 70 eV LiD plasma expanding in a field of 300 G. The behaviour of this plasma in the absence of a magnetic field had earlier been computed by Fader (1968) (see Section (6)). Haught et al. (1970) included a viscous term and calculated the behaviour of a plasma formed from a particle of LiH, 40 μm in diameter, in a magnetic field of 10 kG, by a laser pulse of half-intensity duration 10 ns and peak power 500 MW, focused to a spot of 200 μm diameter. Figure 7.46 shows the results of calculations of the expansion with and without the magnetic field. The laser pulse peak power

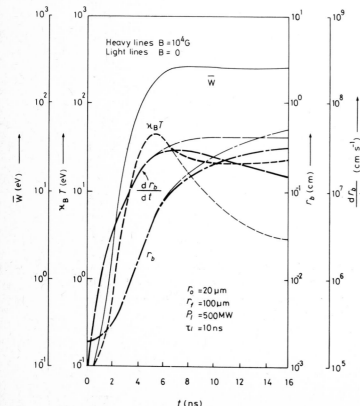

7.46   Calculations of heating and expansions of plasmas formed from small lithium hydride spheres with and without a magnetic field of 10^4 G. (After Haught et al. (1970))

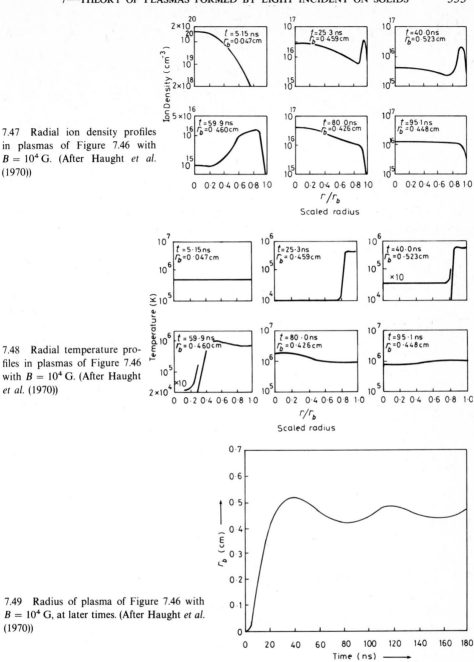

7.47 Radial ion density profiles in plasmas of Figure 7.46 with $B = 10^4$ G. (After Haught *et al.* (1970))

7.48 Radial temperature profiles in plasmas of Figure 7.46 with $B = 10^4$ G. (After Haught *et al.* (1970))

7.49 Radius of plasma of Figure 7.46 with $B = 10^4$ G, at later times. (After Haught *et al.* (1970))

occurred at $t = 10$ ns. The radial density and temperature profiles in the plasma at six different times are shown in Figures 7.47 and 7.48. Figure 7.49 shows the radius of the plasma boundary at later times: the bouncing motion at a radius of about 0·5 cm is similar to that calculated by Bhadra (1968).

The expansion of a uniform spherical plasma into a uniform background plasma in a uniform magnetic field was discussed by Wright (1971, see also Book & Clark (1973), Wright (1973)), who found an analytic solution for the magnetic field configuration during the early stages of the expansion, and investigated its stability.

Haines (1963) showed that departures from azimuthal symmetry can impart a rotation to an isolated plasma in a magnetic field. Bhadra (1972) suggested that a plasma expanding in an asymmetric magnetic field might receive angular momentum as it bounces off the field.

Extremely high magnetic fields have been produced by imploding a conducting cylinder within which a strong field has been initially established parallel to the axis. As the cylinder shrinks the field lines are compressed. By this technique fields of the order of 10^7 G have been obtained. A similar compression of the lines of force may be produced inside a fixed conducting cylinder if a conducting plasma, created on the axis by a laser pulse, expands against the field. This configuration should have a strong confining effect on the plasma, as was suggested by Cavaliere et al. (1967), who carried out approximate calculations of the expansion process. Martineau & Tonon (1969b) investigated theoretically the development of a plasma heated by a laser pulse of 6·6 GW peak power in a magnetic field of 1 MG within a conducting cylinder of radius 2 cm. The plasma was assumed to be perfectly conducting, with an elliptical shape as shown in Figure 7.50. The minor semi-diameter ℓ_2 oscillated rapidly at first whereas the major semi-diameter ℓ_1, parallel to the field lines, increased

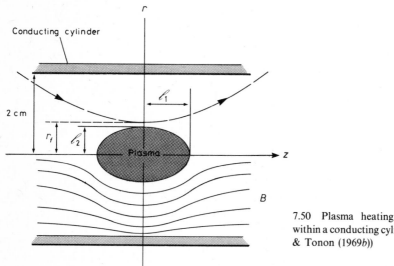

7.50 Plasma heating in a magnetic field within a conducting cylinder. (After Martineau & Tonon (1969b))

monotonically with time. Some numerically computed results are shown in Figure 7.51. The geometry of the laser beam and the time dependence of the laser pulse power were taken into account. The electron–ion equipartition time was also included, and as may be seen the electron and ion temperatures differ, although not very

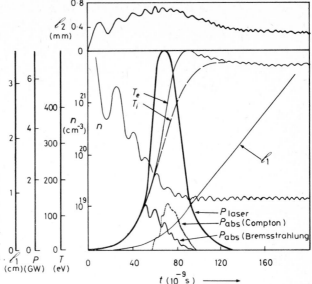

7.51 Calculations of deuterium plasma heating and expansion in the magnetic configuration of Figure 7.50, with $B = 10^6$ G, $N = 3 \times 10^{17}$ ions, $r_f = 0.25$ mm. Plasma is assumed to be perfectly conducting. (After Martineau & Tonon (1969b))

greatly. The contributions of inverse Bremsstrahlung and of induced Compton effect to the absorption of energy from the laser beam were both calculated and found to be roughly equal, as may be seen: the Compton effect is more important near the peak of the laser pulse, when the electron temperature is high and the electron and ion densities are low. Calculations for a similar model but without the conducting cylinder were reported by Beaudry & Martineau (1972) for spherical geometry, and by Martineau (1974) (taking into account the finite conductivity of the plasma) for elliptical geometry. Pashinin & Prokhorov (1971, 1972) have discussed the possibility of using a cylindrical deuterium target inside a heavy shell as the target for laser fusion experiments. The laser light would enter axially, and the heavy shell would assist inertial confinement. A megagauss magnetic field would be applied initially, parallel to the cylinder axis, to limit thermal conduction losses from the deuterium to the walls.

Schwirzke & Tuckfield (1969) have pointed out that if a plasma expands sufficiently in a magnetic field for the collision frequency to become small, two-stream instabilities may develop (see Chapter 1), which will increase the resistivity of the plasma and thus reduce the rate of cooling by thermalizing some of the energy of expansion.

Dunn & Lubin (1970: see also Lubin et al. (1969)) discussed the radiation to be expected from a plasma expanding in a magnetic field. In addition to normal cyclotron radiation from the volume of the plasma, whose intensity varies as B^2 and as T_e, a surface contribution is calculated whose intensity varies as B and as $T_e^{\frac{3}{2}}$. Lubin suggested that the time history and the extent of the interaction of the expanding plasma could be determined from a study of the emission spectrum in the vicinity of the cyclotron frequency.

9.2. SPONTANEOUS MAGNETIC FIELDS

Spontaneous magnetic fields can occur in laser-heated plasmas, unless the geometry of the interaction is perfectly spherical. Stamper *et al.* (1971), and Stamper (1972) have shown that these fields, which are due to pressure and temperature gradients in the plasma, may be as large as 10^5 G with a single 2 GW laser beam focused on a solid target. A simple circuit model has been described by Tidman & Stamper (1973).

Calculations by Chase *et al.* (1973) suggest that such fields could have important effects on the transport coefficients in the plasma, and by impeding the flow of heat may substantially enhance the neutron yields from massive solid deuterium targets heated by a single laser beam. It should be noted, however, that the spontaneous magnetic field energy density is only a small fraction of the electron kinetic energy density.

Computer simulations by Winsor & Tidman (1973) have shown the effect of the spontaneous magnetic field on plasmas formed by neodymium laser pulses incident on massive high-Z targets. Figures 7.52 and 7.53 show contours of several parameters calculated with and without the magnetic field, which reaches a maximum of about

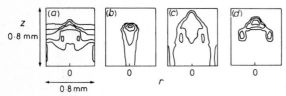

7.52   Computed contours in (r, z) space of plasma parameters for an aluminium target (see text). Contours are at intervals of $\times 10$, the lowest levels being (*a*) 10^{-5} g cm^{-3}, (*b*) 10^{13} W cm^{-3}, (*c*) 10^4 J cm^{-3}, (*d*) 10^{13} W cm^{-3}. Laser light is incident from below: effects of spontaneous magnetic fields are included. (After Winsor & Tidman (1973))

(*a*) Density   (*b*) Laser power absorbed per unit volume
(*c*) Energy density   (*d*) Radiated power per unit volume

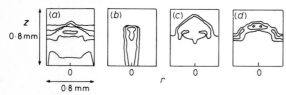

(*a*) Density   (*b*) Laser power absorbed per unit volume
(*c*) Energy density   (*d*) Radiated power per unit volume

7.53   As Figure 7.52, but omitting effects of magnetic fields. (After Winsor & Tidman (1973))

10^6 G. The contours are for an aluminium target heated by a 20 J laser pulse with Gaussian space and time distributions (50 μm half width, 1 ns pulse length), at 1 ns. Figure 7.54 shows parameters plotted as functions of distance into the target from the surface, at a radial distance of 40 μm from the axis of the laser beam. The laser beam is incident from the left. The ripples in the laser absorption curve are ascribed to an acoustic wave resulting from the density dependence of the inverse Bremsstrahlung absorption, and tend to be absent when $B = 0$. Winsor & Tidman suggested

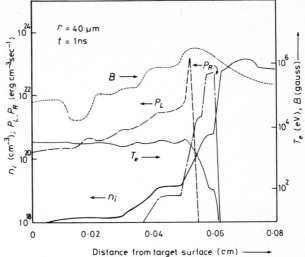

7.54 Profiles of spontaneous magnetic field (B), laser power absorbed per unit volume (P_L), power radiated per unit volume (P_R), ion density (n_i) and electron temperature T_e computed for plasma from an aluminium target (see text). (After Winsor & Tidman (1973))

that a further instability may also arise even in uniformly irradiated targets, because a perturbation in the temperature somewhere in the critical density layer will generate a magnetic field which enhances the perturbation.

9.3. LASER-COMPRESSED MAGNETIC FIELDS

Askaryan *et al.* (1973) have pointed out that if a magnetic field of 10^6 G is established in a highly-conducting target before laser-ablation compression increases the density by a factor of 3×10^4, the field in the compressed material might be in the region of 10^9 G. This is considerably higher than has yet been achieved on earth. The magnetic pressure would then be about 4×10^{10} atmospheres, which is small compared to the plasma pressures discussed in Section (8).

8

Experimental Studies of Plasmas Formed by Light Incident on Solids

1. INTRODUCTION

Much experimental work has been reported on plasmas formed from a wide variety of solid targets by ruby, neodynium and carbon dioxide lasers. We shall find it convenient to group together experiments using a particular target geometry, considering in turn massive solids, thin films, and isolated grains or filaments as targets, in vacuum. A group of experiments of special interest from the point of view of nuclear fusion research involves the use of solid hydrogen isotope targets: this work will be described separately. Within each group the results will be arranged roughly in the order of increasing laser power, which coincides broadly with the chronological order. We shall then consider the effects of magnetic fields on these plasmas, and finally we shall discuss plasmas formed from solid targets in gases.

The techniques which have proved useful in investigating plasmas formed in gases are all equally suitable for studying plasmas formed from solid targets. Two additional diagnostic methods are available for plasmas from solids, provided they are formed in a sufficiently good vacuum. These are far ultraviolet or soft X-ray spectroscopy, and the collection and analysis of charged particles.

2. PLASMAS FORMED FROM MASSIVE TARGETS IN VACUUM

2.1. MILLISECOND LASER PULSES

The earliest experimental studies of the effects of focused laser pulses on solid targets in vacuum were published in 1963. They were made with free-running ruby lasers whose output energies were of the order of one joule, in multiple spikes over a period of up to about a millisecond. Verber & Adelman (1963), Lichtman & Ready (1963), Giori et al. (1963), Honig & Woolston (1963) and Cobb & Muray (1965) all reported the emission of copious electrons from metal targets. Current densities of many hundreds or thousands of amps cm^{-2} were deduced from the current, measured with apparatus similar to that shown in Figure 8.1, and the observed area of the focal spot. Honig & Woolston (1963) and Honig (1963) also measured relatively small currents of thermal ions emitted from various targets, and deduced surface temperatures of 5000 to 10 000 K. The number of charged particles emitted from germanium was found to be less than a thirtieth of that from tantalum under similar conditions, but the holes drilled in germanium targets were ten times deeper. This may be due in part to the absorption length for λ 6943 Å being 1200 Å in germanium, in contrast to 95 Å in tantalum.

8.1 Arrangement for determining electron current from metal target illuminated by laser light. (After Lichtman & Ready (1963))

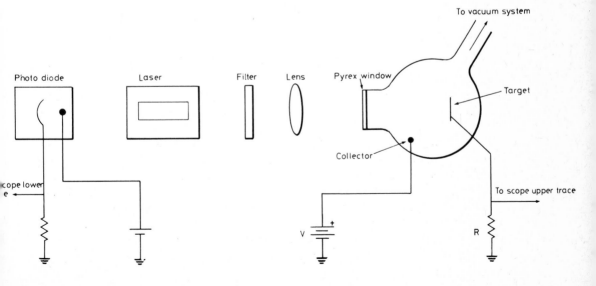

The complex time and space structure of the output of a free-running ruby laser (see for example Röss (1969)) made interpretation difficult. Lichtman & Ready (1963) also measured the electron emission due to light from a free-running neodymium laser producing a few well-separated spikes: the results suggested that the electron emission could be explained as being thermionic: the laser acts purely as a heat source at the surface of the target. Later work (Verber & Adelman (1965)) confirmed this view, at least for pulses delivering an irradiance of up to 10^5 W cm^{-2} for a millisecond. A more energetic ruby laser pulse of 10 J total energy with a duration of about 0·5 ms was found by Arifov et al. (1967) to produce a plasma when focused with a lens of 50 mm focal length on to a target of aluminium, copper, steel or titanium. The irradiance, if we assume a focal spot area of 10^{-3} cm^2, would have been about 2×10^7 W cm^{-2}. The plasma was studied by measuring the absorption of 8 mm microwaves, which showed that electron densities exceeding 10^{13} cm^{-3} were present.

2.2. NANOSECOND LASER PULSES

2.2.1. Laser power up to 100 MW

The invention of the Q-switched ruby laser (McClung & Hellwarth (1962)) immediately raised the available laser pulse power from kilowatts to megawatts. Shortly thereafter, highly-ionized plasmas were being produced from solid targets in several laboratories. The first measurements of ion energies from laser heated plasmas were reported by Linlor (1963), who used a time of flight method. The arrangement is shown in Figure 8.2. With a 5·4 MW laser pulse (0·2 J, 40 ns) focused by a lens of 67 mm focal length, Linlor found that the distribution of the energies of ions emitted

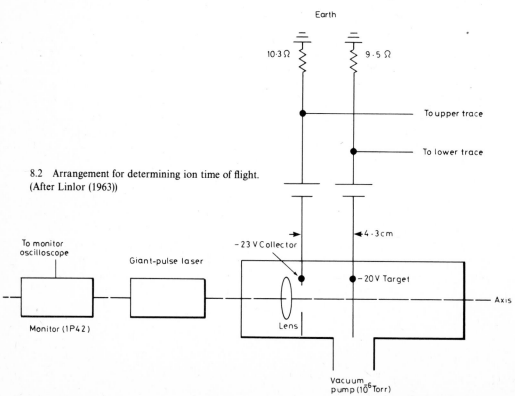

8.2   Arrangement for determining ion time of flight.
(After Linlor (1963))

from metallic targets with widely varying atomic masses (Al to Pb) always had a maximum at about 1 keV. This result was confirmed by Fenner (1966) who found that with a less powerful laser (8×10^{-3} J, 30 ns) ions from targets of different metals all had very similar energy distributions, peaking at 200–300 eV. However, in a composite target, such as gold and aluminium, all the ions were found to have the same velocity distribution, which corresponded to that expected for ions of the heavier element. Isenor (1964a) found from similar experiments that for magnesium the velocity of the fastest ions increased more or less proportionally with the peak power of the laser pulse: the maximum ion energy observed again corresponded to several hundred eV. In a subsequent paper (Isenor (1964b)) more careful measurements over a wider range of laser powers were described which indicated that the energy of the fastest ions was directly proportional to the peak power of the laser pulse and was independent of the atomic mass (in agreement with Linlor). Isenor used a photomultiplier removed from its envelope as the ion detector. His laser pulses, with a peak power of 2 MW (0.1 J, 50 ns) were focused by a lens of 50 mm focal length: the irradiance was probably at least 10^8 W cm^{-2}. Ion energies as high as 180 eV were also reported by Bernal et al. (1966), with a rather lower irradiance of 70 MW cm^{-2}, for ions desorbed from a tungsten target. Namba & Kim (1966, 1968) carried out similar experiments at higher irradiances with a 3 MW neodymium glass laser and a 5 MW ruby laser, both giving pulses of about 50 ns duration. They observed that the positive ion signal had two peaks. The energies of the fast ions increased as the square of the irradiance and reached 100 eV at 3×10^9 W cm^{-2}: the energies of the slow ions only attained 25 eV, and varied directly with the irradiance. Again, the ion energy was found to be independent of the atomic mass.

The momentum imparted to solid targets by laser pulses of about 6 MW peak power (0.3 J, 50 ns) was measured by Neuman (1964) by means of a piezo-electric transducer. When the target material was ejected in vapour form, as was the case with most of the metals investigated, almost the same momentum was delivered to every target. Lead was an exception: molten globules were observed to be ejected as well as vapour, and twice the usual momentum was measured. Again, splinters were ejected from glazed porcelain, and a high target momentum resulted. By defocusing the light on the target, the energy density of the pulse could be reduced at constant total energy. The momentum of the target remained constant until the irradiance fell below 8×10^8 W cm^{-2} (energy per unit area 4×10^{-3} J cm^{-2}): below this level the momentum varied roughly as the square root of the irradiance. This result showed clearly the shielding effect of the dense plasma generated by high fluxes.

Spectroscopic evidence of high plasma electron temperatures in neodymium laser heated carbon plasmas was provided by Archbold & Hughes (1964), who used a method described by Kaufman & Williams (1958) (see also Kunze et al. (1968)) to estimate the electron temperature from the intensity of the CV λ 2271 Å line. A lower limit of about 10 eV was found. The first spectroscopic studies of laser produced plasmas in the vacuum ultraviolet were made by Ehler & Weissler (1966). The spectra emitted by plasmas produced from several different metals by 10 MW ruby laser pulses (0.4 J) were studied in the normal and grazing incidence regions. Tungsten and platinum targets both showed a strong continuum with a peak at about 200 Å, corresponding to a temperature of 30 eV (see Chapter 2, Section (3.3)), while the weaker aluminium and beryllium continua peaked at about 140 Å and 175 Å respectively. The time dependence of the ultraviolet emission followed that of the laser pulse closely. Multiply-ionized atoms were reported by Gibson *et al.* (1968) in plasmas produced by *Q*-switched carbon dioxide laser pulses with peak power as low as 0·35 MW, lasting about 400 ns. Lines of the silicon spectrum of all stages of ionization up to Si IV were recorded photographically when a silicate glass target was used (Figure 8.3). Because of the long wavelength of the carbon dioxide laser radiation the expanding plasma continued to be heated at low densities when it would have been transparent to ruby or neodymium laser light. Thus the spectral lines were less perturbed by Stark effect when the carbon dioxide laser was used. This has important implications for laser applications in spectrochemical analysis. The plasma was a good enough reflector, for the laser light wavelength, to cause oscillation in the amplifier, which generated secondary pulses.

Izawa *et al.* (1968) investigated light scattering from plasmas obtained with 20 MW

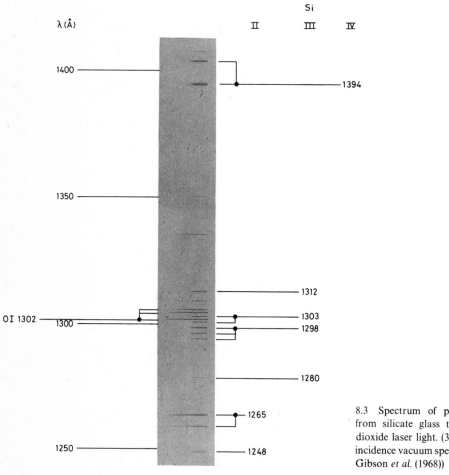

8.3 Spectrum of plasma generated from silicate glass target by carbon dioxide laser light. (3m Hilger normal incidence vacuum spectrograph). (After Gibson *et al.* (1968))

(0·6 J, 30 ns) neodymium glass laser pulses focused by a lens of 70 mm focal length on a carbon target. The beam from a non-Q-switched ruby laser giving 100 kW for 1 ms was passed through the plasma at right angles to the neodymium laser beam, 5 mm from the target surface, being brought to a focus in the plasma by a 300 mm lens. Scattered ruby light was detected at 90° to the incident beam by a photo-multiplier behind an interference filter of 10 Å passband. The signal fell rapidly during the first 4 μs after the neodymium laser pulse, and then rose again. The first peak was assumed to be due to Thomson scattering by plasma electrons (see Chapter 3) and the second to Rayleigh scattering from neutral carbon atoms. On this basis the results shown in Figure 8.4 were obtained. The electron density was also measured by a double electrostatic probe method (see for example Chen (1965)) at a rather

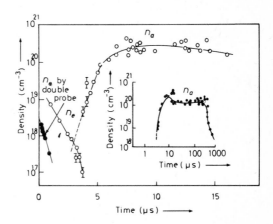

8.4 Time dependence of electron and neutral atom densities 5 mm from carbon target surface, as determined from scattering observations, and also double probe measurements of electron density 7 mm from target. (After Izawa *et al.* (1968))

greater distance (7 mm) from the target surface. The electron density 5 mm from the target attained 10^{19} cm^{-3}, and decayed exponentially with a time constant of about 1 μs. The neutral atom density exceeded 10^{20} cm^{-3} for 300 μs.

An experiment was reported by Carillon *et al.* (1970) in which the extreme ultra-violet *absorption* spectrum of a laser-heated aluminium plasma was studied. The arrangement is shown in Figure 8.5. The laser beam was divided into two equal parts, each of which was focused on to one of two aluminium targets. The laser power in each beam was 25 MW, and the pulse duration about 40 ns. Target I was placed so that continuous radiation emitted from a small region near its surface entered a two-grating, grazing-incidence spectrograph. The position of target II could be varied along the laser beam axis so that light from target I passed through any desired region of the plasma from target II. By comparing intensities with plasma from target I only, from target II only, and from both targets, the absorption in the plasma from target II could be measured. A mean value was obtained over the duration of the source continuum, which was about 40 ns. The relative timing of the arrival of the two laser pulses at their targets could be adjusted by an optical delay line. Figure 8.6 shows the transmittance of the plasma at 98 Å as a function of the distance z' from the hottest part of the plasma, along the laser axis. The hottest region was found to occur at a distance of about 0·5 mm from the target surface. For a delay time of

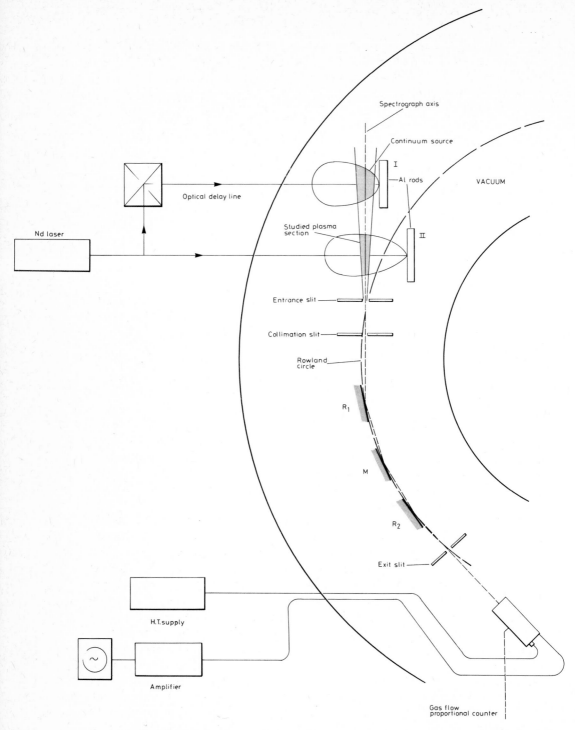

Spectrograph axis

Continuum source

I

Al rods

VACUUM

Optical delay line

Nd laser

Studied plasma
section

II

Entrance slit

Collimation slit

Rowland
circle

R_1

M

R_2

Exit slit

H.T.supply

Amplifier

Gas flow
proportional counter

8.5 Experimental arrangement for measurements of far ultra-violet absorption by laser-heated plasma. R_1, R_2—gratings. M—concave mirror. (After Carillon *et al.* (1970))

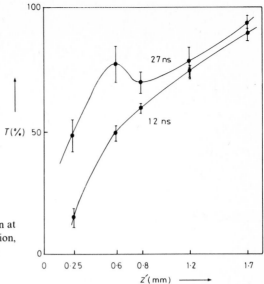

8.6   Transmittance $\mathscr{T}$ of plasma for radiation at 98 Å as function of distance z' from hottest region, along laser axis. (After Carillon *et al* (1970))

12 ns the transmittance increased monotonically with distance, but for a delay of 27 ns the transmittance had a minimum at 0·8 mm. The authors interpreted this minimum as being due to the presence, as a result of recombination, of significant numbers of ions with sufficiently low ionization potential to be photo-ionized by the 98 Å radiation ($\chi \lesssim 125$ eV). Radiative transfer processes thus appear to be quite important in these plasmas.

Allen (1971) has described an experiment by Schlier *et al.* in which a 20–30 keV electron beam of diameter 0·3 mm was used to probe the plasma produced by a 25 MW (1·3 J, 50 ns) ruby laser pulse. The integrated mass density along the beam was derived as a function of time and used to estimate the surface temperature and vaporization rate of the graphite target.

A detailed study of ions emitted from plasmas produced from several different solid target materials was reported by Langer *et al.* (1966, 1968a, see also Ducauze *et al.* (1965), Ducauze & Langer (1966) and Tonon (1966)). Measurements were made of the spatial distribution of electron and ion emission from plasmas produced by a neodymium glass laser of about 30 MW peak power (0·8 J, 30 ns) focused by a 40 or 50 mm lens on to the target. The experimental arrangement is shown in Figure 8.7. Thirty-two Faraday cups, placed equidistant from the focus of the lens, were used to collect ions. Figures 8.8(a) and (b) show the results for a gold target: similar diagrams were obtained from aluminium, beryllium and tungsten targets, but in the case of carbon the emission was almost isotropic in the half-planes. The H and V planes are the horizontal and vertical planes containing the laser beam axis: the S plane is parallel to the target surface, which is perpendicular to the laser beam axis. Varying the direction of polarization of the laser beam had no effect on the shapes of the distributions. Figure 8.9 shows the effect of astigmatism introduced by inclining the lens at an angle of 30° to the laser beam axis. In an analogous experiment

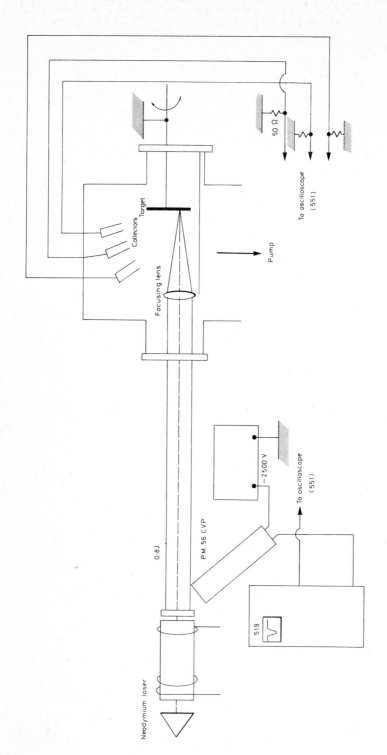

8.7 Arrangement for studying spatial distribution of electron and ion emission. (After Ducauze & Langer (1966))

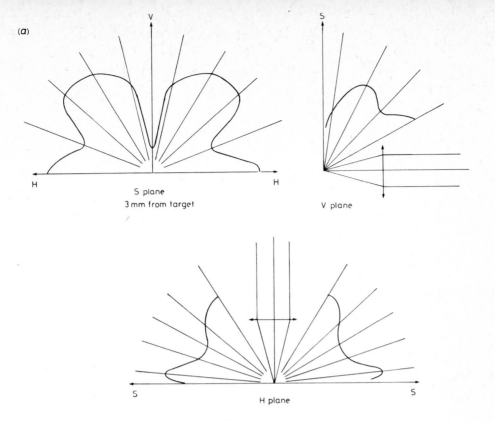

(a)

S plane
3 mm from target

V plane

H plane

Electron distribution

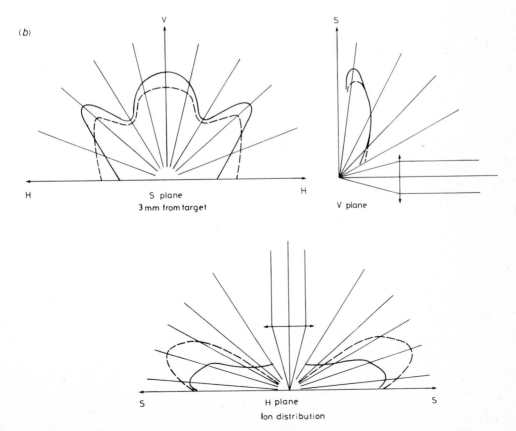

(b)

S plane
3 mm from target

V plane

H plane

Ion distribution

8.8   (a) Electron and (b) ion distributions from gold target (see text). (After Ducauze & Langer (1966))

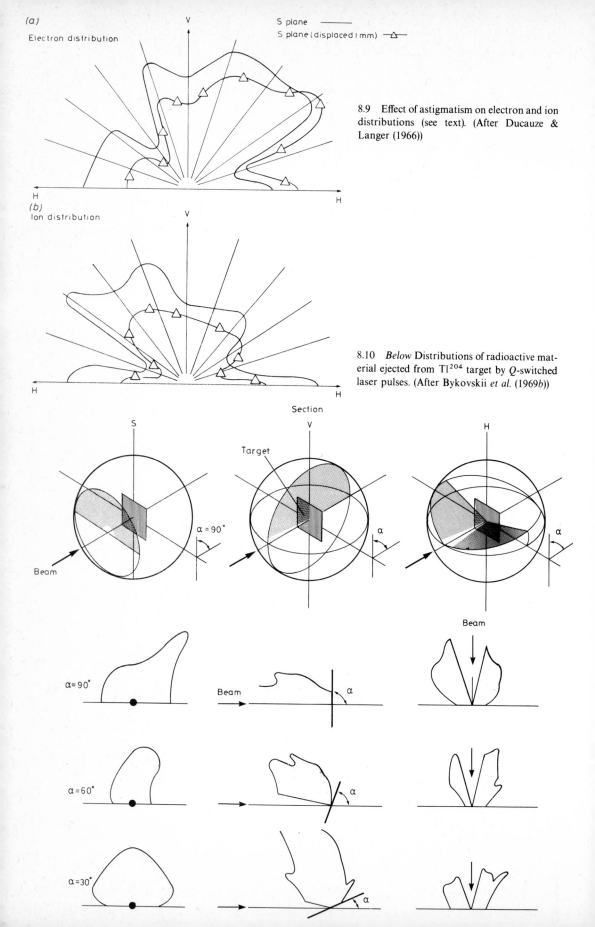

(a)

Electron distribution

V

S plane ————
S plane (displaced 1 mm) —△—

H   H

8.9   Effect of astigmatism on electron and ion distributions (see text). (After Ducauze & Langer (1966))

(b)

Ion distribution

V

H   H

8.10   *Below* Distributions of radioactive material ejected from Tl^{204} target by *Q*-switched laser pulses. (After Bykovskii *et al.* (1969*b*))

Section

S   V   H

Target

$\alpha = 90°$ α α

Beam

Beam

$\alpha = 90°$ Beam α

$\alpha = 60°$ α

$\alpha = 30°$ α

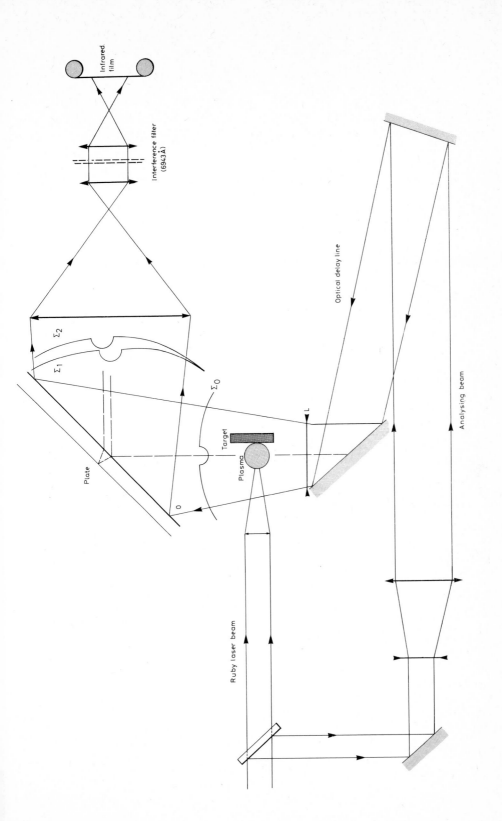

8.11   Interferometer for determining electron density in laser-heated plasma. (After Bobin *et al.* (1967))

Bykovskii *et al.* (1969*b*) used a radio-isotope technique to find the distribution of all the material ejected from the target. Their results for a Tl^{204} target are shown in Figure 8.10. The diagrams varied somewhat from shot to shot.

Experiments on plasmas produced by a ruby laser of maximum power 30 MW (0·8 J, 30 ns) focused on to a beryllium target by lenses of about 50 mm focal length were carried out by Bobin *et al.* (1967) and Langer *et al.* (1968*a*). An interferometric study was made of the average electron density as a function of distance from the target surface along the axis of the laser beam. A part of the laser pulse generating the plasma was used as the light source. The interferometer is shown in Figure 8.11. Optical delays of 20 and 40 ns were used to give the results shown in Figure 8.12.

Langer *et al.* also investigated the energy spectra of ions emitted when the same ruby laser was focused by a 50 mm lens on to targets of beryllium, carbon and

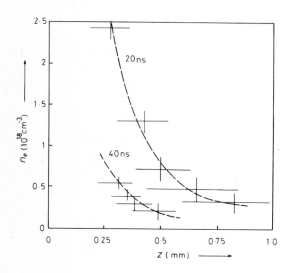

8.12 Average electron density as function of distance from target surface along laser beam axis, with interferometer beam delayed by 20 and 40 ns. (After Bobin *et al.* (1967))

molybdenum. Figure 8.13 shows the arrangement of the electrostatic analyser, which deflected the particles detected through 90° along a circular arc of radius 10 cm, and had an energy resolving power of 100. The theory is straightforward. When a potential difference v is applied to the plates, all ions passing through the analyser must have the same ratio of ion energy w_i to charge Z, and

$$\frac{w_i}{Z} = Sv, \qquad (8.1)$$

S being some constant. If the distance travelled by the ions of mass m_i from the target to the detector is d, the time of flight

$$t = d\left(\frac{m_i}{2w_i}\right)^{\frac{1}{2}}. \qquad (8.2)$$

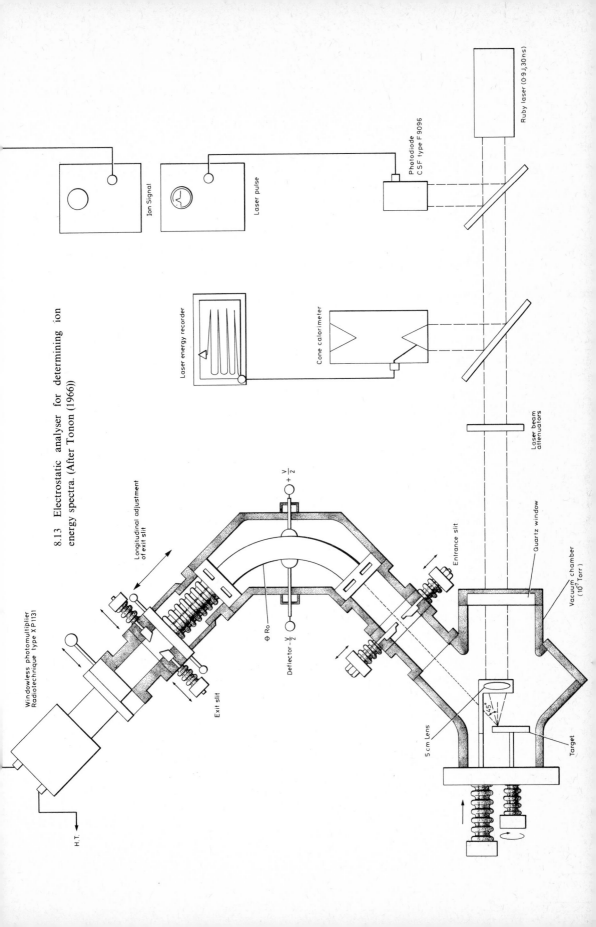

8.13 Electrostatic analyser for determining ion energy spectra. (After Tonon (1966))

With a constant potential difference applied to the deflectors,

$$t = d\left(\frac{m_i}{2Sv}\right)^{\frac{1}{2}}\left(\frac{1}{Z}\right)^{\frac{1}{2}} = \text{const.}\left(\frac{m_i}{Z}\right)^{\frac{1}{2}}. \tag{8.3}$$

If the target is pure, containing only atoms of mass m_i, and the plasma generated from it contains these atoms in several stages of ionization ($Z = 1, 2, \ldots$), then provided the range of energies is sufficiently broad the detector will receive the ions at times

$$t_1 = d\left(\frac{m_i}{2Sv}\right)^{\frac{1}{2}}, t_2 = t_1/\sqrt{2}, \ldots, t_Z = t_1/\sqrt{Z}. \tag{8.4}$$

Figures 8.14(a), (b) and (c) show the energy spectra of beryllium, carbon and molybdenum ions with a laser irradiance of 6×10^{10} W cm^{-2}. As may be seen, the maximum

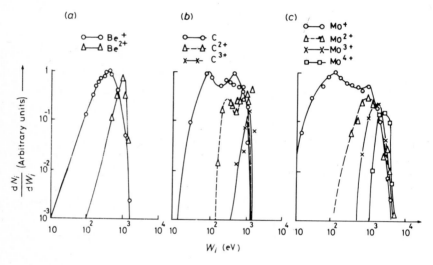

8.14   Energy spectra of ions from beryllium, carbon and molybdenum targets. (After Tonon (1966))

ion energy measured is the same for all ions of the same mass, but the mean ion energy increases with increasing Z. Figure 8.15 shows the ranges of ion energies observed from a carbon target, plotted against the irradiance: the sensitivity of the analyser was much greater for Figure 8.15 than for Figure 8.14(b). An analysis of these results was made by Allen (1970) on the assumption that the observed distributions of ion velocity could be fitted by a Maxwellian superimposed on a drift velocity, a model that had been applied to another plasma by Gorog (1968). The apparent ion temperatures thus obtained varied from 13 eV for C^{2+} to 40 eV for Be^+. These results are probably overestimates, because the ions are emitted over a period of time, with the result that the apparent velocity distributions are broadened.

Similar experimental results were obtained by Paton & Isenor (1968) with 25 MW, 20 ns ruby laser pulses focused to an area of 3×10^{-4} cm^2 on aluminium, copper and

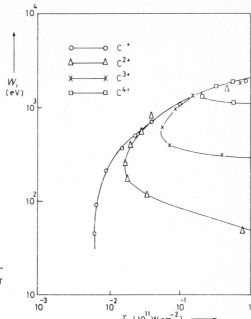

8.15 Ranges of ion energies observed with carbon target as functions of irradiance. (After Tonon (1966))

gold targets. The mean ion energies are shown plotted against the ion charge in Figure 8.16. No allowance was made for recombination.

Ehler (1973) placed an evacuated chamber with an entrance slit in the path of a freely expanding plasma produced from a tungsten target by a 30 MW, 20 ns ruby laser pulse. In this way a well-defined slab of plasma was formed which was passed

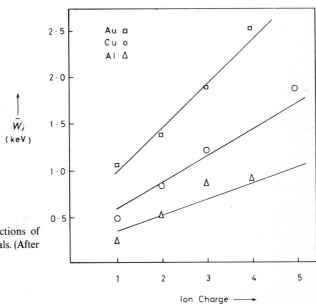

8.16 Mean ion energies as functions of ion charge for three target materials. (After Paton & Isenor (1968))

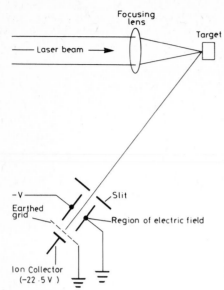

8.17 Arrangement for determining Debye length of plasma. (After Ehler (1973))

through a region of transverse field to a biased ion collector (Figure 8.17). By measuring the field required to eliminate the ion current for various slit widths, the curve shown in Figure 8.18 was obtained. If the plasma thickness, determined by the entrance slit width, was less than about two Debye lengths none of the ions experienced any shielding: in this case it was calculated that all ions would be deflected away from the detector by a field of 42 V cm^{-1}. For slit widths greater than $2\lambda_D$ larger fields were needed. The transition in Figure 8.18 indicates a Debye length of 5×10^{-3} cm. The ion time-of-flight and the peak collector current, with the geometry of the experiment, gave an electron density of 2×10^9 cm^{-3}, from which, with the Debye

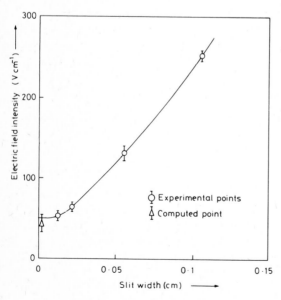

8.18 Electric field required to eliminate ion current, as function of slit width. (After Ehler (1973))

length, an electron temperature of about 0·1 eV was deduced. The ion energy was about 1050 eV.

A 22 MW (2 J, 90 ns) transversely-excited carbon dioxide laser, focused to peak irradiances of up to 8×10^9 W cm^{-2} over a focal spot area of 3×10^{-3} cm^2 was used by Tonon & Rabeau (1972, 1973), to produce plasma from a polyethylene target. Interferograms in neodymium laser light gave electron density profiles along the laser beam axis which are shown in Figure 8.19. As may be seen, an essentially steady flow of plasma is established. It is of particular interest that with this arrangement

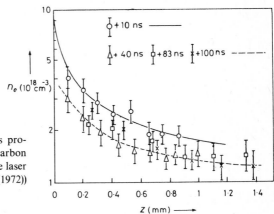

8.19 Electron density profiles in plasmas produced from polyethylene targets by a carbon dioxide laser. Times are measured from the laser pulse maximum. (After Tonon & Rabeau (1972))

regions of the plasma can be studied where the electron density is approaching the critical value (10^{19} cm^{-3}) for 10·6 μm radiation, as a consequence of the shorter wavelength of the neodymium laser. The total number of electrons in the plasma rises from about 1.5×10^{16} at the peak of the laser pulse to about 4×10^{16}. Ion velocity spectra for carbon ions showed a strong peak in C$^+$ at a velocity corresponding to a trough in C^{2+}, and a weaker C$^+$ peak corresponding to a minimum in the C^{3+} spectrum. The C^{4+} spectrum was approximately Maxwellian about a drift velocity of 1.4×10^7 cm s^{-1}. The ion temperature derived from this Maxwellian following Allen (1970) was 30 eV, which was close to the electron temperature giving the best fit to the measured ion drift velocities according to the theory of Mirels & Mullen (1963) (see Chapter 7, Section (6.2)).

The heating time $t(T_i)$ of an ion of charge Z in the hot plasma region was given by Bobin *et al.* (1971):

$$n_e t(T_i) \simeq 10^8 \frac{\mathscr{A} T_e^{\frac{3}{2}}}{Z} \text{ s cm}^{-3}, \qquad (8.5)$$

where T_e is in eV and $\mathscr{A}$ is the atomic mass. The time $t(Z)$ required for an atom to reach an ionization state with charge Z in a plasma at electron density n_e and temperature T_e (see Chapter 2, Section (7)) was calculated, assuming coronal conditions, by Tonon & Rabeau (1973) for all the carbon ions as functions of temperature, as shown

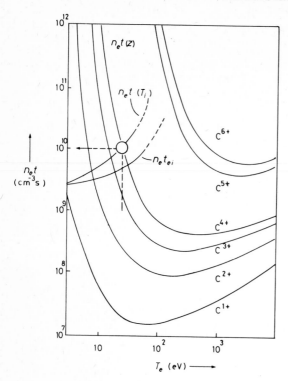

8.20 Ionization, ion heating and electron–ion equipartition times for a carbon plasma, as functions of the electron temperature. (After Tonon & Rabeau (1973))

in Figure 8.20. The intersection of expression (8.5) with the curve for the highest observed value of Z gives an estimate of the maximum temperature in the plasma, assuming that $T_e \simeq T_i$ in the heating region. This assumption should be valid because the electron–ion equipartition time is less than the heating time. The temperature (about 25 eV) derived in this way was in good agreement with that obtained from the ion expansion measurements. The dependence of the total number of particles in the plasma, and the mean plasma velocity, on the irradiance agreed well with self-regulating absorption theory allowing for transverse expansion (see Chapter 7, Section (4.1)).

Mass spectrometric studies were made by Bykovskii et al. (1968), who were able to detect ions from surface impurities in the plasma generated from a tantalum target. This technique was also used to measure the angular distribution of ions of different charges and energies leaving an aluminium plasma, and the energy distributions of ions emerging in different directions (Bykovskii et al. (1971b)). Figure 8.21 shows angular distributions for two different laser irradiances. Ion energy distributions were also measured with LiD and ZrH targets (Bykovskii et al. (1972b)). With several different metal targets the maximum ion energy was found to be about the same for ions of any charge, and was insensitive to atomic mass above $\mathscr{A} = 50$ (Bykovskii et al. (1969a, 1972a)). Results with neodymium laser pulses lasting 20 ns at several values of the irradiance are shown in Figure 8.22. The application of this method to

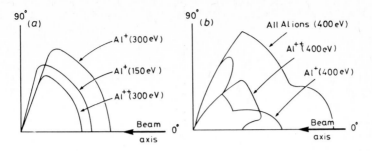

8.21 Angular distributions of ions leaving a plasma generated by a laser pulse incident normally on an aluminium target. (a) $I = 4 \times 10^9$ W cm^{-2}: (b) $I = 10^{10}$ W cm^{-2}. (After Bykovskii *et al.* (1971b))

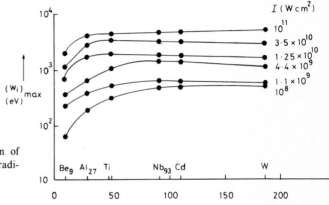

8.22 Maximum ion energy as function of atomic mass for several values of the irradiance. (After Bykovskii *et al.* (1972a))

more energetic plasmas (Bykovskii *et al.* (1971a)) is described later in this chapter.

Schwob *et al.* (1970) and Seka *et al.* (1970a) made detailed spectroscopic studies of plasmas produced from an aluminium target by 30 MW (1 J, 30 ns) neodymium laser pulses focused by a 10 cm lens. A grazing incidence (83°) vacuum spectrograph was used. The spectrum consisted of an intense continuum together with lines from highly-ionized atoms up to Al X. The time dependence of a number of lines from different stages of ionization, and of the continuum at $\lambda\,297$ Å, measured with a Bendix magnetic photomultiplier are shown in Figure 8.23. The electron temperature

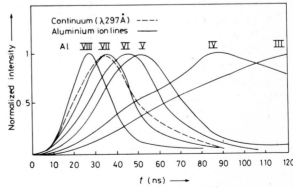

8.23 Time dependence of intensities of lines from aluminium ions, and of continuum at $\lambda297$ Å. (After Schwob *et al.* (1970))

was derived from the wavelength of the maximum of the Bremsstrahlung continuum after allowing for contributions from higher-order spectra. At the moment when the continuous emission reached its maximum intensity, the electron temperature was found to be 25 ± 2.5 eV. The electron temperature in these plasmas was also measured by the method described in Chapter 6, Section (8.5), comparing the intensity of X-rays transmitted by two different absorbing foils. The results depend upon a knowledge of the recombination continua, which were calculated on the assumption of local thermal equilibrium and an electron density of 10^{21} cm^{-3}, treating all aluminium ions as hydrogen-like, and considering only recombination to the ground state. The dependence of the electron temperature derived in this way upon the laser output energy is shown in Figure 8.24.

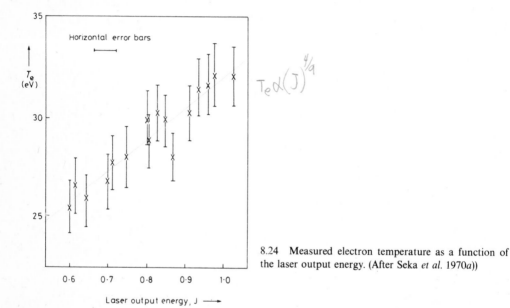

$T_e \propto (J)^{4/9}$

8.24   Measured electron temperature as a function of the laser output energy. (After Seka *et al.* 1970*a*))

Rumsby & Paul (1974) used four different methods to determine the electron density and temperature distributions in expanding plasmas produced from a carbon target by 37 MW, 20 ns ruby laser pulses (irradiance 7×10^{10} W cm^{-2}). Negatively-biased ion collecting probes gave the electron density at distances from 5 mm to 1·3 m from the target. Between 30 mm and 140 mm, the microwave measurements at 4 mm wavelength were used to give n_e and T_e. From 20 to 50 mm, n_e and T_e were obtained from the Thomson scattering of light from another ruby laser, with a scattering angle of 135°. Finally, the electron temperature was measured near the target surface from the relative intensities of two lines of C V, λ 40·27 Å and 34·98 Å, in the grazing incidence vacuum ultraviolet region.

The electron density profiles at times up to 1 μs after the laser pulse are shown in Figure 8.25. The plasma consists of a relatively thin shell expanding rapidly away

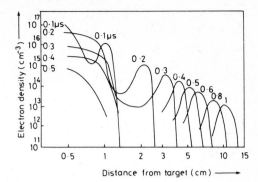

8.25   Electron density profiles at several times. (After Rumsby & Paul (1974))

from the target, together with a denser central core with a steady diameter of about 1 cm. The peak density of the shell is followed in Figure 8.26, and may be seen to vary as t^{-3}, indicating that the decay of electron density due to recombination is negligible compared to the effect of expansion of the shell. The areas under the

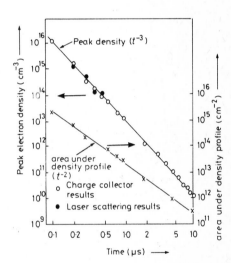

8.26   Peak electron density in plasma shell, and area under density profile, as functions of time. (After Rumsby & Paul (1974))

density profiles are also shown: these vary as t^{-2}, suggesting that Fader's (1968) similarity solution (see Chapter 7, Section (6)) is valid, with the expansion velocity increasing linearly with distance from the target.

The electron temperature was found to be uniform in the plasma shell. Figure 8.27 shows that the time dependence of T_e in the shell is as t^{-1} and not as t^{-2} as might be expected for an adiabatic expansion. This is presumably because the electrons receive energy from 3-body recombination processes (Dawson (1964)). (Electron–ion collisions should cease to heat ions after the shell has expanded to about 20 mm, and the ion temperature would then be expected to fall as t^{-2}.) The electron temperature near the target surface was found from the spectroscopic measurements to be about 30 to 40 eV, in good agreement with Puell's theory (see Chapter 7, Section (4.1)).

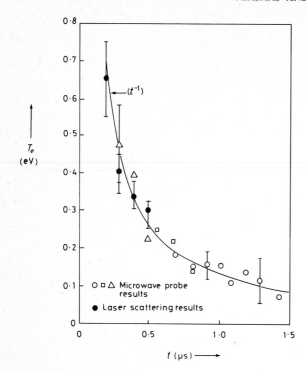

8.27   Time dependence of electron temperature in carbon plasma shell. (After Rumsby & Paul (1974))

Electrostatic single probes, 8 mm microwaves and a high-speed image-converter framing camera were used by Fabre & Vasseur (1966) and Fabre *et al.* (1966) to study plasmas produced from a copper target by focused 50 MW (1 J, 20 ns) ruby laser pulses. In rather brief descriptions of their results they reported that 8 mm microwave cut-off (corresponding to an electron density in excess of 1.5×10^{13} cm^{-3}) was observed for about 7 μs at a distance of 3 cm from the target, and for about 1 μs at 6 cm. The electrostatic probes gave an electron density of 5.4×10^{14} cm^{-3} when the plasma front reached a distance of 12 mm from the target, decreasing roughly as the inverse cube of the distance to 1.7×10^{12} cm^{-3} at 72 mm. The characteristic curve of the probe 12 mm from the target, obtained with many laser shots, indicated an electron temperature of the order of 10 eV. The onset of the probe signal implied an expansion velocity of 6.6×10^{6} cm s^{-1}. This was higher than that observed photographically—about 3.5×10^{6} cm s^{-1} falling to 2.5×10^{6} cm s^{-1}. The interpretation of electrostatic probe signals, from plasmas formed from aluminium and lead targets by 70 MW neodymium laser pulses, has been discussed by Andreev *et al.* (1968).

Plasmas produced under similar conditions from a lithium hydride target were studied by Izawa *et al.* (1969) using a Thomson scattering technique. Their arrangement is shown in Figure 8.28. The plasma was formed by a 50 MW, 100 ns neodymium glass laser pulse. The rotating prism *Q*-switch also switched a 5 MW, 20 ns ruby laser which was used as the source for the scattering measurements. Light scattered

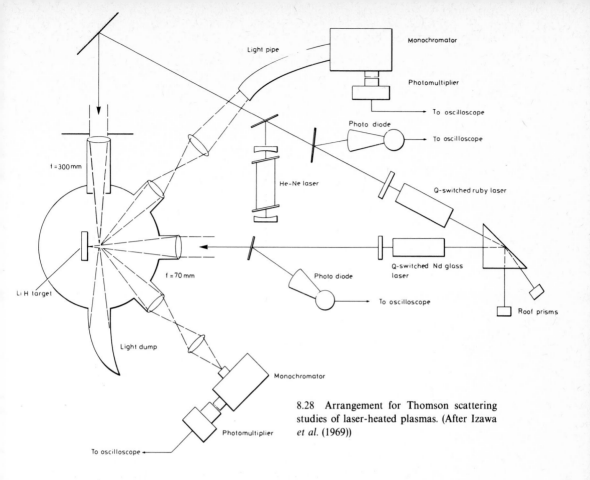

8.28 Arrangement for Thomson scattering studies of laser-heated plasmas. (After Izawa et al. (1969))

through both 45° and 135°, coming from a point 5 mm from the target surface, was collected and analysed: the density was high enough and the temperature low enough to give values of $\alpha \gg 1$ for both directions. Figure 8.29 shows the spectrum of radiation scattered through 45°, 800 ns after the peak of the neodymium laser pulse.

8.29 Scattered light spectrum from lithium hydride plasma. (After Izawa et al. (1969))

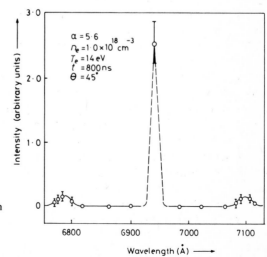

The satellite peaks 155 Å away from the ruby laser wavelength gave values of 1.0×10^{18} cm^{-3} for n_e and 14 eV for T_e. This is in quite good agreement with a rough estimate of 20 eV for the electron temperature in a similar plasma by Ambartsumyan *et al.* (1965).

The maximum electron density, 6×10^{18} cm^{-3}, was observed 200 ns after the neodymium laser pulse peak. Figure 8.30 shows the central 'ion feature' of the scattered light spectrum 100 ns after the neodymium laser pulse peak. The marked central minimum indicates a strong ion acoustic resonance, and thus an electron

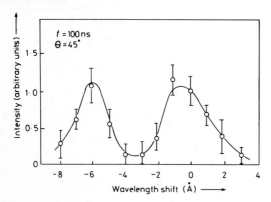

8.30   Central 'ion feature' in scattered light spectrum from lithium hydride plasma. (After Izawa *et al.* (1969))

temperature many times greater than the ion temperature. The separation of almost 6 Å between the peaks corresponds to twice the ion acoustic frequency, from which an anomalously high electron temperature of about 600 eV is derived if it is assumed that $Z = 2$ and the ion mass is taken to be four times the proton mass. The central ion feature also showed a general displacement of 3 Å to shorter wavelengths, which may be ascribed to the expansion of the plasma. The corresponding expansion velocity is 1.8×10^7 cm s^{-1}.

Weichel *et al.* (1967) (see also Weichel & Avizonis (1968)) investigated the spectrum of light scattered from plasmas produced by a 90 MW ruby laser (8 J, 90 ns) focused by a 10 cm lens on a carbon (pyrolytic graphite) target. The laser light scattered from the plasma, in a direction perpendicular to both the axis and the E-vector of the laser beam, entered a spectrograph. The spectrum, recorded photographically, showed the splitting of the central ion feature due to the ion acoustic resonance: the maxima were separated by up to 0.9 Å. On the assumption of an (interferometrically measured) electron density of 10^{19} cm^{-3}, with $\alpha \gg 1$, and with $Z = 3$, this corresponds to an electron temperature of approximately 30 eV.

Considerably higher electron temperatures were measured by Puell *et al.* (1971) using the X-ray relative intensity method, for carbon plasmas generated by neodymium laser pulses of 70 MW power and 17 ns duration, focused by an 8 mm lens. The time dependence of the electron temperature is shown in Figure 8.31. As may be seen, it follows quite closely the theoretical curve, which was calculated assuming that $T(t)$ varies as $[I(t)]^{\frac{4}{5}}$ (see Chapter 7, Section (4.1)).

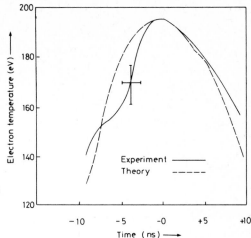

8.31   Time dependence of electron temperature in carbon plasma. (After Puell *et al.* (1971))

2.2.2. Laser power from 100 MW to 1 GW

Interferometric studies have provided much information about particle densities, particularly in the later phases of development of the plasmas. Early work by Bruce *et al.* (1966) and David *et al.* (1966) was extended by David (1967, see also David & Avizonis (1968)), who used an argon ion laser as the source of light at both λ 4880 Å and 5145 Å for two-wavelength Mach–Zehnder interferometry. The arrangement is shown in Figure 8.32. The argon laser beam was brought to a 25 μm diameter focus at a distance from the target surface which could be varied by tilting the quartz plate. The plasma was produced by a 100 MW (7·5 J, 76 ns) ruby laser focused to a spot of 1 mm diameter on a carbon target. At a distance of 1 mm from the target, on the assumption that the plasma diameter traversed by the argon laser light was also 1 mm, the time dependence of the average neutral atom and electron densities were found to be as shown in Figure 8.33. The two wavelengths used are rather close together, and the detailed structure suggested in the diagram for the neutral atom density curve is by no means certain. There appears to be a marked discrepancy between the time dependences of n_e and n_a measured in this way and by Thomson and Rayleigh scattering (Izawa *et al.* (1968)) although the maximum values are in good agreement. Streak photographs of plasmas produced at the same laboratory under very similar conditions (Weichel & Avizonis (1966)) showed that the expansion velocity of the luminous plasma front reached a maximum value of 7×10^6 cm s^{-1} about 60 ns after the peak of the laser pulse, at which time the front was about 0·4 cm away from the target. The effect of radiation pressure on the behaviour of these plasmas was considered by Weichel *et al.* (1968) who suggested that it might restrain the initial expansion of the vapour plume. However, the vapour density assumed in their calculation (10^{18} cm^{-3}) is much too small to cause significant reflection or absorption of light. In any case, radiation pressure is usually negligible compared to kinetic pressures in laser-produced plasmas.

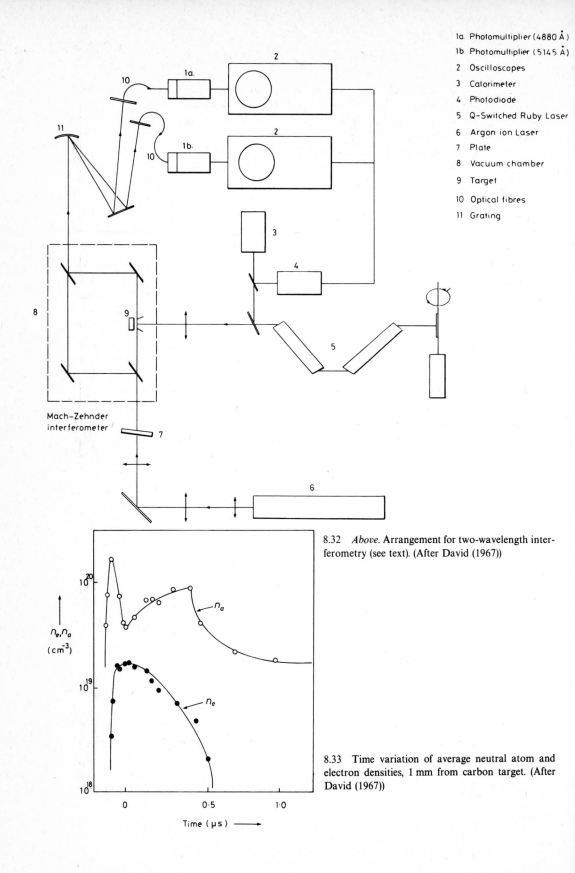

1a. Photomultiplier (4880 Å)
1b. Photomultiplier (5145 Å)
2 Oscilloscopes
3 Calorimeter
4 Photodiode
5 Q-Switched Ruby Laser
6 Argon ion Laser
7 Plate
8 Vacuum chamber
9 Target
10 Optical fibres
11 Grating

Mach–Zehnder
interferometer

8.32 *Above.* Arrangement for two-wavelength interferometry (see text). (After David (1967))

8.33 Time variation of average neutral atom and electron densities, 1 mm from carbon target. (After David (1967))

In a later paper David & Weichel (1969) re-examined these results and described a further experiment in which the carbon target was surrounded by a thin quartz bubble which intercepted the carbon vapour. The bubble was itself surrounded by a silver sphere which acted as a calorimeter. Small holes in the quartz and silver admitted the laser beam to the target. The energy collected by the silver sphere was assumed to be equal to the energy of the plasma electrons, on the grounds that the electron energy was mostly lost by radiation which passed through the quartz sphere. The thickness of the carbon layer deposited on the quartz sphere gave the quantity of carbon ablated, minus that which condensed on the target. From this a value for the energy per ion was deduced. However, it is fairly clear from other experiments (see for example Boland et al. (1968)) that most of the electron energy is converted into radially directed ion kinetic energy: also, even if the electrons did lose significant amounts of energy by radiation most of it would be absorbed by the quartz unless $\hbar\omega \lesssim 6$ eV. Thus in one way or another most of the energy of the plasma was delivered to the quartz, which presumably radiated to the silver calorimeter. Finally, if the Rayleigh scattering result of Izawa et al. (1968) is accepted, much of the carbon reaching the quartz bubble was removed from the target long after the laser pulse ceased, as a cool vapour. The energies per ion or per electron deduced from this experiment are probably much too low.

Burnett & Smy (1970) made related afterglow studies of plasmas produced from aluminium targets by 2 J, 20 ns ruby laser pulses. Using a technique similar to that of Lampis & Brown (1968) for plasmas in gases they measured the absorption of a narrow probe beam from a helium–neon laser (λ 6328 Å), and also the plasma emission near that wavelength, and deduced the electron temperature. Figure 8.34 shows the time variation of the temperature 1·0 mm from the target surface. The arrested decay in temperature after 75 ns may be due to recombination supplying energy equal to the work done in expansion, as predicted by Dawson (1964).

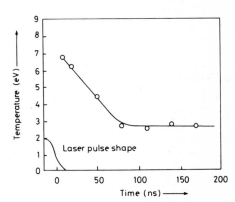

8.34 Time variation of electron temperature 1 mm from aluminium target surface. (After Burnett & Smy (1970))

Two-wavelength interferograms of laser-generated plasmas have also been obtained by Hugenschmidt & Vollrath (1971), using the fundamental and second harmonic frequencies of a ruby laser.

Gregg & Thomas (1966a, b; 1967) carried out a number of experiments with a ruby laser producing a pulse of 7·5 ns duration, which could be focused to produce an irradiance of up to 4×10^{11} W cm^{-2}. They observed mean ion energies up to 2 keV for a wide variety of target materials by a time of flight method, and using a ballistic pendulum estimated the shock pressure produced in the target material to be up to 10^6 atm. They also examined the wavelength distribution of the radiation from the plasma in the range λ 4000 to 10 500 Å and estimated maximum plasma temperatures of up to 8×10^5 K on the basis of L.T.E. Opower & Press (1966) used a ruby laser of 250 MW peak power (3·5 J, 14 ns) to produce a similar irradiance of $2 \cdot 5 \times 10^{11}$ W cm^{-2}. With lithium hydride and with carbon targets the ions collected were found from times of flight to have maximum energies of 1·7 keV and 6·3 keV respectively. On the basis of the theory of Chapter 7, Section (6.2) these correspond to mean energies of 340 eV for a lithium hydride plasma, assuming a mean mass number of 4 and a mean charge number of 2, and 550 eV for a carbon plasma assuming total stripping of the ions. However, as Opower & Press pointed out, the plasma temperatures were probably lower because some expansion occurred during heating. Later work by Opower *et al.* (1967) made use of two dye-cell-switched ruby lasers, whose dye cells were simultaneously bleached by a third laser. Each laser beam had a peak power of 500 MW (3·5 J, 7 ns). This arrangement was used to overcome problems of dielectric breakdown in the ruby. When both beams were focused from slightly different directions on the same target area, an irradiance of 3×10^{12} W cm^{-2} was obtained. The dependence of the maximum ion energy observed with a lithium deuteride target on the irradiance is shown in Figure 8.35: the ion energy varies as (irradiance)$^{0 \cdot 4}$.

Basov *et al.* (1966, see also Afanas'ev *et al* (1966*a*)) used a neodymium glass laser providing pulses of 200 MW peak power and about 15 ns duration, focused by a lens

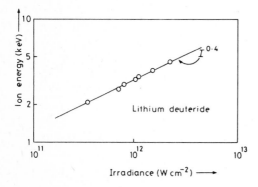

8.35   Dependence of maximum energy of ions from a lithium deuteride target on irradiance. (After Opower *et al.* (1967))

of 60 mm focal length, to produce plasmas from carbon targets. The expansion of the plasma was investigated by means of an electrostatic probe, by high-speed streak photography of the light emitted by the plasma, and by shadow photography of the plasma using some of the neodymium laser radiation (suitably delayed) and a gated image converter. The probe was a small screened Faraday cylinder. It received a potential which was at first negative, corresponding to the halo of electrons surrounding the plasma, then fell to zero and became positive as the rest of the plasma moved

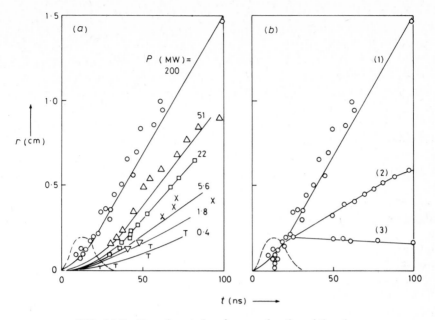

8.36 (a) Position of neutral surface r as function of time for several laser pulse peak powers P. (b) Position of neutral surface (1), self-luminous boundary (2) and boundary of opaque region (3) as functions of time. The laser pulse shape (peak power 200 MW) is also shown (dashed curve). (After Basov et al. (1966))

past. The time at which the signal passed through zero as the polarity reversed could be measured accurately, and was taken as the time of passage of the surface of charge neutrality in the plasma. Results are shown in Figure 8.36(a) for several laser peak powers. The asymptotic velocity of the surface of charge neutrality was found to vary as the fourth root of the peak laser power: this dependence may be compared with the result of Opower et al. (1967) that the ion velocities varied as the fifth root of the power. The shadow photographs showed a sharply defined, roughly hemispherical opaque region extending from the target. Measurements were also made of the absorption of a narrow probe beam passing through the plasma parallel to the target surface. Figure 8.36(b) shows the behaviour of the self-luminous boundary (2), and the boundary of the opaque region as determined by shadow photography (3) for a laser power of 200 MW. Figure 8.37 shows how the position of the boundary of the opaque region varied with time for three different laser powers.

Subsequently Basov et al. (1971a) made detailed interferometric measurements of the plasma produced from a carbon target by an 8 J, 80 ns neodymium laser pulse focused to a spot whose radius increased from 0·05 mm initially to 0·2 mm at the peak of the laser pulse. A Mach–Zehnder interferometer was used, with a ruby laser source, a narrow section of the fringe pattern being selected by a slit and scanned by an image-converter camera to provide time resolution of 0·2 ns. The length of the slit was perpendicular to the target surface and parallel to the beam axis. At the boundary of the opaque zone near the target surface, the electron density reached $5 \times 10^{19} \text{ cm}^{-3}$,

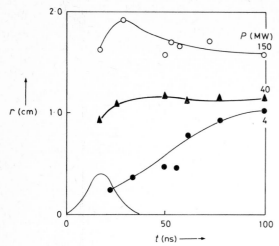

r (cm)

t (ns) ⟶

8.37 Position of boundary of opaque region as function of time for three laser peak powers. The laser pulse shape is also shown. (After Basov *et al.* (1966))

which is well below the critical density, so the apparent opacity was probably due to refraction. Figure 8.38 shows the electron density distribution along the axis of the laser beam at six different times. Careful analysis of the results allowed a detailed picture of the density and velocity profiles to be constructed. Figure 8.39 shows the variation of the total mass of the plasma M, and of the total number of electrons N_e, with time, together with the vaporization rate dM/dt, which reached 4 gm s^{-1}. The laser pulse shape is also shown. The plasma pressure reached a maximum of 2.5×10^6 atm very early in the laser pulse, at the time of maximum evaporation rate.

Demtröder & Jantz (1970) studied the plasmas produced by 100 MW, 30 ns laser pulses focused by a 3·5 cm lens to an irradiance of 5×10^{11} W cm^{-2} on aluminium

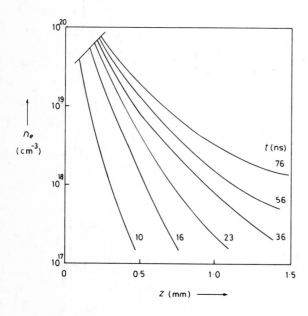

n_e (cm^{-3})

z (mm) ⟶

8.38 Electron density distributions in carbon plasma, as functions of distance z from target surface, along axis of laser beam, at six different times from start of laser pulse. (After Basov *et al.* (1971a))

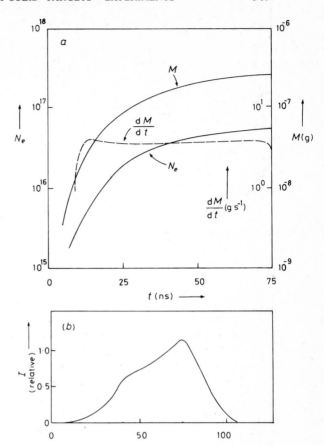

8.39 (a) Total mass of plasma M, total number of electrons N_e, and rate of vaporization dM/dt as functions of time. (b) laser pulse shape. (After Basov et al. (1971a))

and copper targets. They used time-of-flight and retarding potential methods* to determine the energy distributions and relative numbers of ions of different charges. The average ion energy is shown plotted against the charge Z for copper ions in Figure 8.40, before and after allowing for recombination. An electron temperature

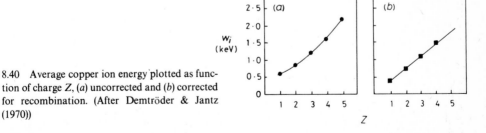

8.40 Average copper ion energy plotted as function of charge Z, (a) uncorrected and (b) corrected for recombination. (After Demtröder & Jantz (1970))

* Details of a retarding potential probe have also been given by Matoba (1972).

of about 10 eV was deduced, so that the number of times the electrons gained and lost energy to the ions may have been about 40 (see Chapter 7, Section (6.2)).

Ions produced by similar laser pulses (up to 100 MW, 3 J, 30 ns) focused to an irradiance of up to 4×10^{12} W cm^{-2} on aluminium or iron targets were investigated by Faure *et al.* (1971, see also Tonon *et al.* (1971a)), using the electrostatic analyser described above (Figure 8.13). Ions with Z up to 11 (Al) and 20 (Fe) were collected: the energy per unit charge was 1040 eV for Al and 1380 for Fe. The maximum energy measured for Fe^{16+} ions was 20 keV. Ion heating times and ionization times (as described earlier—see p. 355) are shown for iron plasmas in Figure 8.41. From the intersection at the highest observed value of Z, a plasma temperature of 300 eV

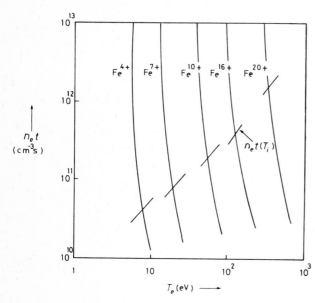

8.41   Ionization and ion heating times for an iron plasma, as functions of the electron temperature. (After Tonon *et al.* (1971a))

was deduced for an irradiance of 4×10^{12} W cm^{-2}, in good agreement with X-ray observations with similar plasmas. Ion velocity spectra observed from plasmas produced from carbon and beryllium targets, interpreted as displaced thermal Maxwellians (Allen (1970)) gave ion temperatures which agreed well with X-ray electron temperatures: for beryllium (Be^{+}), $T_i = 40$ eV when $I = 6 \times 10^{10}$ W cm^{-2}, while for carbon (C^{5+}), $T_i = 110$ eV when $I = 10^{12}$ W cm^{-2}.

The expansion of plasmas generated from a variety of targets by ruby laser pulses with peak powers up to 200 MW (6 J, 30 ns) was investigated by Koopman (1971b) using a cylindrical Langmuir probe and also by a microwave method. Table 8.1 compares probe results for five different target materials at a distance of 32·6 cm from the target. Two probes at 21·5 and 43 cm from a copper target gave charge density values in the ratio 8 to 1 measured at times t_1 and $t_2 = 2t_1$ respectively, confirming the expected $1/r^3$ dependence of density on radius.

Stumpfel *et al.* (1972) studied the grazing incidence vacuum ultraviolet spectra emitted by plasmas produced from a magnesium target by 17 ns ruby laser pulses

TABLE 8.1: Measurements on plasmas expanding in vacuum, at a distance of 32·6 cm from the target

Target	Cu	Ti	Al	C	$(C_2H_4)_n$
Thickness (cm)	Solid	Solid	Solid	0·025	0·0025
Front velocity (cm μs^{-1})	8·6	10·4	14·4	17·5	19·0
Front ion kinetic energy (keV)	2·3	2·6	2·8	1·85	0·84*
Peak electron density (10^{11} cm^{-3})	2·8	3·9	6·8	10·5	14·0
Velocity of electron density peak (cm μs^{-1})	4·9	6·5	9·0	13·5	14·0
Electron density at twice time of peak (10^{11} cm^{-3})	1·8	2·8	2·2	1·8	1·4

* Assuming $m_i/m_p = 4·7$.
After Koopman (1971b).

of up to 340 MW peak power, focused by a 74 mm focal length lens. By inserting a moveable 1 mm aperture in the near field, the time variation of the laser output power from a small region of the laser rod was examined. Usually, a single spike of 5 ns duration was produced, but sometimes a double spike was seen. The superposition of these short pulses in the full laser beam produced the overall pulse of 17 ns duration. It was probable therefore that higher irradiance occurred over a small area of the target, even early on in the pulse, than would be deduced from the overall energy and duration. Both photographic and time resolved measurements were made. Lines of Mg IX and X dominated the line spectrum below λ 200 Å at laser powers above 200 MW: the resonance line of Mg XI at λ 9·168 Å was outside the range of the spectrograph. The foil X-ray absorber method gave an electron temperature of 400 to 500 eV at the highest laser power: the authors regarded this result with some reserve because the contributions of recombination radiation were not taken into account.

For spectroscopic observations in the vacuum ultraviolet, especially at grazing incidence, it is desirable for the target to be within a few cm of the spectrograph slit. The narrow slit tends to be rather quickly blocked by solid particles ejected from the target. It is often possible to use materials which are normally gases or liquids at room temperature as solid targets by freezing them on to a cold surface before the target chamber is evacuated (Holden & Hughes (1967)). Any large particles thrown from the target evaporate rapidly on contact with the slit. A convenient arrangement is shown in Figure 8.42. Figure 8.43 shows some examples of spectra obtained in this way, excited by neodymium laser pulses of a few hundred MW and recorded with a 2m spectrograph (Gabriel et al. (1965)). The O VII lines shown in Figure 8.43(b) were recorded using water as the target: the high electron density is apparent from the relative intensities of the resonance and intercombination lines (Griffin & Peacock, see Kunze et al. (1968)). Figure 8.43(a), obtained with CCl_4 as the target material, shows good chlorine spectra with convenient carbon reference lines.

A neodymium glass laser delivering 660 MW pulses (10 J, 15 ns) was used by Basov et al. (1967b) for spectroscopic studies of highly-ionized atoms. With this

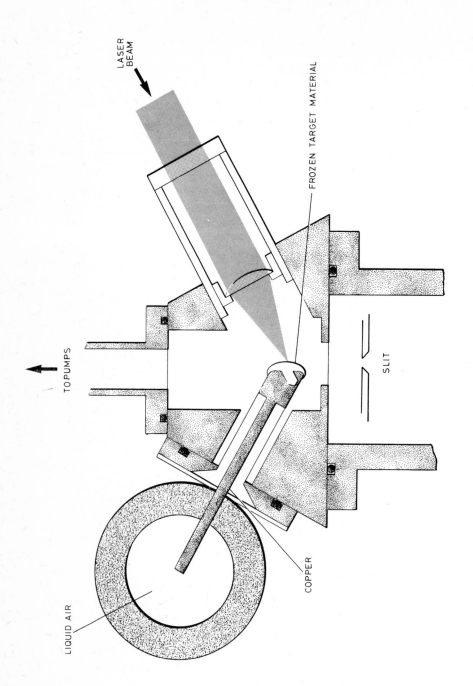

LASER BEAM

TOPUMPS

LIQUID AIR

COPPER

FROZEN TARGET MATERIAL

SLIT

8.42   Arrangement for freezing liquids or gases to form solid targets. (After Holden & Hughes (1967))

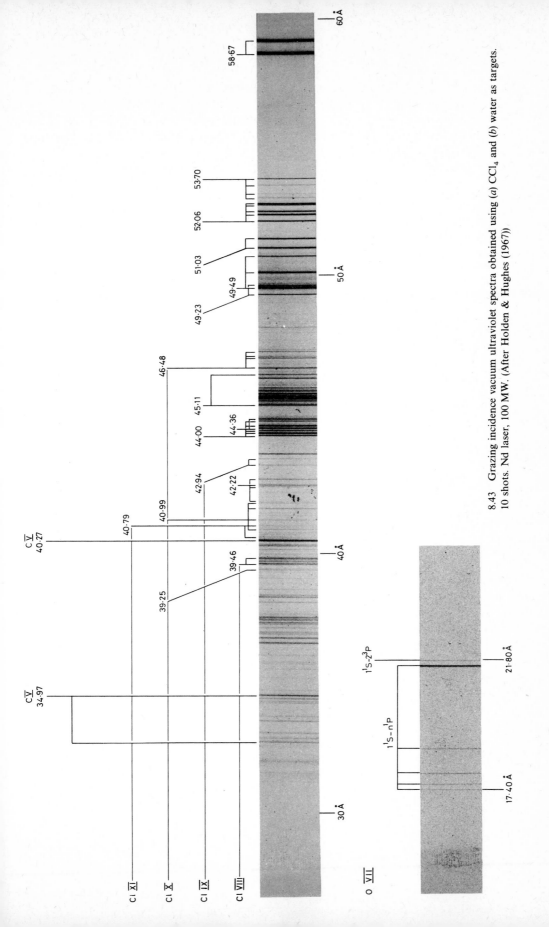

8.43 Grazing incidence vacuum ultraviolet spectra obtained using (a) CCl_4 and (b) water as targets. 10 shots. Nd laser, 100 MW. (After Holden & Hughes (1967))

laser, using an aluminium target and a 100 mm focal length lens, the edge of the opaque region was again about 1·5 mm from the target at the end of the laser pulse. The velocity of expansion of the self-luminous region was 10^7 cm s^{-1}, corresponding to an aluminium ion energy of about 1 keV. With a calcium target, the electron temperature of the plasma was estimated from the presence of lines of Ca XIII and XIV to be about 130 eV in the emitting region. By improving the focus to produce an irradiance of 5×10^{12} W cm^{-2}, it was possible to excite Ca XVI ions (Basov et al. (1967c)), implying an electron temperature of 200–300 eV (the latter after allowing for photo-recombination at excited levels). The electron density was determined in plasmas from a carbon target by interferometry and also by measurements of the Stark broadening of C VI λ 520·6 Å and 3434 Å (Aglisky et al. (1971)). The results are shown in Figure 8.44: the use of the vacuum ultraviolet λ 520·6 Å line permits measurements down to within about 1 mm of the target surface. An expansion

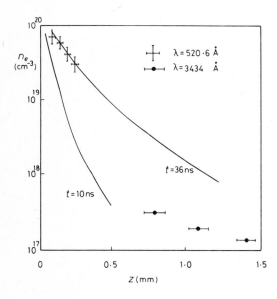

8.44 Electron density distributions in carbon plasma as functions of distance z from target surface, along axis of laser beam, 10 and 36 ns after start of laser pulse. Curves are from interferometric measurements, individual points from Stark broadening measurements. (After Aglitsky et al. (1971))

velocity for the plasma was determined from the Doppler shift of the absorption observed in the C IV resonance lines at λ 1548·2 Å and 1550·8 Å. The very dense inner region of the plasma emits a continuum, which is absorbed by ground state C IV ions in the expanding outer regions. The shift corresponded to a velocity of about 3×10^7 cm s^{-1}. The observed time and space variation of light emitted by ions in the different stages of ionization was interpreted (Basov et al. (1970)) as being consistent with the different species of ions expanding in spherical layers (as we shall see, more detailed observations (Irons et al. (1972)) indicate a different distribution.)

Basov et al. (1967d, 1968b) modified their laser by the addition of a pulse-sharpening Kerr cell to produce 10 J pulses with a rise time of 4 ns and a half-intensity duration of 12 ns. The arrangement is shown in Figure 8.45. The pulse from the main neodymium glass laser oscillator (5) passes through an isolating and pulse-sharpening

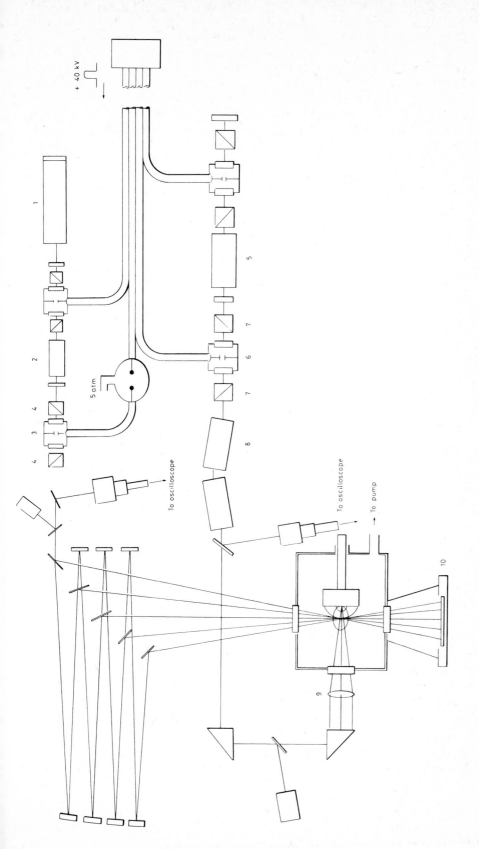

8.45 Arrangement for obtaining multiple shadowgrams of laser-generated plasma (see text). (After Basov *et al.* (1968*b*))

Kerr cell (6) and Glan prism polarizers (7) to the amplifiers (8) and thence to the
target through the lens (9). A pulse from the ruby laser (1), (2) is also sharpened by
the Kerr cell (3) and Glan prisms (4) and divided by reflectors into several beams,
all of which pass through the plasma in slightly different directions, at different
times, on to the photographic plate (10). All four Kerr cells are driven from the same
pulse generator with suitable delay lines. In this way five shadowgrams were obtained
at adjustable intervals with exposure times of 3 ns. The results showed evidence that
the apparent shape of the opaque region was considerably influenced by refraction
of the parallel probe beam caused by a large electron density gradient. Bright regions
could be seen which, if they were due to refracted light, implied a decrease in the
electron density towards the target surface, with a gradient of about 10^{20} cm^{-4}.
This might be due to a rapid decrease in the electron temperature towards the target.
It was possible to obtain a more accurate image of the opaque region by using a
lens to focus it on to the photographic plate: in this way light refracted through
moderate angles is collected.

Ruby lasers producing somewhat slower pulses (rise time about 10 ns, duration
about 20 ns) with peak powers of up to 500 MW were used at Culham to generate
plasmas from a number of target materials. Fawcett et al. (1966) focused the laser
pulse with a 10 cm lens to a spot of 0·25 mm diameter, giving an irradiance of
10^{12} W cm^{-2}. They found that the dominant spectra from carbon plasmas were
C V and VI: from iron, Fe XV and XVI; and from nickel, Ni XVII and XVIII. The
continuous spectra obtained from many solid targets had a maximum intensity at
λ 70 Å, corresponding to an electron temperature of about 100 eV. High-speed
streak photographs of the visible radiation showed that the luminous plasma from
a polyethylene target expanded at 2×10^7 cm s^{-1}, and that from iron at 4×10^6 cm s^{-1}.
Fawcett & Peacock (1967), Fawcett et al. (1967) and Tondello (1969) extended this
work to other target elements and identified many new lines.

Burgess et al. (1967, 1968) carried out a careful study of the vacuum ultraviolet
emission spectra from several targets with the ruby laser power in the region of 450
to 500 MW. Problems of spatial resolution are rather severe at grazing incidence,
and photographic recording did not provide time resolution. However, the most
highly ionized atoms are present only in the hottest region of the plasma, which is
close to the target surface and very short-lived. Thus by studying their emission
spectra useful information may be gained about this most interesting phase in the
life of the plasma. The emission lines observed were often very broad, and as their
breadth varied greatly between different transitions in the same ion, it seems reasonable to suppose that the principal source of broadening was the Stark effect due to
neighbouring electrons and ions. From the profiles of O VI and K IX lines, electron
densities of 10^{20} to 10^{21} cm^{-3} were deduced for the plasma from an oxidized potassium target in the region up to 1 mm from the target surface. The electron temperature
deduced from the relative intensities of the $2p^2P^0$–$4d^2D$ and $2s^2S$–$3p^2P$ transitions
in O VI was $40(-10, +20)$ eV, on the assumption that the upper levels are thermalized
relative to the continuum (as should be the case on McWhirter's criterion—see
Chapter 2, Section (7)). The continuous emission spectrum due to free–bound and
free–free transitions was also investigated. In a carbon plasma the frequency depen-

dence of the C V continuum intensity yielded an electron temperature of 113 ± 20 eV within 0·5 mm of the target surface and an overall average of 15 ± 3 eV when no spatial resolution was employed. The distances from the target surface within which lines were emitted with appreciable intensity from various ion species in the normal incidence region are given in Figure 8.46. It appears that the hottest region of the plasma was only about 0·2 mm thick.

Boland et al. (1968, see also Boland & Irons (1968)) investigated plasmas produced from polyethylene by a similar laser with a peak output power of about 300 MW

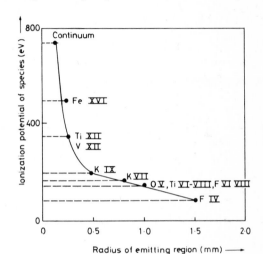

8.46 Regions within which lines of various ion species were emitted with appreciable intensity. (After Burgess et al. (1967))

(5 J, 17 ns). They made detailed space-resolved measurements of both C V and C VI free–bound continuum intensities, and deduced the electron temperature distribution shown in Figure 8.47. The two points at 3 and 5 mm from the target surface were obtained from C VI line intensity ratios. These results are averages over the periods

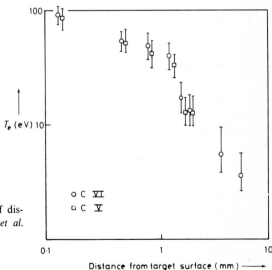

8.47 Electron temperature as function of distance from target surface. (After Boland et al. (1968))

of time during which the C V and C VI ions are emitting, and of course these periods vary with distance from the target. Boland *et al.* also carried out the spectroscopic equivalent of time-of-flight measurements on ions from carbon plasmas, by measuring the time-dependence of line intensities emitted from small elements of the plasma at various distances z from the target surface. The monochromator-photomultiplier detector system had a rise time of about 3 ns. Results at $z = 2, 5$ and 10 mm are shown in Figure 8.48. The distance of the region of peak intensity from the target surface

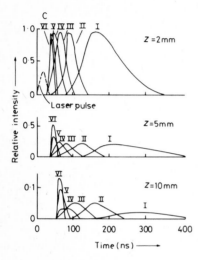

8.48   Line intensities emitted from elements of plasma at three different distances from target surface. (After Boland *et al.* (1968))

is plotted for each ion as a function of time in Figure 8.49. Table 8.2 lists the lines studied and the velocities and energies deduced from observations of the plasma at a distance of 2 mm from the target surface. Most of the radiation emitted in the visible region comes from the less highly ionized atoms, which explains why streak photographs show lower velocities than those obtained from the times of flight of the

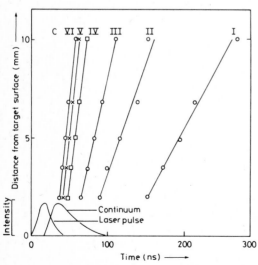

8.49   Distances from target surface of regions of peak intensity of emission, as functions of time, for carbon ions. (After Boland *et al.* (1968))

TABLE 8.2: Spectroscopic measurements of carbon ion energies in plasma 2 mm from a polyethylene target

Line	Transition	Ion to which velocity measurements refer	Velocity (cm s^{-1})	Ion energy w_i (keV)	$\dfrac{w_i}{z}$
C VI 3434	6–7	C^{6+}	$3\cdot3 \times 10^7$	6·7	1·1
C V 4945	6–7	C^{5+}	$3\cdot1 \times 10^7$	5·9	1·2
C IV 2524	$4d\,^2D$–$5f\,^2F$	C^{4+}	$2\cdot5 \times 10^7$	3·9	1·0
C III 3609	$4p\,^3P$–$5d\,^3D$	C^{3+}	$1\cdot6 \times 10^7$	1·6	0·5
C II 4267	$3d\,^2D$–$4f\,^2F$	C^{2+}	$1\cdot0 \times 10^7$	0·62	0·3
C I 2479	$2p\,^1S$–$3s\,^1P$	C^{1+}	$0\cdot7 \times 10^7$	0·26	0·3

After Boland et al. (1968).

fastest ions (see for example Figure 8.36(b)). (Related experimental studies on far ultraviolet spectra from aluminium targets were reported by Dhez et al. (1969).) It should perhaps be emphasized that in observing radiation from an i-times ionized atom, we effectively study the velocity of an $(i+1)$-times ionized atom which has just recombined with a free electron. Calculations show that the recombination rates are so small that their effect on the ion populations between $z = 1$ and $z = 5$ mm may be ignored.

From the absolute intensities (taking into account the depth of the plasma) and electron temperature distribution, and the known transition probabilities of the lines, Boland et al. calculated the electron and ion densities. It was found (Figure 8.50) that throughout the plasma, from the target surface up to a distance z of 5 mm, the electron temperature and density were related by the expression

$$T_e = 2 \times 10^{-11} n_e^{\frac{2}{3}} \tag{8.6}$$

where T_e is in eV and n_e in cm^{-3}. This is consistent with an adiabatic expansion of the electron gas. The total number of ions was found to be $\sim 10^{16}$.

An approximate energy balance for the plasma was drawn up. Radiation losses were negligible: the recoil energy of the target was estimated to be 0·1 J. The thermal energy of the electrons was 0·1 J, and the ionization energy about 0·2 J. The kinetic

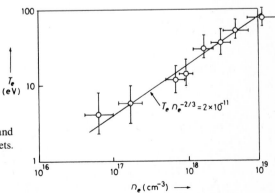

8.50 Relation between electron temperature and density in plasmas from polyethylene targets. (After Boland et al. (1968))

$T_e\, n_e^{-2/3} = 2 \times 10^{11}$

energy of the ions was estimated to be between 1 and 6 J, with a mean of 3·5 J. The laser pulse energy was 5 J, a little of which was probably reflected. Thus the greater part of the laser energy ends up in the directed expansion energy of the ions.

Irons *et al.* (1972) made further spectroscopic studies of plasmas formed by 5 J, 17 ns ruby laser pulses focused on polyethylene targets. The spectrum of light emitted at right angles to the target normal, at different distances from the target surface and from the laser beam axis was studied with time resolution. The spatial distributions of intensity of lines of C I to C VI at different times were determined, and also line profiles which in most cases could be interpreted in terms of the Doppler effect due to streaming velocities of the ions. Velocities along the target normal were also determined for C VI, C V and C IV from the Doppler displacement, and agreed with those found from the spectroscopic time-of-flight measurements of Boland *et al.* (1968) within the experimental uncertainty.

A very detailed study was made of C V λ 2271 Å. Figure 8.51 shows the spatial distribution at $t = 45$ ns, while Figure 8.52 shows the time variation of the intensity

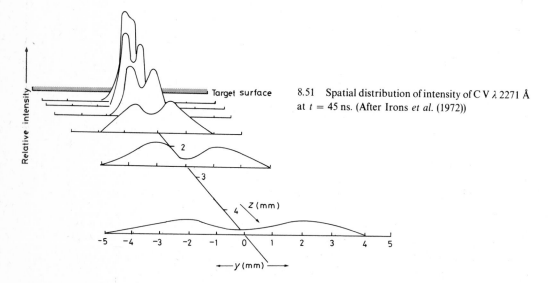

8.51   Spatial distribution of intensity of C V λ 2271 Å at $t = 45$ ns. (After Irons *et al.* (1972))

distribution in the plane parallel to the target surface and 1·6 mm towards the laser, and also the time dependence of the line profile of light from a pencil through the laser beam axis in the same plane. Figure 8.53 shows a sequence of intensity distribu-

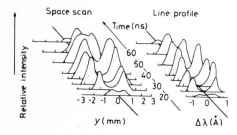

8.52   Time variation of intensity distribution and integrated line profile of C V λ 2271 Å in plane parallel to target, 1·6 mm away from the surface towards the laser (see text). (After Irons *et al.* (1972))

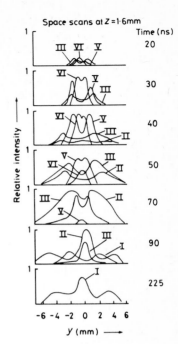

8.53 Sequence of spatial distributions for C I to C VI 1·6 mm away from the target surface. The maximum laser irradiance occurred at $t = 20$ ns. (After Irons *et al.* (1972))

tions in the same plane at different times for C I to C VI. In general, it was found that the various ion species tended to be separated in space, forming conical annular regions surrounding the laser beam axis, the most highly charged ions being on the axis. The mean radial velocity increased with distance from the axis. C V ions were found to travel away from the focal area in roughly straight lines.

Within about 2 mm of the target surface the electron and ion densities are very high, and first-order Stark broadening dominates the spectral line profiles of several hydrogen-like C VI and quasi-hydrogen-like C V lines. Of the possible Stark broadening mechanisms Irons (1973) showed that ion quasi-static (Holtsmark) theory (see Griem (1964)) best fitted the measured profiles of C VI λ 5290 and 3434 Å and C V λ 4945 and 2982 Å, although the theory predicted twice the observed half-intensity breadths. At large distances from the target (12 mm) Doppler broadening dominated the much narrower line profiles. The C V line at λ 2271 Å is not significantly Stark broadened even at very high densities and was found to have the same breadth (ignoring shifts due to radial velocities) all the way from 0·35 to 12 mm from the target. Figure 8.54 shows time-averaged half-intensity breadths at several distances from the target. As is to be expected for the Doppler effect, the line breadths at large distances from the target are proportional to the wavelength for lines from the same ion species.

Irons & Peacock (1974) also investigated spectroscopically the recombination of C^{6+} in carbon plasmas produced by 1 GW neodymium laser pulses. As the plasmas expanded, at greater distances from the target, three-body recombination was found to be progressively more important, while two-body recombination decreased,

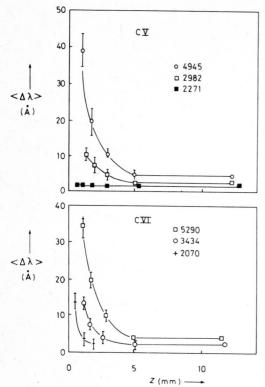

$\langle\Delta\lambda\rangle$
(Å)

$\langle\Delta\lambda\rangle$
(Å)

z (mm) ⟶

8.54   Time-averaged half-intensity breadths of C V and C VI lines as functions of distance z from target surface. (After Irons (1973))

becoming negligible after a few millimetres. L.T.E. was attained for bound states above a principal quantum number which varied from 4, at 1 mm from the target surface, to 6 at 2·8 mm.

Sigel *et al.* (1972) made X-ray absorption measurements of electron temperatures in carbon plasmas using four different beryllium foil absorbers. With neodymium laser pulses of about 30 ns duration and a peak power of 210 MW, the electron temperature was found to be 120 eV, while with 450 MW pulses T_e was 160 eV. The results indicated Maxwellian electron energy distributions with a dependence of T_e on irradiance as $I^{\frac{2}{5}}$. We may, however, compare the results of Waki *et al.* (1972), who measured electron temperatures in plasmas from a polyethylene target by the X-ray absorption method. Their results for the variation of T_e with irradiance, using neodymium laser pulses with 4 ns and 10 ns duration, are shown in Figure 8.55. T_e varies as $I^{\frac{2}{3}}$, as is predicted by deflagration wave theory (Chapter 7, Section (4.2)), assuming that $T_e \propto T_i$. Oscillations at 10^{10} rad s^{-1} appeared in the reflected light intensity when the incident irradiance was about 2×10^{12} W cm^{-2}, corresponding to an electron temperature of about 250 eV. These were ascribed to the presence of ion acoustic waves in the plasma.

Kang *et al.* (1972) measured the velocity distributions of beryllium ions in plasmas

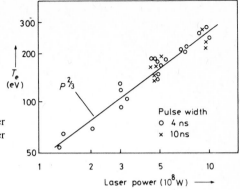

8.55 Electron temperature as function of laser power for plasmas from polyethylene targets. (After Waki *et al.* (1972))

produced by 4 ns neodymium laser pulses with powers up to 625 MW, focused to irradiances up to $8·5 \times 10^{11}$ W cm^{-2}. Figure 8·56 shows that the distributions have approximately Maxwellian shapes with superimposed drift velocities: following Allen (1970), Kang *et al.* ascribed their breadths to ion temperatures, and deduced values of T_i varying from 32 eV for Be^{2+} to 120 eV for Be^{4+}.

Puell *et al.* (1970a) used ruby laser pulses of up to 1 GW peak power with a duration of 7 ns to produce plasmas from carbon and from lithium deuteride targets. The

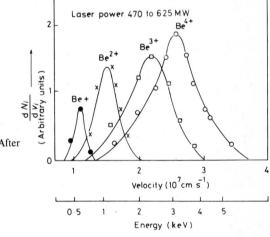

8.56 Velocity spectra of beryllium ions. (After Kang *et al.* (1972))

targets were prepared in the form of rods 100 μm in diameter: the laser beam was focused to a spot of diameter 80 μm on the end of the rod. The plasma electron temperature was measured by the X-ray absorption method using several different foils: the results are plotted against the laser irradiance at the target surface in Figures 8.57(a) and (b). The theoretical predictions of Chapter 7, Section (4.1) allowing for lateral expansion with $r_f = 40$ μm, $Z = 6$ (carbon) and for $r_f = 40$ μm, $\bar{Z} = 2$ (lithium deuteride) are plotted on the assumption that $T_i = T_e$, and also assuming

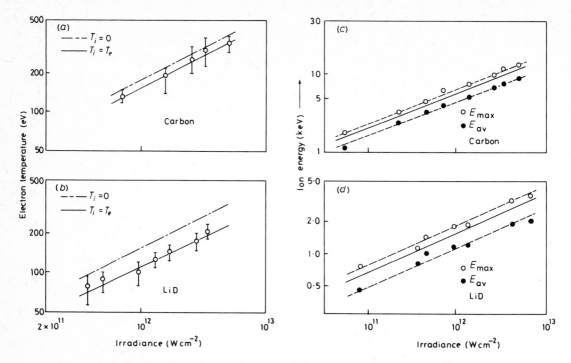

8.57 Experimental and theoretical values of electron temperatures and ion energies as functions of irradiance for carbon and lithium deuteride targets. (After Puell *et al.* (1970*a*))

that $T_i = 0$. As may be seen, agreement is good if $T_i = T_e$. Time-of-flight measurements were also made, giving the total charge and the energy of expansion of the ions in the plasma. The total ion charge collected from a carbon target is shown plotted against the irradiance in Figure 8.58, and agrees well with the theoretical (irradiance)$^{5/9}$ dependence. Measured ion expansion energies are plotted in Figures 8·57(*c*) and (*d*) for carbon and lithium deuteride targets and again show good agreement with the theoretical values.

In another experiment Puell *et al.* (1970*b*) arranged for two plasmas very similar to those described above to collide head-on. The arrangement is shown in Figure

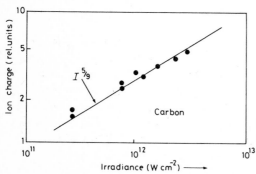

8.58 Total ion charge collected from a carbon target as function of irradiance (After Puell *et al.* (1970*a*))

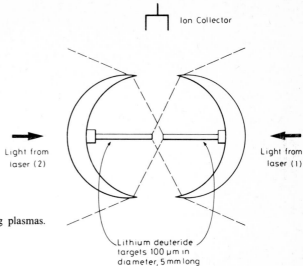

8.59   Arrangement for producing two colliding plasmas.
(After Puell *et al.* (1970*b*))

8.59. The two laser beams originated from the same oscillator with a pulse duration of 5 to 7 ns, each being amplified to 1 GW. The ends of the two lithium deuteride targets were placed 200 μm apart, so that the densities of the plasmas should be sufficiently high when they met (10^{19} cm^{-3}) for a strong interaction to take place. Ions were collected with only one target illuminated, and then with both. The ion energy spectra are shown in Figure 8.60. They were derived from time of flight measurements assuming an average mass of 4·5 proton masses. When both targets were illuminated the total number of ions collected increased by a factor of 4 and the average ion energy increased by a factor of 3. The effects of a collision between two plasmas were described briefly also by Basov *et al.* (1968*d*). The temperature was doubled and the density increased ten times immediately after the collision. The use of an average ion mass is satisfactory for a single freely-expanding plasma because ions of different mass expand with approximately the same velocity. However,

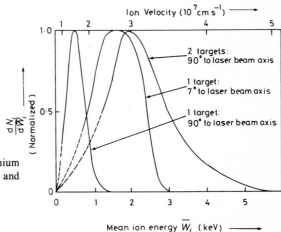

8.60   Energy spectra of ions from lithium deuteride plasmas produced using one and two targets. (After Puell *et al.* (1970*b*))

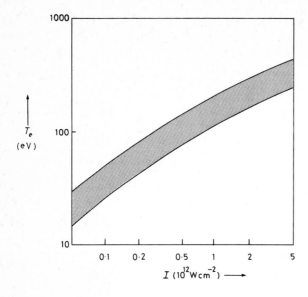

T_e
(eV)

I (10^{12}W cm^{-2}) ⟶

8.61 Composite plot of measured plasma electron temperature as a function of irradiance, for
several target materials and pulse durations

when two plasmas collide their interaction will tend to give all ions equal energies, so that deuterons will gain energy from lithium ions. Puell *et al.* concluded that this is the explanation for the change in the ion spectrum, and that there is no real increase in the mean energy of the plasmas as a result of the interaction.

Figure 8.61 shows a composite plot of plasma electron temperature as a function of ruby or neodymium laser irradiance, as measured by several authors using a variety of target materials and pulse durations. T_e varies approximately as $I^{\frac{2}{3}}$ up to $I \simeq 5 \times 10^{11}$ W cm^{-2}, and as $I^{\frac{1}{4}}$ thereafter.

Dyer *et al.* used carbon dioxide laser pulses with a peak power of up to 0·3 GW and a duration at half maximum power of 60 ns to produce plasmas from a carbon target. Studies of the X-ray emission using the absorption technique with several different pairs of foils (Dyer *et al.* (1974*a*)) indicated that the electron velocity distribution was non-Maxwellian, with a high-energy component having an energy of about 2 keV. The fraction of the laser radiation reflected back into the $f/1\cdot5$ focusing lens was found to be almost constant, below 10%, over the range of target irradiance from 5×10^8 to 4×10^{11} W cm^{-2} (Dyer *et al.* (1974*b*)). It was estimated that less than 10% of the light incident on the target was scattered outside the acceptance angle of the lens. In contrast to this observation, Yamanaka *et al.* (1974) reported that under similar conditions the reflectivity rose with the irradiance to a peak of over 60% at an irradiance of 4×10^{10} W cm^{-2}, and then fell steadily to less than 20% at 8×10^{10} W cm^{-2}. Hall & Negm (1975) observed a very similar dependence of reflectivity on irradiance. These results suggest the onset of a non-linear absorption process, possibly the parametric decay instability. When the reflectivity was a maximum the electron temperature measured by Hall & Negm was about 300 eV. The threshold electric field for the parametric decay instability in a homogeneous plasma under these conditions was estimated to be somewhat higher than that corresponding to an irradiance of 4×10^{10} W cm^{-2} in vacuum, but it should be remembered that the electric field due to the laser irradiance is enhanced near the critical density layer, where the group velocity falls (Thomson *et al.* (1974)). On the other hand, the effect of plasma inhomogeneity is to raise the threshold.

2.2.3. Laser power above 1 GW

The development of large neodymium glass lasers led to the availability of pulses of more than 1 GW peak power. Basov *et al.* (1969*b*) measured the intensity of X-ray emission from the plasma produced with a solid carbon target by a 1·8 GW pulse (27 J, 15 ns) focused by a 50 mm lens. Two scintillation detectors were used, placed behind beryllium foils of mass per unit area 15·5 and 31 mg cm^{-2}. From a comparison of the recorded intensities, using the calibrating data given by Jahoda *et al.* (1960) and Elton & Roth (1967), it was found that the corresponding plasma electron temperature remained between 150 and 220 eV over a period of about 25 ns around the time of arrival of the laser pulse. It was suggested that this might be explained by an increase in the laser beam divergence with increasing power (Basov *et al.* (1967*a*)).

Belland *et al.* (1971*c*) produced plasmas from metal and plastic targets using neodymium laser pulses of 1·8 ns duration and up to 1·5 GW peak power. The incident light had a smooth spectrum with a half-intensity breadth of 37 Å. At irradiances exceeding $1·5 \times 10^{12}$ W cm^{-2} the spectrum of light returned from metal plasmas towards the laser contained between 3 and 5 strong lines on the short wavelength side of the incident light maximum, with a constant uniform spacing of 3 Å. The threshold was higher by a factor of 3 for plastic targets. If the target surface was tilted through more than 12° from the normal to the laser beam axis, so that none of the incident light was specularly reflected back into the focusing lens, the lines were not observed. When a longer laser pulse was used the lines appeared at the same threshold irradiance. Subsequent observations (Belland *et al.* (1972)) showed that the lines were due to free-running oscillation of the two amplifier rods in the laser, feedback being provided by the electro-optic switch and by the target, which was 10 m away from the laser. They were eliminated by inserting a saturable absorber between the amplifiers.

Belland *et al.* repeated their measurements with light from a mode-selected laser whose spectrum consisted of a small number of narrow ($\lesssim 1$ Å) lines 16 Å apart. The reflected light spectrum showed a displacement of each line towards longer wavelengths of 1 to 3 Å (Figure 8.62) and a broadening of about 4 Å. The red shift

8.62 Wavelength shift in reflected light, relative to incident light wavelength, as function of laser power. (After Belland *et al.* (1972))

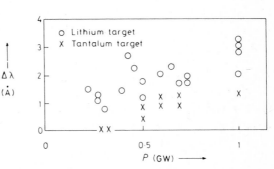

was explained as being due to the Doppler effect: the light is reflected from a plasma layer which moves away from the laser as the target is compressed and ablated.

Zaritskii *et al.* (1972*b*) carried out similar experiments but with a narrow-band neodymium laser ($\Delta\lambda \lesssim 0.05$ Å) consisting of a mode-locked oscillator, with axial mode selection, and six amplifiers. Each laser spike of up to 20 joules lasted 1 ns. A frequency-doubling crystal could be inserted in the beam. The reflected light spectrum at λ 0.53 μm consisted of several equidistant lines 0.23 Å apart, irrespective of the target material. At λ 1.06 μm the spacing increased to 0.46 Å. As in the experiments of Belland *et al.*, it was found that the incident light had weak satellites at these spacings, which were amplified, possibly by a stimulated Raman-type process. The laser light was focused on to the target surface by an $f/1$, 45 mm focal length lens. The fraction of the laser energy reflected back into the lens from an 8 J pulse at λ 1.06 μm was measured as a function of the axial distance of the lens from the target (Basov *et al.* (1972*d*)) with the result shown in Figure 8.63: as may be seen, the energy reflectivity fell by a factor of 2 as the lens was moved through 50 μm. The focal spot diameter was 20 μm, and it seems probable that the curvature of

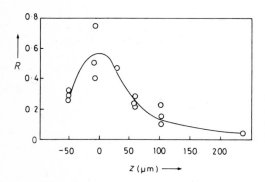

8.63   Fraction of laser energy reflected back into lens focusing light on to a polyethylene target, as function of axial position of focal plane. z is measured from the target surface with $z > 0$ towards the laser ($f/1$ lens, $f = 45$ mm). (After Basov *et al.* (1972*d*))

the critical density surface in the plasma played an important part in determining the spatial distribution of the reflected light. At the second harmonic wavelength the maximum specular energy reflectivity was only 5% from LiD, and 6 and 7% from polyethylene and aluminium respectively. Frequency-doubling can be achieved with greater than 50% efficiency, so the shorter wavelength is more advantageous for plasma heating: higher harmonics may be even better. The polarization of the frequency-doubled pulse was determined and compared to that of the reflected light (Zaritskii *et al.* (1972*a*)). Whereas the incident pulse was 99.8% linearly polarized, the reflected pulse was only 90 to 95% linearly polarized (varying from shot to shot), in the same plane as the incident light. The change in polarization may have been the result of Faraday rotation in the self-induced magnetic field of the plasma (see Chapter 7, Section (9.2)): assuming that $n_e \sim 5 \times 10^{20}$ cm^{-3} and the plasma layer was 100 μm thick, the magnetic field was estimated to be about 30 kG.

Very detailed mass spectrometric studies of plasmas generated by neodymium laser pulses of up to 2 GW peak power were undertaken by Bykovskii *et al.* (1971*a*). They used a variety of metal targets—aluminium, manganese, cobalt, niobium, tantalum, tungsten—and measured the numbers of multiply-charged ions, their energies and angular distributions. At the highest irradiance (about $10^{13}\,\mathrm{W\,cm^{-2}}$), ions with charges up to $Z = 25$ and energies as large as 40 keV were detected. The most energetic ions had the highest charges: they were found to be emitted only in directions close to the normal to the target surface. Figure 8.64 shows smoothed ion energy distribution curves for cobalt ions.

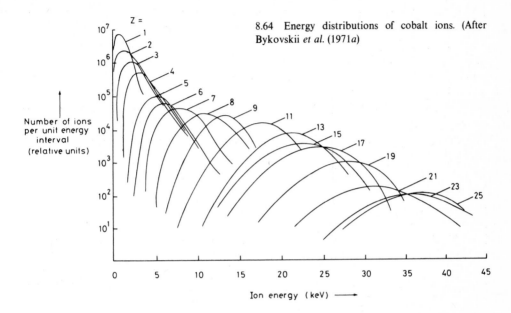

8.64 Energy distributions of cobalt ions. (After Bykovskii *et al.* (1971*a*)

Number of ions per unit energy interval (relative units)

Ion energy (keV)

Plasmas produced from Perspex, nylon and PTFE targets by neodymium laser pulses of up to 2 GW power were investigated by Donaldson *et al.* (1973). The electron temperature was measured from X-ray absorption observations using combinations of aluminium and Mylar foil filters and also from absolute intensities, and was found to vary with the irradiance I as $I^{0.4\pm0.05}$. Grazing-incidence spectrograms were obtained from which the highest stage of ionization detectable could be related to the laser irradiance and the electron temperature.

Basov *et al.* (1971*b*) observed that about 1000 neutrons were emitted from a deuterated polyethylene $((CD_2)_n)$ powder target heated by 14 J neodymium laser pulses of a few nanoseconds duration. X-ray absorption measurements indicated the occurrence of hard radiation quanta with energies of more than 100 keV. These were probably emitted from the walls of the vacuum chamber by fast electrons from the plasma: when the detector was screened from radiation emitted directly from the target region the X-ray flux was not noticeably reduced.

Basov *et al.* (1973) investigated the X-ray emission from plasmas produced by similar laser pulses (20 J, 2 ns) from $(CD_2)_n$ and from iron targets. The X-ray absorption measurements, using several different combinations of foils, again indicated a strongly non-Maxwellian electron energy distribution with an enhanced high-energy tail in plasmas from the $(CD_2)_n$ target, but in the case of the iron target the results were consistent with a thermal electron energy distribution at a temperature of 800 eV.

Preliminary results of experiments on plasmas produced by 10 GW (35 J, 2·5 ns) neodymium laser pulses were described by Olsen et al. (1972, see also Jones et al. (1972)). Using deuterated polyethylene targets electron temperatures of 350–400 eV were deduced from the X-ray absorption measurements, and C^{6+} ion expansion energies as high as 20 keV were observed. The maximum irradiance was 5×10^{14} W cm^{-2}. The reflectivity of the target was very low—less than 0·4% of the incident laser power was reflected specularly into the $f/5$ lens, while an upper limit of $14 \pm 7\%$ of the laser power was diffusely scattered.

Key et al. described two methods for studying X-rays generated by 20 J, 5 ns neodymium laser pulses from a variety of targets, and gave some preliminary results. A cylindrical gold-cathode photodiode was constructed (Key et al. (1974a)) with a resolving time of 1 ns and a signal current sufficient to drive a fast oscilloscope directly. Space-resolved measurements were made using a double pinhole camera with two different filter foils (Key et al. (1974b)): the spatial resolution was about 15 μm.

Olsen et al. (1973) also described a useful new diagnostic device, consisting of a Thomson parabola ion energy analyser with a channeltron electron-multiplier array to convert the ion image into an optical image which can be recorded photographically from a single shot. Ion energy spectra were measured for plasmas produced from LiH targets by 2·5 ns pulses of energy up to 70 J and correlated with other observations. A 53 J pulse produced a fairly flat distribution of Li^{3+} ion energies between 4 and 20 keV, the maximum detectable energy being about 27 keV. The specular reflectivity of the plasma was 2·5%. In contrast, a 22 J pulse produced two clearly distinct groups of Li^{3+} ions, one centred at about 20 keV and extending to 45 keV, the other centred at about 5 keV. In this case the reflectivity of the plasma was higher—about 12%. The emission of hard X-rays with photon energies between 200 and 800 eV correlated positively with the reflectivity of the plasma and negatively with the input laser pulse energy. When lithium deuteride was used as the target material neutrons were detected with about one third of the shots, the maximum yield being about 800 per pulse.

Similar laser pulses (20 to 70 J, 2 to 5 ns) were used by Mead et al. (1972) and Shearer et al. (1972) to generate plasmas from both polyethylene and deuterated polyethylene targets. The maximum irradiance was 2×10^{14} W cm^{-2} at the focus of an $f/7$ lens. Electron temperatures measured by the X-ray method with different pairs of absorbing foils gave inconsistent results, again indicating a non-Maxwellian electron energy distribution with a high-energy tail at more than 40 keV. Neutrons were detected from the CD_2 targets: the total neutron emission is shown plotted as a function of the laser pulse energy in Figure 8.65. The authors warn that this regular relationship was not reproduced in later experiments, but a yield proportional to (energy)^6 is not very far from the predictions of deflagration wave theory (Chapter 7, Section (4.2)). The neutron yield was correlated positively with high target reflectivity and a large flux of high-energy X-rays. Target and focusing geometry seemed to be

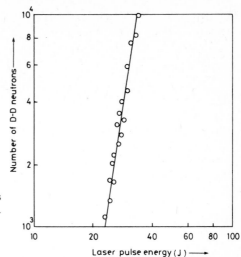

8.65 Number of neutrons emitted from CD_2 targets as function of laser pulse energy. (After Mead *et al.* (1972))

important since repeated shots on the same target point sometimes gave twice the neutron yield of the earlier pulse. Single laser pulses shorter than 2 ns did not produce a detectable number of neutrons, though two short pulses separated by a few nanoseconds did. Measurements of scattered light (Shearer *et al.* (1972)), together with the observation of high-energy non-thermal electrons, have already been mentioned in Chapter 2, Section (11.2) as offering evidence for the occurrence of non-linear interactions.

2.3. PICOSECOND LASER PULSES

The duration of a mode-locked spike from a neodymium glass laser is typically about 10 ps, although this can be increased by limiting the number of modes available to lock together by placing one or more additional Fabry–Perot resonators inside the laser cavity. If a neodymium-doped YAG crystal is used in the laser oscillator, the spike duration is again increased to about 100 ps, because the bandwidth is smaller. When even a small amount of energy is compressed into such a short time, the output power becomes very great: terawatt $(1 \text{ TW} \equiv 10^{12} \text{ W})$ pulses have been generated. Non-linear effects are to be expected in the focal volume of these pulses, and some observations of anomalous reflectivities have already been described in Chapter 2, Section (11.2).

None of the experiments reported so far on plasmas produced from solid targets by picosecond pulses have had a time resolution shorter than the pulse duration. The subsequent evolution of the plasma can, of course, be followed by the methods used for plasmas produced by longer pulses. Thus Belland *et al.* (1971a) have reported holographic interferometric experiments on plasmas produced from an aluminium target by 7 ps neodymium laser pulses of 0·1 to 0·5 J. The total number of free electrons in the plasma produced by a pulse of 0·2 J, at a time 6·5 ns after the plasma was formed, was 5×10^{14}: the plasma boundary velocity normal to the target was 10^7 cm s^{-1}.

Caruso *et al.* (1970*c*) measured the time of flight of ions produced from a nickel target by 100 ps pulses of 0·05 to 2·5 J, with a focal spot radius of 10^{-2} cm. Ion kinetic energies of ~ 20 keV were found with pulses of 1 J: the ion beam was directed predominantly normally to the target surface. Figure 8.66 shows the variation of the number of charges collected in various directions as a function of the laser pulse

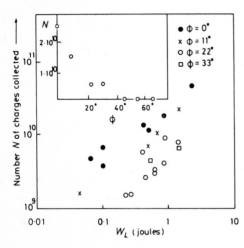

8.66   Number of charges N collected from nickel plasma in various directions relative to the normal to the target surface, as function of laser pulse energy W_L. Inset shows measurements at $W_L = 1$ joule. (After Caruso *et al.* (1970*c*))

energy. Figure 8.67 shows how the ion time of flight (for the probe signal maximum) varied with the pulse energy W_L. The results are in good agreement with the thermal wave theory given by Caruso & Gratton (Chapter 7, Section (4.3)) in that the number of particles emitted varies as $W_L^{\frac{2}{3}}$, while the time of flight varies as $W_L^{-\frac{1}{3}}$. The authors

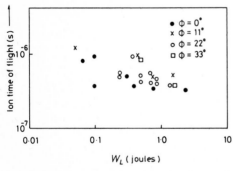

8.67   Ion time of flight over 10 cm, as a function of laser pulse energy W_L, in various directions relative to the normal to the target surface. (After Caruso *et al.* (1970*c*))

point out, however, that the theory is not strictly applicable in this case because the lifetime of the heated layer is shorter than the laser pulse duration.

An important experiment was carried out by Basov *et al.* (1968*c*) with an isolated mode-locked neodymium laser pulse of 2 TW (2×10^{12} W) peak power (20 J, 10 ps), in which neutrons from D–D reactions were detected from a laser-heated plasma for the first time. The laser is shown in Figure 8.68. Five stages of amplification were used,

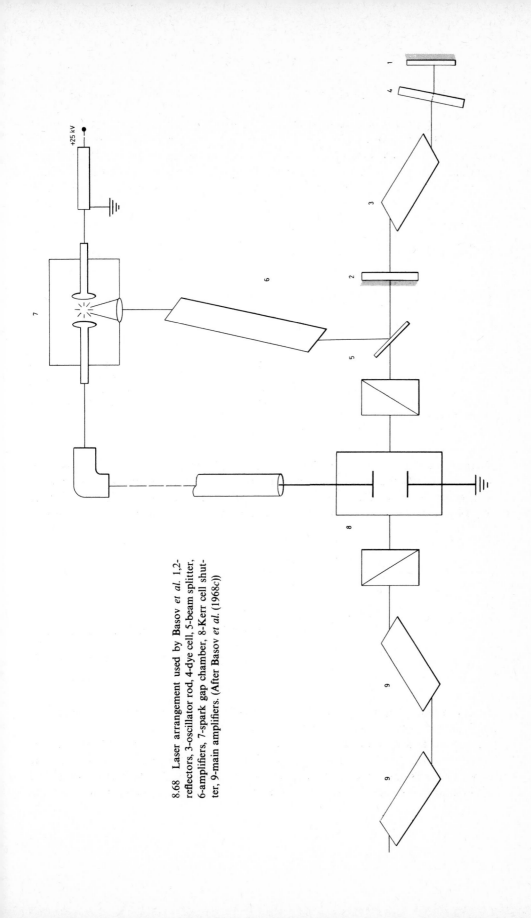

8.68 Laser arrangement used by Basov *et al.* 1,2-reflectors, 3-oscillator rod, 4-dye cell, 5-beam splitter, 6-amplifiers, 7-spark gap chamber, 8-Kerr cell shutter, 9-main amplifiers. (After Basov *et al.* (1968c))

with a gain factor of 10^4. The laser output was focused by a lens of 60 mm focal length on to a target of lithium deuteride prepared in an argon atmosphere to avoid surface contamination (see Langer et al. (1968b)), but placed in vacuum for the experiment as usual. The laser irradiance was about 10^{16} W cm^{-2}. A neutron scintillation counter was arranged to detect neutrons emitted from the target area with an efficiency of 10%. The counter was calibrated using Cs^{137} and Co^{59} sources. Coincidences were observed, with some of the series of laser shots, between pulses from the counter and pulses from the Kerr cell controlling the laser pulse emission. A comparison with observations for natural background radiation showed that coincidences obtained with the lithium deuteride target were twenty times more frequent than would be expected from the background. The pulse height was consistent with the pulses being due to single neutrons.

On the assumption that the deuterons have a Maxwellian velocity distribution, the deuteron temperature may be estimated (see Chapter 4) from the number of nuclear reactions which is given by:

$$N_R = \langle\sigma v\rangle_{T_D}\tau_p V \frac{n_D^2}{2}. \tag{8.7}$$

Here $\langle\sigma v\rangle_{T_D}$ is the average value of the product of the reaction cross section and the relative velocity of the reacting particles, taken over a Maxwellian distribution at temperature T_D, and is a function of T_D only. The duration of the hot plasma is τ_p, its volume is V, and the deuteron density is n_D. The observations suggested that $N_R \sim 1$, which leads to a mean deuteron energy of 2×10^3 eV, in reasonable agreement with an estimate of the energy delivered per particle. The plasma reflected sufficient light back into the laser to cause damage at the input end of the amplifier chain: with a gain of 10^4, the reflectivity need not have been great for this to occur.

This experiment was repeated by Gobeli et al. (1969) with 2–3 ps laser pulses of up to 25 J. In 15 shots, 29 neutron counts were recorded. The average background count for the same 1 ms gating period with the laser operating was 0·72, so the probability that neutrons were indeed observed is $1 - (4 \times 10^{-6})$.

Experiments with picosecond pulses incident on solid hydrogen isotope targets are described in Section (5).

The Bremsstrahlung emission spectra from plasmas produced by 10 J, 25 ps neodymium laser pulses and by 10 J, 1·2 to 1·5 ns CO_2 laser pulses were determined directly by Kephart et al. (1974) using a flat crystal X-ray spectrometer. Graphite and lithium fluoride crystals at various Bragg angles gave measurements at photon energies between 4 and 50 keV. With $(CH_2)_n$ targets the measurements clearly indicated a non-Maxwellian distribution of electron velocities with a high-energy tail, both for the neodymium and for the CO_2 laser pulses.

Flat crystal spectra obtained with potassium acid phthalate crystals were recorded photographically by Nagel et al. (1974) from plasmas produced by 5 J neodymium laser pulses of duration up to 1 ns, focused on a wide variety of targets. The resolution in these experiments was poor, but it was suggested that high-resolution studies of X-ray line profiles might yield useful information about the density attained in compression experiments. Pinhole photographs gave an indication of the size of the emitting plasma: the Bremsstrahlung emission power per unit volume exceeded 10^{13} W cm^{-3}.

3. PLASMAS FORMED FROM THIN SOLID FILMS

If a thin film target is used instead of a massive solid, the later part of a sufficiently long and energetic focused laser pulse will penetrate it. This facilitates investigation of the absorption of light during the pulse, and the rate of ablation of the target material. In this section we shall discuss experiments on plane targets which are too thin to be regarded as semi-infinite, so far as plasma production is concerned.*

The first studies using thin foil targets were made by Linlor (1963, 1964) with a 5 MW ruby laser (0·2 J, 40 ns) focused on to an area of 10^{-3} cm^2. A gold foil of mass per unit area $1·8 \times 10^{-4}$ g cm^{-2} was found to transmit a peak power of only 0·2 MW. A hole was nevertheless produced in the foil as the vaporized material moved away after the laser pulse ended: a second pulse was transmitted through the hole without attenuation. No reflected light was detected. The 5×10^{14} gold atoms present in the focal spot absorbed about 96% of the entire incident laser pulse energy. (With an aluminium foil of mass per unit area $1·6 \times 10^{-4}$ g cm^{-2}, 4×10^{15} atoms absorbed 99% of the incident light). The plasma formed from the gold target therefore received a total energy equivalent to 2·5 keV per initial atom. However, each atom must have been ionized to produce several free electrons, so the mean plasma energy per particle probably amounted to some hundreds of eV, in agreement with other observations.

The charged particles emitted from plasmas formed from thin gold and aluminium foils by 4 J, 25 ns ruby laser pulses were investigated by Fabre & Vasseur (1968). About 10^{16} charges were collected from gold foils of 0·05 to 0·2 μm thickness, and about (4 or 5) $\times 10^{16}$ charges from aluminium foils 2 μm thick. The angular distribution of charge was predominantly along the axis of the laser beam, very little being observed in or near the plane of the target.

A high-power carbon dioxide laser was used to produce plasmas from thin aluminium foils by Pépin et al. (1972) and Dick et al. (1973). The laser pulse shape was rather complex, consisting of a 30 MW pulse of 120 ns half intensity duration followed by a long relatively weak tail lasting several μs with a peak at 1 μs. The total pulse energy was between 3·5 and 7 J, depending on the laser gas mixture. When focused by a 50 mm focal length germanium lens the maximum irradiance was about 10^{10} W cm^{-2}. Streak photographs showed luminous boundaries moving in both directions away from the foil with a velocity of about 10^6 cm s^{-1} for about 100 ns from the beginning of the laser pulse, after which both boundaries came to a halt. After 0·8 μs for the foil of 10 μm thickness, or 1·4 μs for the 20 μm foil, the target was observed to become transparent to the incident light: about this time a further expansion of the boundary furthest from the laser occurred, with a velocity of about

---

* A method of producing very thin metal films has been described by Valenzuela & Eckardt (1971).

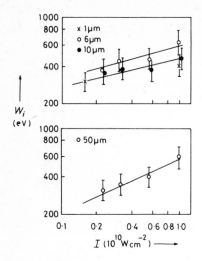

W_i
(eV)

8.69   Mean kinetic energies of ions from aluminium foil targets of four different thicknesses, as functions of irradiance (After Dick *et al.* (1973))

3×10^5 cm s^{-1}. The reflectivity of the plasma was found to be very small, apart from an initial spike lasting a few nanoseconds. Interferograms showed electron densities up to 10^{18} cm^{-3} at a time 150 ns after the laser pulse reached its maximum power. The luminous region of the second expansion was found to consist mainly of neutral atoms. Mean ion energies, determined from time of flight measurements, are shown in Figure 8.69 for four different foil thicknesses. The corresponding electron temperatures at the maximum irradiance varied from 27 eV for a 1 μm foil to 31 eV for a 150 μm foil. The results for the thick foils agreed well with the self-regulating theory of Chapter 7, Section (4.1): the theory of Bonnier & Martineau (1973), which takes into account the increasing transparency of the target during the pulse, fitted the thin foil observations.

Opower *et al.* (1967) compared the incident and transmitted pulse shapes with thin lithium hydride targets, using a ruby laser of 150 MW peak power (1·5 J, 10 ns) focused into an area of 3×10^{-4} cm^2. A target 3 μm thick transmitted a significant amount of light during the laser pulse, but a 5 μm target remained opaque. Thus the laser pulse energy was absorbed by about 9×10^{-8} cm^3 of lithium hydride, containing 5×10^{15} atoms. Assuming an average of two electrons per ion, the mean energy delivered to the plasma per particle was therefore 600 eV.

Thin films of polythene were used as targets for a 200 MW ruby laser by Griffin & Schlüter (1968). The focal area was $2·5 \times 10^{-4}$ cm^2. The laser light transmitted through the target film was collected by an integrating sphere and detected by a photodiode. No significant amount of reflected light was observed. The incident and transmitted pulses for a 70 μm film are shown in Figure 8.70. The transmitted power follows the incident pulse shape initially: the film is then transparent. At time t_1, the fraction of the incident light power transmitted falls rapidly. It does not fall to zero, because the spatial distribution of irradiance in the focal spot falls off at the periphery, where the film remains transparent. However, in the centre of the spot

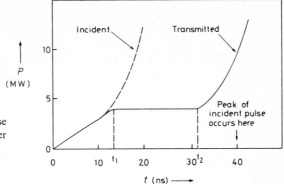

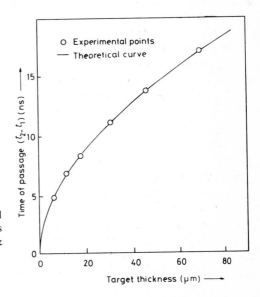

8.70 Incident and transmitted light pulse shapes for thin polythene targets. (After Griffin & Schlüter (1968))

the film becomes strongly absorbing, and remains so until time t_2 when the fraction transmitted increases rapidly. The time t_1 is found from measurements with various laser fluxes to be the time at which the irradiance reaches the value of 10^{10} W cm^{-2}, and is independent of the film thickness. However, the time interval from t_1 to t_2 does depend on the film thickness. Thus t_2 is interpreted as the moment at which the absorbing layer, initially formed at the front surface nearest the laser, reaches the back of the film. The time of passage $t_2 - t_1$ is plotted in Figure 8.71 as a function of the film thickness. The velocity of the absorbing layer as a function of the instantaneous irradiance may be derived from these results: this is plotted in Figure 8.72. All these measurements were made on the rising edge of the laser pulse. The times t_1 and t_2 were clearly defined even with films only 3 μm thick, from which it would appear that the effective absorption length is not greater than about 1 μm. Griffin & Schlüter suggested that although the irradiance was increasing exponentially during the measurements with a rise time of about 3·5 ns, the theory of Caruso *et al.* (Chapter 7,

8.71 Comparison of experimental and theoretical results for time of passage of absorbing layer, as function of target thickness. (After Griffin & Schlüter (1968) and Caruso & Gratton (1968b))

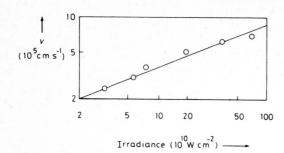

8.72   Velocity of absorbing layer through polythene film targets as function of irradiance. (After Griffin & Schlüter (1968))

Section (4.1)), which assumes a constant irradiance, should apply to their experiments since the velocity of a particle in the absorbing layer, relative to the layer, is almost everywhere greater than the ratio of the absorption length to the laser pulse rise time. Caruso & Gratton (1968b) pointed out that what was actually measured was the velocity of the absorbing layer relative to the shocked material moving in the same direction, and found good agreement with the theoretically predicted dependence of the time of passage on the film thickness (Figure 8.71).

Bruneteau et al. (1970) measured the time of passage of the absorbing layer through polyethylene and also aluminium films of different thicknesses, for a laser pulse of power up to 150 MW with a half-intensity duration of 15 ns, focused to a mean irradiance of up to 2×10^{11} W cm^{-2}. The total number of charges collected was measured for various pulse energies and is shown plotted in Figure 8.73. The slope of the line is about 0·7. With these targets, however, the average stage of ionization increases with the laser pulse energy W_L and Bruneteau et al. estimated that the number of ionized atoms collected probably varied approximately as $W_L^{\frac{1}{2}}$. The velocity of the absorbing layer was found to be 10^6 cm s^{-1}, which is higher than that reported by Griffin & Schlüter for the same laser irradiance. The difference may be due to the difference in laser pulse shapes.

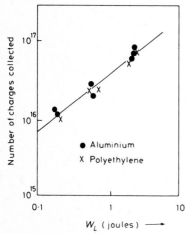

8.73   Number of charges collected from plasmas generated from polyethylene and aluminium targets as function of laser pulse energy W_L. (After Bruneteau et al. (1970))

TABLE 8.3: *Electron temperatures of plasmas produced from thin foil targets by 300 MW, 6 J ruby laser pulses*

Target material	Thickness (μm)	Electron temperature (eV)
Polythene	14	170 ± 15
	100	145 ± 15
	200	85 ± 10
Aluminium	20	170 ± 15
Steel	100	145 ± 20

After Muller & Green (1971).

Seka *et al.* (1971) prepared multi-layer targets of aluminium and polyethylene foils on plastic or aluminium bases and studied the far ultraviolet spectra emitted when they were irradiated by 200 MW, 40 ns neodymium laser pulses. When aluminium targets were used, covered with polyethylene foils of different thicknesses, the relative intensities of aluminium lines from all stages of ionization from Al V to Al XI remained unchanged, but all the lines became fainter as the polyethylene thickness was increased up to 300 μm, when no aluminium lines were observed. It seems that the plasma produced at different depths in the target has much the same characteristics.

Muller & Green (1971) measured the electron temperatures of plasmas produced from various thin foil targets by a 300 MW, 6 J laser ruby focused by a lens of 30 mm focal length, using the X-ray absorption technique. The results are shown in Table 8.3. In the case of polythene foils, the transit time of the absorbing layer through the foil was estimated to be equal to the pulse duration when the foil thickness was about 100 μm. The low temperature observed with the 200 μm foil was ascribed to energy being shared with material ahead of the absorbing layer.

Mode-locked neodymium laser pulses of 10 ps duration were focused on to thin Mylar films by Salzmann (1973). Significant ion emission from the back of a 2 μm film was observed when the pulse energy exceeded 0·5 J, corresponding to 5×10^3 J cm^{-2}. Most of the laser pulse energy was absorbed in the target: the results were consistent with thermal wave theory, as is to be expected at this irradiance ($\sim 5 \times 10^{14}$ W cm^{-2}).

4. PLASMAS FORMED FROM SMALL ISOLATED TARGETS

Small, homogeneous solid targets have been used in many experiments. Another form of target of particular interest is a small, hollow glass sphere into which hydrogen isotopes or other gases may be introduced at high pressure by diffusion: one method of preparation has been given by Lewkowicz (1974).

We shall describe a number of methods for suspending small targets in vacuum, and then go on to describe experiments in which plasmas have been produced with single- and multiple-beam systems.

4.1. METHODS OF SUSPENSION

The simplest methods of placing a small target particle at the focus of a laser beam are to suspend it from a small-diameter filament, as was done for example by Fabre & Lamain (1969), or to place it on the point of a needle (Dolgov-Savel'ev et al. (1970)). However, with these arrangements the target isolation is not complete, and preparation is slow. More sophisticated methods have been devised.

The electrodynamic method of suspension used by Haught & Meyerand, who in 1964 produced the first plasmas with isolated particles, was described by Haught & Polk (1966). It is based on the fact that it is possible to arrange a three-dimensional oscillating electric field in such a way that a charged particle placed in it will move in a small, stable elliptical orbit. This arrangement was developed by Wuerker et al. (1959b) from a 1-dimensional system investigated by Wuerker et al. (1959a). The electrode arrangement is shown in Figure 8.74. Each parallel pair of electrodes, 5 cm apart, was connected to one of the terminals of a star-connected three phase a.c. supply. It was found to be convenient in a later version (Haught & Polk (1970)) to increase the spacing to 10 cm and to use a variable frequency supply (10 to 400 Hz). Bias voltages could be applied to each plate separately. Powdered target material was contained in a box from which an externally operated spring could flip a small number of particles into the central region of the electrode system. A hot filament produced electrons which charged up the powder particles. By careful manipulation of the electrode potentials a single 10 to 20 μm particle could be placed in a stable orbit with a diameter about three times that of the particle and one third of that of the focal spot of the laser beam. The stability of the suspension is discussed, and some useful practical details are given, by Zaritskii et al. (1971).

Consoli et al. (1968) described some other geometries, making use of a number of parallel rod or plate electrodes arranged on the circumference of a cylinder. Such arrangements may have practical advantages in certain experiments, particularly if it is desired to produce the plasma within a resonant microwave cavity. Another electrodynamic suspension geometry was used by Waniek & Jarmuz (1968).

Dolgov-Savel'ev & Karnyushin (1970) used an electrostatic field to attach a particle to a plane electrode. A pulsed field of opposite polarity detached the particle, which then fell 20 cm with a satisfactorily predictable trajectory into the middle of a chamber in which the laser beam was focused.

Another method of positioning an isolated particle is to place it on top of a supporting pedestal, and then to remove the pedestal rapidly, leaving the particle moving downwards relatively slowly under gravity. This was done by Pack et al. (1968), who used as the pedestal a long, thin, vertical metal stem with a wider anvil at the

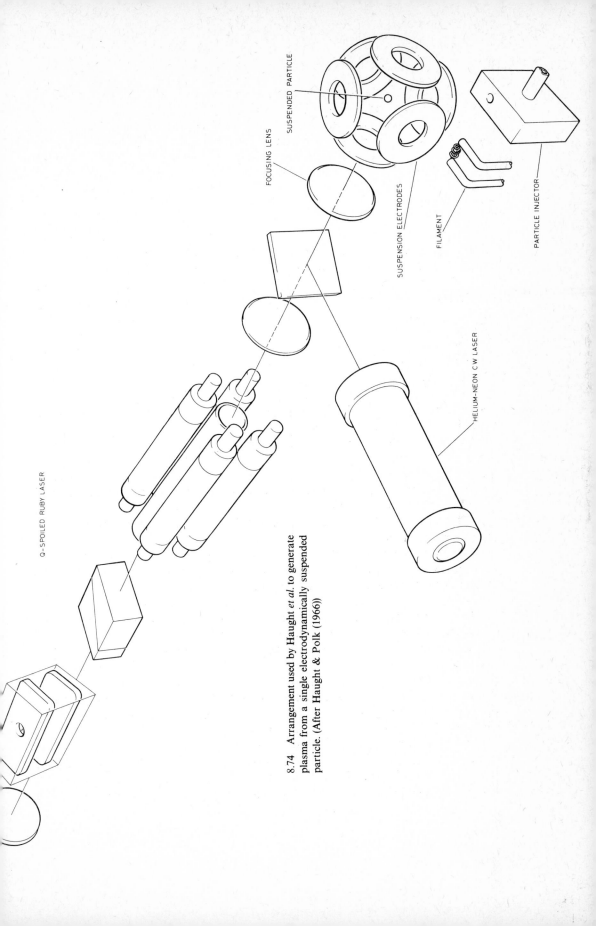

Q-SPOILED RUBY LASER

FOCUSING LENS

SUSPENDED PARTICLE

SUSPENSION ELECTRODES

FILAMENT

PARTICLE INJECTOR

HELIUM-NEON CW LASER

8.74 Arrangement used by Haught *et al.* to generate plasma from a single electrodynamically suspended particle. (After Haught & Polk (1966))

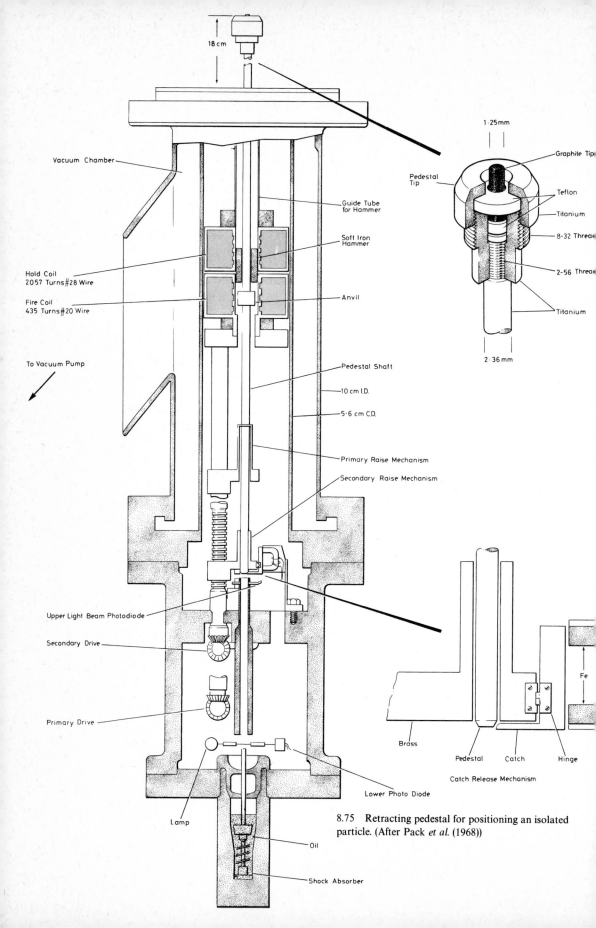

18 cm

Vacuum Chamber

Guide Tube for Hammer

Soft Iron Hammer

Hold Coil
2057 Turns #28 Wire

Fire Coil
435 Turns #20 Wire

Anvil

To Vacuum Pump

Pedestal Shaft

10 cm I.D.

5·6 cm C.D.

Primary Raise Mechanism

Secondary Raise Mechanism

Upper Light Beam Photodiode

Secondary Drive

Primary Drive

Lower Photo Diode

Lamp

Oil

Shock Absorber

1·25mm

Graphite Tip

Pedestal Tip

Teflon

Titanium

8-32 Thread

2-56 Thread

Titanium

2·36 mm

Fe

Brass

Pedestal

Catch

Hinge

Catch Release Mechanism

8.75 Retracting pedestal for positioning an isolated
particle. (After Pack *et al.* (1968))

lower end which was driven downwards by a magnetically-operated hammer. Figure 8.75 shows the arrangement. It was found that the particle fell almost exactly vertically under gravity after the pedestal was removed, the sideways velocity being very small. This fortunate behaviour is probably due to the fact that longitudinal waves propagate more rapidly than transverse waves in the stem. Retracting pedestals have the advantages that they can accommodate a wide range of target dimensions, and that the orientation of the target to the laser beam may be fairly accurately established.

Very small particles have been levitated by the pressure of light from a c.w. argon laser (Yamanaka *et al.* (1974)).

4.2. EXPERIMENTS USING SINGLE LASER BEAMS

Haught & Polk (1966) used a 20 MW, 20 ns ruby laser to form plasmas from 10 to 20 μm diameter lithium hydride pellets. Charge collection experiments indicated that the particles were totally ionized. The asymptotic radial velocity of the plasma particles between 20 and 40 cm from the target was found to be the same for ions as for electrons, and was equal to $2 \cdot 5 \times 10^7$ cm s^{-1}. The time of flight was the same within 20% in three different directions. An average energy of 170 eV per particle was deduced. Reasonably good agreement was obtained with the theory described in Chapter 7, Section (6). Very similar plasmas were produced by Breton *et al.* (1968) from electrodynamically suspended lithium hydride particles 20 μm in diameter at the focus of a 20 MW ruby laser pulse. The plasma became transparent in less than 10 ns. The line profiles of H$_\beta$ and of Li I λ 4603 Å were measured at several different times after the laser pulse, by means of a monochromator and photomultiplier. Many laser shots were needed, and this limited the accuracy with which the profile could be determined. The convolution with the instrumental slit function was allowed for. It was assumed that the H$^+$ and Li^{2+} ions both varied in density exponentially with radius, with the same characteristic radius, and that their velocities were proportional to the radius. The maximum velocities 250 ns after the laser pulse were found to be $(2 \cdot 8 \pm 0 \cdot 5) \times 10^7$ cm s^{-1} for H$^+$ and $(2 \cdot 5 \pm 0 \cdot 5) \times 10^7$ cm s^{-1} for Li^{2+}.

Inoue *et al.* (1971) measured the energies of ions from the plasma produced when the tip of a polyethylene filament was placed at the focus of an 80 MW, 1·2 J ruby laser pulse. Only C^{2+} ions were observed; the ion energy distribution peaked at 70 eV. About a tenth of the laser energy was absorbed in the target.

Lithium hydride particles suspended electrodynamically were used as targets by De Michelis & Ramsden (1967). Their ruby laser produced a 10 ns pulse with 120 MW peak power, which was focused on to the target particle by a lens of 5 cm focal length. The electron temperature of the resulting plasma was measured by the X-ray absorption method and an upper limit of 100 eV was found. Streak photographs of the plasma showed an initial expansion velocity of $(5 \pm 1) \times 10^7$ cm s^{-1}, which fell in 10 to 15 ns to about 10^7 cm s^{-1}. The plasma expanded almost symmetrically.

Sucov *et al.* (1967) reported observations on plasmas produced by a 60 MW (2 J, 30 ns) ruby laser focused on to aluminium targets cemented on to a fine silica fibre 12 μm in diameter. In one group of experiments the targets were evaporated foils 3 μm thick and 400 μm in diameter, corresponding to the diameter of the focal spot using a lens of 80 mm focal length. 70 to 80% of the laser energy was absorbed. The main signal received by an electrostatic probe indicated a plasma expansion velocity of 1.5×10^7 cm s^{-1}. High-speed photography showed a plume moving towards the laser with a velocity of 10^7 cm s^{-1} for 1 μs, followed by a fairly symmetrical plasma of much lower velocity (2×10^6 cm s^{-1}) and lower brightness. It appears that the foil was not penetrated during the laser pulse. Some measurements of 4 mm microwave transmission and phase shift indicated average electron densities of about 10^{13}, which fell to 10^{11} cm^{-3} in about 10 μs. After 2 μs the electron temperature was about 0.03 eV. On replacing the disc target by an aluminium sphere of 125 μm diameter, again supported by a silica fibre, only 25 to 30% of the laser light was absorbed. A more nearly spherical plasma was generated with an expansion velocity of 3×10^6 cm s^{-1} after 100 ns. Microwave measurements gave similar results initially to those obtained with disc targets, but a more precise double-pass technique (Pack *et al.* (1969)) indicated that the average electron temperature after 2 μs was somewhat lower—about 0.01 eV: the electron density was also lower at that time by a factor of 3. Microwave cut-off observations gave the velocity of expansion of the spherical surface in which $n_e = 6 \times 10^{13}$ cm^{-3}. This was found to be 5×10^6 cm s^{-1} at a time 300 ns after the laser pulse. In further experiments (Engelhardt *et al.* (1970)), using a retracting pedestal to hold the target, the proportion of the incident laser light absorbed was measured for several aluminium target sphere radii, and hence the mean energy per atom of the target was deduced. From the expansion velocity of the outermost luminous region, recorded photographically, a value for the maximum ion energy was obtained. These results are shown in Figure 8.76. The ratio between the mean and the maximum energies varies between 1.5 and 2. Some preliminary

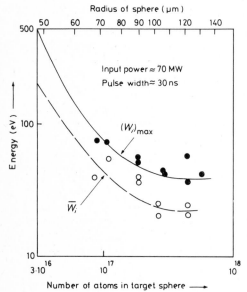

8.76   Maximum and average ion energies as functions of number of aluminium atoms in spherical targets. (After Engelhardt *et al.* (1970))

results of Thomson scattering measurements on similar plasmas were reported by George *et al.* (1970), which showed anomalously high electron temperatures and low electron densities and require confirmation.

Koopman (1967) described studies of plasmas produced from lithium hydride and lithium deuteride targets by 200 MW (3 J, 15 ns) ruby laser pulses. A retracting pedestal was used to position groups of about 10 particles of 25 to 50 μm diameter in the 400 μm diameter of the laser beam focus. An expansion velocity of (6 to 9) $\times$ 10^6 cm s^{-1} was observed with an image-converter streak camera. From the velocity, together with a knowledge of the target mass and the quantity of energy absorbed from the laser beam, it was deduced that initially the plasma was formed at a temperature of about 20 eV from about 10^{16} target molecules. A small number (about 10^{10} to 10^{11}) of fast ions was found to leave the target volume with w_i/Z up to about 200 eV per proton charge.

Fabre & Lamain (1970) studied the variation with target diameter of the total number of charges produced by laser pulses focused on to small spherical paraffin targets suspended by fine silica fibres. The 500 MW laser pulse of duration 12 to 15 ns was focused to a 200 μm diameter spot. The number of free electrons produced was determined interferometrically. For target diameters less than 80 μm, every CH_2 molecule in the target was fully ionized, generating between 5 and 7 free electrons. As the diameter increased this number fell off, and only one electron per molecule was generated when the diameter was 180 μm, corresponding to complete ionization of a layer only 10 μm thick.

Faugeras *et al.* (1968) investigated the plasmas produced from electrodynamically suspended lithium hydride particles of 30 to 60 μm diameter. The neodymium glass laser produced pulses of energy up to 45 J with a half-intensity duration of about 35 ns (1·3 GW). The diameter of the focal spot was about 200 μm. The light emitted from the plasma in the visible and near ultraviolet regions was monitored by a fast photomultiplier. The laser light transmitted through the plasma was studied using a fast photodiode. Charged particle collectors measured the flux and energy of the plasma ions. The mean initial particle energy was deduced to be 20 to 40 eV, the mean ion energy detected being 100 to 350 eV. It was found that all laser pulses of more than three joules total energy (100 MW peak power) produced very similar plasmas. The moment at which light from the plasma was first detected always corresponded to the time at which the rising laser pulse power reached about 1 MW. The shape of the laser pulse was constant, rising exponentially between 100 ns and 20 ns before the time of peak power, with a time constant of about 12 ns. For pulses containing more than about 3 J, the entire process of plasma heating takes place during the period of exponential rise, the plasma being transparent to the remainder of the pulse. Under these conditions, therefore, we have the important result that increasing the pulse energy has no effect on the process of plasma heating, which merely takes place earlier relative to the peak of the laser pulse.

Haught & Polk (1970) continued their experiments with lithium hydride particle targets of 10 to 50 μm radius, using 10 to 15 ns ruby laser pulses of up to 500 MW peak power. The focal spot diameter was about 200 μm: small changes in the positioning of the target particle had a marked effect on the plasma produced. Ion charge collectors at different distances from the plasma origin gave signals whose time dependence scaled linearly with the distance of the collector from the origin. This self-similar behaviour implies a linear variation of velocity with radius, as was assumed in the theory of Chapter 7, Section (6). The radial charge density profile in the expanding plasma was deduced and is shown in Figure 8.77 compared with the theoretical mass density profile. It seems that considerable recombination occurs in the denser regions of the plasma. On integrating over the plasma volume the charge collected was found to be about half that which would lie under the Gaussian curve.

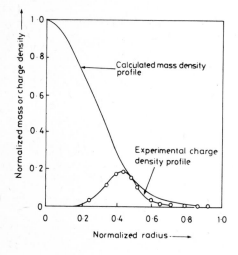

8.77 Experimental charge density and calculated mass density distributions. (After Haught & Polk (1970))

This is consistent with the fact that the total charge collected corresponded to half that which would have been expected from the fully-ionized target particle.

Experimental measurements of the boundary velocity for a wide range of laser pulse conditions are shown in Table 8.4, together with the corresponding average plasma particle energy before expansion, and also particle energies derived from the

TABLE 8.4: Measurements on plasmas produced from lithium deuteride particles by ruby laser pulses

Laser pulse					Experimental			Theoretical
Power (MW)	Rise time (ns)	Half width (ns)	Focal spot radius (μm)	Target particle radius (μm)	Velocity (cm s^{-1})	Average particle energy (eV)	Laser energy absorbed (%)	average particle energy (eV)
30	10	10	400	10	1×10^7	15	1·2	34
20	20	20	60	10	2×10^7	60	3·8	68
300	15	15	100	15	4×10^7	210	3·9	182
315	17	13·5	120	25	$3·6 \times 10^7$	175	16	184
320	20	14	120	40	$3·5 \times 10^7$	160	57	156

After Haught & Polk (1970).

integrated similarity theory. The agreement is very good. Several other diagnostic measurements were made, including a mass-spectrometric analysis which showed no trace of impurity ions. The ratio of hydrogen to lithium ions, at high plasma energies (~ 100 eV) when both atoms are fully ionized, was near unity. Time-resolved measurements showed that the hydrogen and lithium atoms expand at the same rate, maintaining a homogeneous plasma at all times, as is to be expected because of the short mean free path for atomic collisions.

It is apparent from these results that in order to produce hotter plasmas it is desirable to use laser pulses with shorter rise-times. Mattioli & Véron (1971) used a fast electro-optic switch to produce neodymium glass laser pulses rising from zero to 0·3 GW in 1·5 ns. With an irradiance of $1·5 \times 10^{12}$ W cm^{-2} the heating time for a 50 μm diameter lithium hydride particle was found to be 20 ns: about 10% of the laser energy was absorbed in that time. Seka et al. (1970b) investigated spectroscopically the plasma generated from 15 to 25 μm diameter lithium hydride particles by 3 ns, 1 GW neodymium glass laser pulses. From the behaviour of Li ion lines and lines of oxygen present as a surface impurity, they concluded that the plasma consisted of (a) a rapidly expanding ($v \sim 2 \times 10^7$ cm s^{-1}) periphery in which the density fell so rapidly that recombination could not occur to any appreciable extent so that ionization states were 'frozen-in', and (b) a dense, relatively slowly expanding ($v \sim 2 \times 10^6$ cm s^{-1}) core in which recombination was complete in 100 to 150 ns. Ions in this region could not be detected by mass spectrometers or time-of-flight spectrometers, which are usually placed many centimetres from the target.

Plasmas produced from lithium hydride particles were investigated by Lubin et al. (1969) using several experimental techniques. The particles were suspended electrodynamically, and in the initial experiments were heated by a laser of 600 MW peak power with a pulse duration at half intensity of 9 ns. Time-of-flight measurements showed distinct asymmetry: the plasma expanded up to 1·8 times faster towards the laser than away from it. The total charge collected varied erratically from shot to shot, suggesting that the solid particle was incompletely ionized, particularly for particle radii greater than 65 μm. It was therefore decided to preheat the target particle with a pulse of relatively low power, sufficient to vaporize it completely, and then to apply the main heating pulse. For 20 μm diameter lithium hydride particles, a prepulse of about 5 MW lasting 2 ns was used, followed by a main pulse of peak power $1·5 \times 10^9$ W with a half-power duration of 4 ns. The expansion symmetry was greatly improved. By means of charged particle collectors the average energy per particle and the total number of ions in the expanding plasma were determined. Thomson scattering observations were made using a subsidiary 100 MW, 15 ns ruby laser. The scattering parameter α was as high as 5 at the time of maximum temperature, so collective effects dominated the spectrum (see Chapter 3). A seven-channel array of interference filters and photomultipliers was used to resolve and record the scattered light spectrum. In this way the electron temperature and density were determined. Some of the results are shown in Figure 8.78. The curves are theoretical values based on the theory of Haught & Polk (1966). The agreement is good, except perhaps for the average energy per particle for larger target diameters.

In later experiments (Lubin et al. (1972), Goldman et al. (1973)), spherical lithium

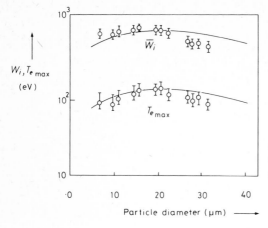

$W_i, T_{e\,max}$ (eV)

Particle diameter (μm)

8.78 Measured and calculated average ion energy and peak electron temperature as functions of lithium hydride particle diameter (see text). (After Lubin *et al.* (1969))

deuteride and deuterated polyethylene targets of diameter 150 to 250 μm were heated by 5 to 40 J neodymium laser pulses of 120 ps duration, following a weak prepulse to improve absorption. Electron temperatures of about 1 keV were measured. More than 10^4 neutrons were reported from LiD targets: none were observed without a prepulse. The specular plasma reflectivity increased with laser pulse energy, saturating at a value of about 10%. At the very high irradiances occurring in these experiments (more than 10^{16} W cm^{-2}) many of the non-linear effects discussed in Chapter 2 are expected to appear.

4.3. EXPERIMENTS USING MULTIPLE LASER BEAMS

Basov *et al.* (1972c) described a nine-beam neodymium laser which produced pulses lasting between 2 and 16 ns with a total energy of 600 to 1300 J. Almost uniform irradiation of a spherical target of radius 50 to 250 μm could be achieved with an irradiance of about 10^{15} W cm^{-2}; the optical paths in the nine beams were adjusted to be almost equal so that the arrival of laser energy at the target from all directions was synchronized to within 30 ps. The ratio of pulse height to background was 10^7. The system was used to heat a small deuterated polyethylene sphere (Basov *et al.* (1972a)). The beams were focused 200 μm away from the target surface to give uniform heating and to reduce reflections. The total pulse duration was 6 ns. Table 8.5 shows the results of electron temperature and neutron emission measurements for four different target radii and total laser energies.

Experiments in which up to four coplanar laser beams were focused on small, hollow cylindrical polyethylene targets were reported by Schirmann *et al.* (1974). The neodymium laser oscillator generated pulses of 3 ns duration which were divided and amplified to a maximum energy of 150 J in each beam. The beams were focused by 117 mm focal length, $f/1.5$ lenses. Interferograms of the plasma were obtained in

TABLE 8.5: *Electron temperature and neutron yield measurements on plasmas produced from deuterated polyethylene spheres by neodymium laser pulses*

Target radius (μm)	Laser energy (J)	Average electron temperature (eV)	Neutron yield
250	600	40	—
125	202	120	—
55	214	840	3×10^6
30	232	4000	—

After Basov *et al.* (1972a).

light at the second harmonic of another neodymium laser using a Jamin interferometer and three image-converter cameras. Figure 8.79 shows three interferograms obtained with different positions of the foci of the four beams relative to the target. All three were obtained under similar conditions at a time 5 ns after the laser pulse began. The symmetry of the plasma improves as the distance d of the focal point beyond the axis of the cylinder is increased, and appears to be quite good when $d = 1.2$ mm. Figure 8.80 shows the results obtained with 1, 2 and 4 beams. The

8.79   Corona structures produced by four coplanar laser beams focused at distance d from the axis of a cylindrical target (see text). (After Schirmann *et al.* (1974), by courtesy of C. Patou)

$E_L = 4 \times 50$ J $t = +5$ ns

$\Delta t = 3$ ns

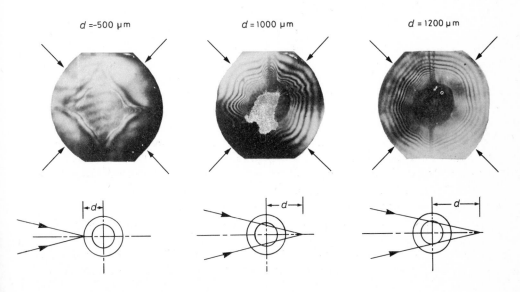

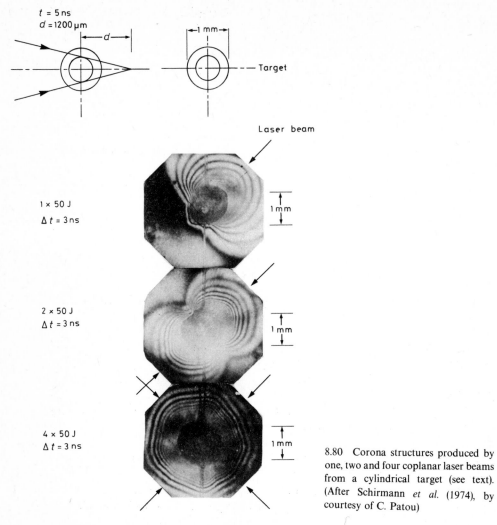

8.80 Corona structures produced by one, two and four coplanar laser beams from a cylindrical target (see text). (After Schirmann *et al.* (1974), by courtesy of C. Patou)

interferograms at λ 0·53 μm only give information about electron densities up to about 10^{20} cm^{-3}. It does not follow from the symmetry of an interferogram that the denser ablation surface is also symmetrical. However, photographs of the plasma luminosity at about λ 0·5 μm, which is strongest near the dense, hot, critical density surface, also indicated good cylindrical symmetry when $d = 1·2$ mm.

Observations of the shock wave within the cylindrical target were frustrated by the appearance of a plasma from the inner wall of the cylinder very early during the later pulse, due to light passing through the target wall before the highly-reflecting critical density layer formed near the outer surface. This internal plasma rapidly became opaque to the interferometric wavelength.

Evidence for laser-driven compression of small gas-filled glass shells was first presented by the K.M.S. Fusion group (Charatis *et al.* (1974)). The experimental

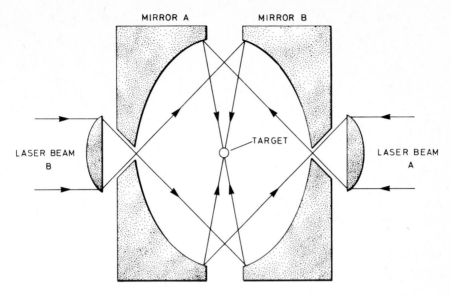

8.81   Optical arrangement for obtaining nearly-uniform illumination of a target sphere from two laser beams. (After Charatis *et al.* (1974))

arrangement, shown in Figure 8.81, ingeniously produced almost uniform illumination of the spherical target using only two laser beams. The 50 mm gap between the reflectors permitted diagnostic observations, for which extensive facilities were provided. Up to 230 J of neodymium laser energy could be delivered to the target: the pulse half-intensity duration could be varied from 30 ps to 1 ns. The targets, with diameters of 30 to 700 μm and wall thicknesses of 0·5 to 12 μm, were filled with deuterium or a deuterium–tritium mixture at pressures up to 100 atm. X-ray pinhole photographs with DT-filled targets indicated considerable compression of the DT atoms to about 100 times the initial density. Neutron yields up to 4×10^5 were obtained, and the temperature of the central compressed core was estimated to be as high as 500 to 700 eV, with total laser energies up to 60 J. The fraction of the

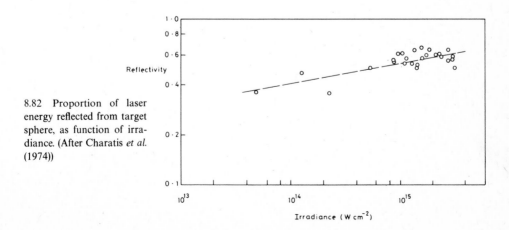

8.82   Proportion of laser energy reflected from target sphere, as function of irradiance. (After Charatis *et al.* (1974))

total laser energy reflected or scattered from the target is shown in Figure 8.82 as a function of the average irradiance incident on the target: it rises very slowly with the irradiance throughout the range from 4×10^{13} to 4×10^{15} W cm^{-2}.

5. PLASMAS FORMED FROM SOLID HYDROGEN ISOTOPE TARGETS

Although the experiments described in this section could properly have been included in the previous sections of this chapter, they are treated separately here because of their special interest from the point of view of nuclear fusion research, and because of the special techniques necessary to prepare the targets. A paradox of this work is the use of cryogenic methods as a preliminary to reaching very high temperatures.

5.1. METHODS OF TARGET PREPARATION

Solid hydrogen and deuterium targets have been prepared in three main forms: extruded filaments, thin discs, and pellets.

Extruded filaments, first used as targets by Ascoli-Bartoli et al. (1966) at Frascati, have been prepared in several laboratories. At Culham, Saunders et al. (1967) extruded filaments $\frac{1}{3}$ mm in diameter and a few mm long. The arrangement used by Bobin et al. (1969) at Limeil is shown in Figure 8.83. Deuterium gas enters the helium-cooled cryostat and condenses in the copper extruder. Square filaments 1 or 2 mm thick have been produced with different extruders.

Thin discs have been produced by Sigel et al. (1969) at Garching with the apparatus shown in Figure 8.84. The cold copper plate shown in the inset is 1 mm thick. To form a hydrogen disc, the bell glass is raised into the position shown, and filled with hydrogen gas at a carefully controlled pressure. The hydrogen begins to condense on the copper plate, and a film of liquid hydrogen covers the plate and fills the hole. As cooling proceeds the liquid hydrogen freezes, the interface moving slowly to the centre of the hole. By controlling the conditions, an optically clear disc of solid hydrogen may be obtained. The bell glass is then removed, exposing the solid hydrogen film to high vacuum. The hydrogen begins to vaporize. The disc, being in poor thermal contact with the copper plate, evaporates (slowly as a result of

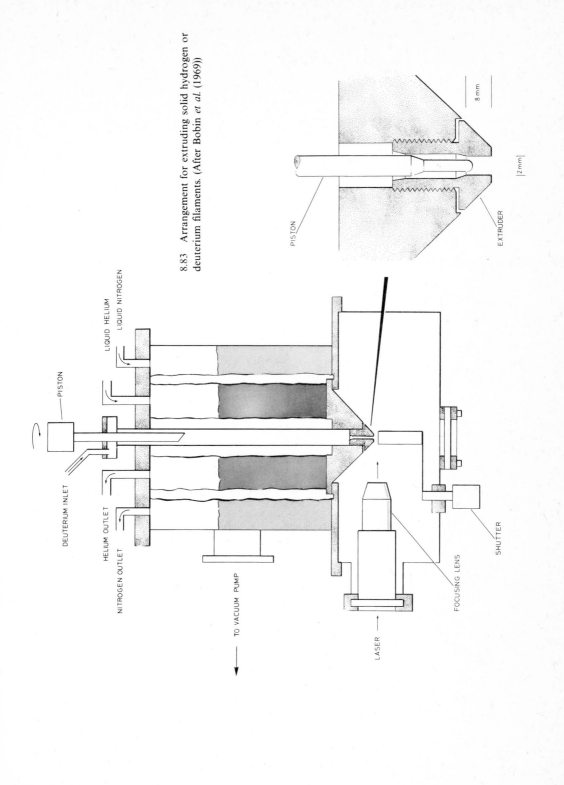

8.83 Arrangement for extruding solid hydrogen or deuterium filaments. (After Bobin *et al.* (1969))

PISTON

EXTRUDER

8 mm

|2 mm|

LIQUID HELIUM

LIQUID NITROGEN

PISTON

DEUTERIUM INLET

HELIUM OUTLET

NITROGEN OUTLET

TO VACUUM PUMP

LASER

FOCUSING LENS

SHUTTER

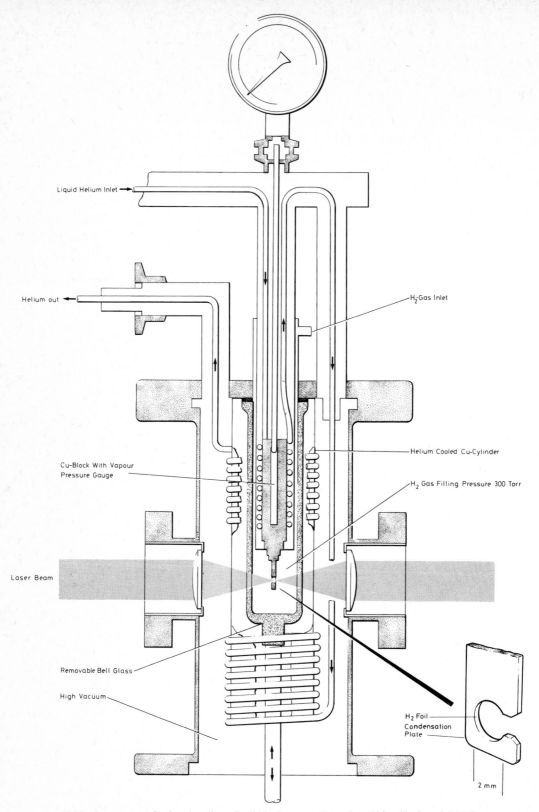

Liquid Helium Inlet →

Helium out →

H₂ Gas Inlet

Helium Cooled Cu-Cylinder

Cu-Block With Vapour
Pressure Gauge

H₂ Gas Filling Pressure 300 Torr

Laser Beam

Removable Bell Glass

High Vacuum

H₂ Foil
Condensation
Plate

2 mm

8.84 Arrangement for forming discs of solid hydrogen or deuterium. (After Sigel *et al.* (1969))

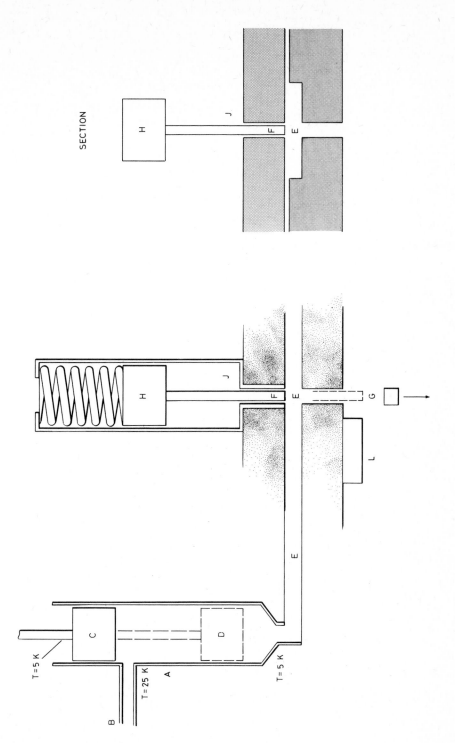

8.85 Arrangement for producing detached pellets of solid hydrogen or deuterium: A—Tube with temperature gradient. B—H$_2$ gas inlet. C—Cold piston (inlet position). D—Cold Piston (extruding position). E—Rectangular extrusion channel. F—Initial position of punch. G—Final position of punch. (After Francis et al. (1967))

evaporative cooling) in about 10 minutes, while the solid film on the copper plate evaporates relatively rapidly.

Three methods of producing pellets have been described. Francis *et al.* (1967) used the punch shown in Figure 8.85, which presses out pellets 0·25 mm in diameter and 0·25 mm long from a rectangular extruded strip of solid hydrogen. The pellet leaves the punch with a velocity of 10^3 cm s^{-1} when the head H hits the stop J. The microphone L provides a synchronizing signal to fire the laser flash tubes. The laser is focused about 2 cm below the point of ejection. Cecchini *et al.* (1968) used an r.f. heating coil to detach a pellet from an extruded filament (Figure 8.86). Deuterium is solidified in the capillary tube. The needle is then inserted and compresses the

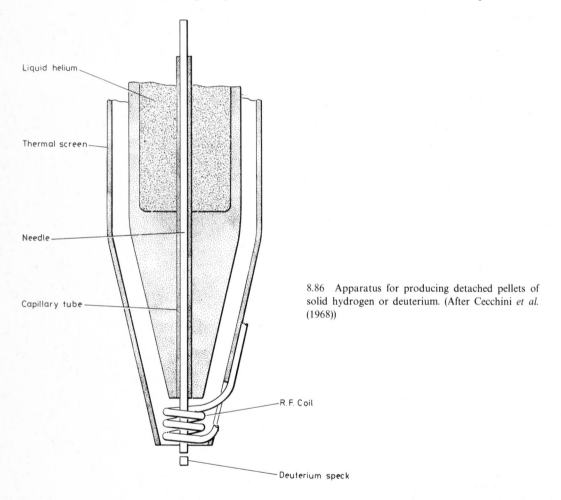

Liquid helium

Thermal screen

Needle

Capillary tube

R.F. Coil

Deuterium speck

8.86　Apparatus for producing detached pellets of solid hydrogen or deuterium. (After Cecchini *et al.* (1968))

deuterium 'snow' against a retractable plug (not shown). The plug is removed and the needle is moved down to the position shown in the diagram, carrying the deuterium pellet which is detached by heating the needle. The system described by

Sigel *et al.* was modified (see Witkowski (1971)) to produce cylindrical pellets 100 μm in diameter and 100 μm long by providing a piston which punches out the hydrogen ice. This arrangement has the advantage that the hydrogen 'ice' is more homogeneous than extruded hydrogen 'snow'.

In each case the position of the pellet in the laser focus was checked by means of sensing light beams before the laser Q-switch was operated, because the pellet trajectories were somewhat variable.

5.2. PLASMAS FORMED FROM FILAMENTS

Solid deuterium filaments 0·4 mm in diameter were used as the targets for 3 J, 10 to 15 ns ruby laser pulses by Ascoli-Bartoli *et al.* (1966). Streak photography, schlieren photography, time-resolved spectroscopy of light scattered in different directions and measurements with charge collectors combined to give a picture of the expansion of the target material, which was tentatively described by the polar velocity diagram of Figure 8.87.

Saunders *et al.* (1967) used a 500 MW ruby laser to produce a plasma from solid hydrogen filaments $\frac{1}{3}$ mm in diameter. Infrared observations showed that hardly

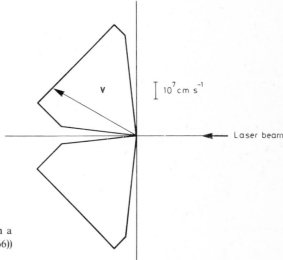

8.87 Velocity distribution of material emitted from a solid deuterium target. (After Ascoli-Bartoli *et al.* (1966))

any radiation was emitted from the plasma at wavelengths above 2·5 μm, the cut-off occurring between 1·8 and 2·5 μm (see Chapter 2) corresponding to an electron density of about 3×10^{20} cm^{-3}. Streak photographs showed the boundary of the plasma to be expanding with a velocity of 4×10^7 cm s^{-1} which is equivalent to a proton energy of 1 keV. The expansion velocity was symmetrical within 20% in the plane, perpendicular to the filament, containing the laser beam axis. Time-of-flight measurements gave similar results to the photographic measurements of velocity, but showed that particles were emitted preferentially towards the laser, in contrast to the pattern of Figure 8.87. The total number of particles in the plasma was estimated to be 3×10^{17}. The initial mean energy of the plasma per particle, derived on the basis of Dawson's theory (Chapter 7, Section (6.2)), was 200 eV. This suggests a higher efficiency of conversion of laser energy to plasma energy (between 30 and 100%) than is to be expected according to the theory of Haught & Polk (1966).

Caruso et al. (1971) formed plasmas from solid deuterium filament targets 1 mm in diameter with neodymium laser pulses of 10 ns duration and of energy 1 to 30 joules. The focal spot radius was of order 10^{-2} cm. The times of flight of ions from the plasma were measured between probes 10 and 15 cm from the target as a function of the laser pulse energy (Figure 8.88). The number of particles reaching the nearer probe,

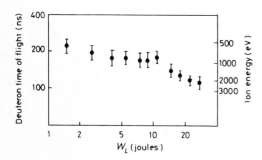

8.88   Deuteron time of flight and equivalent energy as function of laser pulse energy W_L. (After Caruso et al. (1971))

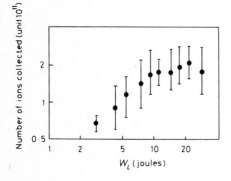

8.89   Number of ions collected by probe from plasmas formed from solid deuterium target, as function of laser pulse energy W_L. (After Caruso et al. (1971))

of effective area 2×10^{-3} cm^2, is shown in Figure 8.89. Above a laser pulse energy of about 10 J, corresponding to an irradiance of about 3×10^{12} W cm^{-2}, the ion energy increased more or less proportionally to the pulse energy, while the number of ions collected remained almost constant. Neutrons were detected when the pulse energy was greater than 4·5 J; up to 1000 neutrons were produced per shot. The deuteron temperature derived from the neutron yield was in the region 200 to 400 eV. The electron temperature may be calculated from a knowledge of the ion temperature and the asymptotic ion expansion energy, which equals $5(T_i + T_e)/2$. For the highest laser pulse energy, the ion expansion energy was about 2 keV, so

$T_i + T_e \sim 800$ eV, and $T_e \sim 400$ to 600 eV. Caruso *et al.* interpreted the change in the behaviour of the plasma for pulse energies above 10 J as being due to a transition from the self-regulating plasma regime to the deflagration wave regime (Chapter 7), when the plasma electron density in the focal plane approaches the critical value n_{ec}.

A series of experiments was carried out at Limeil using square solid hydrogen and deuterium filaments 1 or 2 mm thick and 10 to 15 mm long as targets for gigawatt neodymium glass laser pulses. The spectrum of the light transmitted through the target region was compared with that of the incident laser light by Decroisette *et al.* (1970, 1972) for 10 J laser pulses of about 1 GW peak power, focused to an irradiance of about 10^{12} W cm^{-2}. Figure 8.90 shows the time variation of incident and transmitted intensities at λ 10 575 Å. The line profiles were compared at several times: Figure 8.91 shows the profiles at $t = 6$ ns on the scale of Figure 8.90, while Figure

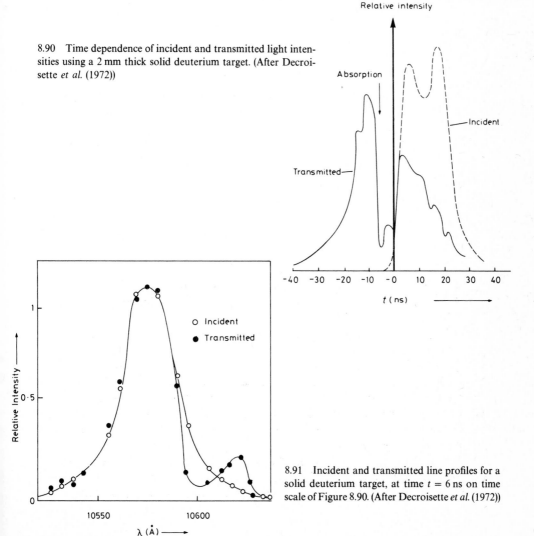

8.90   Time dependence of incident and transmitted light intensities using a 2 mm thick solid deuterium target. (After Decroisette *et al.* (1972))

8.91   Incident and transmitted line profiles for a solid deuterium target, at time $t = 6$ ns on time scale of Figure 8.90. (After Decroisette *et al.* (1972))

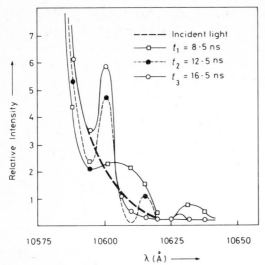

8.92 Long-wavelength wings of incident and transmitted light spectra, for a solid deuterium target. (After Decroisette *et al.* (1972))

8.92 shows the long-wavelength wings at several other times. In each case secondary maxima appear in the transmitted light spectrum on the long wavelength side only. Decroisette *et al.* interpreted their results, which were similar to those reported by Krasyuk *et al.* (1970*a*), as being due to induced Compton scattering (see Chapter 2, Section (11.2)). If this was the case, the energy transferred to the plasma by the effect was small: only a few percent of the incident photons appear to have been noticeably affected, and only a small fraction (approximately equal to $\Delta\lambda/\lambda_0 \sim (2$ to $5) \times 10^{-3}$) of the energy of this group of photons was lost to the plasma by this process under these conditions.

With 2 GW, (75J, 35 ns) neodymium glass laser pulses focused by a 5 cm focal length lens (Colin *et al.* (1968*a*, *b*), Bobin *et al.* (1968*b*)), it was found from streak photographs that breakdown of the deuterium target occurred as long as 100 ns before the laser pulse reached its maximum power, at an irradiance of $(1$ to $2) \times 10^{10}$ W cm^{-2}. About 95% of the laser pulse energy was absorbed in the plasma. Streak photographs showed the expansion velocity of the plasma boundary to be about 2×10^6 cm s^{-1}, corresponding to a temperature of about 1 eV. However, some deuterons were detected with time-of-flight velocities of 3.5×10^7 cm s^{-1}, corresponding to energies of 1.5 keV. Also, X-ray absorption measurements indicated average electron temperatures of 90 to 120 eV. The X-ray intensity reached a maximum 30 to 40 nanoseconds before the peak of the laser pulse. The electron temperature depended on the position of the target surface relative to the laser beam focal plane, being highest when the laser was focused on, or just ($\frac{1}{2}$ mm) inside the target surface.

These measurements were extended (Bobin *et al.* (1969): see also Fauquignon & Floux (1970)) with laser pulses of energy up to 180 J varying in duration from 30 to 80 ns. The penetration of what appeared to be either the shock or the deflagration wave into the target was measured from streak photographs. The position of the interaction zone moved a considerable distance relative to the focal plane of the laser beam during the heating pulse, resulting in a variation in the laser irradiance. The highest temperatures were attained with the laser beam focused 1 mm inside the target surface. The electron temperature derived from X-ray measurements was plotted against the maximum laser power in the high-temperature zone, after allowing for attenuation during penetration into the target, and was found to vary as (power)$^{\frac{2}{3}}$ in agreement with theory (Figure 8.93).

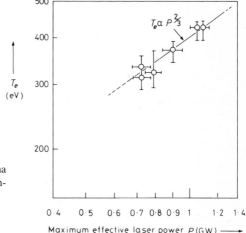

8.93 Electron temperature in a deuterium plasma as function of maximum effective laser power in high-temperature zone. (After Bobin *et al.* (1969))

It is apparent that, with a given amount of energy available from the laser amplifiers, shorter laser pulses will heat the target more efficiently, allowing the thermal wave to penetrate farther and heat more of the target material at high density before the rarefaction wave arrives. Also, to produce high temperatures the laser pulse should be shaped so that the deflagration front in the solid target should reach the focal plane of the laser beam at the moment of maximum laser power. Accordingly, the neodymium glass laser was modified by placing an electro-optic shutter, consisting of a Pockels cell between two Glan prisms, after the first amplifier rod (Floux *et al.* (1969, 1970)). In this way the laser pulse rise time was reduced to less than 5 ns, the half-power duration being 15 ns. The final output energy was about 40 J, the peak power varying from 3 to 5 GW. The shutter also protected the first amplifier and the oscillator from damage by laser light reflected from the plasma and amplified on its return. A 50 mm focal length aspheric lens of aperture $f/1$ was specially designed for the work (Champetier *et al.* (1968)): for an account of the optics of these laser systems see De Metz (1971). The solid hydrogen, or deuterium, target filament was 1 mm × 1 mm in cross section. The best position for the focal plane was at a depth of 100 to 300 μm inside the target. The electron temperature was found to be between 500 and 700 eV. Using large plastic phosphors and photomultipliers, shielded by lead, neutrons were detected coming from the plasma formed from a deuterium target at the time of highest temperature. Their minimum time of flight was consistent with a maximum neutron energy of 2·45 MeV (45 ns per metre) as is to be expected from the D–D reaction (see Chapter 4). The slower particles observed were probably scattered on their way to the detector. The total number of neutrons emitted from the plasma reached a maximum of about 500 per laser shot, and was very sensitive to changes in laser power and focusing conditions. The maximum irradiance was a few times 10^{13} W cm^{-2}. No neutrons were observed when hydrogen was used instead of deuterium for the target. The appearance of harmonics and subharmonics in light scattered through 135° has been mentioned in Chapter 2, Section (11.2).

In later experiments the laser pulse rise time was further reduced to 1·2 ns, and the half-power duration to a minimum of 1·75 ns (Salères *et al.* (1971*a*, *b*; 1972), Floux *et al.* (1972)). The pulses were focused by an $f/1$ lens ($f = 75$ mm) or an $f/1·75$ lens ($f = 117$ mm) to spot diameters of about 100 μm or 150 μm. Considerably larger neutron yields were obtained, which are shown plotted against the absorbed laser energy for 3·5 ns pulses in Figure 8.94. The results at the higher energies are consistent with the hemispherical thermal wave theory of Zakharov *et al.* (1971*b*) described in

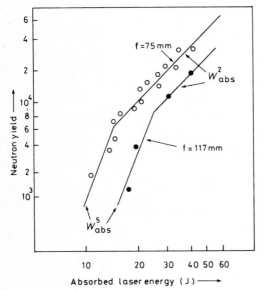

8.94 Neutron yield from deuterium plasma as function of absorbed laser energy. (After Salères *et al.* (1972))

Chapter 7, Section (4.3), while at lower energies they fit the deflagration wave theory (Chapter 7, Section (4.2)). The transition seems to occur at an irradiance of about 5×10^{13} W cm^{-2}.

With the $f/1$ lens the specular energy reflectivity of the plasma increased from 28% at 20 J incident energy to 42% at 60 J, varying as (energy)$^{\frac{1}{2}}$. Both the neutron output and the plasma reflectivity were very sensitive to the focusing of the laser beam, as may be seen from Figure 8.95. In contrast to the result of the earlier experiment with longer laser pulses, the optimum position for the focal plane was now at the target surface. At constant laser power, the reflectivity was independent of the

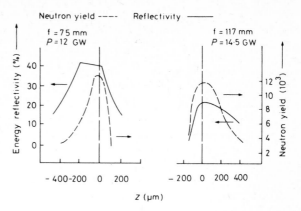

8.95 Dependence of neutron yield and energy reflectivity on position of solid deuterium target relative to focal plane. Laser light is incident from the left. The focal plane is at the target surface when $z = 0$, and inside the target for $z < 0$. (After Salères *et al.* (1972))

pulse duration. The neutron output increased by a factor of 2 as the pulse duration was increased from 1·75 to 7 ns.

At even higher laser powers, up to 22 GW, with a pulse width of 2·8 ns the neutron yield rose to about 3×10^4, the maximum yield being observed when the focal plane was $350\,\mu m$ inside the target, with a subsidiary maximum when the focal plane coincided with the target surface (Salères *et al.* (1973a, b)). The reflectivity at the laser frequency ω_0 had a maximum of 57% when the focal plane was $140\,\mu m$ inside the target. A similar pattern (though more sharply peaked) was observed for light collected at $\frac{3}{2}\omega_0$, but the light energy collected at $2\omega_0$ followed the pattern of the neutron yield, with two maxima. The X-ray emission was studied (Floux *et al.* (1973)) with two scintillation detectors behind different absorbers: one recorded 'soft' X-rays between 6 and 20 Å, the other 'hard' X-rays up to 5 Å. As the position of the focal plane relative to the target surface was varied, the 'soft' X-ray energy varied in much the same way as the neutron yield. On the other hand, the 'hard' X-rays reached a maximum where the plasma reflectivity was greatest, midway between the two neutron yield maxima. The conditions responsible for high reflectivity and para-metric conversion to $\frac{3}{2}\omega_0$ also appear to generate high-energy electrons.

The reflectivity and neutron yield measurements were repeated with the laser beam frequency-doubled by a KDP crystal (Carion *et al.* (1973)). With a pulse energy at $\lambda\ 0.53\,\mu m$ of about 7·5 J in a 2·8 ns pulse, the variation of the neutron yield with focal plane position again showed two maxima (of about 500 neutrons per pulse), near the target surface and about $150\,\mu m$ inside the target, while the reflectivity peaked (at 10%) at an intermediate position.

Yamanaka *et al.* (1972a, b, see also Yamanaka (1972)) described experiments using 2 ns neodymium laser pulses of up to 20 GW peak power, focused by a lens of 50 mm focal length on to deuterium filaments 2 mm in diameter. They reported an abrupt change in the plasma heating process when the irradiance was increased above $10^{13}\ W\ cm^{-2}$. Below this threshold, the X-ray electron temperature varied with the laser power, P_0, as $P_0^{\frac{2}{3}}$ and time-of-flight measurements showed that the ions arrived at the collector as a single group. No neutrons were observed with laser irradiances

in this region. However, above the threshold the dependence of the electron temperature on the laser power became erratic, varying roughly as $P_0^{\frac{4}{3}}$, while two groups of
ions could be distinguished. Plasma ion temperatures were derived from the energies
obtained from time-of-flight measurements. Above threshold, the ion temperature
corresponding to the fast ion energy was very close to the measured electron temperature, varying as $P_0^{\frac{4}{3}}$, while the ion temperature derived from the slow ion energy
was much lower and varied as $P_0^{\frac{1}{3}}$. As the electron temperature rose, the number of
ions in the fast group increased while that in the slow group fell. Experimental
results are shown in Table 8.6. The neutron yield is shown plotted against the laser
pulse energy in Figure 8.96, and the straight line inserted by Yamanaka *et al.* has a
slope of about 4·5. If the results giving total neutron yields below 100 are ignored,

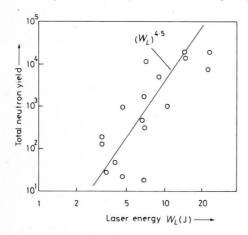

8.96 Neutron yield from deuterium plasma as
function of laser pulse energy. (After Yamanaka
(1972))

quite a good fit is obtained with a straight line indicating a yield proportional to the
square of the absorbed energy. The results at these higher energies, which correspond
to irradiances greater than about 2×10^{13} W cm^{-2}, are quite close to those reported
by Salères *et al.* (Figure 8.94) for similar energies and pulse durations. Thermal wave
theory appears to be applicable to these conditions. Regarding the absorption

TABLE 8.6: *Measurements on plasmas formed from deuterium filaments by 2 ns neodymium laser pulses*

| Input energy (J) | Electron temperature (keV) | Fast ions | | Slow ions | | Neutrons per shot (approx) |
		Time-of-flight energy (keV)	Ion temperature (keV)	Time-of-flight energy (keV)	Ion temperature (eV)	
3	∼0·2	—	—	0·6	70	—
5	∼0·5	9	0·8	0·9	80	300
12	2	18	2·1	1·5	90	5000
20	4	27	4	2	140	2×10^4

After Yamanaka (1972).

mechanism, Yamanaka *et al.* concluded that the change occurring at $10^{13}\,\mathrm{W\,cm^{-2}}$ was due to the onset of the parametric instability (see Chapter 2).

Caruso *et al.* (1969) used a mode-locked neodymium laser giving single pulses of 0·05 to 1·3 J in about 10 ps, focused to a spot of diameter about 0·25 mm. The targets were small cylinders of solid deuterium, 0·2 mm in diameter and 1 mm long. The alignment was thus rather critical. Figure 8.97 shows the ion energy, measured from the ion time of flight, as a function of the laser pulse energy. The total number

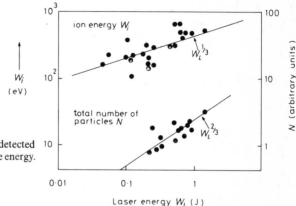

8.97 Ion energy, and total number of particles detected by an ion probe, plotted as functions of laser pulse energy. (After Caruso *et al.* (1969))

of particles detected by an ion probe is also plotted. In agreement with thermal wave theory (see Chapter 7, Section (4.3)) the ion energy varies as $W_L^{\frac{1}{3}}$ while the number of ions produced varies as $W_L^{\frac{2}{3}}$. The ion energies are consistent with a rather large fraction of the laser pulse energy being absorbed: there was some evidence that more than a half and perhaps as much as nine tenths of the energy was gained by the plasma. Such a strong absorption implies a non-linear effect, which was probably present since even the lowest pulse energy used produced an irradiance exceeding $10^{13}\,\mathrm{W\,cm^{-2}}$.

Carbon dioxide laser pulses of 300 MW peak power and 60 ns duration were focused on to solid deuterium filaments by Martineau *et al.* (1974). The maximum irradiance at the target was $5 \times 10^{11}\,\mathrm{W\,cm^{-2}}$. The mean electron energy determined from X-ray absorption measurements varied between 20 and 35 eV as the irradiance was varied from 5×10^{10} to $5 \times 10^{11}\,\mathrm{W\,cm^{-2}}$. These results agreed with deductions from time-of-flight measurements. However, when the incident laser power was greater than 200 MW, for certain positions of the focal plane relative to the target surface it was observed that X-rays emitted with energies greater than 500 eV were over a hundred times more intense than they should have been for a Maxwellian electron velocity distribution at 30 eV.

5.3. PLASMAS FORMED FROM THIN SOLID FILMS

Solid hydrogen films, prepared as described in Section (5.1) above, were used by Sigel *et al.* (1968, see also Sigel (1970)) as targets for ruby laser pulses of up to 500 MW peak power. Figure 8.98 shows the time of passage of the absorbing front as a function of the film thickness for 2·9 J pulses of 18 ns half-intensity duration. Sigel (1969, 1970)

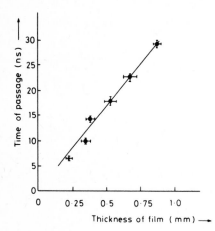

8.98   Times of passage of absorbing front through hydrogen films of different thicknesses. (After Sigel (1970))

also described a holographic study of plasmas generated using similar pulses. The number of electrons generated was found to be 4×10^{16}. In these experiments strong molecular Raman Stokes scattering was observed at λ 9755 Å in hydrogen and λ 8750 Å in deuterium (Mennicke (1971*b*), see also Sigel *et al.* (1972)).

The electron temperature of plasmas produced by 1 GW neodymium laser pulses (30 J, 30 ns) from deuterium discs was investigated by the X-ray absorption method using several different pairs of absorbing foils (Sigel *et al.* (1972)). The results were inconsistent with a Maxwellian electron velocity distribution and implied an excess of high-energy electrons. (Results using a carbon target have already been mentioned in Section (2.2.2): in that experiment a thermal distribution was indicated, which gives confidence in the method.) The X-ray intensity was positively correlated to the intensity of the reflected light. Ion probes (Büchl *et al.* (1972*a*)) gave the distributions of asymptotic energies and numbers shown in Figure 8.99. About half the laser pulse energy appeared as ion kinetic energy (Sigel (1970)).

When 10 J, 10 ns neodymium laser pulses were focused on deuterium discs with a simple lens of focal length 15 cm, neutrons were detected (Büchl *et al.* (1971)). With a special aspheric lens of focal length 5 cm neutrons appeared sometimes with 5 J pulses and in 80% of shots when the energy was raised to 15–20 J. The total number of neutrons per shot was a few hundred: they appeared over a period of about 100 ns, presumably as the result of scattering. The longer laser pulse length probably explains the smaller neutron yield for a given pulse energy, as compared

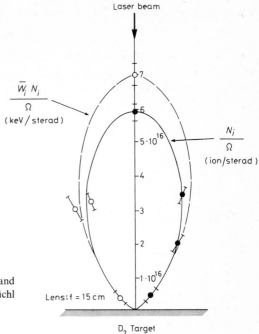

8.99 Polar diagram of distributions of ion numbers and asymptotic energies from a solid deuterium target. (After Büchl et al. (1972a)

to the observations of Floux et al. and Yamanaka et al. described in the previous sub-section.

Büchl et al. (1972b) studied the spectrum of light reflected from the plasma generated from solid deuterium by 10 J, 10 ns neodymium glass laser pulses, and compared it with the spectrum of the incident light. The irradiance at the focus was 5×10^{13} W cm^{-2}. The reflected light spectrum was similar in general shape to that of the incident light, but lacked much of the fine structure. The entire reflected light spectrum was displaced by about 2·5 Å to longer wavelengths which, interpreted as a Doppler shift on reflection, corresponds to a velocity of the reflecting region away from the laser, into the target, of about 3.5×10^6 cm s^{-1}. This result is in good agreement with the theoretical value of the absorbing layer velocity u_1 (see Chapter 7, Section (4.1)) which for $I = 5 \times 10^{12}$ W cm^{-2} is equal to 3.7×10^6 cm s^{-1}.

Mode-locked pulses of duration less than 10 ps with energies up to 3 J were also used by Büchl et al. (1971) and Salzmann (1973) to heat deuterium discs. With picosecond pulses X-ray absorption measurements indicated Maxwellian distributions of electron velocity, in contrast to the results obtained with nanosecond pulses. An electron temperature of 550 eV was attained with a 1 J pulse. The high-energy tail observed with nanosecond pulses may be due to an instability, which cannot grow significantly during a 10 ps pulse.

Büchl (1971, 1972) also reported observations on plasmas produced from solid hydrogen targets 1 mm thick by 1 MW pulses from a transversely excited, high-pressure carbon dioxide laser. The peak irradiance was 5×10^9 W cm^{-2}. The results were in good agreement with the self-regulating plasma theory of Chapter 7, Section (4.1), provided recombination is taken into account. The electron temperature was between 10 and 25 eV and the total number of electrons in the plasma, measured interferometrically, was up to 4×10^{16}. The velocity of the shock wave through the target was $1 \cdot 1 \times 10^5$ cm s^{-1}. During the slow decay of the laser pulse the plasma was found to be strongly reflecting.

5.4. PLASMAS FORMED FROM ISOLATED PELLETS

Only two experiments have been reported so far in which isolated pellets of solid hydrogen or deuterium have been used as targets. Accurate positioning is difficult to achieve.

Francis et al. (1967) focused ruby laser pulses of up to 500 MW on to cylindrical hydrogen pellets $\frac{1}{4}$ mm in diameter and $\frac{1}{4}$ mm long. When the laser beam diameter was equal to that of the pellet, the plasma formed by 150 MW pulses was always asymmetrical, probably because of uneven heating of the pellet. The diameter of the laser beam was therefore doubled. With a pulse power of 500 MW the plasma then expanded symmetrically with boundary velocities between $(1$ and $2) \times 10^7$ cm s^{-1}. Measurement of the Bremsstrahlung emission from the plasma in a wavelength region free from spectral lines, together with a knowledge of the plasma radius and an approximate value of the electron temperature, showed that the number of ions in the plasma exceeded 5×10^{16}. The number of hydrogen atoms in a pellet would have been about 5×10^{17} if the pellet was fully consolidated by the extruder, so at least 10% of the atoms in the pellet were ionized.

Very similar pellets (diameter 0·2 mm, length 0·25 mm) were used as targets by Ascoli-Bartoli et al. (1969) but formed of deuterium instead of hydrogen. A neodymium laser produced pulses of energy 50 J with a peak power of 3·4 GW and a half-intensity duration of 30 ns, focused to a diameter of 0·5 mm. Observations of the incident and transmitted pulse shapes showed that the central interval of the laser pulse, over a period of 30 to 40 ns, interacted strongly with the plasma, being attenuated by up to 50%. The initial and final parts of the pulse were unaffected. Streak photographs showed a strongly asymmetrical luminous region expanding with a velocity of about 2×10^6 cm s^{-1} and moving preferentially away from the laser, surrounded by a low-intensity region expanding at a velocity of a few times 10^7 cm s^{-1}. Measurements of ion times of flight indicated ion velocities of 1 to 2×10^7 cm s^{-1}, which correspond to the expansion velocity of the low-intensity region and are equivalent to ion energies of 100 to 400 eV.

6. EFFECTS OF MAGNETIC FIELDS ON PLASMAS FORMED FROM SOLID TARGETS IN VACUUM

Several experiments have been reported in which plasmas have been produced by laser pulses focused on targets placed in vacuum in magnetic fields. We shall describe them in two groups, divided according to the magnetic field strength. The division at 10^5 G (10 T) is entirely arbitrary.

6.1. LOW MAGNETIC FIELDS ($B < 10^5$ G)

The first experiments on the effects of magnetic fields on laser produced plasmas were made by Linlor (1963, 1964). The plasmas were generated from massive solid targets by ruby laser pulses of up to 5 MW peak power with 40 ns duration, in uniform magnetic fields up to 1200 G. A loop conductor round the plasma gave a signal when the plasma was formed which showed that the magnetic field was being displaced by the expanding conducting plasma. The sign of this diamagnetic signal changed when the field direction was reversed. No signal was observed when the field was absent. The radius r of the luminous region, recorded photographically without time resolution, decreased as the field B increased: the product r^2B was reasonably constant, suggesting that the region was limited to a tube of flux. On the other hand from the data the quantity $r^{\frac{3}{2}}B$ could equally well be constant, which would agree better with Bhadra's theory (Bhadra (1968); see Chapter 7, Section (9)).

Schwarz (1971) measured energies of ions emitted from metal targets heated by ruby laser pulses as a function of the strength of a magnetic field applied perpendicular to the target surface. Two groups of ions were distinguished, the slower group having energies which varied linearly with the laser power up to 8·5 eV at 30 MW in the absence of a magnetic field. When a field was applied, the slow ion energies w_i increased by δw_i, such that $w_i^3 \delta w_i \simeq 6 \times 10^{-11} B^4$, where B is in gauss ($B < 2000$ G) and w_i in eV. The result was interpreted in terms of a theory involving pondero-motive forces (see for example Hora (1969b, 1971, 1972)).

Mattioli & Véron (1969) studied the effect of uniform magnetic fields up to 5 kG on plasmas produced from isolated lithium hydride particles 50 μm in diameter by 35 ns neodymium laser pulses of up to 1 GW peak power. They noted that whereas in the absence of the magnetic field all singly-ionized particles collected at a distance from the plasma, including oxygen and carbon present as impurities, had very similar velocities ($(5·0$ to $5·6) \times 10^6$ cm s^{-1}), in the presence of a field of 2·2 kG all these particles had roughly the same energy (135 to 195 eV), suggesting thermalization of the directed energy of expansion as a result of interaction with the field. They also observed that the intensity of recombination radiation was strongly reduced by the field.

Haught et al. (1970) produced very similar plasmas in a variety of magnetic field configurations. Three orthogonal pairs of parallel coils were arranged round the target. Mirror, cusp, or 'minimum-B' fields (see for example Krall & Trivelpiece

(1973)) could be created by passing currents in the same direction through one pair of coils, or in opposite directions through one pair, or in the same directions through one pair and in opposite directions through each of the other two pairs respectively. The stopping time of the plasma, as indicated by a magnetic probe coil, is plotted as a function of the magnetic field in Figure 8.100, and varies as $B^{-\frac{2}{3}}$, in agreement with Bhadra's theory and the theoretical model used by Haught *et al.* With either a simple

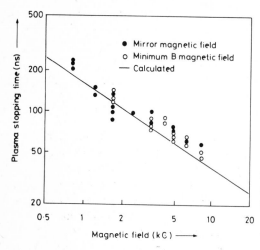

8.100 Dependence of stopping time of plasma on magnetic field for two configurations. (After Haught *et al.* (1970))

mirror or a 'minimum B' field, charge was collected from the plasma over a longer time when the magnetic field was present, and much more of the charge moved along the axis of the field than orthogonally to it. The maximum ion energies observed were lower, and nearer the initial average plasma energy per particle. This was good evidence for a significant degree of trapping of the plasma. Since most of the plasma escaped through the loss cones of the field configuration, collisional scattering into lossy trajectories was the major loss mechanism. There was some evidence for the presence of flute instabilities with simple mirror or 'minimum-B' fields below 5 kG, but none in 'minimum-B' fields above 5 kG. From other measurements it was concluded that a plasma with a mean energy of up to 200 eV had been contained at a density of more than 3×10^{13} cm^{-3} for up to 150 μs.

Plasmas were produced from silica filaments 20 μm in diameter, in mirror fields up to 1 200 G, with 20 ns, 150 MW neodymium glass laser pulses by Tuckfield & Schwirzke (1969), The development of the plasma was studied with a framing image-converter camera having an exposure time of 50 ns. The diameter of the plasma was measured and is plotted as a function of time in Figure 8.101, which shows several bounces of the plasma against the magnetic field. The field had very little effect on the early stages of development of the plasma, because the kinetic pressure of the plasma then greatly exceeded the magnetic pressure. The plasma boundary attained very nearly its asymptotic velocity of $1 \cdot 3 \times 10^7$ cm s^{-1} before any significant inter-action took place. According to Bhadra's theory for spherical symmetry the bounce

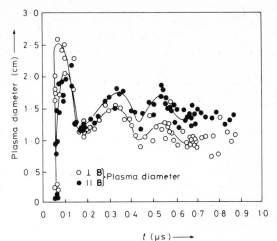

8.101 Variation of radial (⊥**B**) and axial (∥**B**) plasma diameters with time. Target at centre of 2:1 magnetic mirror, with axis parallel to laser beam axis. Field at target 1200 gauss. (After Tuckfield & Schwirzke (1969))

period of about 0·2 μs should correspond to a radius of 2·6 cm, which is about twice the measured value. However, the plasma diameter actually falls between bounces, which shows the marked effect of (axial) loss of plasma in the real geometry: with perfect containment the radius of a resistive plasma should increase from bounce to bounce. Further studies of plasmas produced under similar conditions (Schwirzke & Tuckfield (1969)) were interpreted as implying the occurrence of two-stream instabilities in the interface between the plasma and the magnetic field.

Lubin *et al.* (1969) and Dunn & Lubin (1970) measured the intensity of far infrared radiation (70 μm < λ < 8 mm) from plasmas formed from lithium hydride particles in uniform magnetic fields of up to 10 kG, and found that the intensity increased with the magnetic field. The results are shown in Figure 8.102. The dependence of intensity

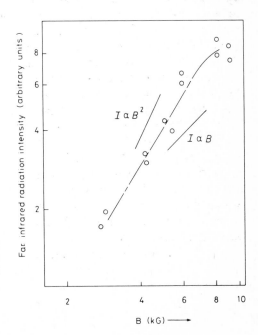

8.102 Variation of far infrared (70 μm < λ < 8 mm) radiation intensity from lithium hydride plasmas with magnetic field. (After Lubin *et al.* (1969))

on magnetic field appears to be intermediate between that expected for volume and that for surface cyclotron emission (see Chapter 7, Section (9.1)).

Sucov *et al.* (1967) studied plasmas produced from a variety of aluminium targets in a 14–6–14 kG mirror field, whose axis was perpendicular to the laser beam axis. The rate of loss of ions from the plasma was consistent with scattering into the loss cones of the mirror field. The broad plasma plume normally observed without a magnetic field narrowed when the field was present. The duration of luminosity of the central core increased from 1 to 5 μs when the field was applied. There was a tendency for the plasma to break up in the field. The boundary moved across the magnetic field lines towards the laser almost as rapidly as it did with no field. Sucov *et al.*, following Baker & Hammel (1965), suggested that this was the result of charge separation creating an electric field which cancelled out the $\mathbf{v} \times \mathbf{B}$ force.

Following early work by Fabre *et al.* (1966) with copper targets, in which an increase in emitted light intensity and a decrease in plasma diameter were noted when plasmas were formed in a magnetic field of 13 kG, Fabre & Lamain (1969) studied plasmas produced from small paraffin spheres of 20 to 100 μm radius. The target spheres were suspended from a fine filament at the focus of 500 MW, 13 ns laser pulses in a uniform magnetic field of 60 kG perpendicular to the laser beam axis. The diameter of the focal spot was about 400 μm. Interferograms showed that with 80 μm radius targets a jet of plasma was always produced, perpendicular to the magnetic field, whose direction depended on the exact position of the sphere in the laser beam: the jet always moved normally to the most strongly heated area of the particle surface. The reaction to the jet tended to push the target particle out of the focal volume of the laser pulse, and the transmitted light signal then rose prematurely. With small target particles of 20 μm radius, no jet was observed: the entire particle was heated, in agreement with the observations of Haught *et al.* (1970).

Bruneteau *et al.* (1970) studied the plasmas produced from plane aluminium and polyethylene targets. The target surface was normal to the axis of a laser beam focused to an irradiance of up to 2×10^{11} W cm^{-2}. The uniform magnetic field of 60 kG was parallel to the target surface. Again, a narrow jet of plasma was observed to flow across the magnetic field towards the laser, in agreement with the observations of Sucov *et al.* (1967). The particle energies were measured: Figure 8.103 shows the energies of the fastest particles as functions of the laser pulse energy in a constant

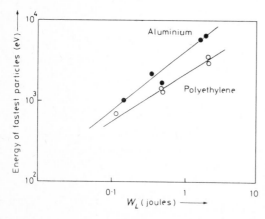

8.103 Energies of fastest particles produced from aluminium and polyethylene targets, in a magnetic field of 55 kG, as function of laser pulse energy W_L. (After Bruneteau *et al.* (1970))

magnetic field of 55 kG. At the highest laser irradiance aluminium ions were observed with a velocity of 2×10^7 cm s^{-1}. The plasma jet was explained as outlined by Sucov *et al.*: initially, the hot plasma produced near the target surface has a kinetic pressure normally to the surface which is much greater than the magnetic pressure. The condition for penetration (Tuck (1959): see also Rose & Clark (1961), p. 416) is that

$$\frac{B^2}{2\mu_0} < \frac{3n_i m_i v_i^2}{2},$$

(8.8)

v_i being the ion velocity in the jet. This is easily satisfied with the irradiances used in these experiments.

In later experiments (Bruneteau *et al.* (1971), Fabre *et al.* (1973)), the time at which the plasma reached its maximum diameter was measured for two orientations of the uniform magnetic field. A ruby laser giving 3 to 5 J, 15 to 20 ns pulses was used to produce the plasma. The results are shown in Figure 8.104 for polyethylene targets. For fields below $B = 15$ kG, the stopping time varied approximately as $B^{-\frac{2}{3}}$ for both orientations (the two curves do not coincide because the experimental conditions

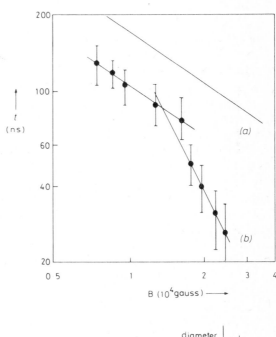

8.104   Time at which plasma from a polyethylene target reached its maximum diameter, as a function of the magnetic field, for two orientations of the field. (After Fabre *et al.* (1973))

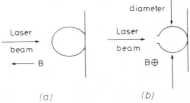

were different). At higher fields, losses along the normal to the target surface combined with losses along the magnetic field to reduce the confinement time in case (b).

6.2. HIGH MAGNETIC FIELDS ($B \geqslant 10^5$ G)

Ascoli-Bartoli et al. (1969) produced plasmas from solid deuterium pellets in a magnetic
field of 250 kG parallel to the axis of a 3·4 GW, 30 ns laser beam. Conducting cylinders
of 5 mm internal diameter arranged axially prevented the rapid escape of magnetic
field lines displaced by the plasma, resulting in field compression. The main reported
effect of the field was to increase by a factor of 10 the brightness of the low-density,
rapidly expanding outer region of the plasma. A bright axial core 1 mm in diameter
persisted for over 150 ns.

Schirmann et al. (1970: see also Tonon et al. (1971b)) used similar laser pulses to
generate plasmas from polyethylene foil at the centre of a magnetic mirror whose
axis was perpendicular to that of the laser beam. The field at the target was varied
up to 210 kG and interferograms were recorded with a framing camera having an
exposure time of 1 ns. Electron density distributions along the laser beam axis from
the target surface were derived from the interferograms taken 25 ns after the laser
pulse maximum; they are shown in Figure 8.105. There is a marked rise in electron
density at the boundary of the plasma. As the field increases, the distance travelled
by the boundary up to the moment when the interferogram was recorded is reduced,
being approximately inversely proportional to B. The density distributions resemble
those calculated by Haught et al. (1970) (Figure 7.47, Chapter 7) except that the interaction occurs earlier and at higher density with the higher magnetic field. Further
information about the conditions in the shock boundary region would be very interesting. X-ray absorption measurements of the electron temperature were made with
and without magnetic field; the results are shown in Figure 8.106. The magnetic field
appeared to make no significant difference. This is only to be expected, because this

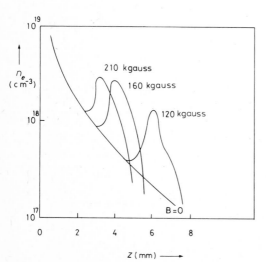

8.105 Electron density distributions along laser
beam axis in plasmas formed from polyethylene in
magnetic fields. (After Schirmann et al. (1970))

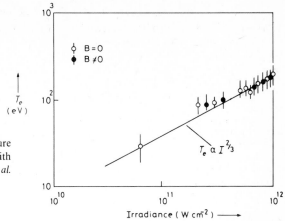

8.106 Dependence of electron temperature in a polyethylene plasma on irradiance, with and without magnetic field. (After Tonon *et al.* (1971*b*))

measurement represents the electron temperature very early in the development of the plasma, when the kinetic pressure is much greater than the magnetic pressure.

7. PLASMAS FORMED FROM SOLID TARGETS IN GASES

7.1. PLASMAS FORMED BY RUBY AND NEODYMIUM LASERS

Some of the earliest work on plasmas produced by focused laser beams was done using targets in atmospheric air. Harris (1963) published colour ciné photographs of the plume of hot vapour produced by focusing free-running ruby laser light on to a solid target. Ready (1963) described high-speed photography of the plume produced by a 30 MW Q-switched laser pulse from a carbon target, and found the velocity of the boundary to be about 2×10^6 cm s^{-1}.

Time-resolved spectroscopic studies of plasmas produced by pulses from a small Q-switched ruby laser focused on to targets in atmospheric air were reported by Archbold *et al.* (1964). With a germanium target lines of neutral and up to three times ionized atoms were recorded, together with a strong continuum. Several first spectrum lines corresponding to transitions to the ground state or to a low-lying metastable level were found to be reversed (Figure 8.107). The continuum was attenuated at the centres of these lines, so a substantial density of neutral atoms must have been present to absorb light between the target and the spectrograph before the continuum was excited.

Basov *et al.* (1967*e*) studied the expansion of the plasma generated by a 6 J, 15 ns neodymium glass laser pulse from a carbon target in air at a pressure of 2 Torr.

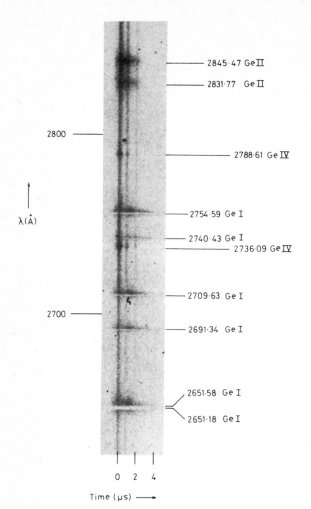

2845·47 Ge II

2831·77 Ge II

2800

2788·61 Ge IV

λ(Å)

2754·59 Ge I

2740·43 Ge I
2736·09 Ge IV

2709·63 Ge I

2700

2691·34 Ge I

2651·58 Ge I

2651·18 Ge I

0   2   4

Time (μs) ⟶

8.107 Time-resolved spectrum of plasma from a germanium target heated by a multi-phase Q-switched ruby laser. (After Archbold et al. (1964))

Shadow photographs showed that a strong, almost spherical shock wave was formed, whose velocity at a radius of 1 cm was $1·8 \times 10^7$ cm s^{-1}. The later motion corresponded closely to that of a blast wave. The development of the wave was not affected by the angle of incidence of the laser beam on the target surface: the wave moved symmetrically relative to the target normal. Interferograms of shock waves were obtained 90 ns after the laser pulse ended and were analysed in detail (on the assumption of cylindrical symmetry about the laser beam axis) to give the electron density distribution (Basov et al. (1968b)). The results for a shock in air at a pressure of 1 Torr are shown in Figures 8.108 and 8.109. The temperature behind the shock front was found to be about 40 eV. Schlieren photographs showed that the front was not more than 0·6 mm thick under these conditions, falling to less than 0·3 mm when the pressure exceeded 4 Torr. Similar electron density distributions were observed by Ashmarin et al. (1971) using 0·5 J, 20 ns ruby laser pulses focused on a copper target.

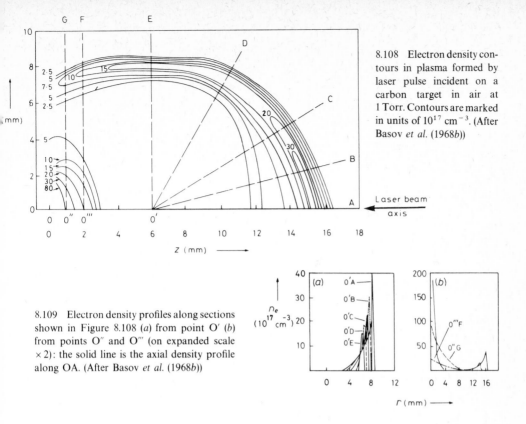

8.108 Electron density contours in plasma formed by laser pulse incident on a carbon target in air at 1 Torr. Contours are marked in units of 10^{17} cm^{-3}. (After Basov et al. (1968b))

8.109 Electron density profiles along sections shown in Figure 8.108 (a) from point O′ (b) from points O″ and O‴ (on expanded scale × 2): the solid line is the axial density profile along OA. (After Basov et al. (1968b))

Figure 8.110 shows distributions in air at a pressure of 5 Torr, 50 and 90 ns after the laser pulse, as derived from holographic interferograms. The average shock wave velocity between these two times was 3×10^6 cm s^{-1}.

Bobin et al. (1968a) reported measurements on shock waves produced by the expansion of laser-heated plasmas from a beryllium target in argon and in air at pressures p from 0·2 to 3 Torr, and found that the velocity of the front varied as

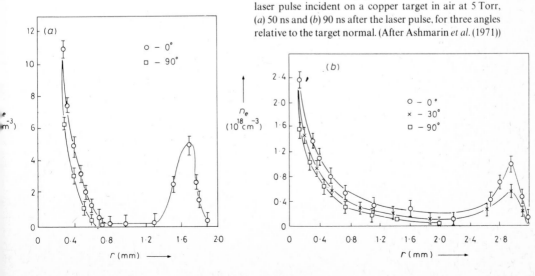

8.110 Electron density profiles in plasmas formed by laser pulse incident on a copper target in air at 5 Torr, (a) 50 ns and (b) 90 ns after the laser pulse, for three angles relative to the target normal. (After Ashmarin et al. (1971))

$p^{-0.2}$ in agreement with blast wave theory (Chapter 6, Section (5)). Hall (1969*b*) studied the development of the shock waves produced by plasmas from a tantalum target in argon at pressures above 0·6 Torr using a high-speed camera and electro-static probes. The results were again in good agreement with the predictions of blast wave theory.

Emmony & Irving (1968, 1969) produced plasmas from carbon targets in argon, air and helium using 40 ns, 0·35 J neodymium glass laser pulses. As may be seen from Figure 8.111, they found that although blast wave theory ($r \propto t^{0.4}$) fitted the later stages of the motion of the shock front, the early expansion, even after the laser

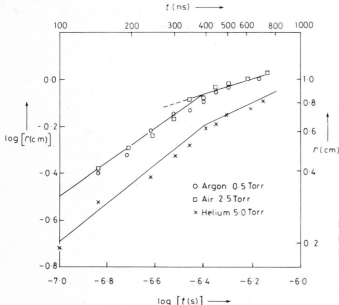

8.111   Time dependence of plasma radius, for plasmas produced from carbon targets in three different gases. (After Emmony & Irving (1969))

pulse ended, was more rapid ($r \propto t^{0.6 \text{ to } 0.7}$), suggesting that additional energy was being delivered to the blast wave system from the heated carbon plasma, perhaps as the result of recombination.

In experiments by Basov *et al.* (1972*c*) on plasmas produced from solid targets by a spherical array of nine laser beams, a small amount of background gas was present in which a shock wave produced by the expanding target material could be observed (related work by Mead (1970) is described in Chapter 6, Section (9.1.3)). Blast wave theory was satisfactory from 50 to 250 ns after the laser pulse: at longer times the radius varied as $t^{0.3}$ or $t^{0.2}$ (Basov *et al.* (1972*b*)) instead of $t^{0.4}$, probably because at low densities the plasma behind the wave became transparent to recom-bination radiation and thus lost energy. Measurements in the blast wave phase provide a useful method of measuring the amount of laser energy absorbed by the target. Basov *et al.* (1968*b*) have pointed out the possibility of using a weak laser pre-pulse to produce a vapour atmosphere in which a plasma produced by a powerful pulse could expand. This method might give useful information about the thermo-dynamic properties of vapours of refractory materials.

Arifov et al. (1967, 1968a, b) used microwave and probe techniques to study the behaviour of plasmas produced by a Q-switched ruby laser focused on to a solid target in air and hydrogen at several pressures. At low pressures they observed an ionized halo surrounding the plasma from the target. They concluded that this was probably caused by photo-ionization of the ambient gas by ultraviolet radiation from the plasma near the target surface. At atmospheric pressure the mean free path for this radiation would be only a small fraction of a millimetre. The photo-ionization of gases by radiation from laser-heated solids was also investigated by Bruneteau & Fabre (1972) using 200 MW pulses focused on a lead target. In air at low pressures ($p \lesssim 0.1$ Torr), the electron density varied inversely as the square of the distance from the target. At a distance of 7 cm the maximum electron density observed was 10^{13} cm^{-3}. The energies of the ionizing photons were estimated to be between 85 and 130 eV (150 Å $\geq \lambda \geq$ 100 Å).

Koopman (1972a), using microwave and Langmuir probe techniques, observed fast ionization fronts ahead of laser-produced plasmas generated from a carbon target in low-pressure hydrogen by 6 J, 30 ns ruby laser pulses. In a vacuum, the plasma boundary expanded at a velocity of 2.75×10^7 cm s^{-1}. When hydrogen at a density of 10^{15} molecules cm^{-3} ($p = 30$ m Torr) was present, rapid photo-ionization of the gas was detected due to radiation from the high-density carbon plasma. The electron density in the gas at radius r from the target, about 50 ns after the laser pulse, was given by

$$n_e(r) = 4 \times 10^{-2} \frac{n_{H_2}}{r^2} \text{ cm}^{-3}, \qquad (8.9)$$

the density of hydrogen molecules n_{H_2} being in cm^{-3} and r in cm. This result is consistent with the observations of Bruneteau & Fabre in air: for a further discussion of photo-ionization see Koopman (1972b). At a later time t_F the electron density was observed to rise, increasing by a factor of 3 up to a time t_P. It then remained steady until the arrival of the main laser plasma boundary at t_L. These observations were made with Langmuir probes at several distances from the target and also by microwave interferometry. A microwave signal recorded 33 cm from the target is shown in Figure 8.112, and the times t_P, t_F and t_L are plotted as functions of distance in Figure 8.113. The ionizing front contained little energy or momentum: the main laser-generated plasma boundary velocity was not affected by the presence of the gas.

8.112 Signal from X-band microwave interferometer, 33 cm from a carbon target in low-pressure hydrogen (30 mTorr). (After Koopman (1972a))

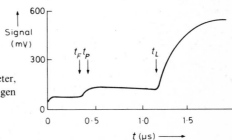

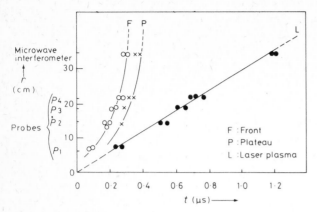

8.113 Langmuir probe observations of t_F, t_P and t_L (see Figure 8.112) at several distances r from a carbon target in hydrogen. (After Koopman,(1972a))

Koopman suggested that the front arose from a balance of electron diffusion and space charge coupled to a basically collisional ionization mechanism. At somewhat higher background hydrogen gas pressures ($p \gtrsim 0.1$ Torr) the interaction with the laser plasma became much stronger (Koopman (1972b)). Ion probes indicated the presence of an ion precursor front.

After a laser-produced plasma has expanded to a radius of about 1 mm, most of the energy is in the form of radially-directed ordered flow. At sufficiently large distances and times the mean free path for ion–ion collisions becomes large compared to other dimensions of interest and the plasma is described as 'collisionless'. Most of the plasma kinetic energy is carried by the heavier ions, since charge neutrality requires that the flow velocities of electrons and ions be equal. Thus if the flow of the plasma is to be significantly affected some interaction involving the ions must occur. Even in a collisionless plasma a number of collective interactions can affect the ion momentum. If the laser-produced plasma is flowing into a region containing a cold low-density plasma, the relative velocity of the two plasmas can give rise to streaming instabilities which broaden the ion velocity distributions until the random velocities become comparable with the drift velocity. If this happens over a short distance the laser-produced plasma boundary collects up the low-density plasma ahead of it like a snow-plough. Behind the boundary is a hot mixture of laser target material and background gas. The interaction is described as a 'collisionless ion wave shock'. Stability criteria for streaming instabilities have been discussed by Koopman (1971a): magnetic field effects are important and will be discussed in Section (7.3). For a general review of collisionless shocks the reader is referred to a paper by Paul (1970).

Finally, in this sub-section we mention some experiments with very long pulses by Batanov et al. (1972). Plasmas were generated from a bismuth target in helium at several atmospheres by 0.8 ms neodymium laser pulses of 2.3 to 3.6 kJ, focused to a diameter of 8 mm. The bismuth plasma extended over 4 to 6 cm along the beam axis, detaching itself from the target, and absorbed 80 to 90% of the laser light. High-speed photography provided evidence for self-focusing in the plasma.

7.2. PLASMAS FORMED BY CARBON DIOXIDE LASERS

Pirri *et al.* (1972) and Lowder *et al.* (1973) investigated the plasmas formed by carbon dioxide lasers focused on solid targets in atmospheric air. At the lowest irradiances ($\sim 10^7$ W cm^{-2}) the plasmas were close to the target surface and more or less spherical. Above about 5×10^7 W cm^{-2}, the plasma boundary nearest the laser moved up the laser beam with velocities of 10^5 to 10^6 cm s^{-1} which were consistent with laser-supported detonation wave theory (Chapter 6, Section (3)), for $\gamma = 1.27$. The later development of the plasma, and the impulse received by the target, were consistent with cylindrical blast wave theory, and were independent of the nature of the target. The spectrum of radiation emitted from the plasma was recorded by Wei & Hall (1973). At the higher irradiances the spectrum consisted almost entirely of lines of oxygen and nitrogen atoms superimposed on a continuum. The target appears to act simply as an initiator of a plasma in air—the normal breakdown threshold is about 10^9 W cm^{-2} (see Chapter 5, Section (5.3)). Measurements of thresholds for ignition of the gaseous plasma with several targets in air at S.T.P., and further details of the development of the plasma, were reported by Maher *et al.* (1974).

Barchukov *et al.* (1973) confirmed this general picture. In their experiments the carbon dioxide laser was focused several millimetres in front of the solid target. Breakdown occurred at the focus, not at the target surface, and the plasma expanded away from the focal plane in both directions along the laser beam axis, irrespective of the orientation of the axis relative to the surface. The threshold irradiance was about 5×10^6 to 10^7 W cm^{-2}. No craters were produced in the target material and the nature of the target had little effect, unless it was very highly reflecting, in which case the breakdown threshold rose. With a highly-polished copper surface the threshold was four times higher than with an unpolished surface, and repeated shots on the same area raised the threshold above the available laser output. Microscopic examination showed that 10 to 50 μm damage spots had been formed, suggesting that the initial breakdown had been due to very small strongly-absorbing particles on the surface, which were removed by the first laser pulse. Measurements of the impulse received by the target were again consistent with blast wave theory.

7.3. MAGNETIC EFFECTS

When a laser-produced plasma expands into a low-density background plasma generated (for example) by photo-ionization, the collisionless momentum coupling described in Section (7.1) can be strongly enhanced by the presence of a magnetic field. Fields which are not strong enough to affect the ion motion significantly can allow streaming instabilities to develop under conditions which would be stable in the absence of the field. This was demonstrated experimentally by Cheung *et al.* (1973) with laser-produced copper plasmas expanding into argon at pressures up to 1 mTorr. Figure 8.114 shows electron densities measured by a microwave method

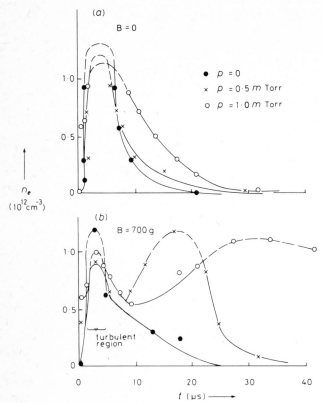

8.114 Electron densities in plasmas produced from a copper target in low-pressure argon, measured 13 cm from the target, as functions of time (a) without magnetic field (b) with 700 G magnetic field transverse to plasma flow. (After Cheung et al. (1973))

at a distance of 13 cm from the target. The presence of the photo-ionized background plasma had little effect on the behaviour of the copper plasma in the absence of the magnetic field. When a 700 G field was applied, a second density maximum was observed to move past the measurement point. Momentum conservation calculations indicate that the laser plasma was 'snowploughing' the ionized particles in the background gas.

Conduction currents arising spontaneously in laser-heated plasmas from solid targets were first reported by Askaryan et al. (1967b). These currents generate magnetic fields (see Chapter 7, Section (9.2)) which were measured by Stamper et al. (1971) in plasmas produced from a variety of target materials in background gases. Small (diameter <1 mm) magnetic probe coils recorded fields of more than 1 kG a few millimetres from the target region, in directions azimuthal to the laser beam axis. The field depends on the background gas pressure, since this affects the pressure gradients in the laser-produced plasma. The pressure dependence was studied by Bird et al. (1973) using a 300 MW neodymium laser focused on Mylar and aluminium targets. The maximum azimuthal field is shown plotted against the pressure of nitrogen, at a radius of 3 mm, 4 mm away from the target surface, in Figure 8.115. For pressures above 250 mTorr and at axial distances greater than 1 cm from the

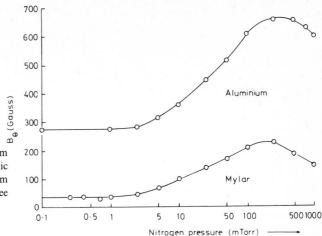

8.115 Dependence of maximum spontaneous azimuthal magnetic field on gas pressure for aluminium and Mylar targets in nitrogen (see text). (After Bird *et al.* (1973))

surface, azimuthal fields were generated at the front of an expanding aluminium plasma which were in the opposite direction to that of the initial field. This has not been explained. At 5 mTorr, a glass plate placed at an axial distance of 1·15 cm from the surface produced field reversal by reversing the direction of the pressure gradient.

The spatial distribution of the spontaneous field was mapped by Schwirzke & McKee (1972) at several times relative to the laser pulse.

Dean *et al.* (1971) obtained spectroscopic evidence for momentum transfer to the background gas by an expanding laser-heated plasma. The plasma was produced from a lucite fibre in nitrogen. At a pressure of 100 mTorr, Doppler shift observations of N II (Figure 8.116) indicated ion velocities comparable with the velocity of the boundary of the laser plasma. The dependence of the later expansion velocity of the laser plasma boundary on gas pressure was consistent with blast wave theory, again indicating strong momentum transfer. Dean *et al.* estimated that the interaction must be collisionless and suggested that it was due to a streaming instability occurring

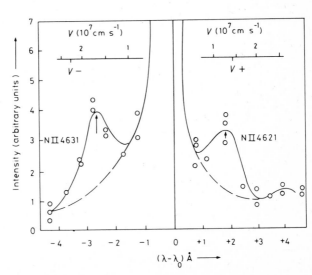

8.116 Doppler shift observations for N II lines from plasma generated from a lucite target in nitrogen at 100 mTorr. The line of sight was perpendicular to the laser beam axis and 8 mm from the target surface. Arrows show centres of satellites after subtracting dashed background. (After Dean *et al.* (1972))

as the result of the presence of spontaneous magnetic fields. Wright (1972, but see also Dean *et al.* (1972)) and Koopman (1972*b*) suggested that the collision cross-sections used by Dean *et al.* were too small, and that the interaction may instead have been collisional. Koopman has also remarked that substantial momentum transfer takes place in regions where no significant spontaneous magnetic field is observed.

Appendix A

CONVERSION FACTORS

ENERGY

$$1 \text{ eV} \equiv 1{\cdot}602 \times 10^{-12} \text{ erg} \equiv 1{\cdot}602 \times 10^{-19} \text{ J}$$
$$1 \text{ erg} \equiv 10^{-7} \text{ J} \equiv 6{\cdot}241 \times 10^{11} \text{ eV}$$
$$1 \text{ J} \equiv 6{\cdot}241 \times 10^{18} \text{ eV} \equiv 10^{7} \text{ erg}.$$

TEMPERATURE

$$\text{If } \kappa_B T = 1 \text{ eV}, \qquad T = 1{\cdot}160 \times 10^{4} \text{ K}.$$

WAVELENGTH

$$1 \text{ Å} \equiv 10^{-10} \text{ m} \equiv 0{\cdot}1 \text{ nm}$$

PRESSURE

$$1 \text{ Torr } [\equiv 1 \text{ mm Hg}] \equiv 133{\cdot}3 \text{ Pa}$$
$$1 \text{ atm} \equiv 760 \text{ Torr} \equiv 1{\cdot}013 \times 10^{5} \text{ Pa}$$
$$(1 \text{ Pa} \equiv 1 \text{ J m}^{-3})$$

IRRADIANCE AND RADIATION ELECTRIC FIELD

In vacuum, an irradiance of I W cm^{-2} corresponds to a peak radiation electric field of $27\cdot45I^{\frac{1}{2}}$ V cm^{-1}, or an r.m.s. electric field of $19\cdot41I^{\frac{1}{2}}$ V cm^{-1} (I W m^{-2} corresponds to a peak electric field of $27\cdot45I^{\frac{1}{2}}$ V m^{-1}, or an r.m.s. electric field of $19\cdot41I^{\frac{1}{2}}$ V m^{-1}.)

Appendix B

PHOTON WAVELENGTH, WAVENUMBER AND FREQUENCY

Laser	Wavelength (λ) Å	μm	Wavenumber (k) m^{-1}	Angular frequency (ω) rad s^{-1}	Frequency Hz
Ruby	6 943	0·6943	$1·440 \times 10^6$	$2·713 \times 10^{15}$	$4·318 \times 10^{14}$
Neodymium	10 600	1·06	$9·43 \times 10^5$	$1·78 \times 10^{15}$	$2·83 \times 10^{14}$
Carbon dioxide		10·6	$9·43 \times 10^4$	$1·78 \times 10^{14}$	$2·83 \times 10^{13}$

PHOTON ENERGY

Laser	Photon energy ($\hbar\omega$) eV	J	erg	Number of photons per J
Ruby	1·786	$2·861 \times 10^{-19}$	$2·861 \times 10^{-12}$	$3·495 \times 10^{18}$
Neodymium	1·17	$1·87 \times 10^{-19}$	$1·87 \times 10^{-12}$	$5·34 \times 10^{18}$
Carbon dioxide	0·117	$1·87 \times 10^{-20}$	$1·87 \times 10^{-13}$	$5·34 \times 10^{19}$

CRITICAL ELECTRON DENSITY

	Critical electron density (n_{ec})	
Laser	m^{-3}	cm^{-3}
Ruby	2.313×10^{27}	2.313×10^{21}
Neodymium	9.92×10^{26}	9.92×10^{20}
Carbon dioxide	9.92×10^{24}	9.92×10^{18}

References

Numbers in square brackets denote pages in the text where references appear.

AARON, J. M. *et al.* (1974) (with IRELAND, C. L. M. & GREY MORGAN, C.). Aberration effects in the interaction of focused laser beams with matter, *J. Phys. D.*, **7**, 1907–17 (1974). [171]

ABRAMOV, A. A. *et al.* (1969) (with LUGOVOI, V. N. & PROKHOROV, A. M.). Self-focusing of ultrashort laser pulses, *Zh. Eksp. i Teor. Fiz. Pis'ma*, **9**, 675–9 (1969). (Transl: *JETP Lett.*, **9**, 419–22 (1969)). [196]

ABRIKOSOVA, I. I. & BOCHKOVA, O. M., (1969). Breakdown of liquid and gaseous helium by a laser beam, and observation of stimulated Mandel'shtam–Brillouin scattering in liquid helium, *Zh. Eksp. i Teor. Fiz. Pis'ma*, **9**, 285–9 (1969). (Transl: *JETP Lett.*, **9**, 167–9 (1969).) [177]

ABRIKOSOVA, I. I. & SHCHERBINA-SAMOILOVA, M. B. (1968). Measurement of the threshold parameters for the breakdown of liquid and gaseous helium by a laser beam, *Zh. Eksp. i Teor. Fiz. Pis'ma*, **7**, 305–8 (1968). (Transl: *JETP Lett.*, **7**, 238–40 (1968).) [177]

AFANAS'EV, YU. V. & KROKHIN, O. N. (1967). Vaporization of matter exposed to laser emissions, *Zh. Eksp. i Teor. Fiz.*, **52**, 966–75 (1967). (Transl: *Sov. Phys. JETP*, **25**, 639–45 (1967).) [287]

AFANAS'EV, YU. V. & ROZANOV, V. B. (1972). Energy spectrum of multiply charged ions in a laser-produced plasma, *Zh. Eksp. i Teor. Fiz.*, **62**, 247–52 (1972). (Transl: *Sov. Phys. JETP*, **35**, 133–5 (1972).) [318]

AFANAS'EV, YU. V. *et al.* (1966a) (with KROKHIN, O. N. & SKLIZKOV, G. V.). Evaporation and heating of a substance due to laser radiation, *I.E.E.E. J. Quantum Electron.*, **QE-2**, 483–5 (1966). [287, 366]

AFANAS'EV, YU. V. *et al.* (1966b) (with KROL', V. M., KROKHIN, O. N. & NEMCHINOV, I. V.). Gasdynamic processes in heating of a substance by laser radiation, *Prikl. Mat. i Mekh.*, **30**, 1022–8 (1966). (Transl: *Appl. Math & Mech.*, **30**, 1218–25 (1966).) [287]

AFANAS'EV, YU. V. *et al.* (1968) (with BELENOV, E. M. & KROKHIN, O. N.). Cascade ionization of gas in a high-intensity light field, *Zh. Eksp. i Teor. Fiz. Pis'ma*, **8**, 209–11 (1968). (Transl: *JETP Lett.*, **8**, 126–7 (1968).) [170]

449

AFANAS'EV, YU. V. *et al.* (1969*a*) (with BELENOV, E. M. & KROKHIN, O. N.). Cascade ionization of a gas by a powerful ultrashort pulse of light, *Zh. Eksp. i Teor. Fiz.*, **56**, 256–63 (1969). (Transl: *Sov. Phys. JETP*, **29**, 141–4 (1969).) [170]

AFANAS'EV, YU. V. *et al.* (1969*b*) (with BELENOV, E. M., KROKHIN, O. N. & POLUEKTOV, I. A.). Ionization processes in a laser plasma, *Zh. Eksp. i Teor. Fiz. Pis'ma*, **10**, 553–7 (1969). (Transl: *JETP Lett.*, **10**, 353–6 (1969).) [214]

AFANAS'EV, YU. V. *et al.* (1969*c*) (with BELENOV, E. M., KROKHIN, O. N. & POLUEKTOV, I. A.). Cascade ionization of a gas during optical breakdown in a wide range of radiation fluxes, *Zh. Eksp. i Teor. Fiz.*, **57**, 580–4 (1969). (Transl: *Sov. Phys. JETP*, **30**, 318–20 (1970).) [170]

AFANAS'EV, YU. V. *et al.* (1970) (with BELENOV, E. M., KROKHIN, O. N. & POLUEKTOV, I. A.). Self-consistent regime of heating of matter by a laser pulse under conditions of nonequilibrium ionization, *Zh. Eksp. i Teor. Fiz.*, **60**, 73–82 (1970). (Transl: *Sov. Phys. JETP*, **33**, 42–6 (1971).) [300]

AFANAS'EV, YU. V. *et al.* (1972) (with BELENOV, E. M. & POLUEKTOV, I. A.). Optical breakdown of molecular gases, *Zh. Eksp. i Teor. Fiz. Pis'ma*, **15**, 60–3 (1972). (Transl: *JETP Lett.*, **15**, 41–4 (1972).) [167]

AGLITSKY, E. V. *et al.* (1971) (with BASOV, N. G., BOIKO, V. A., GRIBKOV, V. A., ZAKHAROV, S. A., KROKHIN, O. N. & SKLIZKOV, G. V.). Determination of electron density, velocity and gas-dynamic pressure in laser plasma, *X International Conference on phenomena in ionized gases, Oxford (1971). Contributed papers*, p. 229. Oxford, Donald Parsons & Co. Ltd. (1971). [374]

AGOSTINI, P. *et al.* (1968) (with BARJOT, G., BONNAL, J. F., MAINFRAY, G., MANUS, C. & MORELLEC, J.). Multiphoton ionization of hydrogen and rare gases, *I.E.E.E. J. Quantum Electron*, **QE-4**, 667–9 (1968). [157]

AGOSTINI, P. *et al.* (1970*a*) (with BARJOT, G., MAINFRAY, G., MANUS, C. & THEBAULT, J.). Multiphoton ionization of rare gases at 1.06μ and 0.53μ, *I.E.E.E. J. Quantum Electron*, **QE-6**, 782–8 (1970). [157, 158]

AGOSTINI, P. *et al.* (1970*b*) (with BARJOT, G., MAINFRAY, G., MANUS, C. & THEBAULT, J.) Multiphoton ionization of rare gases at 1.06μ and 0.53μ, *Phys. Lett.*, **31A**, 367–8 (1970). [157]

AGOSTINI, P. *et al.* (1970*c*) (with BARJOT, G., MAINFRAY, G., MANUS, C. & THEBAULT, J.). Probabilités d'ionisation multiphotonique des gaz rares à 1.06 et 0.53μ, *C.R. Acad. Sci., Paris*, **270B**, 1566–8 (1970). [157]

AGOSTINI, P. *et al.* (1971*a*) (with LU VAN, M. & MAINFRAY, G.). Multiphoton ionization and dissociation of the hydrogen molecule at 1.06μ, *Phys. Lett.*, **36A**, 21–2 (1971). [160]

AGOSTINI, P. *et al.* (1971*b*) (with MAINFRAY, G., MANUS, C. & MORELLEC, J.). Étude experimentale de l'ionisation multiphotonique des gaz rares et des atomes alcalins, *J. Physique*, Suppl. C5B to **32**, 142–4 (1971). [157, 158]

AHLBORN, B. (1971). Trigger criterion for steady fusion detonations in D–T ice, *Phys. Lett.*, **37A**, 227–8 (1971). [142]

AHLBORN, B. (1972). Decaying laser sparks, *Phys. Lett.*, **40A**, 289–90 (1972). [220]

AHLBORN, B. & STRACHAN, J. D., (1973). Dynamics of step heat waves in gases and plasmas, *Can. J. Phys.*, **51**, 1416–27 (1973). [220]

AHMAD, N. & KEY, M. H. (1969). Enhanced plasma heating in laser-induced gas breakdown at the shock front of a blast wave, *Appl. Phys. Lett.*, **14**, 243–5 (1969). [244]

AHMAD, N. & KEY, M. H. (1972). Plasma temperature in laser pulse induced gas breakdown, *J. Phys. B.*, **5**, 866–77 (1972). [232, 256]

AHMAD, N. *et al.* (1969*a*) (with GALE, B. C. & KEY, M. H.). Experimental and theoretical studies of the time and space development of plasma parameters in a laser induced spark in helium, *Proc. Roy. Soc.*, **A310**, 231–52 (1969). [231, 242, 243]

AHMAD, N. *et al.* (1969*b*) (with GALE, B. C. & KEY, M. H.). Scattering of laser light in radiation-driven breakdown waves in gases, *J. Phys. B.*, **2**, 403–9 (1969). [242, 244, 245]

AKCASU, A. Z. & WALD, L. M. (1967). Bremsstrahlung of slow electrons in neutral gases and free–free absorption of microwaves, *Phys. Fluids*, **10**, 1327–35 (1967). [48]

AKHIEZER, A. I. *et al.* (1957) (with PROKHODA, I. G. & SITENKO, A. G.). Scattering of electromagnetic waves in a plasma, *Zh. Eksp. i Teor. Fiz.*, **33**, 750–7 (1957). (Transl: *Sov. Phys. JETP*, **6**, 576–82 (1958).) [87]

AKHIEZER, A. I. *et al.* (1967) (with AKHIEZER, I. A., POLOVIN, R. V., SITENKO, A. G. & STEPANOV, K. N.). *Collective oscillations in a plasma.* Oxford, Pergamon. (1967). (Transl: from *Kollektivnye kolebaniya v plasme.* Moscow, Atomizdat (1964).) [18, 63]

AKHMANOV, S. A. *et al.* (1965) (with KOVRIGIN, A. I., STRUKOV, M. M. & KHOKHLOV, R. V.). Frequency dependence of the threshold of optical breakdown in air, *Zh. Eksp. i Teor. Fiz. Pis'ma*, **1**, 42–7 (1965). (Transl: *JETP Lett.*, **1**, 25–9 (1965).) [181]

ALCOCK, A. J. (1972). Experiments on self-focusing in laser-produced plasmas. In *Laser interaction and related plasma phenomena*, (ed. H. J. Schwarz and H. Hora), Vol. 2, pp. 155–75. New York, Plenum (1972). [198, 247]

ALCOCK, A. J. & RAMSDEN, S. A. (1966). Two wavelength interferometry of a laser-induced spark in air, *Appl. Phys. Lett.*, **8**, 187–8 (1966). [229, 240]

ALCOCK, A. J. & RICHARDSON, M. C. (1968). Creation of a spark by a single subnanosecond laser pulse, *Phys. Rev. Lett.*, **21**, 667–70 (1968). [186, 263]

ALCOCK, A. J. & RICHARDSON, M. C. (1972). Comment on 'New observations of dielectric breakdown in air induced by a focused Nd^{3+} glass laser with various pulse widths' by C. C. Wang and L. I. Davis, *Phys. Rev.*, **A5**, 1566–7 (1972). [188]

ALCOCK, A. J. *et al.* (1966*a*) (with PANARELLA, E. & RAMSDEN, S. A.). An interferometric study of laser induced breakdown in air. *VII International Conference on phenomena in ionized gases, Belgrade (1965)*, Vol. 3, pp. 224–7. Belgrade, Gradevinska Knjiga (1966). [227, 239]

ALCOCK, A. J. *et al.* (1966*b*) (with PASHININ, P. P. & RAMSDEN, S. A.). Temperature measurements of a laser spark from soft-X-ray emission, *Phys. Rev. Lett.*, **17**, 528–30 (1966). [232, 256, 258]

ALCOCK, A. J. *et al.* (1968*a*) (with DE MICHELIS, C. & HAMAL, K.). Subnanosecond schlieren photography of laser-induced gas breakdown, *Appl. Phys. Lett.*, **12**, 148–50 (1968). [225, 254, 255]

ALCOCK, A. J. *et al.* (1968*b*) (with DE MICHELIS, C., HAMAL, K. & TOZER, B. A.). A mode-locked laser as a light source for schlieren photography, *I.E.E.E. J. Quantum Electron*, **4**, 593–7 (1968). [225, 254, 255]

ALCOCK, A. J. *et al.* (1968*c*) (with DE MICHELIS, C., HAMAL, K. & TOZER, B. A.). Expansion mechanism in a laser-produced spark, *Phys. Rev. Lett.*, **20**, 1095–7 (1968). [255]

ALCOCK, A. J. *et al.* (1968*d*) (with DE MICHELIS, C. & RICHARDSON, M. C.). Production of a spark by a train of mode-locked laser pulses, *Phys. Lett.*, **28A**, 356–7 (1968). [185, 264]

ALCOCK, A. J. *et al.* (1969*a*) (with DE MICHELIS, C., KOROBKIN, V. V. & RICHARDSON, M. C.). Preliminary evidence for self-focusing in gas breakdown produced by picosecond laser pulses, *Appl. Phys. Lett.*, **14**, 145–6 (1969). [199, 264]

ALCOCK, A. J. *et al.* (1969*b*) (with DE MICHELIS, C., KOROBKIN, V. V. & RICHARDSON, M. C.). Frequency broadening in laser-induced sparks, *Phys. Lett.*, **29A**, 475–6 (1969). [198]

ALCOCK, A. J. *et al.* (1969*c*) (with DE MICHELIS, C. & RICHARDSON, M. C.). Wavelength dependence of laser-induced gas breakdown using dye lasers, *Appl. Phys. Lett.*, **15**, 72–3 (1969). [184]

ALCOCK, A. J. *et al.* (1970) (with DE MICHELIS, C. & RICHARDSON, M. C.). Breakdown and self-focusing effects in gases produced by means of a single-mode ruby laser, *I.E.E.E. J. Quantum Electron*, **QE-6**, 622–9 (1970). [198, 247]

ALEXANDROV, A. F. *et al.* (1968) (with MITSUK, V. E. & TIMOFEEV, J. B.). On the scattering of light by plasma in the presence of collisions, *VIII International Conference on phenomena in ionized gases, Vienna (1967). Contributed papers*, p. 519. Vienna, IAEA (1968). [99]

ALIEV, YU. M. *et al.* (1972) (with GRADOV, O. M. & KIRII, A. YU). Parametric resonance in an inhomogeneous plasma, *Zh. Tekh. Fiz.*, **42**, 1811–17 (1972). (Transl: *Sov. Phys. Tech. Phys.*, **17**, 1453–7 (1973).) [64]

ALLEN, C. W. (1962). *Astrophysical Quantities*, 2nd edition. London, Athlone Press (1962). [85]

ALLEN, F. J. (1970). Method of determining ion temperatures in laser-produced plasmas, *J. Appl. Phys.*, **41**, 3048–51 (1970). [352, 370, 383]

ALLEN, F. J. (1971). Surface temperature and disposition of beam energy for a laser-heated target, *J. Appl. Phys.*, **42**, 3145–9 (1971). [345]

ALLEN, F. J. (1972). Production of high-energy ions in laser-produced plasmas, *J. Appl. Phys.*, **43**, 2169–75 (1972). [318]

AMANO, T. & OKAMOTO, M. (1969). Parametric effects of an alternating field on inhomogeneous plasmas, *J. Phys. Soc. Jap.*, **26**, 529–40 (1969). [64, 69, 70]

AMBARTSUMYAN, R. V. *et al.* (1965) (with BASOV, N. G., BOIKO, V. A., ZUEV, V. S., KROKHIN, O. N., KRYUKOV, P. G., SENAT-SKII, YU. V. & STOILOV, YU. YU.). Heating of matter by focused laser radiation, *Zh. Eksp. i Teor. Fiz.*, **48**, 1583–7 (1965). (Transl: *Sov. Phys. JETP*, **21**, 1061–4 (1965).) [257, 361]

AMBARTSUMYAN, R. V. *et al.* (1971) (with KALININ, V. N. & LETOKHOV, V. S.). Two-step selective photoionization of rubidium atoms by laser radiation, *Zh. Eksp. i Teor. Fiz. Pis'ma*, **13**, 305–7 (1971). (Transl: *JETP Lett.*, **13**, 217–19 (1971).) [160]

ANDELFINGER, C. *et al.* (1966) (with DECKER, G., FÜNFER, E., HEISS, A., KEILHACKER, M., SOMMER, J. & ULRICH, M.). ISAR I—a fast megajoule theta-pinch experiment with extremely high compression fields, *Conference on plasma physics and controlled fusion research, Culham (1965)*, Vol. I, pp. 249–60. Vienna, IAEA (1966). [115]

ANDREEV, N. E. *et al.* (1969) (with KIRYI, A. YU. & SILIN, V. P.). Parametric excitation of longitudinal oscillations in a plasma, *Zh. Eksp. i Teor. Fiz.*, **57**, 1024–39 (1969). (Transl: *Sov. Phys. JETP*, **30**, 559–66 (1970).) [62]

ANDREEV, S. I. *et al.* (1968) (with DYMSHITS, YU. I., KAPORSKII, L. N. & MUSATOVA, G. S.). Thermionic emission and expansion of a plasma formed by focused single-pulse laser radiation on a solid target, *Zh. Tekh. Fiz.*, **38**, 875–83 (1968). (Transl: *Sov. Phys. Tech. Phys.*, **13**, 657–62 (1968).) [360]

ANDREWS, A. J. (1970). Private communication. [264]

ANDREWS, A. J. *et al.* (1972) (with HALL, T. A. & HUGHES, T. P.). Absorption of picosecond light pulses by solid targets, *Vth European Conference on controlled fusion and plasma physics, Grenoble (1972)*, Vol. I, p. 62. Grenoble, Ass. Euratom-CEA (1972). [77]

ANISIMOV, S. I. (1970). Effect of very short laser pulses on absorbing substances, *Zh. Eksp. i Teor. Fiz.*, **58**, 337–40 (1970). (Transl: *Sov. Phys. JETP*, **31**, 181–2 (1970).) [306]

ANISIMOV, S. I. *et al.* (1966) (with BONCH-BRUEVICH, A. M., EL'YASHEVICH, M. A., IMAS, YA. A., PAVLENKO, N. A. & ROMANOV, G. S.). Effect of powerful light fluxes on metals, *Zh. Tekh. Fiz.*, **36**, 1273–84 (1966). (Transl: *Sov. Phys. Tech. Phys.*, **11**, 945–52 (1967).) [277]

ARCHBOLD, E. & HUGHES, T. P. (1964). Electron temperature in a laser-heated plasma, *Nature*, **204**, 670 (1964). [341]

ARCHBOLD, E. *et al.* (1964) (with HARPER, D. W. & HUGHES, T. P.). Time-resolved spectroscopy of laser-generated microplasmas, *Br. J. Appl. Phys.*, **15**, 1321–7 (1964). [278, 279, 435, 436]

ARIFOV, T. U. *et al.* (1967) (with ASKAR'YAN, G. A., RABINOVICH, M. S., RAEVSKII, I. M. & TARASOVA, N. M.). Plasma produced by a beam from a non-Q-switched laser acting on a medium, *Zh. Eksp. i Teor. Fiz. Pis'ma*, **6**, 681–2 (1967). (Transl: *JETP Lett.*, **6**, 166–8 (1967).) [340, 439]

ARIFOV, T. U., *et al.* (1968*a*) (with ASKAR'YAN, G. A., RAEVSKII, I. M. & TARASOVA, N. M.). Laser-induced current pulses from a target in a gas, *Zh. Eksp. i Teor. Fiz.*, **55**, 385–8 (1968). (Transl: *Sov. Phys. JETP*, **28**, 201–2 (1969).) [439]

ARIFOV, T. U. *et al.* (1968*b*) (with ASKAR'YAN, G. A. & TARASOVA, N. M.). Ionizing action of radiation due to heating of substance in the focus of a laser beam and production of a plasma with high degree of ionization, *Zh. Eksp. i Teor. Fiz. Pis'ma*, **8**, 128–32 (1968). (Transl: *JETP Lett.*, **8**, 77–9 (1968).) [439]

ARKHIPENKO, V. I. *et al.* (1971) (with BUDNIKOV, V. N. & OBUKHOV, A. A.). Nonlinear effect in the absorption of microwaves by a plasma, *Zh. Tekh. Fiz.*, **41**, 2334–6 (1971). (Transl: *Sov. Phys. Tech. Phys.*, **16**, 1852–4 (1972).) [76]

ARNUSH, D. & KENNEL, C. F. (1973). Refraction by the electromagnetic pump of parametrically generated electrostatic waves, *Phys. Rev. Lett.*, **30**, 597–600 (1973). [71]

ARSLANBEKOV, T. U. *et al.* (1971) (with BAKOS, J., KISS, A., NAGAEVA, M. L., PETROSIAN, K. B. & ROSA, K.). Multiphoton ionization of the excited helium atoms, *X International Conference on Phenomena in ionized gases, Oxford (1971). Contributed papers*, p. 43. Oxford, Donald Parsons & Co. Ltd. (1971). [160]

ARUTYUNYAN, I. N. *et al.* (1970) (with ASKAR'YAN, G. A. & POGOSYAN, V. A.). Multiphoton processes in the focus of an intense laser beam with expansion of the interactive region taken into consideration, *Zh. Eksp. i Teor. Fiz.*, **58**, 1020–4 (1970). (Transl: *Sov. Phys. JETP*, **31** (3), 548–50 (1970).) [157]

ASCOLI-BARTOLI, U. (1970). Plasma diagnostics based on refractivity. *Physics of Hot Plasmas*, (ed. B. J. Rye & J. C. Taylor), pp. 404–55. Edinburgh, Oliver & Boyd (1970). [223]

ASCOLI-BARTOLI, U. *et al.* (1964) (with KATZENSTEIN, J. & LOVISETTO, L.). Forward scattering of light from a laboratory plasma, *Nature*, **204**, 672–3 (1964). [118]

ASCOLI-BARTOLI, U. *et al.* (1965) (with KATZENSTEIN, J. & LOVISETTO, L.). Spectrum of laser light scattered from a single giant pulse in a laboratory plasma, *Nature*, **207**, 63–4 (1965). [118]

ASCOLI-BARTOLI, U. *et al.* (1966) (with DE MICHELIS, C. & MAZZUCATO, E.). The hot-ice experiment, *Conference on plasma physics and controlled nuclear fusion research, Culham (1965)*, Vol. 2, pp. 941–51. Vienna, IAEA (1966). [412, 417]

ASCOLI-BARTOLI, U. *et al.* (1969) (with BRUNELLI, B., CARUSO, A., DE ANGELIS, A., GATTI, G., GRATTON, R., PARLANGE, F. & SALZMANN, H.). Production of a dense deuterium plasma in a strong magnetic field by a giant laser pulse (Hot-ice experiment). *Conference on plasma physics and controlled nuclear fusion research, Novosibirsk (1968)*, Vol. I, pp. 917–24. Vienna, IAEA (1969). [428, 434]

ASHKIN, M. (1966). Radiative absorption cross section of an electron in the field of an argon atom, *Phys. Rev.*, **141**, 41–4 (1966). [48]

ASHMARIN, I. I. *et al.* (1971) (with BYKOVSKII, Yu. A., DEGTYARENKO, N. N., ELESIN, V. F., LARKIN, A. I. & SIPAILO, I. P.). Pulsed-hologram investigation of gas breakdown in front of a laser-produced plasma, *Zh. Tekh. Fiz.*, **41**, 2369–77 (1971). (Transl: *Sov. Phys. Tech. Phys.*, **16**, 1881–7 (1971).) [436, 437]

ASKAR'YAN, G. A. (1962). Effects of the gradient of a strong electro-magnetic beam on electrons and atoms. *Zh. Eksp. i Teor. Fiz.*, **42**, 1567–70 (1962). (Transl: *Sov. Phys. JETP*, **15**, 1088–90 (1962).) [195]

ASKAR'YAN, G. A. (1964). Excitation and dissociation of molecules in an intense light field, *Zh. Eksp. i Teor. Fiz.*, **46**, 403–5 (1964). (Transl: *Sov. Phys. JETP*, **19**, 273–4 (1964).) [154]

ASKAR'YAN, G. A. (1966). Self-focusing of a light beam upon excitation of the atoms and molecules of the medium in the beam, *Zh. Eksp. i Teor. Fiz. Pis'ma*, **4**, 400–3 (1966). (Transl: *JETP Lett.*, **4**, 270–2 (1966).) [196]

ASKAR'YAN, G. A. & MOROZ, E. M. (1962). Pressure on evaporation of matter in a radiation beam, *Zh. Eksp. i Teor. Fiz.*, **43**, 2319–20 (1962). (Transl: *Sov. Phys. JETP*, **16**, 1638–9 (1963).) [277]

ASKAR'YAN, G. A. & TARASOVA, N. M. (1971). Initial stage of optical explosion of a material particle in an intense light flux, *Zh. Eksp. i Teor. Fiz.*, **60**, 617–20 (1971). (Transl: *Sov. Phys. JETP*, **33**, 336–7 (1971).) [313]

ASKAR'YAN, G. A. *et al.* (1965a) (with RABINOVICH, M. S., SAVCHENKO, M. M. & SMIRNOVA, A. D.). Observation of a fast photoionization aureole and of a concentrated long lived ionization cloud due to a shock wave from a spark in a laser beam, *Zh. Eksp. i Teor. Fiz. Pis'ma*, **1** (6), 18–23 (1965). (Transl: *JETP Lett.*, **1**, 162–4 (1965).) [232, 241, 252]

ASKAR'YAN, G. A. *et al.* (1965b) (with RABINOVICH, M. S., SAVCHENKO, M. M. & SMIRNOVA, A. D.). Light spark in a magnetic field, *Zh. Eksp. i Teor. Fiz. Pis'ma*, **1** (1), 9–15 (1965). (Transl: *JETP Lett.*, **1**, 5–8 (1965).) [271]

ASKAR'YAN, G. A. *et al.* (1966) (with RABINOVICH, M. S., SAVCHENKO, M. M. & STEPANOV, V. K.). Fast overlap of microwave radiation by an ionization aureole of a spark in a laser beam, *Zh. Eksp. i Teor. Fiz. Pis'ma*, **3**, 465–8 (1966). (Transl: *JETP Lett.*, **3**, 303–5 (1966).) [232, 241, 252]

ASKAR'YAN, G. A. *et al.* (1967*a*) (with RABINOVICH, M. S., SAVCHENKO, M. M. & STEPANOV, V. K.). Optical breakdown 'fireball' in the focus of a laser beam, *Zh. Eksp. i Teor. Fiz. Pis'ma*, **5**, 150–4 (1967). (Transl: *JETP Lett.*, 5, 121–4 (1967).) [271]

ASKAR'YAN, G. A. *et al.* (1967*b*) (with RABINOVICH, M. S., SMIRNOVA, A. D. & STUDENOV, V. B.). Currents produced by light pressure when a laser beam acts on matter, *Zh. Eksp. i Teor. Fiz. Pis'ma*, **5**, 116–18 (1967). (Transl: *JETP Lett.*, **5**, 93–5 (1967).) [442]

ASKAR'YAN, G. A. *et al.* (1968) (with RABINOVICH, M. S., SAVCHENKO, M. M., SMIRNOVA, A. D., STEPANOV, V. K. & STUDENOV, V. B.). Diamagnetic moment of the 'fireball' of the light spark in the laser focus, *VIII International Conference on phenomena in ionized gases, Vienna (1967). Contributed papers*, p. 264. Vienna, IAEA (1968). [271]

ASKAR'YAN. G. A. *et al.* (1969) (with SAVCHENKO, M. M. & STEPANOV, V. K.). Investigation of light spark and of other optical effects in focusing of light by a lens with a channel on the axis, *Zh. Eksp. i Teor. Fiz. Pis'ma*, **10**, 161–5 (1969). (Transl: *JETP Lett.*, **10**, 101–3 (1969).) [224]

ASKAR'YAN, G. A. *et al.* (1970) (with SAVCHENKO, M. M. & STEPANOV, V. K.). Diamagnetic moment of a strong shock wave of a high-temperature light explosion in gases, *Zh. Eksp. i Teor. Fiz.*, **59**, 1133–45 (1970). (Transl: *Sov. Phys. JETP*, **32**, 617–23 (1971).) [271]

ASKAR'YAN, G. A. *et al.* (1972) (with KAITMAZOV, S. D. & MEDVEDEV, A. A.). Light flash from a strong shock wave of a laser spark. The effect of a strong external magnetic field, *Zh. Eksp. i Teor. Fiz.*, **62**, 918–23 (1972). (Transl: *Sov. Phys. JETP*, **35**, 487–9 (1972).) [271]

ASKAR'YAN, G. A. *et al.* (1973) (with NAMIOT, V. A. & RABINOVICH, M. S.). Supercompression of matter by reaction pressure to obtain microcritical masses of fissioning matter, to obtain ultrastrong magnetic fields, and to accelerate particles, *Zh. Eksp. i Teor. Fiz. Pis'ma*, **17**, 597–600 (1973). (Transl: *JETP Lett.*, **17**, 424–6 (1973).) [143, 337]

ÅSTRÖM, E. (1950). On waves in an ionized gas, *Ark. Fys.*, **2**, 443–57 (1950). [17]

BABUEL-PEYRISSAC, J. P. (1972). Induced Compton effect in a plasma, *Phys. Lett.*, **41A**, 143–4 (1972). [79]

BABUEL-PEYRISSAC, J. P. *et al.* (1969) (with FAUQUIGNON, C. & FLOUX, F.). Effect of powerful laser pulse on low Z solid material, *Phys. Lett.*, **30A**, 290–1 (1969) [301, 302]

BABYKIN, M. V. & STARYKH, V. V. (1971). Optimum temperatures and critical parameters of a self-sustained reaction in a solid thermonuclear fuel, *Zh. Tekh. Fiz.*, **41**, 1618–23 (1971). (Transl: *Sov. Phys. Tech. Phys.*, **16**, 1273–7 (1972).) [142, 316]

BACONNET, J. P. *et al.* (1969) (with CESARI, G., COUDEVILLE, A. & WATTEAU, J. P.). 90° laser light scattering by a dense plasma focus, *Phys. Lett.*, **29A**, 19–20 (1969). [112]

BAKER, D. A. & HAMMEL, J. E. (1965). Experimental studies of the penetration of a plasma stream into a transverse magnetic field, *Phys. Fluids*, **8**, 713–22 (1965). [432]

BAKOS, J. *et al.* (1972*a*) (with KISS, A., SZABO, L. & TENDLER, M.). Resonance multiphon ionization of the triplet metastable He atoms, *Phys. Lett.*, **39A**, 283–4 (1972). [160]

BAKOS, J. *et al.* (1972*b*) (with KISS, A., SZABO, L. & TENDLER, M.). Light intensity dependence of the multiphoton ionization probability in the resonance case, *Phys. Lett.*, **41A**, 163–4 (1972). [160]

BARAVIAN, G. *et al.* (1970) (with BENATTAR, R., BRETAGNE, J., GODART, J. L. & SULTAN, G.). Multiphoton ionization of neon: experimental study of a resonance, *Appl. Phys. Lett.*, **16**, 162 (1970). [158]

BARAVIAN, G. *et al.* (1971) (with BENATTAR, R., BRETAGNE, J., CALLEDE, G., GODART, J. L. & SULTAN, G.). Experimental determination of the cross section for multiphoton ionization of neon near a resonance, *Appl. Phys. Lett.*, **18**, 387–9 (1971). [158]

BARAVIAN, G. *et al.* (1972) (with BENATTAR, R., BRETAGNE, J., GODART, J. L. & SULTAN, G.). Electron–ion recombination in a helium plasma produced by laser, *Z. Phys.*, **254**, 218–31 (1972). [252, 253]

BARCHUKOV, A. I. *et al.* (1973) (with BUNKIN, F. V., KONOV, V. I. & PROKHOROV, A. M.). Low-threshold breakdown of air near a target by CO_2 radiation, and the associated large recoil momentum, *Zh. Eksp. i Teor. Fiz. Pis'ma*, **17**, 413–16 (1973). (Transl: *JETP Lett.*, **17**, 294–6 (1973).) [441]

BARHUDAROVA, T. M. *et al.* (1968) (with VORONOV, G. S., DELONE, G. A., DELONE, N. B. & MARTAKOVA, N. K.). Multiphoton ionization of the noble gas atoms, *VIII International Conference on phenomena in ionized gases, Vienna (1967). Contributed papers*, p. 266. Vienna, IAEA (1968). [157]

BARTHELEMY, C. *et al.* (1968) (with LEBLANC, M. & BOUCAULT, M. T.). Variation of the breakdown threshold of air as a function of the wavelength of laser radiation, *C.R. Acad. Sci., Paris*, **266B**, 1234–5 (1968) [182, 183]

BARYNIN, V. A. & KHOKLOV, R. V. (1966). The mechanism of the optical breakdown in a gas, *Zh. Eksp. i Teor. Fiz.*, **50**, 472–3 (1966). (Transl: *Sov. Phys. JETP*, **23**, 314–15 (1966).) [169]

BASOV, N. G. & KROKHIN, O. N. (1964). Conditions for heating up of a plasma by the radiation from an optical generator, *Zh. Eksp. i Teor. Fiz.*, **46**, 171–5 (1964). (Transl: *Sov. Phys.*, **19**, 123–5 (1964).) [310]

BASOV, N. G. *et al.* (1966) (with BOIKO, V. A., DEMENT'EV, V. A., KROKHIN, O. N. & SKLIZKOV, G. V.). Heating and decay of plasma produced by a giant laser pulse focused on a solid target, *Zh. Eksp. i. Teor. Fiz.*, **51**, 989–1000 (1966). (Transl: *Sov. Phys. JETP*, **24**, 659–66 (1967).) [285, 366, 367, 368]

BASOV, N. G. *et al.* (1967a) (with BOIKO, V. A., KROKHIN, O. N. & SKLIZKOV, G. V.). Formation of a long spark in air by weakly focused laser radiation, *Dokl. Akad. Nauk. SSSR*, **173**, 538–41 (1967). (Transl: *Sov. Phys. Dokl.*, **12**, 248–50 (1967).) [197, 387]

BASOV, N. G. *et al.* (1967b) (with BOIKO, V. A., VOINOV, YU. P., KONONOV, E. YA., MANDEL'SHTAM, S. L. & SKLIZKOV, G. V.). Production of spectra of multiply charged ions by focusing laser radiation on a solid target, *Zh. Eksp. i Teor. Fiz. Pis'ma*, **5**, 177–80 (1967). (Transl: *JETP Lett.*, **5**, 141–3 (1967).) [371]

BASOV, N. G. *et al.* (1967c) (with BOIKO, V. A., VOINOV, YU. P., KONONOV, E. YA., MANDEL'SHTAM, S. L. & SKLIZKOV, G. B.). Spectra of calcium ions Ca XV, Ca XVI obtained by focusing laser emission on a target, *Zh. Eksp. i Teor. Fiz. Pis'ma*, **6**, 849–51 (1967). (Transl: *JETP Lett.*, **6**, 291–3 (1967).) [374]

BASOV, N. G. *et al.* (1967d) (with KROKHIN, O. N. & SKLIZKOV, G. V.). Laser application for the production and diagnosis of pulsed plasma, *Appl. Opt.*, **6**, 1814–17 (1967). [374]

BASOV, N. G. *et al.* (1967e) (with KROKHIN, O. N. & SKLIZKOV, G. V.). Formation of shock waves with the aid of powerful laser radiation, *Zh. Eksp. i Teor. Fiz. Pis'ma*, **6**, 683–4 (1967). (Transl: *JETP Lett.*, **6**, 168–71 (1967).) [435]

BASOV, N. G. *et al.* (1968a) (with BOIKO, V. A., KROKHIN, O. N. & SKLIZKOV, G. V.). Influence of laser radiation focusing on the air breakdown, *VIII International Conference on phenomena in ionized gases, Vienna (1967). Contributed papers*, p. 261. Vienna, IAEA (1968). [197]

BASOV, N. G. *et al.* (1968b) (with GRIBKOV, V. A., KROKHIN, O. N. & SKLIZKOV, G. V.). High-temperature effects of intense laser emission focused on a solid target, *Zh. Eksp. i Teor. Fiz.*, **54**, 1073–87 (1968). (Transl: *Sov. Phys. JETP*, **27**, 575–82 (1968).) [374, 375, 436, 437 438]

BASOV, N. G. *et al.* (1968c) (with KRIUKOV, P. G., ZAKHAROV, S. D., SENATSKY, YU. V. & TCHEKALIN, S. V.). Experiments on the observation of neutron emission at a focus of high-power laser radiation on a lithium deuteride surface, *IEEE J. Quantum Electron*, **4**, 864–7 (1968). [392, 393]

BASOV, N. G. *et al.* (1968d) (with KROKHIN, O. N. & SKLIZKOV, G. V.). Laser application for investigations of the high-temperature and plasma phenomena, *IEEE J. Quantum Electron*, **4**, 988–91 (1968). [385]

BASOV, N. G. *et al.* (1969a) (with BOIKO, V. A., GRIBKOV, V. A., ZAKHAROV, S. M., KROKHIN, O. N. & SKLIZKOV, G. V.). Kinetics of heating and 'freezing' of ionization state of hot plasma formed at interaction of laser radiation and solid target, *IX International Conference on phenomena in ionized gases, Bucharest (1969). Contributed papers*, p. 333. Bucharest, Ed. Acad. R.S. Romania (1969). [318]

BASOV, N. G. *et al.* (1969b) (with BOIKO, V. A., GRIBKOV, V. A., ZAKHAROV, S. M., KROKHIN, O. N. & SKLIZKOV, G. V.). Measurement of the time variation of the temperature of the plasma of a laser flare by means of its X-radiation, *Zh. Eksp. i Teor. Fiz. Pis'ma*, **9**, 520–3 (1969). (Transl: *JETP Lett.*, **9**, 315–16 (1969).) [387]

BASOV, N. G. *et al.* (1970) (with BOIKO, V. A., DROZHBIN, YU. A., ZAKHAROV, S. M., KROKHIN, O. N., SKLIZKOV, G. V. & YAKOVLEV, V. A.). Study of the initial stage of gas-dynamic expansion of a laser flare plasma, *Dokl. Akad. Nauk. SSSR*, **192**, 1248–50 (1970). (Transl: *Sov. Phys. Dokl.*, **15**, 576–8 (1970).) [374]

BASOV, N. G. *et al.* (1971*a*) (with BOIKO, V. A., GRIBKOV, V. A., ZAKHAROV, S. M., KROKHIN, O. N. & SKLIZKOV, G. V.). Gas dynamics of laser plasma in the course of heating, *Zh. Eksp. i Teor. Fiz.*, **61**, 154–61 (1971). (Transl: *Sov. Phys. JETP*, **34**, 81–4 (1972).) [367, 368, 369]

BASOV, N. G. *et al.* (1971*b*) (with BOIKO, V. A., ZAKHAROV, S. M., KROKHIN, O. N. & SKLIZKOV, G. V.). Generation of neutrons in a laser CD_2 plasma heated by pulses of nanosecond duration, *Zh. Eksp. i Teor. Fiz. Pis'ma*, **13**, 691–4 (1971). (Transl: *JETP Lett.*, **13**, 489–91 (1971).) [389]

BASOV, N. G. *et al.* (1972*a*) (with IVANOV, YU. S., KROKHIN, O. N., MIKHAILOV, YU. A., SKLIZKOV, G. V. & FEDOTOV, S. I.). Neutron generation in spherical irradiation of a target by high-power laser radiation, *Zh. Eksp. i Teor. Fiz. Pis'ma*, **15**, 589–92 (1972). (Transl: *JETP Lett.*, **15**, 417–19 (1972).) [408, 409]

BASOV, N. G. *et al.* (1972*b*) (with KROKHIN, O. N. & SKLIZKOV, G. V.). Heating of laser plasmas for thermonuclear fusion, In *Laser interaction and related plasma phenomena*, (ed H. J. Schwarz and H. Hora) Vol. 2, pp. 389–408. New York, Plenum (1972). [258, 438]

BASOV, N. G. *et al.* (1972*c*) (with KROKHIN, O. N., SKLIZKOV, G. V., FEDOTOV, S. I. & SHIKANOV, A. S.). A powerful laser setup and investigation of the efficiency of high temperature heating of a plasma, *Zh. Eksp. i Teor. Fiz.*, **62**, 203–12 (1972). (Transl: *Sov. Phys. JETP*, **35**, 109–14 (1972).) [408, 438]

BASOV, N. G. *et al.* (1972*d*) (with ZARITSKII, A. R., ZAKHAROV, S. D., KROKHIN, O. N., KRYUKOV, P. G., MATVEETS, YU. A., SENATSKII, YU. V. & FEDOSIMOV, A. I.). Generation of high-power light pulses at wave-lengths 1·06 and $0·53\mu$ and their application in plasma heating. I. Experimental investigations of reflection of light of two wavelengths in laser heating of plasma *Kvantovaya Electron.*, **5** (11), 63–71 (1972). (Transl: *Sov. J. Quantum Electron.*, **2**, 439–44 (1973).) [388]

BASOV, N. G. *et al.* (1973) (with BOIKO, V. A., ZAKHAROV, S. M., KROKHIN, O. N., SKLIZKOV, G. A. & FAENOV, A. YA.). Spectrum of continuous X-ray emission from a laser plasma and deviations of the electron distribution function from Maxwellian form, *Kvantovaya Elektron.*, **5** (17), 126–8 (1973). (Transl: *Sov. J. Quantum Electron.*, **3**, 444–5 (1974).) [389]

BATANOV, V. A. *et al.* (1972) (with BUNKIN, F. V., PROKHOROV, A. M. & FEDOROV, V. B.). Self focusing of light in a plasma and supersonic ionization wave in a laser beam, *Zh. Eksp. i Teor. Fiz. Pis'ma*, **16**, 378–82 (1972). (Transl: *JETP Lett.*, **16**, 266–9 (1972).) [440]

BATENIN, V. M. & CHINNOV, V. F. (1971). Electron bremsstrahlung in the field of argon or helium atoms, *Zh. Eksp. i Teor. Fiz.*, **61**, 56–63 (1971). (Transl: *Sov. Phys. JETP*, **34**, 30–3 (1972).) [53]

BATES, D. R. *et al.* (1962*a*) (with KINGSTON, A. E. & MCWHIRTER, R. W. P.). Recombination between electrons and atomic ions I. Optically thin plasmas, *Proc. Roy. Soc.*, **A267**, 297–312 (1962). [38]

BATES, D. R. *et al.* (1962*b*) (with KINGSTON, A. E. & MCWHIRTER, R. W. P.). Recombination between electrons and atomic ions II. Optically thick plasmas, *Proc. Roy. Soc.*, **A270**, 155–67 (1962). [38]

BEACH, A. D. (1967). A 12-channel Doppler-profile spectrophotometer for plasma-scattered laser light, *J. Sci. Instrum.*, **44**, 690–2 (1967). [116]

BEACH, A. D. *et al.* (1969) (with BODIN, H. A. B., BUNTING, D. A., DANCY, D. J., HEYWOOD, G. C. H., KENWARD, M. R., MCCARTAN, J., NEWTON, A. A., PASCO, I. K., PEACOCK, R. & WATSON, J. L.). Temperature and density measurements in the midplane of a long theta pinch, *Nucl. Fusion*, **9**, 215–22 (1969). [115]

BEAUDRY, G. & MARTINEAU, J. (1972). L'effet de la dynamique d'une impulsion-laser sur le chauffage de plasmas confinés, *Can. J. Phys.*, **50**, 594–9 (1972). [335]

BEAUDRY, G. & MARTINEAU, J. (1973). Plasma heating by beating of two laser beams, *Phys. Lett.*, **43A**, 331–2 (1973). [70]

BEAUDRY, G. et al. (1971) (with MARTINEAU, J. & PÉPIN, H.). Influence of laser pulse dynamics on the heating of a confined plasma, X International Conference on phenomena in ionized gases, Oxford (1971). Contributed papers, p. 226. Oxford, Donald Parsons & Co. Ltd. (1971). [73]

BEBB, H. B. (1967). Theory of three-photon ionization of the alkali atoms, Phys. Rev., 153, 23–8 (1967). [152]

BEBB, H. B. & GOLD, A., (1966a). Multiphoton ionization of hydrogen and rare-gas atoms, Phys. Rev., 143, 1–24 (1966). [152, 153, 154]

BEBB, H. B. & Gold, A. (1966b). Multiphoton ionization of rare gas and hydrogen atoms. In Physics of Quantum Electronics (ed. P. L. Kelley, B. Lax and P. E. Tannenwald), pp. 489–98. New York, McGraw-Hill (1966). [152]

BEKEFI, G. (1966). Radiation processes in plasmas. New York, J. Wiley (1966). [1, 9, 19, 29, 47, 88]

BELLAND, P. et al. (1971a) (with DE MICHELIS, C. & MATTIOLI, M.). Holographic interferometry of laser produced plasmas using picosecond pulses, Opt. Commun., 3, 7–8 (1971). [391]

BELLAND, P. et al. (1971b) (with DE MICHELIS, C. & MATTIOLI, M.). Self-focusing in laser induced gas breakdown, Opt. Commun., 4, 50–3 (1971). [178, 198]

BELLAND, P. et al. (1971c) (with DE MICHELIS, C., MATTIOLI, M. & PAPOULAR, R.). Spectral analysis of the backscattered light from a laser produced plasma, Appl. Phys. Lett., 18, 542–4 (1971). [387]

BELLAND, P. et al. (1972) (with DE MICHELIS, C., MATTIOLI, M. & PAPOULAR, R.). Spectral analysis of backscattered light from laser-produced plasma: reexamined, Appl. Phys. Lett., 21, 32–3 (1972). [387]

BEREZHETSKAYA, N. K. et al. (1970) (with VORONOV, G. S., DELONE, G. A., DELONE, N. B. & PISKOVA, G. K.). Effect of a strong optical-frequency electromagnetic field on the hydrogen molecule, Zh. Eksp. i Teor. Fiz., 58, 753–9 (1970). (Transl: Sov. Phys. JETP, 31, 403–6 (1970).) [160]

BERGER, P. J. & SMITH, D. C. (1972). Gas breakdown in the laser as the limitation of pulsed high-pressure CO_2 lasers, Appl. Phys. Lett., 21, 167–70 (1972). [193]

BERGQVIST, T. & KLEMAN, B. (1966). Breakdown in gases by 10 600 Å laser radiation, Ark. Fys., 31, 178–88 (1966). [177]

BERNAL, G. E. et al. (1966) (with READY, J. F. & LEVINE, L. P.). Ion emission from laser irradiated tungsten, IEEE J. Quantum Electron., 2, 480–2 (1966). [341]

BERNEY, A. (1973). Scattering of laser light from a pulsed arc plasma, Plasma Phys., 15, 699–704 (1973). [117]

BERNSTEIN, I. B. & FADER, W. J. (1968). Expansion of a resistive spherical plasma in a magnetic field, Phys. Fluids, 11, 2209–17 (1968). [332]

BERNSTEIN, I. B. & TREHAN, S. K. (1960). Plasma oscillations (I). Nuclear Fusion, 1, 3–41 (1960). [23]

BERNSTEIN, I. B. et al. (1964) (with TREHAN, S. K. & WEENINK, M. P. H.). Plasma oscillations: II. Kinetic theory of waves in plasmas, Nucl. Fusion, 4, 61–104 (1964). [23, 87, 88]

BERRY, M. et al. (1964) (with DURAND, Y., NELSON, P. & VEYRIE, P.). Étude expérimentale et théorique du claquage de l'air sous l'action d'un faisceau laser, C.R. Acad. Sci., Paris, 259, 2401–3 (1964). [234]

BESPALOV, V. I. & TALANOV, V. I. (1966). Filamentary structure of light beams in nonlinear liquids, Zh. Eksp. i Teor. Fiz. Pis'ma, 3, 471–6 (1966). (Transl: JETP Lett., 3, 307–10 (1966).) [196]

BETHE, H. A. & SALPETER, E. E. (1957). Quantum mechanics of one- and two-electron systems. In Handbuch der Phys. (ed. S. Flugge), Vol. 35, p. 324. Berlin, Springer (1975). [147]

BEZZERIDES, B. & WEINSTOCK, J. (1972). Nonlinear saturation of parametric instabilities, Phys. Rev. Lett., 28, 481–4 (1972). [65]

BHADRA, D. K. (1968). Expansions of a resistive plasmoid in a magnetic field, Phys. Fluids, 11, 234–9 (1968). [331, 333, 429]

BHADRA, D. K. (1972). Dynamics of a resistive plasmoid in a magnetic field. In Laser interaction and related plasma phenomena (ed. H. J. Schwarz and H. Hora), Vol. 2, pp. 291–300. New York, Plenum (1972). [334]

BIBERMAN, L. M. & NORMAN, G. E. (1967). Continuous spectra of atomic gases and plasma, *Usp. Fiz. Nauk.*, **91**, 193–246 (1967). (Transl: *Sov. Phys. Usp.*, **10**, 52–90 (1967).) [27, 43, 48]

BICKERTON, R. J. (1973). Thermal conduction limitations in laser fusion, *Nucl. Fusion*, **13**, 457–8 (1973). [306]

BILLMAN, K. W. *et al.* (1972) (with STALLCOP, J. R., ROWLEY, P. D. & PRESLEY, L. L.). Comparisons of measured and theoretical inverse bremsstrahlung and photoionization absorption of infrared radiation in a H–He plasma, *Phys. Rev. Lett.*, **28**, 1435–8 (1972). [53, 54]

BIRD, R. S. *et al.* (1973) (with MCKEE, L. L., SCHWIRZKE, F. & COOPER, A. W.). Pressure dependence on self-generated magnetic fields in laser-produced plasmas, *Phys. Rev.*, **A7**, 1328–31 (1973). [442, 443]

BIRMINGHAM, T. *et al.* (1965) (with DAWSON, J. & OBERMAN, C.). Radiation processes in plasmas, *Phys. Fluids*, **8**, 297–307 (1965). [49]

BIZE, D. *et al.* (1967) (with CONSOLI, T., SLAMA, L., STEVENIN, P. & ZYMANSKI, M.). Mesure de la densité d'un plasma en évolution obtenu par ionisation laser a l'aide d'l'interféromètre Fabry-Pérot, *C.R. Acad. Sci., Paris*, **264B**, 1235–8 (1967). [232, 241]

BOBIN, J. L. (1971). Flame propagation and overdense heating in a laser created plasma, *Phys. Fluids*, **14**, 2341–54 (1971). [294, 299]

BOBIN, J. L. (1972). Laser created plasmas and controlled thermonuclear fusion, *V European Conference on Controlled Fusion and Plasma Physics, Grenoble (1972)*, Vol. 2, p. 171. Grenoble, Ass. EURATOM-CEA (1972). [300]

BOBIN, J. L. *et al.* (1967) (with BUGES, J. C., LANGER, P. & TONON, G.). Profil de densité électronique dans un plasma de béryllium créé par le faisceau laser, *C.R. Acad. Sci., Paris*, **265B**, 1400–3 (1967). [349, 350]

BOBIN, J. L. *et al.* (1968a) (with CANTO, C., FLOUX, F., GUYOT, D., REUSS, J. & VEYRIE, P.). Gas breakdown with nanosecond pulses, *IEEE J. Quantum Electron*, **11**, 923–31 (1968). [262, 263, 437]

BOBIN, J. L. *et al.* (1968b) (with FLOUX, F., LANGER, P. & PIGNEROL, H.). X-rays from a laser-created deuterium plasma, *Phys. Lett.*, **28A**, 398–9 (1968). [420]

BOBIN, J. L. *et al.* (1969) (with DELOBEAU, F., DE GIOVANNI, G., FAUQUIGNON, C. & FLOUX, F.). Temperature in laser-created deuterium plasmas, *Nucl. Fusion*, **9**, 115–20 (1969). [412, 413, 420, 421]

BOBIN, J. L. *et al.* (1971) (with FLOUX, F. & TONIN, G.). Application à la fusion controlée des plasmas denses créés par laser, *IV Conference on plasma physics and controlled nuclear fusion research, Madison (1971)*, pp. 657–72. Vienna, IAEA (1971). [355]

BOBIN, J. L. *et al.* (1972) (with COLOMBANT, D. & TONON, G.). Fusion by laser-driven flame propagation in solid DT-targets, *Nucl. Fusion*, **12**, 445–52 (1972). [300]

BOBIN, J. L. *et al.* (1973) (with DECROISETTE, M., MEYER, B. & VITEL, Y.). Harmonic generation and parametric excitation of waves in a laser-created plasma, *Phys. Rev. Lett.*, **30**, 594–7 (1973). [77, 78]

BOCKASTEN, K. (1961). Transformation of observed radiances into radial distribution of the emission of a plasma, *J. Opt. Soc. Am.*, **51**, 943–7 (1961). [229]

BODNER, S. E. *et al.* (1973) (with CHAPLINE, G. F. & DE GROOT, J.). Anomalous ion heating in a laser heated plasma, *Plasma Phys.*, **15**, 21–7 (1973). [68, 69]

BOGEN, P. & RUSBÜLDT, D. (1968). Bremsstrahlung of a fully ionized plasma in the infrared, *Phys. Fluids*, **11**, 2022–4 (1968). [52]

BOHM, D. & GROSS, E. P. (1949). Theory of plasma oscillations, *Phys. Rev.*, **75**, 1851–64 and 1864–76 (1949). [20]

BOLAND, B. C. & IRONS, F. E. (1968). Diagnostic measurements on a laser generated plasma, *VIII International Conference on phenomena in ionized gases, Vienna (1967). Contributed papers*, p. 452. Vienna, IAEA (1968). [377]

BOLAND, B. C. *et al.* (1968) (with IRONS, F. E. & MCWHIRTER, R. W. P.). A spectroscopic study of the plasma generated by a laser from polyethylene, *J. Phys. B.*, **1**, 1180–91 (1968). [318, 365, 377, 378, 379, 380]

BONCH-BRUEVICH, A. M. *et al.* (1968). (with IMAS, YA. A., ROMANOV, G. S., LIBENSON, M. N. & MAL'TSEV, L. N.). Effect of a laser pulse on the reflecting power of a metal, *Zh. Tekh. Fiz.*, **38**, 851–5 (1968). (Transl: *Sov. Phys. Tech. Phys.*, **13**, 640–3 (1968).) [276]

BONNIER, A. & MARTINEAU, J. (1972). Temperature laws for laser created plasmas from thin targets, *Phys. Lett.*, **38A**, 199–200 (1972). [315]

BONNIER, A. & MARTINEAU, J. (1973). Temperature laws for laser-thin-target interaction, *J. Appl. Phys.*, **44**, 3626–30 (1973). [315, 396]

BONNIER, A. *et al.* (1971) (with MARTINEAU, J. & PÉPIN, H.). Hydrodynamic behaviour of in-homogeneous deuterium plasma under heating by a CO_2 laser, *X International Conference on phenomena in ionized gases, Oxford (1971). Contributed papers*, p. 223. Oxford, Donald Parsons & Co. Ltd. (1971). [315]

BOOK, D. L. & CLARK, R. W. (1973). Comments on 'Early-time model of laser plasma expansion', *Phys. Fluids*, **16**, 341–2 (1973). [334]

BORDÉ, M. C. *et al.* (1966a) (with HENRY, A. & HENRY, M. L.). Emission du gaz ammoniac excité par le rayonnement d'un laser à gaz carbonique, *C.R. Acad. Sci., Paris*, **262B**, 1389–90 (1966). [163]

BORDÉ, M. C. *et al.* (1966b) (with HENRY, A. & HENRY, M. L.). Comportement de differents gaz soumis au rayonnement d'un laser à gaz carbonique, *C.R. Acad. Sci., Paris*, **263B**, 619–20 [163]

BORNATICI, M. *et al.* (1969) (with CAVALIERE, A. & ENGELMANN, F.). Enhanced scattering and anomalous absorption of light due to decay into plasma waves, *Phys. Fluids*, **12**, 2362–73 (1969). [70]

BOTTOMS, P. J. & EISNER, M. (1966). Thermalization of plasma electrons, *Phys. Rev. Lett.*, **17**, 902–3 (1966). [111, 115]

BOWLES, K. L. (1958). Observation of vertical-incidence scatter from the ionosphere at 41 Mc/s, *Phys. Rev. Lett.*, **1**, 454–5 (1958). [82]

BOYD, T. J. M. & SANDERSON, J. J. (1969). *Plasma dynamics*. London, Nelson (1969). [21]

BOYD, T. J. M. & TURNER, J. G. (1972). Lagrangian studies of plasma wave interactions, *J. Phys. A.*, **5**, 881–96 (1972). [70]

BOYD, T. J. M. *et al.* (1971) (with EVANS, D. E. & GARDNER, G. A.). Numerical calculation of the frequency spectrum of light scattered by a magnetised plasma, *X International Conference on phenomena in ionized gases, Oxford (1971). Contributed papers*, p. 412. Oxford, Donald Parsons & Co. Ltd. (1971). [98]

BRADLEY, D. J. & SIBBETT, W. (1973). Streak-camera studies of picosecond pulses from a mode-locked Nd: glass laser, *Opt. Commun.*, **9**, 17–20 (1973). [231]

BRADLEY, D. J. *et al.* (1971) (with LIDDY, B. & SLEAT, W. E.). Direct linear measurement of ultra short light pulses with a picosecond streak camera, *Opt. Commun.*, **2**, 391–5 (1971). [223]

BRAERMAN, W. F. *et al.* (1969) (with STUMPFEL, C. R. & KUNZE, H. J.). Spectroscopic studies of a laser-produced plasma in helium, *J. Appl. Phys.*, **40**, 2549–54 (1969). [257, 258]

BREENE, R. G. & NARDONE, M. C. (1960). Free–free continuum of oxygen, *J. Opt. Soc. Am.*, **50**, 1111–14 (1960). [48]

BREENE, R. G. & NARDONE, M. C. (1961). Free–free continuum of nitrogen, *J. Opt. Soc. Am.*, **51**, 692 (1961). [48]

BREENE, R. G. & NARDONE, M. C. (1963). Free–free continuum of oxygen. II. Effect of polarization and exchange, *J. Opt. Soc. Am.*, **53**, 924–8 (1963). [48]

BREHME, H. (1971). Laser-induced multiphoton processes in e^-–p scattering, *Phys. Rev.*, **C3**, 837–40 (1971). [47]

BRETON, C. *et al.* (1965) (with CAPET, M., CHALMETON, V., NGUYEN QUANG, D. & PAPOULAR, R.). Observations des effets de la focalisation d'un faisceau lumineux intense dans un gaz, *J. Physique.* **26**, 490–3 (1965). [231, 250]

BRETON, C. *et al.* (1968) (with CAPET, M. & COURBIN, C.). Étude par effet Doppler de l'expansion d'un microplasma, *C.R. Acad. Sci., Paris*, **266B**, 52–5 (1968). [403]

BREWER, R. G. & LEE, C. H. (1968). Self-trapping with picosecond light pulses, *Phys. Rev. Lett.*, **21**, 267–70 (1968). [196]

BREWER, R. G. *et al.* (1968) (with LIFSITZ, J. R., GARMIRE, E., CHIAO, R. Y. & TOWNES, C. H.). Small-scale trapped filaments in intense laser beams, *Phys. Rev.*, **166**, 326–31 (1968). [198]

BRIGGS, R. J. (1964). *Electron-stream interaction with plasmas*, Research Monograph No. 29. Cambridge, Mass., M.I.T. Press (1964). [18, 63]

BRODSKII, A. M. & GUREVICH, YU. YA. (1971). Theory of electron photoproduction in a strong electromagnetic field with inclusion of the final-state interaction, *Zh. Eksp. i Teor. Fiz.*, **60**, 1452–64 (1971). (Transl: *Sov. Phys. JETP*, **33**, 786–92 (1971).) [150]

BROWN, S. C. (1967). *Basic Data of Plasma Physics*, 2nd Edition. Cambridge, Mass., M.I.T. Press (1967). [48]

BROWN, R. T. & SMITH, D. C. (1973). Laser-induced gas breakdown in the presence of pre-ionization, *Appl. Phys. Lett.*, **22**, 245–7 (1973). [189]

BROWNE, P. F. (1965). Mechanism of gas breakdown by lasers, *Proc. Phys. Soc.*, **86**, 1323–32 (1965). [164]

BRUCE, C. W. *et al.* (1966) (with DEACON, J. & VONDERHAAR, D. F.). Time and spatially resolved interferometry on pulsed-laser-induced plasmas, *Appl. Phys. Lett.*, **9**, 164–6 (1966). [363]

BRUECKNER, K. A. (1973). Laser driven fusion, *IEEE Trans. Plasma Sci.*, **1**, 13–26 (1973). [319, 320, 321]

BRUNETEAU, J. & FABRE, E. (1972). Photoionisation d'un gaz par le rayonnement ultra violet extrême d'un plasma créé par laser, *Phys. Lett.*, **39A**, 411–12 (1972). [439]

BRUNETEAU, J. *et al.* (1970) (with FABRE, E., LAMAIN, H. & VASSEUR, P.). Experimental investigation of the production and containment of a laser produced plasma, *Phys. Fluids*, **13**, 1795–1801, (1970). [398, 432]

BRUNETEAU, J. *et al.* (1971) (with COLBURN, S., FABRE, E., POQUERUSSE, A. & STENZ, C.). Interaction avec un champ magnetique d'un plasma créé par irradiation laser de solide, *J. Physique*, **32**, Suppl. **C5B**, 136–8 (1971). [433]

BRUSSAARD, P. J. & VAN DER HULST, H. O. (1962). Approximation formulas for nonrelativistic bremsstrahlung and average Gaunt factors for a Maxwellian electron gas, *Rev. Mod. Phys.*, **34**, 507–19 (1962). [40, 43]

BÜCHL, K. (1971). Production of plasma by a CO_2 TEA laser from solid hydrogen targets, *X International Conference on phenomena in ionized gases, Oxford (1971). Contributed papers*, p. 224. Oxford, Donald Parsons & Co. Ltd. (1971). [427]

BÜCHL, K. (1972). Production of plasma with a CO_2 TEA laser from solid hydrogen targets, *J. Appl. Phys.*, **43**, 1032–7 (1972). [427]

BÜCHL, K. *et al.* (1968) (with HOHLA, K., WIENECKE, R. & WITKOWSKI, S.). Investigation of the blast wave from a laser produced gas breakdown, *Phys. Lett.*, **26A**, 248–9 (1968). [225, 255, 256]

BÜCHL, K. *et al.* (1971) (with EIDMANN, K., MULSER, P., SALZMANN, H., SIGEL, R. & WITKOWSKI, S.). Investigation of laser-produced plasmas in the keV temperature range, *Proc. 4th Conference on Plasma Physics and Controlled Nuclear Fusion Research, Madison (1971)*, Vol. I, pp. 645–55. Vienna, IAEA (1971). [77, 290, 426, 427]

BÜCHL, K. *et al.* (1972*a*) (with EIDMANN, K., MULSER, P., SALZMANN, H. & SIGEL, R.). Plasma production with a Nd. laser and non-thermal effects. In *Laser interaction and related plasma phenomena* (ed. H. J. Schwarz and H. Hora), Vol. 2, pp. 503–14. New York, Plenum (1972). [427]

BÜCHL, K. *et al.* (1972*b*) (with EIDMANN, K., SALZMANN, H. & SIGEL, R.). Spectral investigation of light reflected from a laser-produced deuterium plasma, *Appl. Phys. Lett.*, **20**, 3–4 (1972). [427]

BUGES, J. C. & TERNEAUD, A. (1970). Holographic cameras in the nanosecond range, *IEEE J. Quantum Electron*, **6**, 5–6 (1970). [230]

BUNKIN, F. V. (1969). Self-reflection of ultrashort light pulses from condensed media, *Zh. Eksp. i. Teor. Fiz. Pis'ma*, **10**, 561–4 (1969). (Transl: *JETP Lett.*, **10**, 358–60 (1969).) [300]

BUNKIN, F. V. & FEDOROV, M. V. (1965). Bremsstrahlung in a strong radiation field, *Zh. Fksp. i Teor. Fiz.*, **49**, 1215–21 (1965). (Transl: *Sov. Phys. JETP*, **22**, 844–7 (1966).) [47, 152]

BUNKIN, F. V. & KAZAKOV, A. E. (1970a). Electron-heating and incoherent hard radiation produced by interaction of ultrashort intense laser pulses with matter, *Zh. Eksp. i Teor. Fiz.*,
59, 2233–43 (1970). (Transl: *Sov. Phys. JETP*, **32**, 1208–13 (1971).) [73, 75, 300]

BUNKIN, F. V. & KAZAKOV, A. E. (1970b). Compton mechanism of electron gas heating by a laser,
Dokl. Akad. Nauk SSSR, **192**, 71–3 (1970). (Transl: *Sov. Phys. Dokl.*, **15**, 468–70 (1970).)
[75]

BUNKIN, F. V. & PROKHOROV, A. M. (1964). The excitation and ionization of atoms in a strong
radiation field, *Zh. Eksp. i Teor. Fiz.*, **46**, 1090–7 (1964). (Transl: *Sov. Phys. JETP*, **19**,
739–43 (1964).) [152]

BUNKIN, F. V. & PROKHOROV, A. M. (1967). Some features of the interaction between short
laser radiation pulses and matter, *Zh. Eksp. i Teor. Fiz.*, **52**, 1610–15 (1967). (Transl: *Sov.
Phys. JETP*, **25**, 1072–5 (1967).) [169]

BUNKIN, F. V. et al. (1964) (with KARAPETYAN, R. V. & PROKHOROV, A. M.). Dissociation of
molecules in a strong radiation field, *Zh. Eksp. i Teor. Fiz.*, **47**, 216–20 (1964). (Transl:
Sov. Phys. JETP, **20**, 145–8 (1965).) [154]

BUNKIN, F. V. et al. (1971) (with KRASYUK, I. K., MARCHENKO, V. M., PASHININ, P. P. & PROK-
HOROV, A. M.). Structure of a spark produced by a picosecond laser pulse focused in a gas,
Zh. Eksp. i Teor. Fiz., **60**, 1326–31 (1971). (Transl: *Sov. Phys. JETP*, **33**, 717–20 (1971).)
[199]

BUNKIN, F. V. et al. (1972) (with PASHININ, P. P. & PROKHOROV, A. M.). Concerning one possible
use of the IR lasers for high-temperature heating of a superdense plasma, *Zh. Eksp. i Teor.
Fiz. Pis'ma*, **15**, 556–9 (1972). (Transl: *JETP Lett.*, **15**, 394–7 (1972).) [18]

BURGESS, D. D. et al. (1967) (with FAWCETT, B. C. & PEACOCK, N. J.). Vacuum ultra-violet
emission spectra from laser-produced plasmas, *Proc. Phys. Soc.*, **92**, 805–16 (1967). [376,
377]

BURGESS, D. D. et al. (1968) (with FAWCETT, B. C., LONG, J. W. & PEACOCK, N. J.). The extreme
vacuum ultraviolet emission from laser produced plasma sources, *VIII International
Conference on phenomena in ionized gases, Vienna (1967). Contributed papers*, p. 449. Vienna,
IAEA (1968). [376]

BURNETT, N. H. (1972). Absorption of light at oblique incidence on a plasma layer, *Can. J. Phys.*,
50, 3184–92 (1972). [75]

BURNETT, N. H. & SMY, P. R. (1970). Temperature decay of a laser-produced aluminium plasma,
Can. J. Phys., **48**, 1421–5 (1970). [365]

BUSCHER, H. T. et al. (1965) (with TOMLINSON, R. G. & DAMON, E. K.). Frequency dependence
of optically induced gas breakdown, *Phys. Rev. Lett.*, **15**, 847–9 (1965). [179, 182, 183,
184]

BUTLER, S. T. & BUCKINGHAM, M. J. (1962). Energy loss of a fast ion in a plasma, *Phys. Rev.*,
126, 1–4 (1962). [135]

BYKOVSKII, YU. A. et al. (1968) (with DOROFEEV, V. I., DYMOVICH, V. I., NIKOLAEV, B. I., RYZHIKH,
S. V. & SIL'NOV, S. M.). Mass-spectrometer investigation of ions formed by interaction be-
tween laser radiation and matter, *Zh. Tekh. Fiz.*, **38**, 1194–6 (1968). (Transl: *Sov. Phys.
Tech. Phys.*, **13**, 986–8 (1968).) [356]

BYKOVSKII, YU. A. et al. (1969a) (with DEGTYARENKO, N. N., DYMOVICH, V. I., ELESIN, V. F.,
KOZYREV, YU. P., NIKOLAEV, B. I., RYZHIKH, S. V. & SIL'NOV, S. M.). Energy distribution
of the ions produced by a large laser pulse on a solid target, *Zh. Tekh. Fiz.*, **39**, 1694–6 (1969).
(Transl: *Sov. Phys. Tech. Phys.*, **14**, 1269–71 (1970).) [356]

BYKOVSKII, YU. A. et al. (1969b) (with DUDOLADOV, A. G., DEGTYARENKO, N. N., ELESIN, V. F.,
KOZYREV, YU. P. & NIKOLAEV, I. N.). Angular distribution of material vaporized by a
laser beam, *Zh. Eksp. i Teor. Fiz.*, **56**, 1819–22 (1969). (Transl: *Sov. Phys. JETP*, **29**, 977–8
(1969).) [348, 350]

BYKOVSKII, YU. A. et al. (1971a) (with DEGTYARENKO, N. N., ELESIN, V. F., KOZYREV, YU. P. &
SIL'NOV, S. M.). Mass spectrometer study of laser plasma, *Zh. Eksp. i Teor. Fiz.*, **60**, 1306–19
(1971). (Transl: *Sov. Phys. JETP*, **33**, 706–12 (1971).) [357, 388, 389]

BYKOVSKII, YU. A. *et al.* (1971*b*) (with GRYUKHANOV, M. F., DEGTYAREV, V. G., DEGTYARENKO, N. N., ELESIN, V. F., LAPTEV, I. D. & NEVOLIN, V. N.). Angular distribution of the ions of a laser plasma, *Zh. Eksp. i Teor. Fiz. Pis'ma*, **14**, 238–42 (1971). (Transl: *JETP Lett.*, **14**, 157–9 (1971).) [356, 357]

BYKOVSKII, YU. A. *et al.* (1972*a*) (with DEGTYAREV, V. G., DEGTYARENKO, N. N., ELESIN, V. F., LAPTEV, I. D. & NEVOLIN, V. N.). Ion energies in a laser-produced plasma, *Zh. Tekh. Fiz.*, **42**, 658–61 (1972). (Transl: *Sov. Phys. Tech. Phys.*, **17**, 517–20 (1972).) [356, 357]

BYKOVSKII, YU. A. *et al.* (1972*b*) (with VASIL'EV, N. N., DEGTYARENKO, N. N., ELESIN, V. F., LAPTEV, I. D. & NEVOLIN, V. N.). Formation of the energy spectrum of the ions of a laser plasma, *Zh. Eksp. i Teor. Fiz. Pis'ma*, **15**, 308–11 (1972). (Transl: *JETP Lett.*, **15**, 217–20 (1972).) [356]

BYSTROVA, T. B. *et al.* (1967) (with VORONOV, G. S., DELONE, G. A. & DELONE, N. B.). Multiphoton ionization of xenon and krypton atoms at wavelength $\lambda = 1.06\mu$, *Zh. Eksp. i Teor. Fiz. Pis'ma*, **5**, 223–5 (1967). (Transl: *JETP Lett.*, **5**, 178–9 (1967).) [157]

CALLEN, H. B. & WELTON, T. A. (1951). Irreversibility and generalized noise, *Phys. Rev.*, **83**, 34–40 (1951). [49]

CANAVAN, G. H. & NIELSEN, P. E. (1973). Focal spot size dependence of gas breakdown induced by particulate ionization, *Appl. Phys. Lett.*, **22**, 409–10 (1973). [191]

CANTO, C. *et al.* (1968) (with REUSS, J. D. & VEYRIE, P.). Étude théorique de l'ionisation du deutérium sous l'action d'un laser à impulsion courte, *C.R. Acad. Sci., Paris*, **267B**, 878–81 (1968) [203]

CARILLON, A. *et al.* (1970) (with JAEGLE, P. & DHEZ, P.). Extreme-ultraviolet continuum absorption by a laser-generated aluminium plasma, *Phys. Rev. Lett.*, **25**, 140–3 (1970). [343, 344, 345]

CARION, A. *et al.* (1973) (with LANCELOT, J., DE METZ, J. & SALÈRES, A.). Fusion reactions in a plasma created by the second harmonic of a neodymium glass laser, *Phys. Lett.*, **45A**, 439–40 (1973). [423]

CAROLAN, P. G. & EVANS, D. E. (1971*a*). Comparison between calculated and observed spectra of laser light scattered by a magnetized plasma, *X International Conference on phenomena in ionized gases, Oxford (1971). Contributed papers*, p. 413. Oxford, Donald Parsons & Co. Ltd. (1971). [97,129, 130]

CAROLAN, P. G. & EVANS, D. E. (1971*b*). Influence of symmetry about the magnetic vector on the spectrum of light scattered by a magnetized plasma, *Plasma Phys.*, **13**, 947–53 (1971). [97]

CARUSO, A. & GRATTON, R. (1968*a*). Some properties of the plasmas produced by irradiating light solids by laser pulses, *Plasma Phys.*, **10**, 867–77 (1968). [287, 293, 307].

CARUSO, A. & GRATTON, R. (1968*b*). On the ionization of thin films by giant laser pulses, *Phys. Lett.*, **27A**, 48–9 (1968). [397, 398]

CARUSO, A. & GRATTON, R. (1969). Interaction of short laser pulses with solid materials, *Plasma Phys.*, **11**, 839–47 (1969). [301, 302]

CARUSO, A. *et al.* (1966) (with BERTOTTI, B. & GIUPPONI, P.). Ionization and heating of solid material by means of a laser pulse, *Nuovo Cimento*, **45B**, 176–88 (1966). [287]

CARUSO, A. *et al.* (1969) (with DE ANGELIS, A., GATTI, G., GRATTON, R. & MARTELLUCCI, S.). Deuterium plasma produced by subnanosecond laser pulses, *Phys. Lett.*, **29A**, 316–17 (1969). [425]

CARUSO, A. *et al.* (1970*a*) (with DE ANGELIS, A., GATTI, G., GRATTON, R. & MARTELLUCCI, S.). Light scattered in the interaction between subnanosecond laser pulses and solid targets, *Phys. Letts.*, **33A**, 320–1 (1970). [77]

CARUSO, A. *et al.* (1970*b*) (with DE ANGELIS, A., GATTI, G., GRATTON, R. & MARTELLUCCI, S.). Second-harmonic generation in laser produced plasmas, *Phys. Lett.*, **33A**, 29–30 (1970). [77]

CARUSO, A. *et al.* (1970*c*) (with DE ANGELIS, A., GATTI, G., GRATTON, R. & MARTELLUCCI, S.). Energetic ions produced by subnanosecond laser pulses, *Phys. Lett.*, **33A**, 336–7 (1970). [392]

CARUSO, A. *et al.* (1971) (with DE ANGELIS, A., GATTI, G., GRATTON, R. & MARTELLUCCI, S.). Change in the scaling laws for laser produced plasmas through the critical regime, *Phys. Lett.*, **35A**, 279–80 (1971). [418]

CARUSOTTO, S. *et al.* (1968) (with FORNACA, G. & POLACCO, E.). Multiphoton absorption and coherence, *Phys. Rev.*, **165**, 1391–8 (1968). [154]

CAVALIERE, A. *et al.* (1967) (with GIUPPONI, P. & GRATTON, R.). Plasma irradiated in a magnetic field, *Phys. Lett.*, **25A**, 636–7 (1967). [334]

CECCHINI, A. *et al.* (1968) (with DE ANGELIS, A., GRATTON, R. & PARLANGE, F.). Suitable targets for the production of a deuterium plasma cloud in a vacuum environment with a laser pulse, *J. Phys. E.*, **1**, 1040–2 (1968). [416]

CHALMETON, V. (1968). Décharges laser dans les gaz, *J. Physique*, **29**, *Suppl.* C3, 100–2 (1968). [175]

CHALMETON, V. (1969*a*). Étude de la phase initiale du claquage d'un gaz par un laser, *J. Physique*, **30**, 687–99 (1969). [163]

CHALMETON, V. (1969*b*). Detection d'électrons libres dans un gaz éclairé par un laser en dessous du seuil de claquage, *C.R. Acad. Sci., Paris*, **269B**, 197–200 (1969). [163]

CHALMETON, V. & PAPOULAR, R. (1966). Absorption non linéaire de la lumière par les gaz neutres, *C.R. Acad. Sci., Paris*, **262B**, 117–19 (1966). [162]

CHALMETON, V. & PAPOULAR, R. (1967). Émission de lumière par un gaz sous l'effet d'un rayonnement laser intense, *C.R. Acad. Sci., Paris*, **264B**, 213–16 (1967). [162, 163]

CHAMPETIER, J. L. (1965). Interprétation théorique de l'évolution du plasma créé par focalisation d'un faisceau laser dans l'air, *C.R. Acad. Sci., Paris*, **261**, 3954–7 (1965). [203]

CHAMPETIER, J. L. *et al.* (1968) (with MARIOGE, J. P., DE METZ, J., MILLET, F. & TERNEAUD, A.). Utilisation de lentilles asphériques pour l'obtention d'éclairments élevés, *C.R. Acad. Sci., Paris*, **266B**, 838–41 (1968). [421]

CHAN, C. H. *et al.* (1973) (with MOODY, C. D. & McKNIGHT, W. B.). Significant loss mechanisms in gas breakdown at 10·6μ, *J. Appl. Phys.*, **44**, 1179–88 (1973). [167, 179, 189, 190, 191]

CHAN, P. W. (1971*a*). Evidence of density inhomogeneity broadening of plasma satellites from light scattering, *Phys. Fluids*, **14**, 2787–8 (1971). [118]

CHAN, P. W. (1971*b*). Magnetic probe signals of a laser induced spark in a DC magnetic field, *Phys. Lett.*, **35A**, 3–4 (1971). [271]

CHAN, P. W. & NODWELL, R. A. (1966). Collective scattering of laser light by a plasma, *Phys. Rev. Lett.*, **16**, 122–4 (1966). [118]

CHAN, P. W. *et al.* (1968) (with DE MICHELIS, C. & KRONAST, B.). Laser-produced sparks in a 200 kG magnetic field, *Appl. Phys. Lett.*, **13**, 202–3 (1968). [194, 195, 271]

CHANDRASEKHAR, S. (1961). *Hydrodynamic and hydromagnetic stability*. Oxford, Clarendon (1961). [328]

CHANG, C. S. & STEHLE, P. (1973). Theory of resonant multiphoton ionization, *Phys. Rev. Lett.*, **30**, 1283–5 (1973). [152, 160]

CHANG, C. T. *et al.* (1973) (with CHUU, D. S., LEE, P. S. & LEE Y. C.). Collisional effect on the saturation amplitude of nonlinearly excited plasma waves, *Phys. Lett.*, **42A**, 479–80 (1973). [70]

CHANG, R. P. H. *et al.* (1972) (with PORKOLAB, M. & GREK, B.). Parametric instability of plasma waves in a magnetic field due to high-frequency electric fields, *Phys. Rev. Lett.*, **28**, 206–9 (1972). [76]

CHAPELLE, J. *et al.* (1973) (with DUBREUIL, B., DRAWIN, H. W. & EMARD, F.). Lowering of excited state populations of HeI in a He-plasma under the influence of laser radiation, *Phys. Lett.*, **44A**, 201–2 (1973). [269]

CHARATIS, G. *et al.* (1974) (with DOWNWARD, J., GOFORTH, R., GUSCOTT, B., HENDERSON, T., HILDUM, S., JOHNSON, R., MONCUR, K., LEONARD, T., MAYER, F., SEGALL, S., SIEBERT, L., SOLOMON, D. & THOMAS, C.). Experimental study of laser driven compression of spherical glass shells. Paper presented at *5th IAEA Conference on plasma physics and controlled nuclear fusion research, Tokyo* (*1974*). [411]

CHASE, J. B. *et al.* (1973) (with LE BLANC, J. M. & WILSON, J. R.). Role of spontaneous magnetic fields in a laser-created deuterium plasma, *Phys. Fluids*, **16**, 1142–8 (1973). [336]

CHEN, F. F. (1965). Electric probes. In *Plasma diagnostic techniques* (ed. R. H. Huddleston and S. L. Leonard) pp. 113–200. New York, Academic Press (1965). [343]

CHEUNG, A. Y. *et al.* (1973) (with GOFORTH, R. R. & KOOPMAN, D. W.). Magnetically induced collisionless coupling between counter-streaming laser-produced plasmas, *Phys. Rev. Lett.*, **31**, 429–32 (1973). [441, 442]

CHIAO, R. Y. *et al.* (1964) (with GARMIRE, E. & TOWNES, C. H.). Self-trapping of optical beams, *Phys. Rev. Lett.*, **13**, 479–82 (1964). [195]

CHIN, S. L. (1970). Direct experimental evidence of multiphoton ionization of impurities as the initiation process of laser-induced gas breakdown, *Can. J. Phys.*, **48**, 1314–17 (1970). [162]

CHIN, S. L. (1971). Multiphoton ionization of molecules, *Phys. Rev.*, **A4**, 992–6 (1971). [162]

CHIN, S. L. (1972). Effect of local laser-beam intensity fluctuations in multiphoton ionization, *Phys. Rev.*, **A5**, 2303–5 (1972). [154]

CHIN, S. L. & ISENOR, N. R. (1967). Effect of an electron-acceptor gas on the optical breakdown of argon, *Phys. Rev.*, **158**, 93–4 (1967). [181]

CHIN, S. L. & ISENOR, N. R. (1970). Multiphoton ionization in atomic gases with depletion of neutral atoms, *Can. J. Phys.*, **48**, 1445–7 (1970). [157]

CHIN, S. L. *et al.* (1969) (with ISENOR, N. R. & YOUNG, M.). Multiphoton ionization of Hg and Xe, *Phys. Rev.*, **188**, 7–8 (1969). [157]

CHOUDHURY, B. J. (1973*a*). On the validity of momentum translation approximation, *J. Phys. B.*, **6**, L100–2 (1973). [153]

CHOUDHURY, B. J. (1973*b*). Low intensity and high intensity limits of the momentum translation approximation, *J. Phys. B.*, **6**, L103–4 (1973). [153]

CHU, M. S. (1972). Thermonuclear reaction waves at high densities, *Phys. Fluids*, **15**, 413–22 (1972). [135, 142, 316]

CHU, T. K. & HENDEL, H. W. (1972). Measurements of enhanced absorption of electromagnetic waves and effective collision-frequency due to parametric decay instability, *Phys. Rev. Lett.*, **29**, 634–8 (1972). [76]

CILLIE, G. (1932). Hydrogen emission in gaseous nebulae, *Mon. Not. R. Astron. Soc.*, **92**, 820–30 (1932). [42]

CLARKE, J. S. *et al.* (1973) (with FISHER, H. N. & MASON, R. J.). Laser-driven implosion of spherical DT targets to thermonuclear burn conditions, *Phys. Rev. Lett.*, **30**, 89–92 (1973): *corrigendum, ibid.*, **30**, 249 (1973). [323, 325, 326, 327]

COBB, J. K. & MURAY, J. J. (1965). Laser beam-induced electron and ion emission from metal foils, *Brit. J. Appl. Phys.*, **16**, 271–3 (1965). [339]

COHEN, B. I. *et al.* (1972) (with KAUFMAN, A. N. & WATSON, K. M.). Beat heating of a plasma, *Phys. Rev. Lett.*, **29**, 581–4 (1972). [70]

COHN, D. R. & LAX, B. (1972). Magnetic-field-enhanced heating of plasmas with CO_2 lasers, *Appl. Phys. Lett.*, **21**, 217–20 (1972). [222]

COHN, D. R. *et al.* (1972*a*) (with CHASE, C. E., HALVERSON, W. & LAX, B.). Magnetic-field-dependent breakdown of CO_2-laser produced plasma, *Appl. Phys. Lett.*, **20**, 225–7 (1972). [194, 195]

COHN, D. R. *et al.* (1972*b*) (with HALVERSON, W., LAX, B. & CHASE, C. E.). Effect of magnetic field on electron density growth during laser-induced gas breakdown, *Phys. Rev. Lett.*, **29**, 1544–7 (1972). [195]

COLIN, C. *et al.* (1968*a*) (with DURAND, Y., FLOUX, F., GUYOT, D. & LANGER, P.). Étude expérimentale du plasma créé par laser sur une cible cryogénique, *J. Physique*, **29**, *Suppl.*, C3, 59–62 (1968). [420]

COLIN, C. *et al.* (1968*b*) (with DURAND, Y., FLOUX, F., GUYOT, D., LANGER, P. & VEYRIE, P.). Laser-produced plasmas from solid deuterium targets, *J. Appl. Phys.*, **29**, 2991–3 (1968). [420]

COLOMBANT, D. & TONON, G. F. (1973). X-ray emission in laser-produced plasmas, *J. Appl. Phys.*, **44**, 3524–37 (1973). [38, 39, 300]

CONSOLI, T. *et al.* (1966) (with GORMEZANO, C. & SLAMA, L.). Diffusion de la lumière d'un laser à rubis par un plasma dense, *Phys. Lett.*, **20**, 267–8 (1966). [112, 115]

CONSOLI, T. *et al.* (1968) (with GRELOT, P. & SLAMA, L.). Ionization of microparticles by laser beam, *VIII International Conference on phenomena in ionized gases, Vienna (1967). Contributed papers*, p. 265. Vienna, IAEA (1968). [400]

COURANT, R. & FRIEDRICHS, K. O. (1948). *Supersonic flow and shock waves.* New York, Interscience (1948). [209, 294]

DAEHLER, M. & RIBE, F. L. (1967). Cooperative light scattering from θ-pinch plasmas, *Phys. Rev.*, **161**, 117–25 (1967). [110, 111, 112, 121, 124]

DAEHLER, M. *et al.* (1969) (with SAWYER, G. A. & THOMAS, K. S.). Coordinated measurements of plasma density and cooperative light scattering in a theta-pinch plasma, *Phys. Fluids*, **12**, 225–9 (1969). [111, 121, 123]

DAIBER, J. W. & THOMPSON, H. M. (1967). Laser-driven detonation waves in gases, *Phys. Fluids*, **10**, 1162–9 (1967). [210, 253, 257]

DAIBER, J. W. & THOMPSON, H. M. (1970). X-ray temperature from laser-induced breakdown plasmas in air, *J. Appl. Phys.*, **41**, 2043–7 (1970). [258]

DAIBER, J. W. & WINANS, J. G. (1968). Radiation from laser-heated plasmas in nitrogen and argon, *J. Opt. Soc. Am.*, **58**, 76–80 (1968). [259]

DALGARNO, A. & KINGSTON, A. E. (1960). The refractive indices and Verdet constants of the inert gases, *Proc. Roy. Soc.*, **A259**, 424–9 (1960). [85]

DALGARNO, A. & LANE, N. F. (1966). Free–free transitions of electrons in gases, *Astrophys. J.*, **145**, 623–33 (1966). [48]

DAMON, E. K. & TOMLINSON, R. G. (1963). Observation of ionization of gases by a ruby laser, *Appl. Opt.*, **2**, 546–7 (1963). [154]

DAUGHNEY, C. C. *et al.* (1970) (with HOLMES, L. S. & PAUL, J. W. M.). Measurement of spectrum of turbulence within a collisionless shock by collective scattering of light, *Phys. Rev. Lett.*, **25**, 497–9 (1970). [126, 127]

DAVID, C. D. (1967). Two-wavelength interferometry of a laser-produced carbon plasma, *Appl. Phys. Lett.*, **11**, 394–6 (1967). [363, 364]

DAVID, C. D. & AVIZONIS, P. V. (1968). Optical interferometry, *VIII International Conference on phenomena in ionized gases, Vienna (1967). Contributed papers*, p. 511. Vienna, IAEA (1968). [363]

DAVID, C. D. & WEICHEL, H. (1969). Temperature of a laser-heated carbon plasma, *J. Appl. Phys.*, **40**, 3674–9 (1969). [365]

DAVID, C. D. *et al.* (1966) (with AVIZONIS, P. V., WEICHEL, H., BRUCE, C. & PYATT, K. D.). Density and temperature of a laser induced plasma, *IEEE J. Quantum Electron*, **QE2**, 493–9 (1966). [363]

DAVID, J. P. *et al.* (1968) (with FLORET, F., KAFTANDJIAN, V., MILLET, J. & PASCALE, J.). Contribution a l'étude théorique et expérimentale de la photoionisation du césium et de la recombinaison électron-ion césium, *J. Physique*, **29**, Suppl. C3, 15–22 (1968). [159]

DAVIES, W. E. R. & RAMSDEN, S. A. (1964). Scattering of light from the electrons in a plasma, *Phys. Lett.*, **8**, 179–80 (1964). [113, 114]

DAVYDKIN, V. A. *et al.* (1971) (with ZON, B. A., MANAKOV, N. L. & RAPOPORT, L. P.). Quadratic stark effect on atoms, *Zh. Eksp. i Teor. Fiz.*, **60**, 124–31 (1971). (Transl: *Sov. Phys. JETP*, **33**, 70–73 (1971).) [152, 159]

DAWSON, J. M. (1964). On the production of plasma by giant pulse lasers, *Phys. Fluids*, **7**, 981–7 (1964). [310, 312, 317, 331, 365]

DAWSON, J. & OBERMAN, C. (1962). High-frequency conductivity and the emission and absorption coefficients of a fully ionized plasma, *Phys. Fluids*, **5**, 517–24 (1962). [50, 64]

DAWSON, J. & OBERMAN, C. (1963). Effect of ion correlations on high-frequency plasma conductivity, *Phys. Fluids*, **6**, 394–7 (1963). [50]

DAWSON, J. M. & SHANNY, R. (1968). Some investigations of nonlinear behaviour in one-dimensional plasmas, *Phys. Fluids*, **11**, 1506–23 (1968). [66]

DAWSON, J. *et al.* (1968) (with KAW, P. & GREEN, B.). Optical absorption and expansion of laser-produced plasmas, *Phys. Fluids*, **12**, 875–81 (1968). [75, 281]

DAWSON, J. M. *et al.* (1971) (with HERTZBERG, A., KIDDER, R. E., VLASSES, G. C., AHLSTROM, H. G. & STEINHAUER, L. C.). Long-wavelength, high powered lasers for controlled thermonuclear fusion, *Proc. 4th Conference on plasma physics and controlled nuclear fusion research, Madison (1971)*, Vol. I, p. 673–87. Vienna, IAEA (1971). [135, 139]

DEAN, S. O. *et al.* (1971) (with McLEAN, E. A., STAMPER, J. A. & GRIEM, H. R.). Demonstration of collisionless interactions between interstreaming ions in a laser-produced-plasma experiment, *Phys. Rev. Lett.*, **27**, 487–90 (1971). [443]

DEAN, S. O. *et al.* (1972) (with McLEAN, E. A., STAMPER, J. A. & GRIEM, H. R.). Reasons for the collisionless nature of interactions in a laser-produced plasma experiment, *Phys. Rev. Lett.*, **29**, 569–73 (1972). [444]

DECROISETTE, M. *et al.* (1970) (with PIAR, G. & FLOUX, F.). Observation of induced Compton scattering in laser created plasma, *Phys. Lett.*, **32A**, 249–50 (1970). [78, 419]

DECROISETTE, M. *et al.* (1972) (with PEYRAUD, J. & PIAR, G.). Induced Compton scattering and nonlinear propagation in laser-created plasmas, *Phys. Rev.*, **A5**, 1391–6 (1972). [78, 419, 420]

DE GROOT, J. S. & KATZ, J. I. (1973). Anomalous plasma heating induced by a very strong high-frequency electric field, *Phys. Fluids*, **16**, 401–7 (1973). [62, 65, 66, 67]

DELONE, G. A. & DELONE, N. B. (1968). Role of bound states in the process of multiphoton ionization of atoms, *Zh. Eksp. i Teor. Fiz.*, **54**, 1067–8 (1968). (Transl: *Sov. Phys. JETP*, **27**, 570–1 (1968).) [157]

DELONE, G. A. & DELONE, N. B. (1969). Influence of multiphoton resonance on the multiphoton ionization process, *Zh. Eksp. i Teor. Fiz. Pis'ma*, **10**, 413–16 (1969). (Transl: *JETP Lett.*, **10**, 265–7 (1969).) [159]

DELONE, G. A. *et al.* (1969) (with DELONE, N. B., DONSKAYA, N. P. & PETROSYAN, K. B.). Role of field intensity and structure of the atom in the process of multiphoton ionization, *Zh. Eksp. i Teor. Fiz. Pis'ma*, **9**, 103–7 (1969). (Transl: *JETP Lett.*, **9**, 59–61 (1969).) [160]

DELONE, G. A. *et al.* (1971) (with DELONE, N. B. & PISKOVA, G. K.). The role of a multiphoton resonance in multiphoton ionization, *X International Conference on phenomena in ionized gases, Oxford (1971)*. *Contributed papers*, p. 39. Oxford, Donald Parsons & Co. Ltd (1971). [151]

DELONE, G. A. *et al.* (1972) (with DELONE, N. B. & PISKOVA, G. K.). Multiphoton resonance ionization of atoms, *Zh. Eksp. i Teor. Fiz.*, **62**, 1272–83 (1972). (Transl: *Sov. Phys. JETP*, **35**, 672–7 (1972).) [151, 159]

DE METZ, J. (1971). Optical design of a laser system for nuclear fusion research, *Appl. Opt.*, **10**, 1609–14 (1971). [421]

DE MICHELIS, C. (1970). Gas breakdown produced by a train of mode-locked laser pulses, *Opt. Commun.*, **2**, 255–6 (1970). [186]

DE MICHELIS, C. & RAMSDEN, S. A. (1967). Plasma production by laser beam irradiation of a single solid particle, *Phys. Lett.*, **25A**, 162–3 (1967). [403]

DEMTRÖDER, W. & JANTZ, W. (1970). Investigation of laser-produced plasmas from metal-surfaces, *Plasma Phys.*, **12**, 691–703 (1970). [318, 368, 369]

DE SILVA, A. W. & STAMPER, J. A. (1967). Observation of anomalous electron heating in plasma shock waves, *Phys. Rev. Lett.*, **19**, 1027–30 (1967). [127]

DE SILVA, A. W. *et al.* (1964) (with EVANS, D. E. & FORREST, M. J.). Observation of Thomson and co-operative scattering of ruby laser light by a plasma, *Nature*, **203**, 1321–2 (1964). [112, 113, 117]

DEWHURST, R. J. *et al.* (1971) (with PERT, G. J. & RAMSDEN, S. A.). Breakdown in gases with a mode locked train and single picosecond pulse at 1·06μm, *X International Conference on phenomena in ionized gases, Oxford (1971)*. *Contributed papers*, p. 220. Oxford, Donald Parsons & Co. Ltd. (1971). [186, 187, 188]

DE WITT, R. N. (1973a). Circular versus linear polarization in multiphoton ionization of hydrogen, *J. Phys. B.*, **6**, L93–5 (1973). [153]

DE WITT, R. N. (1973b). Application of the momentum translation method to multiphoton ionization of hydrogen by intense electromagnetic fields, *J. Phys. B.*, **6**, 803–8 (1973). [153]

DHEZ, P. *et al.* (1969) (with JAEGLE, P., LEACH, S. & VELGHE, M.). Some characteristics of the extreme ultraviolet spectrum emitted by the plasma produced by laser impact on an aluminium target, *J. Appl. Phys.*, **40**, 2545–8 (1969). [379]

DICK, K. *et al.* (1973) (with PÉPIN, H., MARTINEAU, J., PARBHAKAR, K. & THIBAUDEAU, A.). Plasma creation from thin aluminium targets by a TEA-CO$_2$ laser, *J. Appl. Phys.*, **44**, 3284–93 (1973). [395, 396]

DIMOCK, D. & MAZZUCATO, E. (1968). Normal and anomalous conductivity in a toroidal discharge from Thomson scattering measurements, *Phys. Rev. Lett.*, **20**, 713–14 (1968). [115]

DOLGOV-SAVEL'EV, G. G. & KARNYUSHIN, V. N. (1970). Lithium hydride particle injector, *Pribory i Tekh. Eksp.* (3) (May–June), 220–2 (1970). (Transl: *Instrum. & Exp. Tech.*, **3**, 889–90 (1970).) [400]

DOLGOV-SAVEL'EV, G. G. *et al.* (967) (with ISKOLDSKII, A. M., KRUGLYAKOV, E. P. & MALINOVSKII, V. K.). Electron-optical method for recording the Thomson scattering of ruby laser light by a plasma, *Novosibirsk Preprint* (1967). (Culham Translation, CTO/410 (1967).) [111, 115]

DOLGOV-SAVEL'EV, G. G. *et al.* (1970) (with KARNYUSHIN, V. N. & SEKERIN, V. I.). Investigation of a laser microplasma in the focus of two laser beams, *Zh. Eksp. i Teor. Fiz.*, **58**, 535–40 (1970). (Transl: *Sov. Phys. JETP*, **31**, 287–9 (1970).) [400]

DONALDSON, T. P. *et al.* (1973) (with HUTCHEON, R. J. & KEY, M. H.). Electron temperature and ionization state in laser produced plasmas, *J. Phys. B.*, **6**, 1525–44 (1973) [389]

DOUGAL, A. A. & GILL, D. H. (1968). Breakdown mechanisms for laser-induced discharges in super-high pressure gases, *VIII International Conference on phenomena in ionized gases, Vienna (1967). Contributed papers*, p. 262, Vienna, IAEA (1968). [179]

DOUGHERTY, J. P. & FARLEY, D. T. (1960). A theory of incoherent scattering of radio waves by a plasma, *Proc. Roy. Soc.*, **A259**, 79–99 (1960). [87]

DOUGHERTY, J. P. & FARLEY, D. T. (1963). A theory of incoherent scattering of radio waves by a plasma. 3. Scattering in a partially ionized gas, *J. Geophys. Res.*, **68**, 5473–86 (1963). [99]

DREICER, H. (1964). Kinetic theory of an electron–photon gas, *Phys. Fluids*, **7**, 735–53 (1964). [73]

DREICER, H. *et al.* (1971) (with HENDERSON, D. B. & INGRAHAM, J. C.). Anomalous microwave absorption near the plasma frequency, *Phys. Rev. Lett.*, **26**, 1616–20 (1971). [76]

DREICER, H. *et al.* (1973) (with ELLIS, R. F. & INGRAHAM, J. C.). Hot-electron production and anomalous microwave absorption near the plasma frequency, *Phys. Rev. Lett.*, **31**, 426–9 (1973). [76]

DRUYVESTEYN, M. J. & PENNING, F. M. (1940). The mechanisms of electrical discharges in gases of low pressure, *Rev. Mod. Phys.*, **12**, 87–174 (1940): *corrigendum*; *ibid.*, **13**, 72 (1941). [168]

DU BOIS, D. F. & GILINSKY, V. (1964). Incoherent scattering of radiation by plasmas. II: Effect of Coulomb collisions on classical plasmas, *Phys. Rev.*, **133**, 1317–22 (1964). [98]

DU BOIS, D. F. & GOLDMAN, M. V. (1965). Radiation-induced instability of electron plasma oscillations, *Phys. Rev. Lett.*, **14**, 544–6 (1965). [60]

DU BOIS, D. F. & GOLDMAN, M. V. (1967). Parametrically excited plasma fluctuations, *Phys. Rev.*, **164**, 207–22 (1967). [62, 65]

DU BOIS, D. F. & GOLDMAN, M. V. (1972a). Nonlinear saturation of parametric instability: basic theory and application to the ionosphere, *Phys. Fluids*, **15**, 919–27 (1972). [65, 76]

DU BOIS, D. F. & GOLDMAN, M. V. (1972b). Spectrum and anomalous resistivity for the saturated parametric instability, *Phys. Rev. Lett.*, **28**, 218–21 (1972). [65]

DUCAUZE, A. & LANGER, P. (1966). Répartition des particles émises par focalisation d'un faisceau laser sur une cible solide, *C.R. Acad. Sci., Paris*, **262B**, 1398–1401 (1966). [345, 346, 347, 348]

DUCAUZE, A. *et al.* (1965) (with TONON, G. & VEYRIE, P.). Étude de l'énergie des ions émis par une cible métallique frappée par le faisceau d'un laser, *C.R. Acad. Sci., Paris*, **261**, 4039–41 (1965). [345]

DUGUAY, M. A. & HANSEN, J. W. (1969). An ultrafast light gate, *Appl. Phys. Lett.*, **15**, 192–4 (1969). [223]

DUNN, H. S. & LUBIN, M. J. (1970). Electromagnetic radiation from a laser produced plasma expanding into a uniform magnetic field, *J. Plasma Phys.*, **4**, 573–83 (1970). [335, 431]

DURAND, Y. & VEYRIE, P. (1966). Étude spectroscopique d'un plasma d'helium créé par le faisceau d'un laser, *C.R. Acad. Sci., Paris*, **262B**, 1283–6 (1966). [231, 241]

DYER, P. E. *et al.* (1974*a*) (with JAMES, D. J., RAMSDEN, S. A. & SKIPPER, M. A.). X-ray emission from a CO_2-laser-produced plasma, *Appl. Phys. Lett.*, **24**, 316–17 (1974). [386]

DYER, P. E. *et al.* (1974*b*) (with JAMES, D. J., RAMSDEN, S. A. & SKIPPER, M. A.). Reflection from a CO_2-laser produced plasma, *Phys. Lett.*, **48A**, 311–12 (1974). [386]

DYSHKO, A. L. *et al.* (1967) (with LUGOVOI, V. N. & PROKHOROV, A. M.). Self-focusing of intense light beams, *Zh. Eksp. i Teor. Fiz. Pis'ma*, **6**, 655–9 (1967). (Transl: *JETP Lett.*, **6**, 146–8 (1967).) [196]

EDEN, M. J. & SAUNDERS, P. A. H. (1971). Triggering requirements for pulsed fusion reactors, *Nucl. Fusion*, **11**, 37–41 (1971). [140]

EHLER, A. W. & WEISSLER, G. L. (1966). Vacuum ultraviolet radiation from plasmas formed by a laser on metal surfaces, *Appl. Phys. Lett.*, **8**, 89–91 (1966). [341]

EHLER, W. (1973). Measurement of Debye length in laser-produced plasma, *Phys. Fluids*, **16**, 339–40 (1973). [353, 354]

ELTON, R. C. & ROTH, N. V. (1967). Plasma spectroscopy in the vacuum ultraviolet and soft X-ray regions, *Appl. Opt.*, **6**, 2071–8 (1967). [232, 387]

ELWERT, G. (1939). Intensity and polarization in the continuous X-ray spectrum, *Ann. Physik.*, **34**, 178–208 (1939). [41]

ELWERT, G. (1954). The soft X-radiation from the undisturbed solar corona, *Z. Naturforsch.*, **9A**, 637–53 (1954). [42]

EMMONY, D. C. & IRVING, J. (1968). Observation of laser induced shocks near the surface of a carbon target, *VIII International Conference on phenomena in ionized gases, Vienna (1967). Contributed papers*, p. 58. Vienna, IAEA (1968). [438]

EMMONY, D. C. & IRVING, J. (1969). Strong shock wave generation by the interaction of a giant laser pulse with a solid target, *J. Phys. D.*, **2**, 1186–8 (1969). [438]

ENGELHARDT, A. G. (1963). The generation of dense high temperature plasmas by means of coherent optical radiation. (*Westinghouse research report*, 63-128-113-R2). Pittsburgh, Westinghouse Research Laboratories (1963). [310]

ENGELHARDT, A. G. *et al.* (1964) (with PHELPS, A. V. & RISK, C. G.). Determination of momentum transfer and inelastic collision cross sections for electrons in nitrogen using transport coefficients, *Phys. Rev.*, **135A**, 1566–74 (1964). [168]

ENGELHARDT, A. G. *et al.* (1970) (with GEORGE, T. V., HORA, H. & PACK, J. L.). Linear and nonlinear behaviour of laser produced aluminium plasmas, *Phys. Fluids*, **13**, 212–14 (1970). [404]

ENGELHARDT, A. G. *et al.* (1971) (with FUCHS, V., NEUFELD, R. & RICHARD, C.). Heating of a θ-pinch plasma by CO_2 laser radiation, *Kvantovaya Elektronika*, **1**, 105–8 (1971). (Transl: *Sov. J. Quantum Electron.*, **1**, 516–8 (1972).) [222]

ENGELHARDT, A. G. *et al.* (1972) (with FUCHS, V., NEUFELD, C. R., RICHARD, C. & DÉCOSTE, R.). Evidence for the heating of a high-density θ pinch when transversely irradiated by a pulsed CO_2 laser beam, *Appl. Phys. Lett.*, **20**, 425–8 (1972). [269]

EREMIN, V. I. *et al.* (1971) (with NORINSKII, L. V. & PRYADEIN, V. A.). Frequency dependence of the threshold of optical breakdown in air in the ultra-violet band, *Zh. Eksp. i Teor. Fiz. Pis'ma*, **13**, 433–6 (1971). (Transl: *JETP Lett.*, **13**, 307–10 (1971).) [182]

EUBANK, H. P. (1971). Threshold electric field for excitation of parametric instabilities in inhomogeneous plasmas, *Phys. Fluids*, **14**, 2551–2 (1971). [76]

EVANS, D. E. (1970). The effect of impurities on the spectrum of laser light scattered by a plasma, *Plasma Phys.*, **12**, 573–84 (1970) [100, 101]

EVANS, D. E. & CAROLAN, P. G. (1970). Measurement of magnetic field in a laboratory plasma by Thomson scattering of laser light, *Phys. Rev. Lett.*, **25**, 1605–8 (1970). [98, 129]

EVANS, D. E. & KATZENSTEIN, J. (1969). Laser light scattering in laboratory plasmas, *Rep. Prog. Phys.*, **32**, 207–71 (1969). [88, 90, 94, 105]

EVANS, D. E. *et al.* (1966*a*) (with FORREST, M. J., KACHEN, G. I., KATZENSTEIN, J. & REYNOLDS, J. A.). The cooperative scattering of laser light in a thetatron plasma: a progress report, *VII Conference on phenomena in ionized gases, Belgrade* (*1965*), Vol. 3, pp. 196–9. Belgrade, Gradevinska Knjiga (1966). [111, 112]

EVANS, D. E. *et al.* (1966*b*) (with FORREST, M. J. & KATZENSTEIN, J.). Co-operative scattering of laser light by a thetatron plasma, *Nature*, **211**, 23–4 (1966). [119]

EVANS, D. E. *et al.* (1966*c*) (with FORREST, M. J. & KATZENSTEIN, J.). Asymmetric co-operative scattered light spectrum in a thetatron plasma, *Nature*, **212**, 21–3 (1966). [95, 120, 121]

EVANS, L. R. & GREY MORGAN, C. (1968). Multiple collinear laser-produced sparks in gases, *Nature*, **219**, 712–13 (1968). [171]

EVANS, L. R. & GREY MORGAN, C. (1969). Lens aberration effects in optical frequency breakdown of gases, *Phys. Rev. Lett.*, **22**, 1099–1101 (1969). [171, 247]

EVANS, R. G. & THONEMANN, P. C. (1972*a*). Laser produced dissociation and ionization of residual hydrocarbons in a high vacuum system, *Phys. Lett.*, **38A**, 398–400 (1972). [162]

EVANS, R. G. & THONEMANN, P. C. (1972*b*). Resonant multiphoton ionisation of caesium using a ruby laser, *Phys. Lett.*, **39A**, 133–5 (1972). [160]

EVTUSHENKO, T. P. *et al.* (1966*a*) (with MALYSHEV, G. M., OSTROVSKAYA, G. V., SEMENOV, V. V. & CHELIDZE, T. YA.). Investigation of air sparks by two synchronized lasers, *Zh. Tekh. Fiz.*, **36**, 1115–7 (1966). (Transl: *Sov. Phys. Tech. Phys.*, **11**, 818–20 (1966).) [223, 225]

EVTUSHENKO, T. P. *et al.* (1966*b*) (with ZAIDEL, A. N., OSTROVSKAYA, G. V. & CHELIDZE, T. YA.). Spectroscopic studies of a laser spark. I. Laser spark in helium, *Zh. Tekh. Fiz.*, **36**, 1506–13 (1966). (Transl: *Sov. Phys. Tech. Phys.*, **11**, 1126–30 (1967).) [231, 242]

FABRE, E. & LAMAIN, H. (1969). Plasma créé par irradiation laser de cibles sphériques, *Phys. Lett.*, **29A**, 497–8 (1969). [400, 432]

FABRE, E. & LAMAIN, H. (1970). Destruction de cibles sphériques par irradiation laser, *Phys. Lett.*, **31A**, 203–4 (1970). [405]

FABRE, E. & VASSEUR, P. (1966). Comparaison entre deux techniques de mesure sur un plasma en expansion rapide, *C.R. Acad. Sci., Paris*, **262B**, 923–5 (1966). [360]

FABRE, E. & VASSEUR, P. (1968). Expansion d'un plasma produit par irradiation laser de cibles minces, *J. Physique*, **29**, *Suppl.* **C3**, 123–7 (1968). [395]

FABRE, E. *et al.* (1966) (with VASSEUR, P. & BEVERNAGE, G.). Mesures sur un plasma produit par laser, *Phys. Lett.*, **20**, 381–2 (1966). [360, 432]

FABRE, E. *et al.* (1973) (with STENZ, C. & COLBURN, S.), Étude expérimentale de l'interaction avec un champ magnétique d'un plasma créé par irradiation laser de solides, *J. Physique*, **34**, 323–31 (1973). [433]

FADER, W. J. (1968). Hydrodynamic model of a spherical plasma produced by Q-spoiled laser irradiation of a solid particle, *Phys. Fluids*, **11**, 2200–8 (1968). [313, 332, 359]

FAISAL, F. H. M. (1972*a*). Multiphoton ionization of hydrogenic atoms and the breakdown of perturbative calculations at high intensities, *J. Phys. B.*, **5**, L196–8 (1972). [153]

FAISAL, F. H. M. (1972*b*). Multiphoton ionization by circularly and linearly polarized light: II., *J. Phys. B.*, **5**, L233–6 (1972). [153]

FARKAS, GY. *et al.* (1972) (with HORVATH, Z. GY. & KERTESZ, I.). Influence of optical field emission on the nonlinear photoelectric effect induced by ultrashort laser pulses, *Phys. Lett.*, **39A**, 231–2 (1972). [279]

FARLEY, D. T. *et al.* (1961) (with DOUGHERTY, J. P. & BARRON, D. W.). A theory of incoherent scattering of radio waves by a plasma. II. Scattering in a magnetic field, *Proc. Roy. Soc.*, **A263**, 238–58 (1961). [97]

FAUGERAS, P. E. *et al.* (1968) (with MATTIOLI, M. & PAPOULAR, R.). Optical measurements on laser-produced plasmas, *Plasma. Phys.*, **10**, 939–49 (1968). [405]

FAUQUIGNON, C. & FLOUX, F. (1970). Hydrodynamic behaviour of solid deuterium under laser heating, *Phys. Fluids*, **13**, 386–91 (1970). [286, 294, 308, 420]

FAURE, C. *et al.* (1971) (with PEREZ, A., TONON, G., AVENEAU, B. & PARISOT, D.). Highly multiply-charged heavy ions souce by laser impact, *Phys. Lett.*, **34A**, 313–14 (1971). [370]

FAWCETT, B. C. & PEACOCK, N. J. (1967). Highly ionized spectra of the transition elements, *Proc. Phys. Soc.*, **91**, 973–5 (1967). [376]

FAWCETT, B. C. *et al.* (1966) (with GABRIEL, A. H., IRONS, F. E., PEACOCK, N. J. & SAUNDERS, P. A. H.). Extreme ultra-violet spectra from laser-produced plasmas, *Proc. Phys. Soc.*, **88**, 1051–3 (1966). [376]

FAWCETT, B. C. *et al.* (1967) (with BURGESS, D. D. & PEACOCK, N. J.). Inner-shell transitions in Ca. XII, XIII, XIV and in isoelectronic ions of K, Sc and Ti in laser-produced plasmas, *Proc. Phys. Soc.*, **91**, 970–2 (1967). [376]

FAY, S. W. & JASSBY, D. L. (1972). Transverse and longitudinal heat flow in a laser-heated magnetically confined plasma, *Phys. Lett.*, **42A**, 261–2 (1972). [272]

FECAN, J. C. *et al.* (1966) (with FLOUX, F. & VEYRIE, P.). Étude expérimentale de l'ionisation des gaz par un faisceau lumineux très intense. Mesure de l'absorption de l'énergie dans la zone ionisée, *C.R. Acad. Sci., Paris*, **262B**, 1613–16 (1966). [234]

FEJER, J. A. (1960). Scattering of radio waves by an ionized gas in thermal equilibrium, *Can. J. Phys.*, **38**, 1114–33 (1960). [87]

FEJER, J. A. (1961). Scattering of radio waves by an ionized gas in thermal equilibrium in the presence of a uniform magnetic field, *Can. J. Phys.*, **39**, 716–40 (1961). [97, 100]

FEJER, J. A. & KUO, Y. Y. (1973). Structure in the non-linear saturation spectrum of parametric instabilities, *Phys. Fluids*, **16**, 1490–6 (1973). [65]

FENNER, N. C. (1966). Ion energies in the plasma produced by a high power laser, *Phys. Lett.*, **22**, 421–2 (1966). [341]

FIOCCO, G. & THOMPSON, E. (1963). Thomson scattering of optical radiation from an electron beam, *Phys. Rev. Lett.*, **10**, 89–91 (1963). [112].

FIRSOV, O. B. & CHIBISOV, M. I. (1960). Bremsstrahlung of slow electrons decelerated by neutral atoms, *Zh. Eksp. i Teor. Fiz.*, **39**, 1770–6 (1960). (Transl: *Sov. Phys. JETP*, **12**, 1235–9 (1961).) [48]

FISHER, R. K. & HIRSHFIELD, J. L. (1973). Phase effects in nonlinear wave–wave interactions, *Phys. Fluids*, **16**, 567–9 (1973). [70]

FLOUX, F. *et al.* (1969) (with COGNARD, D., BOBIN, J. L., DELOBEAU, F. & FAUQUIGNON, C.). Réactions nucléaires de fusion engendrées par un faisceau laser. *C.R. Acad. Sci., Paris*, **269B**, 697–700 (1969). [421]

FLOUX, F. *et al.* (1970) (with COGNARD, D., DENOEUD, L.-G., PIAR, G., PARISOT, D., BOBIN, J. L., DELOBEAU, F. & FAUQUIGNON, C.). Nuclear fusion reactions in solid-deuterium laser-produced plasma, *Phys. Rev.*, **A1**, 821–4 (1970). [421]

FLOUX, F. *et al.* (1972) (with BENARD, J. F., COGNARD, D. & SALÈRES, A.). Nuclear DD reactions in solid deuterium laser created plasma. In *Laser interaction and related plasma phenomena*, (ed. H. J. Schwarz and H. Hora) Vol. 2, pp. 409–31. New York, Plenum (1972). [422]

FLOUX, F. *et al.* (1973) (with COGNARD, D., SALÈRES, A. & REDON, D.). X-ray emission from laser created plasmas, *Phys. Lett.*, **45A**, 483–4 (1973). [423]

FORSLUND, D. W. *et al.* (1970) (with MORSE, R. L. & NIELSON, C. W.). Electron cyclotron drift instability, *Phys. Rev. Lett.*, **25**, 1266–70 (1970). [127]

FORSLUND, D. W. *et al.* (1972) (with KINDEL, J. M. & LINDMAN, E. L.). Parametric excitation of electromagnetic waves, *Phys. Rev. Lett.*, **29**, 249–52 (1972). [60]

FORSLUND, D. W. *et al.* (1973) (with KINDEL, J. M. & LINDMAN, E. L.). Nonlinear behaviour of stimulated Brillouin and Raman scattering in laser-irradiated plasmas, *Phys. Rev. Lett.*, **30**, 739–43 (1973). [71, 72]

FOX, R. A. *et al.* (1971) (with KOGAN, R. M. & ROBINSON, E. J.). Laser triple-quantum photo-ionization of cesium, *Phys. Rev. Lett.*, **26**, 1416 (1971). [159]

FRALEY, G. S. *et al.* (1974a) (with LINNEBUR, E. J., MASON, R. J. & MORSE, R. L.). Thermonuclear burn characteristics of compressed deuterium–tritium microspheres, *Phys. Fluids*, **17**, 474–89, (1974). [141, 326]

FRALEY, G. S. *et al.* (1974b) (with GULA, W. P., HENDERSON, D. B., McCRORY, R. L., MALONE, R. C., MASON, R. J. & MORSE, R. L.). Implosion, stability, and burn of multishell fusion targets, Paper IAEA-CN-33/F5-5, *5th IAEA Conference on plasma physics and controlled nuclear fusion research, Tokyo* (1974). [326, 329]

FRANCIS, G. *et al.* (1967) (with ATKINSON, D. W., AVIVI, P., BRADLEY, J. E., KING, C. D., MILLAR, W., SAUNDERS, P. A. H. & TAYLOR, A. F.). Laser produced plasmas from isolated solid hydrogen pellets, *Phys. Lett.*, **25A**, 486–7 (1967). [415, 416, 428]

FRANÇON, M. (1966). *Optical Interferometry*. New York, Academic (1966). [227]

FRANKLIN, R. N. *et al.* (1971) (with HAMBERGER, S. M., LAMPIS, G. & SMITH, G. J.). Nonlinear damping of electron plasma waves by induced decay into ion waves, *Phys. Rev. Lett.*, **27**, 1119–23 (1971). [76]

FRANKLIN, R. N. *et al.* (1972) (with HAMBERGER, S. M., IKEZI, H., LAMPIS, G. & SMITH, G. J.). Nature of the instability caused by electrons trapped by an electron plasma wave, *Phys. Rev. Lett.*, **28**, 1114–17 (1972). [76]

FRANZEN, D. L. (1972). CW gas breakdown in argon using $10 \cdot 6\mu m$ laser radiation, *Appl. Phys. Lett.*, **21**, 62–4 (1972). [189, 270]

FRANZEN, D. L. (1973). Continuous laser-sustained plasmas, *J. Appl. Phys.* **44**, 1727–32 (1973). [270]

FRASER, A. R. (1958). Radiation fronts, *Proc. Roy. Soc.*, **A245**, 536–45 (1958). [294]

FREIDBERG, J. P. & MARDER, B. M. (1971). High-frequency electrostatic plasma instabilities, *Phys. Rev.*, **A4**, 1549–53 (1971). [60, 61]

FREIDBERG, J. P. *et al.* (1972) (with MITCHELL, R. W., MORSE, R. L. & RUDSINSKI, L. I.). Resonant absorption of laser light by plasma targets, *Phys. Rev. Lett.*, **28**, 795–9 (1972). [75, 283]

FRIED, B. D. & CONTE, S. D. (1961). *The plasma dispersion function*. New York, Academic (1961). [19]

FROST, L. S. & PHELPS, A. V. (1964). Momentum-transfer cross-sections for slow electrons in He, Ar, Kr, & Xe from transport coefficients, *Phys. Rev.*, **136A**, 1538–45 (1964). [168]

FUCHS, V. (1973). Heating of θ-pinch plasmas by pulsed CO_2 lasers, *J. Appl. Phys.*, **44**, 1168–73 (1973). [222]

FUCHS, V. *et al.* (1973) (with NEUFELD, C. R., TEICHMANN, J. & ENGELHARDT, A. G.). Nonlinear absorption of radiation by optical mixing in a plasma, *Phys. Rev. Lett.*, **31**, 1110–13 (1973). [70]

FULLER, A. L. & GROSS, R. A. (1968). Thermonuclear detonation wave structure, *Phys. Fluids*, **11**, 534–44 (1968). [142]

FÜNFER, E. *et al.* (1963) (with KRONAST, B. & KUNZE, H. J.). Experimental results on light scattering by a θ-pinch plasma using a ruby laser, *Phys. Lett.*, **5**, 125–7 (1963). [112, 117]

GABOR, D. *et al.* (1965) (with STROKE, G. W., RESTRICK, R., FUNKHAUSER, A. & BRUMM, D.). Optical image synthesis (complex amplitude addition and subtraction) by holographic Fourier transformation, *Phys. Lett.*, **18**, 116–18 (1965). [229]

GABRIEL, A. H. *et al.* (1965) (with SWAIN, J. R. & WALLER, W. A.). A two-metre grazing-incidence spectrometer for use in the range 5–950 Å, *J. Sci. Instr.*, **42**, 94–7 (1965). [371]

GALEEV, A. A. & SAGDEEV, R. Z. (1973). Parametric phenomena in a plasma, *Nuclear Fusion*, **13**, 603–21 (1973). [62]

GALEEV, A. A. *et al.* (1972) (with ORAEVSKII, V. N. & SAGDEEV, R. Z.). Anomalous absorption of electromagnetic radiation at double the plasma frequency, *Zh. Eksp. i Teor. Fiz. Pis'ma*, **16**, 194–7 (1972). (Transl: *JETP Lett.*, **16**, 136–9 (1972).) [69]

GALEEV, A. A. *et al.* (1973) (with LAVAL, G., O'NEIL, T. M., ROSENBLUTH, M. N. & SAGDEEV, R. Z.). Parametric back scattering of a linear electromagnetic wave in a plasma, *Zh. Eksp. i Teor. Fiz. Pis'ma*, **17**, 48–52 (1973). (Transl: *JETP Lett.*, **17**, 35–8 (1973).) [71]

GARDNER, J. W. (1966a). Statistical models for laser-induced ionization of gases, *Electron. Lett.*, **2**, 297–8 (1966). [154]

GARDNER, J. W. (1966b). Photon correlations in ionising laser beams, *Electron. Lett.*, **2**, 397–8 (1966). [154]

GAUNT, J. A. (1930). Continuous absorption, *Proc. Roy. Soc.*, **A126**, 654–60 (1930): *Phil. Trans. Roy. Soc.*, **A229**, 163–204 (1930). [40]

GEKKER, I. R. & SIZUKHIN, O. V. (1969). Anomalous absorption of a powerful electromagnetic wave in a collisionless plasma, *Zh. Eksp. i Teor. Fiz. Pis'ma*, **9**, 408–12 (1969). (Transl: *JETP Lett.*, **9**, 243–6 (1969).) [76]

GENERALOV, N. A. *et al.* (1970*a*) (with ZIMAKOV, V. P., KOZLOV, G. I., MASYUKOV, V. A. & RAIZER, YU. P.). Gas breakdown under the influence of long-wave infrared radiation of a CO_2 beam, *Zh. Eksp. i Teor. Fiz. Pis'ma*, **11**, 343–6 (1970). (Transl: *JETP Lett.*, **11**, 228–31 (1970).) [189]

GENERALOV, N. A. *et al.* (1970*b*) (with ZIMAKOV, V. P., KOZLOV, G. I., MASYUKOV, V. A. & RAIZER, YU. P.). Continuous optical discharge, *Zh. Eksp. i Teor. Fiz. Pis'ma*, **11**, 447–9 (1970). (Transl: *JETP Lett.*, **11**, 302–4 (1970).) [269]

GENERALOV, N. A. *et al.* (1971) (with ZIMAKOV, V. P., KOZLOV, G. I., MASYUKOV, V. A. & RAIZER, YU. P.). Experimental investigation of a continuous optical discharge, *Zh. Eksp. i Teor. Fiz.*, **61**, 1434–46 (1971). (Transl: *Sov. Phys. JETP*, **34**, 763–9 (1972).) [269, 270]

GEORGE, E. V. *et al.* (1971) (with BEKEFI, G. & YA'AKOBI, B.). Structure of the plasma fireball produced by a CO_2 laser, *Phys. Fluids*, **14**, 2708–13 (1971). [267]

GEORGE, T. V. *et al.* (1965) (with GOLDSTEIN, L., SLAMA, L. & YOKOYAMA, M.). Molecular scattering of ruby-laser light, *Phys. Rev.*, **137A**, 369–80 (1965). [112]

GEORGE, T. V. *et al.* (1970) (with ENGELHARDT, A. G. & DE MICHELIS, C.). Thomson scattering diagnostics of laser-produced aluminium plasmas, *Appl. Phys. Lett.*, **16**, 248–51 (1970). [405]

GERRY, E. T. & PATRICK, R. M. (1965). Thomson scattering computations for laboratory plasmas, *Phys. Fluids*, **8**, 208–10 (1965). [91, 92]

GERRY, E. T. & ROSE, D. J. (1966). Plasma diagnostics by Thomson scattering of a laser beam, *J. Appl. Phys.*, **37**, 2715–24 (1966). [109, 110, 112, 113, 115, 116]

GIBSON, A. F. *et al.* (1968) (with HUGHES, T. P., KIMMITT, M. F. & HALLIN, R.). Plasmas produced by focused CO_2 laser radiation, *Phys. Lett.*, **27A**, 470–1 (1968). [342]

GILL, D. H. & DOUGAL, A. A. (1965). Breakdown minima due to electron-impact ionization in super-high-pressure gases irradiated by a focused giant-pulse laser, *Phys. Rev. Lett.*, **15**, 845–6 (1965). [172, 173]

GINZBURG, V. L. (1970). *The propagation of electromagnetic waves in plasmas.* 2nd ed. Oxford, Pergamon (1970). Transl. from second edition of *Rasprostranenie élektromagnitnykh voln v plazme.* Moscow, Nauka (1967). [1, 75]

GIORI, F. *t al.* (1963) (with MACKENZIE, L. A. & MCKINNEY, E. J.). Laser-induced thermionic emission, *Appl. Phys. Lett.*, **3**, 25–7 (1963). [339]

GLOCK, E. (1966). A monochromator of great flexibility for laser light scattered by high temperature plasmas, *VII International Conference on phenomena in ionized gases, Belgrade (1965)*, Vol. 3, pp. 194–6. Belgrade, Gradevinska Knjiga (1966). [111]

GOBELI, G. W. *et al.* (1969) (with BUSHNELL, J. C., PEERCY, P. S. & JONES, E. D.). Observation of neutrons produced by laser irradiation of lithium deuteride, *Phys. Rev.*, **188**, 300–2 (1969). [394]

GODWIN, R. P. (1972). Optical mechanism for enhanced absorption of laser energy incident on solid targets, *Phys. Rev. Lett.*, **28**, 85–7 (1972). [75]

GOLD, A. & BEBB, H. B. (1965). Theory of multiphoton ionization, *Phys. Rev. Lett.*, **14**, 60–3 (1965). [152]

GOLD, A. & BEBB, H. B. (1966). Multiphoton ionization of hydrogen and rare-gas atoms, *Phys. Rev.* **143**, 1–24 (1966). [159]

GOLDMAN, E. B. (1973). Numerical modeling of laser produced plasmas: the dynamics and neutron production in dense spherically symmetric plasmas, *Plasma Phys.*, **15**, 289–310 (1973). [320]

GOLDMAN, L. M. *et al.* (1973) (with SOURES, J. & LUBIN, M. J.). Saturation of stimulated back-scattered radiation in laser plasmas, *Phys. Rev. Lett.*, **31**, 1184–7 (1973). [407]

GOLDMAN, M. V. (1966). Parametric plasmon-photon interaction I. Threshold for amplification of plasmons. II. Analysis of plasmon propagator and correlation functions, *Ann. Phys.*, **38**, 95–116 and 117–69 (1966). [60, 62, 69]

GOLDMAN, M. V. & DU BOIS, D. F. (1965). Stimulated incoherent scattering of light from plasmas, *Phys. Fluids*, **8**, 1404–5 (1965). [70]

GONDHALEKAR, A. M. & KRONAST, B. (1973). Relativistic effect in the light-scattering spectrum of a θ-pinch plasma, *Phys. Rev.*, **A8**, 441–5 (1973). [117]

GONDHALEKAR, A. M. *et al.* (1970) (with KRONAST, B. & BENESCH, R.). Deviations from Maxwellian in the electron velocity distribution function of a θ-pinch plasma, *Phys. Fluids*, **13**, 2623–5 (1970). [116]

GONTIER, Y. & TRAHIN, M. (1968*a*). Multiphoton ionization of atomic hydrogen in the ground state, *Phys. Rev.*, **172**, 83–7 (1968). [152]

GONTIER, Y. & TRAHIN, M. (1968*b*). Ionisation multiphotonique d'un atome hydrogénoïde, *J. Physique*, **29**, *Suppl.* **C3**, 23–6 (1968). [152]

GONTIER, Y. & TRAHIN, M. (1971*a*). Bound–bound multiphoton processes in atomic hydrogen, *X International Conference on phenomena in ionized gases, Oxford (1971). Contributed papers*, p. 44. Oxford, Donald Parsons and Co. Ltd. (1971). [152]

GONTIER, Y. & TRAHIN, M. (1971*b*). On the multiphoton absorption in atomic hydrogen, *Phys. Lett.*, **36A**, 463–4 (1971). [152]

GORBENKO, B. Z. *et al.* (1969) (with DROZHBIN, YU. A., KAITMAZOV, S. D., MEDVEDEV, A. A., PROKHOROV, A. M. & TOLMACHOV, A. M.). Investigation of optical breakdown in air induced by ultrashort pulses, using moving-image photography and an image converter, *Dokl. Akad. Nauk. SSSR*, **187**, 772–4 (1969). (Transl: *Sov. Phys.-Dokl.*, **14**, 764–6 (1970).) [265]

GORBUNOV, L. M. & SILIN, V. P. (1969). Stability of a magnetized plasma in a strong high-frequency field, *Zh. Tekh. Fiz.*, **39**, 3–17 (1969). (Transl: *Sov. Phys. Tech. Phys.*, **14**, 1–9 (1969).) [60]

GOROG, I. (1968). On the expansion anisotropy of laser produced plasmas, *Phys. Lett.*, **28A**, 371–2 (1968). [352]

GOWERS, C. W. *et al.* (1971) (with JONES, P. A. & WATSON, J. L.). An experiment to detect the shift in the Thomson scattered spectrum due to the electron drift in a theta pinch, *X International Conference on phenomena in ionized gases, Oxford (1971). Contributed papers*, p. 418. Oxford, Donald Parsons & Co. Ltd. (1971). [117]

GRANT, I. P. (1958). Calculation of Gaunt factors for free–free transitions near positive ions, *Mon. Not. R. Astron. Soc.*, **118**, 241–57 (1958). [40, 41]

GRAVEL, M. *et al.* (1971) (with ROBERTSON, W. J., ALCOCK, A. J., BÜCHL, K. & RICHARDSON, M. C.). Forward going filament in sparks induced by 10·6μ laser radiation, *Appl. Phys. Lett.*, **18**, 75–7 (1971). [265, 266]

GREEN, T. S. (1970). Ionization in a laser produced plasma, *Phys. Lett.*, **32A**, 530–1 (1970). [287]

GREENE, J. (1959). Bremsstrahlung from a Maxwellian gas, *Astrophys. J.*, **130**, 693–701 (1959). [42, 44]

GREGG, D. W. & THOMAS, S. J. (1966*a*). Momentum transfer produced by focused laser giant pulses, *J. Appl. Phys.*, **37**, 2787–9 (1966). [365]

GREGG, D. W. & THOMAS, S. J. (1966*b*). Kinetic energies of ions produced by laser giant pulses, *J. Appl. Phys.*, **37**, 4313–16 (1966). [365]

GREGG, D. W. & THOMAS, S. J. (1967). Plasma temperatures generated by focused laser giant pulses, *J. Appl. Phys.*, **38**, 1729–31 (1967). [365]

GREWAL, M. S. (1964). Effects of collisions on electron density fluctuations in plasmas, *Phys. Rev.*, **A134**, 86–93 (1964). [98]

GRIEM, H. (1964). *Plasma Spectroscopy*. New York, McGraw-Hill (1964). [26, 231, 241, 242, 381]

GRIFFIN, W. G. & SCHLUTER, J. (1968). The formation mechanism of a laser-produced plasma, *Phys. Lett.*, **26A**, 241–2 (1968). [396–397]

GROSS, R. A. (1964). Continuum radiation behind a blastwave, *Phys. Fluids*, **7**, 1078–80 (1964). [220]

GROSS, R. A. (1971). The physics of strong shock waves in gases. In *Physics of high energy density* (ed. P. Caldirola & H. Knoepfel), pp. 245–77. New York, Academic (1971). [211]

GRYZINSKI, M. (1958). Fusion chain reaction, *Proc. II U.N. Conference on peaceful uses of atomic energy, Geneva*, **31**, 270–4. Geneva, United Nations (1958). [136]

GUCCIONI-GUSH, R. *et al.* (1967) (with GUSH, H. P. & VAN KRANENDONK, J.). Theory of two-photon absorption, *Can. J. Phys.*, **45**, 2513–24 (1967). [154]

GUDERLEY, G. (1942). Starke kugelige und zylindrische Verdichtungsstösse in der Nähe des Kugel-mittelpunktes bzw. der Zylinderachse, *Luftfahrtforschung*, **19**, 302–12 (1942). [319]

GUDZENKO, L. I. *et al.* (1969) (with KAITMAZOV, S. D., MEDVEDEV, A. A. & SHKLOVSKII, E. I.). Observation of dynamic spectrum of radiation from a highly ionized dense plasma, *Zh. Eksp. i Teor. Fiz. Pis'ma*, **9**, 561–4 (1969). (Transl: *JETP Lett.*, **9**, 341–3 (1969).) [265]

GUENTHER, A. H. & PENDLETON, W. K. (1972). Laser-produced gaseous deuterium plasmas. In *Laser interaction and related plasma phenomena*, (ed. H. J. Schwarz and H. Hora) Vol. 2, pp. 97–145. New York, Plenum (1972). [174, 232, 259]

HAGFORS, T. (1961). Density fluctuations in a plasma in a magnetic field, with applications to the ionosphere, *J. Geophys. Res.*, **66**, 1699–1712 (1961). [97, 98]

HAINES, M. G. (1963). Hall rotation in theta-pinches, *Phys. Lett.*, **6**, 313–14 (1963). [334]

HALL, J. L. (1966). Two-quantum photoionization of Cs and I⁻, *IEEE J. Quantum Electron.*, **2**, 361–3 (1966). [159]

HALL, R. B. (1969*a*). Microwave measurements on laser-produced blast waves, *J. Appl. Phys.*, **40**, 36–43 (1969). [232, 241]

HALL, R. B. (1969*b*). Laser production of blast waves in low pressure gases, *J. Appl. Phys.*, **40**, 1941–5 (1969). [438]

HALL, T. A. & NEGM, Y. K. (1975). Private communication. [386]

HAMAL, K. *et al.* (1969) (with DE MICHELIS, C. & ALCOCK, A. J.). Schlieren photography using a mode-locked laser as the light source, *Opt. Acta.*, **16**, 463–9 (1969). [225, 226, 255]

HANCOX, R. (1972). Fusion reactor studies in the United Kingdom, *Trans. Am. Nucl. Soc.*, **15**, 626 (1972). [144]

HANCOX, R. & SPALDING, I. J. (1973). Reactor implications of laser-ignited fusion, *Nucl. Fusion*, **13**, 385–91 (1973). [144]

HARDING, G. N. & ROBERTS, V. (1962). Spectroscopic investigation of plasma in the wavelength range 0·1–2·0 mm, *Nucl. Fusion Suppl.*, **3**, 883–7 (1962). [52]

HARDING, G. N. *et al.* (1961) (with KIMMITT, M. F., LUDLOW, J. H., PARTEAU, P., PRIOR, A. C. & ROBERTS, V.). Emission of sub-millimetre electromagnetic radiation from hot plasma in ZETA, *Proc. Phys. Soc.*, **77**, 1069–75 (1961). [52]

HARRIS, E. G. (1970). Plasma Instabilities. In *Physics of Hot Plasmas*, (ed. B. J. Rye and J. C. Taylor), pp. 145–201. Edinburgh, Oliver & Boyd (1970). [8]

HARRIS, T. J. (1963). High-speed photographs of laser-induced heating, *IBM J. Res. & Develop.*, **7**, 342–4 (1963). [435]

HARTMAN, T. E. (1962). Tunneling of a wave packet, *J. App. Phys.*, **33**, 3427–33 (1962). [150]

HAUGHT, A. F. & POLK, D. H. (1966). High-temperature plasmas produced by laser beam irradiation of single solid particles, *Phys. Fluids*, **9**, 2047–56 (1966). [313, 317, 400, 401, 403, 407, 418]

HAUGHT, A. F. & POLK, D. H. (1970). Formation and heating of laser irradiated solid particle plasmas, *Phys. Fluids*, **13**, 2825–41 (1970). [313, 314, 400, 405, 406]

HAUGHT, A. F. *et al.* (1966) (with MEYERAND, R. G. & SMITH, D. C.). Electrical breakdown of gases by optical frequency radiation. In *Physics of Quantum Electronics* (ed. P. L. Kelley, B. Lax and P. E. Tannenwald), pp. 509–19. New York, McGraw-Hill (1966). [166, 177, 178, 179, 181, 182]

HAUGHT, A. F. *et al.* (1970) (with POLK, D. H. & FADER, W. J.). Magnetic field confinement of laser irradiated solid particle plasmas, *Phys. Fluids*, **13**, 2842–57 (1970). [332, 333, 429, 430, 432, 434]

HAYNES, G. W. *et al.* (1971) (with FRIEDRICH, O. M. & DOUGAL, A. A.). Analysis of threshold breakdown fields for laser produced discharges in high pressure helium, *X International Conference on phenomena in ionized gases, Oxford (1971). Contributed papers*, p. 221. Oxford, Donald Parsons & Co. Ltd. (1971). [178]

HEALD, M. A. & WHARTON, G. B. (1965). *Plasma diagnostics with microwaves*. New York, Wiley (1965). [1, 28, 42, 50]

HEFLINGER, L. O. *et al.* (1966) (with WUERKER, R. F. & BROOKS, R. E.). Holographic interferometry, *J. Appl. Phys.*, **37**, 642–9 (1966). [229]

HELD, B. *et al.* (1971a) (with MAINFRAY, G., MANUS, C. & MORELLEC, J.). Multi-photon ionization of cesium and potassium atoms at 1·06μ and 0·53μ, *X International Conference on phenomena in ionized gases, Oxford (1971). Contributed papers*, p. 45. Oxford, Donald Parsons & Co. Ltd. (1971). [160].

HELD, B. *et al.* (1971b) (with MAINFRAY, G., MANUS, C. & MORELLEC, J.). Multiphoton ionization of cesium and potassium atoms at 1·06μ and 0·53μ, *Phys. Lett.*, **35A**, 257–8 (1971). [160]

HELD, B. *et al.* (1972a) (with MAINFRAY, G., MANUS, C. & MORELLEC, J.). Molecular cesium component in multiphoton ionization of a cesium atomic beam by a Q-switched neodymium-glass laser at 1·06μ, *Phys. Rev. Lett.*, **28**, 130–1 (1972). [160]

HELD, B. *et al.* (1972b) (with MAINFRAY, G. & MORELLEC, J.). Multiphoton ionization probability of potassium atoms at 1·06μ and 0·53μ, *Phys. Lett.*, **39A**, 57–8 (1972). [160]

HELD, B. *et al.* (1973) (with MAINFRAY, G., MANUS, C., MORELLEC, J. & SANCHEZ, F.). Resonant multiphoton ionization of a cesium atomic beam by a tunable-wavelength Q-switched neodymium-glass laser, *Phys. Rev. Lett.*, **30**, 423–6 (1973). [160]

HENDEL, H. W. & FLICK, J. T. (1973). Stochastic ion heating due to the parametric ion-acoustic decay instability, *Phys. Rev. Lett.*, **31**, 199–202 (1973). [76]

HENDERSON, D. B. & MORSE, R. L. (1974). Symmetry of laser-driven implosions, *Phys. Rev. Lett.*, **32**, 355–8 (1974). [326]

HENDERSON, D. B. *et al.* (1974) (with McCRORY, R. L. & MORSE, R. L.). Ablation stability of laser-driven implosions, *Phys. Rev. Lett.*, **33**, 205–8 (1974). [328]

HENNEBERGER, W. C. (1968). Perturbation method for atoms in intense light beams, *Phys. Rev. Lett.*, **21**, 838 (1968). [152]

HERCHER, M. (1965). Single-mode operation of a Q-switched ruby laser, *Appl. Phys. Lett.*, **7**, 39–41 (1965). [109]

HILL, G. A. *et al.* (1972) (with JAMES, D. J. & RAMSDEN, S. A.). Breakdown thresholds in rare and molecular gases using pulsed 10·6μm radiation, *J. Phys. D. (Appl. Phys.)*, **5**, L97–9 (1972). [190]

HIRSCHBERG, J. G. (1960). Nine-channel photoelectric Fabry–Perot interferometer, *J. Opt. Soc. Am.*, **50**, 514 (1960). [111]

HIRSCHBERG, J. G. & PLATZ, P. (1965). A multichannel Fabry–Perot interferometer, *Appl. Opt.*, **4**, 1375–81 (1965). [111]

HOHLA, K. *et al.* (1969) (with BÜCHL, K., WIENECKE, R. & WITKOWSKI, S.). Energiebestimmung der Stosswelle eines laserinduzierten Gasdurchbruchs, *Z. Naturforsch.*, **24A**, 1244–9 (1969). [225]

HOLDEN, P. & HUGHES, T. P. (1967). Unpublished. [371, 372, 373]

HOLSTEIN, T. (1965). Low-frequency approximation to free–free transition probabilities, *Westinghouse Research Laboratories, Pittsburgh, Scientific Paper 65-1E2-Gases-P2* (1965). [168]

HOLZER, W. *et al.* (1971) (with RANSON, P. & PERETTI, P.). Optical breakdown in gases of medium ionization potential, *IEEE J. Quantum Electron.*, **7**, 204–5 (1971). [174]

HONIG, R. E. (1963). Laser-induced emission of electrons and positive ions from metals and semiconductors, *Appl. Phys. Lett.*, **3**, 8–11 (1963). [339]

HONIG, R. E. & WOOLSTON, J. R. (1963). Laser-induced emission of electrons, ions and neutral atoms from solid surfaces, *Appl. Phys. Lett.*, **2**, 138–9 (1963). [339]

HORA, H. (1969a). Self-focusing of laser beams in a plasma by ponderomotive forces, *Z. Phys.*, **226**, 156–9 (1969). [197, 198]

HORA, H. (1969b). Nonlinear confining and deconfining forces associated with the interaction of laser radiation with plasma, *Phys. Fluids*, **12**, 182–91 (1969). [429]

HORA, H. (1971). Nonlinear effect of expansion of laser produced plasmas. In *Laser interaction and related plasma phenomena*, (ed. H. J. Schwarz and H. Hora) Vol. 1, pp. 386–426. New York, Plenum (1971). [429]

HORA, H. (1972). Theoretical aspects of ion energy increase in laser-produced plasmas due to static magnetic fields. In *Laser interaction and related plasma phenomena* (ed. H. J. Schwarz and H. Hora), Vol. 2, pp. 307–13. New York, Plenum (1972). |429|

HORA, H. & WILHELM, H. (1970). Optical constants of fully ionized hydrogen plasma for laser radiation, *Nucl. Fusion*, **10**, 111–16 (1970). |45|

HUBBARD, J. (1961). The friction and diffusion coefficients of the Fokker–Planck equation in a plasma, *Proc. Roy. Soc.*, **A260**, 114–26 (1961). [49]

HUGENSCHMIDT, M. (1970). Evolution temporelle et spatiale de la densité électronique dans un plasma de xénon créé par le faisceau d'un laser à rubis, *C.R. Acad. Sci., Paris*, **271B**, 757–60 (1970). [228, 229, 248]

HUGENSCHMIDT, M. & VOLLRATH, K. (1971). Analyse par interférométrie à deux longueurs d'onde de plasmas produits par focalisation d'un faisceau laser au néodyme, *C.R. Acad. Sci., Paris*, **272B**, 36–9 (1971). [365]

HUGHES, T. P. (1962). A new method for the determination of plasma electron temperature and density from Thomson scattering of an optical maser beam, *Nature*, **194**, 268–9 (1962). [107]

HUGHES, T. P. (1964). Vaporization by laser beams. Proceedings of a conference on *Lasers and their applications* sponsored by the I.E.E. Electronics and Science divisions, London, pp. 5–1,2 (1964). [277]

HUGHES, T. P. & NICHOLSON-FLORENCE, M. B. (1968). Intensity dependence of the inverse bremsstrahlung absorption coefficient in hot plasmas, *J. Phys. A.*, **1**, 588–95 (1968). [47, 316]

HULL, R. J. *et al.* (1972) (with LENCIONI, D. E. & MARQUET, L. C.). Influence of particles on laser induced air breakdown). In *Laser interaction and related plasma phenomena* (ed. H. J. Schwarz and H. Hora), Vol. 2, pp. 147–52. New York, Plenum (1972). [191]

HUNKLINGER, S. & LEIDERER, P. (1971). Influence of impurities on the laser induced breakdown in liquid He^4, *Z. Naturforsch.*, **26A**, 587–8 (1971). [177]

INNES, D. J. & BLOOM, A. L. (1966). Design of optical systems for use with laser beams. *Spectra-physics Laser Technical Bulletin No. 5* (Mountain View, Cal., Spectra-physics Inc, 1966). [171]

INOUE, N. *et al.* (1971) (with KAWASUMI, Y. & MIYAMOTO, K.). Expansion of a laser produced plasma into a vacuum, *Plasma Phys.*, **13**, 84–7 (1971). [403]

IRELAND, C. L. M. & GREY MORGAN, C. (1973). Gas breakdown by a short laser pulse, *J. Phys. D.*, **6**, 720–9 (1973). [169, 187]

IRELAND, C. L. M. & GREY MORGAN, C. (1974) Gas breakdown by single $\sim$ 20 ps, 1·06 μm and 0·53 μm laser pulses, *J. Phys. D.*, **7**, L87–90 (1974). [187]

IRONS, F. E. (1973). Stark broadening of high quantum number $\Delta n = 1$ transitions of carbon V and VI in a laser-produced plasma, *J. Phys. B.*, **6**, 1562–81 (1973). [381, 382]

IRONS, F. E. & PEACOCK, N. J. (1974). A spectroscopic study of the recombination of C^{6+} to C^{5+} in an expanding laser-produced plasma, *J. Phys. B.*, **7**, 2084–99 (1974). [381]

IRONS, F. E. *et al.* (1972) (with MCWHIRTER, R. W. P. & PEACOCK, N. J.). The ion and velocity structure in a laser produced plasma, *J. Phys. B.*, **5**, 1975–87 (1972). [374, 380, 381]

ISENOR, N. R. (1964a). Metal ion emission velocity dependence on laser giant pulse height, *Appl. Phys. Lett.*, **4**, 152–3 (1964). [341]

ISENOR, N. R. (1964b). High-energy ions from a Q-switched laser, *Can. J. Phys.*, **42**, 1413–16 (1964). [341]

ISENOR, N. R. & RICHARDSON, M. C. (1971a). Dissociation and breakdown of molecular gases by pulsed CO_2 laser radiation, *Appl. Phys. Lett.*, **18**, 224–6 (1971). See also *erratum, Appl. Phys. Lett.*, **18**, 586 (1971). [163]

ISENOR, N. R. & RICHARDSON, M. C. (1971b). Light emission from molecular gases irradiated by 10·6 μ laser pulses, *X International Conference on phenomena in ionized gases, Oxford (1971). Contributed papers*, p. 27. Oxford, Donald Parsons & Co. Ltd. (1971). [163]

IZAWA, Y. *et al.* (1968) (with YAMANAKA, T., TSUCHIMORI, N., ONISHI, M. & YAMANAKA, C.). Density measurements of the laser produced plasma by laser light scattering, *Jap. J. Appl. Phys.*, 7, 954 (1968). [342, 343, 363, 365]

IZAWA, Y. *et al.* (1969) (with YOKOYAMA, M. & YAMANAKA, C). Collective scattering of laser light from laser produced LiH plasma, *Jap. J. Appl. Phys.*, 8, 965 (1969). [360, 361, 362]

IZNATOV, A. B. *et al.* (1971) (with KOMISSAROVA, I. I., OSTROVSKAYA, G. V. & SHAPIRO, L. L.). Holograph studies of a laser spark. III. Spark in hydrogen and helium, *Zh. Tekh. Fiz.*, 41, 701–8 (1971). (Transl: *Sov. Phys. Tech. Phys.*, 16, 550–56 (1971).) [248, 249]

JACKSON, E. A. (1967). Parametric effects of radiation on a plasma, *Phys. Rev.*, 153, 235–44 (1967). [60, 69]

JAHODA, F. C. (1971). Pulse laser holographic interferometry. In *Modern optical methods in gas dynamic research* (ed. D. S. Dosanjh), pp. 137–54. New York, Plenum (1971). [229]

JAHODA, F. C. *et al.* (1960) (with LITTLE, E. M., QUINN, W. E., SAWYER, G. A. & STRATTON, T. F.). Continuum radiation in the X-ray and visible regions from a magnetically compressed plasma (Scylla), *Phys. Rev.* 119, 843–56 (1960). [232, 387]

JASSBY, D. L. (1972). Infrared-laser heating of dense arc plasmas, *Phys. Fluids* 15, 2442–4 (1972). [222]

JASSBY, D. L. & MARHIC, M. E. (1972). Thermal decay of an infrared-laser-heated arc plasma, *Phys. Rev. Lett.*, 29, 577–80 (1972). [269]

JAVAN, A. & KELLEY, P. L. (1966). Possibility of self-focusing due to intensity dependent anomalous dispersion, *IEEE J. Quantum Electron.*, 2, 470–3 (1966). [196]

JOHN, P. K. (1972). Laser-beam-scattering measurement of ion temperature in a θ-pinch plasma and evidence for thermonuclear reations, *Phys. Rev.*, A6, 756–64 (1972). [89, 121, 122]

JOHN, P. K. *et al.* (1971) (with IRISAWA, J. & NG, K. H.). Observation of enhanced plasma waves by laser scattering, *Phys. Lett.*, 36A, 277–8 (1971). [117]

JOHNSON, R. R. & HALL, R. B. (1971). Production of net fusion energy from laser-heated target plasmas, *J. Appl. Phys.*, 42, 1035–9 (1971). [316]

JOHNSTON, R. R. (1967). Free–free radiative transitions—a survey of theoretical results, *J. Quant. Spectrosc. & Radiat. Transfer*, 7, 815–35 (1967). [40]

JOHNSTON, T. W. & DAWSON, J. M. (1973). Correct value for high-frequency power absorption by inverse bremsstrahlung in plasmas, *Phys. Fluids* 16, 722 (1973). [50]

JONES, E. D. *et al* (1972) (with GOBELI, G. W. & OLSEN, J. N.). Nanosecond and picosecond laser irradiation of solid targets, In *Laser interaction and related plasma phenomena* (ed. H. J. Schwarz and H. Hora), Vol. 2, pp. 469–80. New York, Plenum (1972). [390]

JOYCE, G. & SALAT, A. (1971). Light scattering in a marginally stable plasma, *Plasma Phys.*, 13, 359–64 (1971). [102]

KADOMTSEV, B. B. (1965). *Plasma turbulence.* (Transl. L. C. Ronson; ed. M. G. Rusbridge). New York, Academic (1965). [65, 127]

KAINER, S. *et al.* (1972) (with DAWSON, J. & COFFEY, T.). Alternating current instability produced by the two-stream instability, *Phys. Fluids*, 15, 2419–22 (1972). See also *erratum*, *Phys. Fluids*, 16, 1382 (1973). [67]

KAITMAZOV, S. D. *et al.* (1968) (with MEDVEDEV, A. A. & PROKHOROV, A. M.). Investigation of optical breakdown in air by a laser operating in the mode-synchronization regime, *Dokl. Akad. Nauk. SSSR*, 180, 1092–3 (1968). (Transl: *Sov. Phys. Dokl.*, 13, 581–2 (1968).) [264]

KAITMAZOV, S. D. *et al.* (1971) (with MEDVEDEV, A. A. & PROKHOROV, A. M.). Effect of a 400-kOe magnetic field on a laser-spark plasma, *Zh. Eksp. i Teor. Fiz. Pis'ma*, 14, 314–16 (1971). (Transl: *JETP Lett.*, 14, 208–10 (1971).) [271]

KAKOS, A. *et al.* (1966) (with OSTROVSKAYA, G. V., OSTROVSKII, YU. I. & ZAIDEL, A. N.). Interferometry holographic investigation of a laser spark, *Phys. Lett.*, 23, 81–3 (1966). [229]

KANG, H.-B. *et al.* (1972) (with YAMANAKA, T., YOSHIDA, K., WAKI, M. & YAMANAKA, C.). Ion temperature of laser-produced plasma, *Jap. J. Appl. Phys.*, 11, 765–6 (1972). [382, 383]

KARLOV, N. V. *et al.* (1970) (with PETROV, YU. N., PROKHOROV, A. M. & STEL'MAKH, O. M.). Dissociation of boron trichloride molecules by CO_2 laser radiation, *Zh. Eksp. i Teor. Fiz. Pis'ma*, **11**, 220–2 (1970). (Transl: *JETP Lett.*, **11**, 135–7 (1970).) [163]

KARLOV, N. V. *et al.* (1972) (with KOMISSAROV, V. M., KUZMIN, G. P. & PROKHOROV, A. M.). Effect of plasma mirror in the breakdown of air in a CO_2 laser cavity, *Zh. Eksp. i Teor. Fiz. Pis'ma*, **16**, 95–8 (1972). (Transl: *JETP Lett.*, **16**, 65–7 (1972).) [269]

KARLOV, N. V. *et al.* (1973) (with KARPOV, N. A., PETROV, YU. N. & STEL'MAKH, O. M.). Self-focusing of CO_2 laser radiation in resonantly absorbing gases, *Zh. Eksp. i Teor. Fiz. Pis'ma*, **17**, 337–40 (1973). (Transl: *JETP Lett.* **17**, 239–41 (1973).) [199]

KARZAS, W. J. & LATTER, R. (1961). Electron radiative transitions in a Coulomb field, *Astrophys. J. Suppl. No. 55*, **6**, 167–212 (1961). [41]

KAS'YANOV, V. & STAROSTIN, A. (1965). On the theory of bremsstrahlung of slow electrons on atoms, *Zh. Eksp. i Teor. Fiz.*, **48**, 295–302 (1965). (Transl: *Sov. Phys. JETP*, **21**, 193–8 (1965).) [48, 53]

KATO, M. (1972). Fluctuation spectrum of plasma with few electrons in Debye volume observed by collective light scattering, *Phys. Fluids*, **15**, 460–3 (1972). [124, 125]

KATZENSTEIN, J. (1965). The Axicon-scanned Fabry–Perot spectrometer, *Appl. Opt.*, **4**, 263–6 (1965). [111, 118]

KAUFMAN, A. N. & COHEN, B. I. (1973). Nonlinear interaction of electromagnetic waves in a plasma density gradient, *Phys. Rev. Lett.*, **30**, 1306–9 (1973). [70]

KAUFMAN, S. & WILLIAMS, R. V. (1958). The electron temperature in Sceptre III, *Nature*, **182**, 557–8 (1958). [341]

KAW, P. (1969). Non-linear effects of laser propagation in dense plasmas, *Appl. Phys. Lett.*, **15**, 16–18 (1969). [75]

KAW, P. & DAWSON, J. M. (1969). Laser-induced anomalous heating of a plasma, *Phys. Fluids*, **12**, 2586–91 (1969). [60, 62, 64]

KAW, P. & DAWSON, J. M. (1970). Relativistic non-linear propagation of laser beams in cold overdense plasmas, *Phys. Fluids*, **13**, 472–81 (1970). [75]

KAW, P. *et al.* (1970) (with VALEO, E. & DAWSON, J. M.). Interpretation of an experiment on the anomalous absorption of an electromagnetic wave in a plasma, *Phys. Rev. Lett.*, **25**, 430–3 (1970). [62, 76]

KAW, P. *et al.* (1971) (with DAWSON, J., KRUER, W., OBERMAN, C. & VALEO, E.). Anomalous heating of plasma by laser irradiation, *Kvantovaya Elektron.*, **3**, 3–14 (1971). (Transl: *Sov. J. Quantum Electron.*, **1**, 205–12 (1971).) [76]

KAZAKOV, A. E. *et al.* (1971) (with KRASYUK, I. K., PASHININ, P. P. & PROKHOROV, A. M.). Experimental observation of the amplification of laser radiation in the interaction of colliding laser beams in a plasma, *Zh. Eksp. i Teor. Fiz. Pis'ma*, **14**, 416–8 (1971). (Transl: *JETP Lett.*, **14**, 280–2 (1971). [73, 79]

KEEN, B. E. & FLETCHER, W. H. W. (1971). Decay instability at the ion-sound frequency induced by a large amplitude Bernstein mode in a plasma, *J. Phys. A.*, **4**, L67–73 (1971). [76]

KEEN, B. E. & FLETCHER, W. H. M. (1973). Plasma heating effects in the presence of a parameteric decay instability, *J. Phys. A.*, **6**, L24–9 (1973). [76]

KEILHACKER, M. & STEUER, K. H. (1971). Time-resolved light-scattering measurements of the spectrum of turbulence within a high-β collisionless shock wave, *Phys. Rev. Lett.*, **26**, 694–7 (1971). [127]

KELDYSH, L. V. (1964). Ionization in the field of a strong electromagnetic wave, *Zh. Eksp. i Teor. Fiz.*, **47**, 1945–57 (1964). (Transl: *Sov. Phys. JETP*, **20**, 1307–14 (1965).) [149, 151]

KELLERER, L. (1970a). Modulated electron spectrum of the light scattered by a magnetized arc plasma, *Z. Phys.*, **232**, 415–17 (1970). [128]

KELLERER, L. (1970b). Measuring magnetic fields in plasmas by means of light scattering, *Z. Phys.*, **239**, 147–61 (1970). [128]

KELLEY, P. L. (1965). Self-focusing of optical beams, *Phys. Rev. Lett.*, **15**, 1005–8 (1965). [195]

KELLEY, P. L. & GUSTAFSON, T. K. (1973). Backward stimulated light scattering and the limiting diameters of self-focused light beams, *Phys. Rev.* **8A**, 315–8 (1973). [196]

KENNARD, E. H. (1938). *Kinetic theory of gases*. New York, McGraw-Hill (1938). [37]

KEPHART, J. F. *et al.* (1974) (with GODWIN, R. P. & McCALL, G. H.). Bremsstrahlung emission from laser-produced plasmas, *Appl. Phys. Lett.*, **25**, 108–9 (1974). [394]

KEY, M. H. (1969). Ionization effects in a hydrodynamic model of radiation-driven breakdown wave propagation, *J. Phys. B.*, **2**, 544–9 (1969). [46, 212, 213, 243]

KEY, M. H. *et al.* (1970) (with PRESTON, D. A. & DONALDSON, T. P.). Self-focusing in gas breakdown by laser pulses, *J. Phys. B.*, **3**, L88–92 (1970). [196, 198, 248]

KEY, M. H. *et al.* (1974a) (with EIDMANN, K., DORN, C. & SIGEL, R.). Time-resolved absolute X-ray measurements on laser-produced plasmas, *Appl. Phys. Lett.*, **25**, 335–7 (1974). [390]

KEY, M. H. *et al.* (1974b) (with EIDMANN, K., DORN, C. & SIGEL, R.). Space resolved measurements on laser produced plasmas, *Phys. Lett.* **48A**, 121–2 (1974). [390]

KIDDER, R. E. (1968). Application of lasers to the production of high-temperature and high-pressure plasma, *Nucl. Fusion*, **8**, 3–12 (1968). [220, 221, 286, 290, 302, 319, 320]

KIDDER, R. E. (1971). Interaction of intense photon and electron beams with plasmas. In *Physics of high energy density* (ed. P. Caldirola and H. Knoepfel), pp. 306–52. New York, Academic (1971). [280]

KIDDER, R. E. & ZINK, J. W. (1972). Decoupling of corona and core of laser-heated pellets. *Nucl. Fusion*, **12**, 325–8 (1972). [329, 330]

KIMMITT, M. F. (1970). *Far-infra-red techniques*. London, Pion (1970). [111]

KIMMITT, M. F. & NIBLETT, G. B. F. (1963). Infra-red emission from the θ-pinch, *Proc. Phys. Soc.*, **82**, 938–46 (1963). [52]

KIMMITT, M. F. *et al.* (1961) (with PRIOR, A. C. & SMITH, P. G.). Observation of far-infra-red radiation from shock waves, *Nature*, **190**, 599–601 (1961). [52]

KINDEL, J. M. *et al.* (1972) (with OKUDA, H. & DAWSON, J. M.). Parametric instabilities and anomalous heating of plasmas near the lower hybrid frequency, *Phys. Rev. Lett.*, **29**, 995–7 (1972). [60]

KIVEL, B. (1967a). Neutral atom bremsstrahlung, *J. Quant. Spectrosc. & Radiat. Transfer*, **7**, 27–49 (1967). [48]

KIVEL, B. (1967b). Bremsstrahlung in air, *J. Quant. Spectrosc. & Radiat. Transfer*, **7**, 51–60 (1967). [48, 53]

KLARSFELD, S. & MAQUET, A. (1972). Circular versus linear polarization in multiphoton ionization, *Phys. Rev. Lett.*, **29**, 79–81 (1972). [153]

KLEIN, H. H. *et al.* (1973) (with MANHEIMER, W. M. & OTT, E.). Effect of side-scatter instabilities on the propagation of an intense laser beam in an inhomogeneous plasma, *Phys. Rev. Lett.*, **31**, 1187–90 (1973). [71]

KLEWE, R. C. & RIZZO, J. E. (1968). Optical breakdown of potassium, *VIII International Conference on phenomena in ionized gases, Vienna (1967). Contributed papers*, p. 263. Vienna, IAEA (1968). [176]

KOMISSAROVA, I. I. *et al.* (1968) (with OSTROVSKAYA, G. V. & SHAPIRO, L. L.). Holograph investigation of laser sparks, *Zh. Tekh. Fiz.*, **38**, 1369–73 (1968). (Transl: *Sov. Phys. Tech. Phys.*, **13**, 1118–21 (1968).) [229, 248]

KOMISSAROVA, I. I. *et al.* (1969) (with OSTROVSKAYA, G. V., SHAPIRO, L. L. & ZAIDEL, A. N.). Two-wavelength holography of a laser spark, *Phys. Lett.*, **29A**, 262–3 (1969). [230, 249]

KOMISSAROVA, I. I. *et al.* (1970) (with OSTROVSKAYA, G. V. & SHAPIRO, L. L.). Holographic studies of a laser spark. II Double interferometer at long wavelength, *Zh. Tekh. Fiz.*, **40**, 1072–80 (1970). (Transl: *Sov. Phys. Tech. Phys.*, **15**, 827–31 (1970).) [214, 230, 249, 250]

KOOPMAN, D. W. (1967). Ion emission from laser-produced plasmas, *Phys. Fluids*, **10**, 2091–3 (1967). [405]

KOOPMAN, D. W. (1971a). Laser-generation of rarefied plasma flows. In *Modern optical methods in gas dynamic research* (ed. D. S. Dosanjh), pp. 177–96. New York, Plenum (1971). [440]

KOOPMAN, D. W. (1971b). Langmuir probe and microwave measurements of the properties of streaming plasma generated by focused laser pulses, *Phys. Fluids*, **14**, 1707–16 (1971). [370, 371]

KOOPMAN, D. W. (1972a). Precursor ionization fronts ahead of expanding laser-plasmas, *Phys. Fluids*, **15**, 56–62 (1972). |439, 440|

KOOPMAN, D. W. (1972b). Momentum transfer interaction of a laer-produced plasma with a low-pressure background, *Phys. Fluids*, **15**, 1959–69 (1972). |439, 440, 444|

KORNHERR, M. *et al.* (1972). (with DECKER, G., KEILHACKER, M., LINDENBERGER, F. & RÖHR, H.). CO_2 laser scattering measurements of turbulence in a high-β collisionless shock wave, *Phys. Lett.*, **39A**, 95–7 (1972). |111, 127|

KOROBKIN, V. V. & ALCOCK, A. J. (1968). Self-focusing effects associated with laser-induced air breakdown, *Phys. Rev. Lett.*, **21**, 1433–35 (1968). |197, 245|

KOROBKIN, V. V. & SEROV, R. V. (1966). Investigation of the magnetic field of a spark produced by focusing laser radiation, *Zh. Eksp. i Teor. Fiz. Pis'ma*, **4**, 103–6 (1966). (Transl: *JETP Lett.* **4**, 70–2 (1966).) |272|

KOROBKIN, V. V. *et al.* (1967) (with MANDEL'SHTAM, S. L., PASHININ, P. P., PROKHINDEEV, A. V., PROKHOROV, A. M., SUKHODREV, N. K. & SHCHELEV, M. YA.). Investigation of the air 'spark' produced by focused laser radiation. III, *Zh. Eksp. i Teor. Fiz.*, **53**, 116–23 (1967). (Transl: *Sov. Phys. JETP*, **26**, 79–85 (1968).) |197, 214, 224, 236, 252, 255, 258|

KOTOVA, L. P. & TERENTEV, M. V. (1967). Resonance ionization of atoms in the field of a strong electromagnetic wave, *Zh. Eksp. i Teor. Fiz.*, **52**, 732–41 (1967). (Transl: *Sov. Phys. JETP*, **25**, 481–7 (1967).) |151|

KOVRIZHNYKH, L. M. *et al.* (1966) (with LIPEROVSKII, V. A. & TSYTOVICH, V. N.). Nonlinear generation of plasma waves by a beam of transverse waves. II, *Zh. Tekh. Fiz.*, **36**, 1339–50 (1966). (Transl: *Sov. Phys. Tech. Phys.*, **11**, 1000–6 (1967).) |60|

KRALL, N. A. & TRIVELPIECE, A. W. (1973). *Principles of plasma physics*. New York, McGraw-Hill (1973). |19, 21, 52, 113, 429|

KRAMERS, H. A. (1923). Theory of X-ray absorption and of the continuous X-ray spectrum, *Phil. Mag.*, **46**, 836–71 (1923). |40|

KRASYUK, I. K. & PASHININ, P. P. (1972). Breakdown in argon and nitrogen under the influence of a $0·35\,\mu$ picosecond laser pulse, *Zh. Eksp. i Teor. Fiz. Pis'ma*, **15**, 471–3 (1972). (Transl: *JETP Lett.*, **15**, 333–4 (1972).) |187|

KRASYUK, I. K. *et al.* (1969) (with PASHININ, P. P. & PROKHOROV, A. M.). Investigation of breakdown in N_2 under the influence of a picosecond ruby-laser pulse, *Zh. Eksp. i Teor. Fiz. Pis'ma*, **9**, 581–4 (1969). (Transl: *JETP Lett.*, **9**, 354–6 (1969).) |187, 188|

KRASYUK, I. K. *et al.* (1970a) (with PASHININ, P. P. & PROKHOROV, A. M.). Experimental observation of stimulated Compton absorption of laser radiation in a spark, *Zh. Eksp. i Teor. Fiz. Pis'ma*, **12**, 439–42 (1970).) (Transl: *JETP Lett.*, **12**, 305–7 (1970).) |78, 420|

KRASYUK, I. K. *et al.* (1970b) (with PASHININ, P. P. & PROKHOROV, A. M.). Investigation of breakdown in argon and helium produced by a picosecond ruby laser light pulse, *Zh. Eksp. i Teor. Fiz.*, **58**, 1606–8 (1970). (Transl: *Sov. Phys. JETP*, **31**, 860–1 (1970).) |187, 188|

KRASYUK, I. K. *et al.* (1973) (with PASHININ, P. P. & PROKHOROV, A. M.). Role of stimulated Compton scattering in the interaction of laser radiation with a superdense plasma, *Zh. Eksp. i Teor. Fiz. Pis'ma*, **17**, 130–2 (1973). (Transl: *JETP Lett.*, **17**, 92–4 (1973). |73, 74|

KROKHIN, O. N. (1964). 'Matched' plasma heating mode using laser radiation, *Zh. Tekh. Fiz.*, **34**, 1324–7 (1964). (Transl: *Sov. Phys. Tech. Phys.*, **9**, 1024–6 (1965).) |286|

KROKHIN, O. N. (1965). Self-regulating regime of plasma heating by laser radiation, *Z. Angew. Math. & Phys.*, **16**, 123–4 (1965). |286|

KROKHIN, O. N. (1971). High-temperature and plasma phenomena induced by laser radiation. In *Physics of high energy density* (ed. P. Caldirola and H. Knoepfel), pp. 278–305. New York, Academic (1971). |277,287|

KROKHIN, O. N. & ROZANOV, V. B. (1972). Escape of α particles from a laser-pulse-initiated thermonuclear reaction, *Kvantovaya Elektron.*, **4**, 118–20 (1972). (Transl: *Sov. J. Quantum Electron.*, **2**, 393–4 (1973).) |135|

KROLL, N. & WATSON, K. M. (1972). Theoretical study of ionization of air by intense laser pulses, *Phys. Rev.*, **A5**, 1883–1905 (1972). |169, 191|

KROLL, N. M. *et al.* (1964) (with RON, A. & ROSTOKER, N.). Optical mixing as a plasma density probe, *Phys. Rev. Lett.*, **13**, 83–6 (1964). [70, 78, 102, 127]

KRONAST, B. & BENESCH, R. (1971). Observation of a line structure in the light scattering spectrum of electron plasma waves in a magneto plasma, *X International Conference on phenomena in ionized gases, Oxford (1971). Contributed papers*, p. 415. Oxford, Donald Parsons & Co. Ltd. (1971). [131]

KRONAST, B. & PIETRZYK, Z. A. (1971). Discrepancy between electron-drift velocities obtained from the ion and electron features of light scattering in a z-pinch plasma, *Phys. Rev. Lett.*, **26**, 67–9 (1971). [117]

KRONAST, B. *et al.* (1966) (with RÖHR, H., GLOCK, E., ZWICKER, H. & FÜNFER, E.). Measurements of the ion and electron temperature in a theta-pinch plasma by forward scattering, *Phys. Rev. Lett.*, **16**, 1082–5 (1966). [112, 119, 120]

KRUER, W. L. (1972). Efficient energy transfer between fast and slow electron plasma oscillations, *Phys. Fluids*, **15**, 2423–6 (1972). [67]

KRUER, W. L. & DAWSON, J. M. (1970). Anomalous damping of large-amplitude electron-plasma oscillations, *Phys. Rev. Lett.*, **25**, 1174–6 (1970). [67]

KRUER, W. L. & DAWSON, J. M. (1972). Anomalous high-frequency resistivity of a plasma, *Phys. Fluids*, **15**, 446–53 (1972). [59, 65]

KRUER, W. L. & VALEO, E. J. (1973). Nonlinear evolution of the decay instability in a plasma with comparable electron and ion temperatures, *Phys. Fluids*, **16**, 675–82 (1973). [65]

KRUER, W. L. *et al.* (1970) (with KAW, P. K., DAWSON, J. M. & OBERMAN, C.). Anomalous high-frequency resistivity and heating of a plasma, *Phys. Rev. Lett.*, **24**, 987–90 (1970). [65]

KUNZE, H. J. (1965). Messung der lokalen Elektronentemperatur und Elektronendichte in einem θ-pinch mittels der Streuung eines Laserstrahls, *Z. Naturforsch*, **20A**, 801–13 (1965). [117]

KUNZE, H. J. *et al.* (1964) (with FÜNFER, E., KRONAST, B. & KEGEL, W. H.). Measurement of the spectral distribution of light scattered by a θ-pinch plasma, *Phys. Lett.*, **11**, 42–3 (1964). [117, 119]

KUNZE, H. J. *et al.* (1968) (wuth GABRIEL, A. H. & GRIEM, H. R.). Electron temperatures and densities from transient helium-like ion line emission, *Phys. Fluids*, **11**, 662–8 (1968). [341, 371]

KUO, Y. Y. & FEJER, J. A. (1972). Spectral-line structures of saturated parametric instabilities, *Phys. Rev. Lett.*, **29**, 1667–70 (1972). [65]

LAEGREID, N. & DAHLGREN, S. D. (1973). Controlled thermonuclear reactor first wall sputtering and wall life estimates, *J. Appl. Phys.* **44**, 2093–6 (1973). [144]

LAMBROPOULOS, P. (1972a). Multiphoton ionization of one-electron atoms with circularly polarized light, *Phys. Rev. Lett.*, **29**, 453–5 (1972). [153]

LAMBROPOULOS, P. (1972b). Effect of light polarization on multiphoton ionization of atoms, *Phys. Rev. Lett.*, **28**, 585–7 (1972). [153]

LAMBROPOULOS, P. (1972c). Effect of spatio-temporal laser light structure on multiphonon ionization, *Phys. Lett.*, **40A**, 199–200 (1972). [154]

LAMPIS, G. & BROWN, S. C. (1968). Afterglow measurements of a laser breakdown plasma, *Phys. Fluids*, **11**, 1137–46 (1968). [233, 234, 251, 252, 365]

LANCZOS, C. (1931). Decrease in intensity of spectral lines in strong electric fields, *Z. Phys.* **68**, 204–32 (1931). [147]

LANDAU, L. D. & LIFSHITZ, E. M. (1959). *Statistical physics.* Oxford, Pergamon (1959). [323]

LANDAU, L. D. & LIFSHITZ, E. M. (1960). *Electrodynamics of continuous media, v.* p. 242. Oxford, Pergamon 1960). [197]

LANDAU, L. D. & LIFSHITZ, E. M. (1965). *Quantum Mechanics*, 2nd ed. *v.* p. 275. Oxford, Pergamon (1965). [147]

LANGER, P. *et al.* (1966) (with TONON, G., FLOUX, F. & DUCAUZE, A.). Laser induced emission of electrons, ions and X-rays from solid targets, *IEEE J. Quantum Electron.*, **2**, 499–506 (1966). [345]

LANGER, P. *et al.* (1968*a*) (with TONON, G., DURAND, Y. & BUGES, J. C.). Étude chronologique du plasma de béryllium créé par un laser, *J. Physique, Suppl.* **C3** *to* **29**, 132–6 (1968). [345, 350]

LANGER, P. *et al.* (1968*b*) (with PIN, B. & TONON, G.). Utilization du laser dans l'analyse isotopique d'un échantillon de DLi, *Revue Phys. Appl.*, **3**, 405–13 (1968). [394]

LASHMORE-DAVIES, C. N. (1973). The coupled mode approach to non-linear wave interactions and parametric instabilities, *UKAEA research group preprint*, CLM-P342. Abingdon, Culham Laboratory (1973). [55]

LAWSON, J. D. (1957). Some criteria for a power producing thermonuclear reactor, *Proc. Phys. Soc.*, **B70**, 6-10 (1957). [137]

LAX, B. *et al.* (1950) (with ALLIS, W. P. & BROWN, S. C.). The effect of magnetic field on the breakdown of gases at microwave frequencies, *J. Appl. Phys.* **21**, 1297–304 (1950). [193]

LEDENEV, V. I. *et al.* (with SUKHORUKOV, A. P. & KHACHATRYAN, A. M.). Changes in the structure of a focal region due to spatial self-focusing of short pulses, *Kvantovaya Elektron.*, **2** (8), 90–4 (1972). (Transl: *Sov. J. Quantum Electron.*, **2**, 163–6 (1972).) [196]

LEE, P. S. *et al.* (1973) (with LEE, Y. C., CHANG, C. T. & CHUU, D. S.). Saturation amplitude of non-linearly excited plasma waves, *Phys. Rev. Lett.*, **30**, 538–40 (1973). [70]

LEE, Y. C. & SU, C. H. (1966). Theory of parametric coupling in plasmas, *Phys. Rev.*, **152**, 129–35 (1966). [60]

LEHNER, G. & POHL, F. (1970). On the possibility of measuring magnetic fields by scattered light, *Z. Phys.*, **232**, 405–14 (1970). [96]

LENCIONI, D. E. (1973). The effect of dust on 10·6 μm laser-induced air breakdown, *Appl. Phys. Lett.*, **23**, 12–14 (1973). [191]

LEONARD, T. A. & BACH, D. R. (1973). Light scattering from an exploded lithium wire plasma, *J. Appl. Phys.*, **44**, 2555–65 (1973). [126]

LEPECHINSKI, D. (1967). Effet des collisions sur le spectre de diffusion incohérente d'un plasma de températures électronique et ionique différentes, *C.R. Acad. Sci., Paris*, **265B**, 1085–8 (1967). [99]

LEWKOWICZ, I. (1974). Spherical hydrogen targets for laser-produced fusion, *J. Phys. D.* **7**, L61-2 (1974). [400]

LIBENSON, M. N. *et al.* (1968) (with ROMANOV, G. S. & IMAS, YA. A.). Temperature dependence of the optical constants of a metal in heating by laser radiation, *Zh. Tech. Fiz.*, **38**, 1116–19 (1968). (Transl: *Sov. Phys. Tech. Phys.*, **13**, 925–8 (1969).) [276]

LICHTMAN, D. & READY, J. F. (1963). Laser beam induced electron emission, *Phys. Rev. Lett.*, **10**, 342–4 (1963). [339, 340]

LINHART, J. G. (1968). A note on power production by pulsed fusion reactors, *Nucl. Fusion*, **8**, 273–4 (1968). [141, 143]

LINHART, J. G. (1970). Very-high-density plasmas for thermonuclear fusion, *Nucl. Fusion*, **10**, 211–34 (1970). [141]

LINLOR, W. I. (1963). Ion energies produced by laser giant pulse, *Appl. Phys. Lett.*, **3**, 210–11 (1963). [340, 395, 429]

LINLOR, W. I. (1964). Some properties of plasma produced by laser giant pulse, *Phys. Rev. Lett.*, **12**, 383–5 (1964). [395, 429]

LINNEBUR, E. J. & DUDERSTADT, J. J. (1973*a*). Theory of light scattering from dense plasmas, *Phys. Fluids*, **16**, 665–74 (1973). [98, 99]

LINNEBUR, E. J. & DUDERSTADT, J. J. (1973*b*). A microscopic theory of density fluctuations in partially ionized gases, *Plasma Phys.*, **15**, 647–58 (1973). [99]

LITVAK, A. G. & TRAKHTENGERTS, V. YU. (1971). Induced scattering of waves and plasma heating by coherent radiation, *Zh. Eksp. i Teor. Fiz.*, **60**, 1702–13 (1971). (Transl: *Sov. Phys. JETP*, **33**, 921–6 (1971).) [71]

LITVAK, M. M. & EDWARDS, D. F. (1966). Electron recombination in laser-produced hydrogen plasma, *J. Appl. Phys.* **37**, 4462–74 (1966). [53, 194, 231, 242]

LIU, C. S. *et al.* (1973) (with ROSENBLUTH, M. N. & WHITE, R. B.). Parametric scattering instabilities in inhomogeneous plasmas, *Phys. Rev. Lett.*, **31**, 697–700 (1973). [71]

LOUIS-JACQUET, M. (1970). Étude de l'ionisation des gaz par faisceau laser au moyen d'un compteur proportionnel, *C.R. Acad. Sci., Paris*, **270B**, 548–51 (1970). [157]

LOWDER, J. E. *et al.* (1973) (with LENCIONI, D. E., HILTON, T. W. & HULL, R. J.). High-energy pulsed CO_2-laser-target interactions in air, *J. Appl. Phys.*, **44**, 2759–62 (1973). [441]

LUBIN, M. J. *et al.* (1969) (with DUNN, H. S. & FRIEDMAN, W.). Heating and confinement studies of laser-irradiated solid-particle plasmas, *International Conference on plasma physics and controlled nuclear fusion research. Novosibirsk (1968)*, pp. 945–65. Vienna, IAEA (1969). [335, 407, 408, 431]

LUBIN, M. *et al.* (1972) (with SOURES, J., GOLDMAN, E. & BRISTOW, T.). Laser heated overdense plasmas for thermonuclear fusion. In *Laser interactions and related plasma phenomena* (ed. H. J. Schwarz and H. Hora), Vol. 2, pp. 433–67. New York, Plenum (1972). [320, 407]

LUDWIG, D. & MAHN, C. (1971). Anomalous scattering of laser light from a magnetized arc plasma, *Phys. Lett.*, **35A**, 191–2 (1971). [130, 131]

LUGOVOI, V. N. & PROKHOROV, A. M. (1968). A possible explanation of the small-scale self-focusing filaments, *Zh. Eksp. i Teor. Fiz. Pis'ma*, **7**, 153–5 (1968). (Transl: *JETP Lett.*, **7**, 117–19 (1968).) [196, 197]

LU VAN, M. & MAINFRAY, G. (1972). Multiphoton ionization and dissociation of the nitrogen molecule at $\lambda = 0\cdot53\mu$, *Phys. Lett.*, **39A**, 21–2 (1972). [161]

LU VAN, M. *et al.* (1972) (with MAINFRAY, G., MANUS, C. & TUGOV, I.). Multiphoton ionization and dissociation of molecular hydrogen at $1\cdot06\,\mu$m, *Phys. Rev. Lett.*, **29**, 1134–7 (1972). [154, 160, 161]

LU VAN, M. *et al.* (1973) (with MAINFRAY, G., MANUS, C. & TUGOV, I.). Multiphoton ionization of atomic and molecular hydrogen at $0\cdot53\,\mu$, *Phys. Rev.*, **A7**, 91–8 (1973). [154, 161]

McCLUNG, F. J. & HELLWARTH, R. W. (1962). Giant optical pulsations from ruby, *J. Appl. Phys.*, **33**, 828–9 (1962). [340]

MACDONALD, A. D. (1966). *Microwave breakdown in gases.* New York, Wiley (1966). [164, 166, 167, 172, 174, 193, 194]

MACHALEK, M. D. & NIELSEN, P. (1973). Light-scattering measurements of turbulence in a normal shock, *Phys. Rev. Lett.*, **31**, 439–42 (1973). [127]

McWHIRTER, R. W. P. (1965). Spectral intensities. In *Plasma diagnostic techniques* (ed. R. H. Huddlestone and S. L. Leonard), pp. 201–64. New York, Academic (1965). [37, 39]

MAHER, W. E. *et al.* (1974) (with HALL, R. B. & JOHNSON, R. R.). Experimental study of ignition and propagation of laser-supported detonation waves, *J. Appl. Phys.*, **45**, 2138–45 (1974). [441]

MAINFRAY, G. *et al.* (1972) (with MANUS, C. & TUGOV, I.). Multiphoton dissociation, predissociation, and autoionization of the hydrogen molecule, *Zh. Eksp. i Teor. Fiz. Pis'ma*, **16**, 19–23 (1972). (Transl: *JETP Lett.*, **16**, 12–15 (1972).) [161]

MAKER, P. D. *et al.* (1964) (with TERHUNE, R. W. & SAVAGE, C. M.). Optical third harmonic generation, *Proc. 3rd International Conference on quantum electronics, Paris (1963)*, Vol. 2, pp. 1559–76. New York, Columbia University Press (1964). [145]

MAKHANKOV, V. G. & TSYTOVICH, V. N. (1973). Anomalous heating of dense plasma by laser radiation, *Phys. Scripta*, **7**, 234–40 (1973). [60]

MALYSHEV, G. M. (1965). Plasma diagnostics by light scattering on electrons, *Zh. Tekh. Fiz.*, **35**, 2129–42 (1965). (Transl: *Sov. Phys. Tech. Phys.*, **10**, 1633–43 (1966).) [113]

MALYSHEV, G. M. *et al.* (1966a) (with OSTROVSKAYA, G. V. & CHELIDZE, T. YA.). Shadow projections of a spark in air occurring when laser radiation is brought to focus, *Opt. i Spektrosk.*, **20**, 374–5 (1966). (Transl: *Opt. & Spectrosc.*, **20**, 207–8 (1966).) [223, 224]

MALYSHEV, G. M. *et al.* (1966b) (with OSTROVSKAYA, G. V., RAZDOBARIN, G. T. & SOKOLOVA, L. V.). Measurement of electron temperature and concentration in an arc plasma via Thomson scattering of laser light, *Dokl. Akad. Nauk SSSR*, **168**, 554–5 (1966). (Transl: *Sov. Phys. Dokl.* **11**, 441–2 (1966).) [112, 113]

MALYSHEV, G. M. *et al.* (1972) (with RAZDOBARIN, G. T. & SEMENOV, V. V.). Determination of parameters of laser-induced plasma in air by a scattering method, *Zh. Tekh. Fiz.*, **42**, 1429–31 (1972). (Transl: *Sov. Phys. Tech. Phys.*, **17**, 1137–9 (1973).) [251]

MANAKOV, N. L. *et al.* (1971*a*) (with RAPOPORT, L. P. & ZON, B. A.). Enlightenment of the absorbtive medium induced by stark shift of multiphoton resonance, *X International Conference on phenomena in ionized gases, Oxford (1971). Contributed papers*, p. 38. Oxford, Donald Parsons & Co. Ltd. (1971). [152]

MANAKOV, N. L. *et al.* (1971*b*) (with RAPOPORT, L. P. & ZON, B. A.). The perturbation theory for multiphoton ionisation of atoms, *X International Conference on phenomena in ionized gases, Oxford (1971). Contributed papers*, p. 46. Oxford, Donald Parsons & Co. Ltd. (1971). [152]

MANDEL'SHTAM, S. L. *et al.* (1964) (with PASHININ, P. P., PROKHINDEEV, A. V., PROKHOROV, A. M. & SUKHODREV, N. K.). Study of the 'spark' produced in air by focused laser radiation, *Zh. Eksp. i Teor. Fiz.*, **47**, 2003–5 (1964). (Transl: *Sov. Phys. JETP*, **20**, 1344–6 (1964).) [250]

MANDEL'SHTAM, S. L. *et al.* (1965) (with PASHININ, P. P., PROKHOROV, A. M., RAIZER, YU. P. & SUKHODREV, N. K.). Investigation of the spark discharge produced in air by focusing laser radiation II, *Zh. Eksp. i Teor. Fiz.*, **49**, 127–34 (1965). (Transl: *Sov. Phys. JETP*, **22**, 91–6 (1966).) [231, 235, 250]

MANLEY, J. M. & ROWE, H. E. (1956). Some general properties of nonlinear elements—Part I. General energy relations, *Proc. IRE*, **44**, 904–13 (1956). [70]

MANLEY, J. M. & ROWE, H. E. (1959). General energy relations in nonlinear reactances, *Proc. IRE*, **47**, 2115–16 (1959). [70]

MARQUET, L. C. *et al.* (1972) (with HULL, R. J. & LENCIONI, D. E.). Studies in breakdown in air induced by a pulsed CO_2 laser, *IEEE J. Quantum Electron.*, **8**, 564 (1972). [191]

MARTINEAU, J. (1971). CO_2 laser heating of an inhomogeneous slab of deuterium resistif θ-pinch plasma, *Phys. Lett.*, **36A**, 67–8 (1971). [222]

MARTINEAU, J. (1974). Laser plasma heating in a strong magnetic field, *Phys. Fluids*, **17**, 131–6 (1974) [335]

MARTINEAU, J. & PÉPIN, H. (1971). Heating of a θ-pinch plasma by a CO_2 laser, *Can. J. Phys.*, **49**, 1685–7 (1971). [222]

MARTINEAU, J. & PÉPIN, H. (1972). CO_2 laser heating of a θ-pinch plasma by inverse bremsstrahlung and induced Compton processes, *J. Appl. Phys.*, **43**, 917–22 (1972). [73]

MARTINEAU, J. & TONON, G. (1969*a*). Study of ionization of a laser irradiated deuterium target, *Phys. Lett.*, **30A**, 32–3 (1969). [279]

MARTINEAU, J. & TONON, G. (1969*b*). Evaluation of a magnetically confined laser created plasma, *Phys. Lett.*, **28A**, 710–11 (1969). [334, 335]

MARTINEAU, J. *et al.* (1974) (with REPOUX, S., RABEAU, M., NIERAT, G. & ROSTAING, M.). Reflection measurements in CO_2 laser solid deuterium target interaction, *Opt. Commun.*, **12**, 307–11 (1974). [425]

MARTONE, M. & SEGRE, S. E. (1969). Cooperative laser scattering at 90° from a cold dense plasma, *Phys. Lett.*, **28A**, 610–11 (1969). [126]

MATOBA, T. (1972). Measurements of laser-produced plasmas from lithium targets by retarding potential probe, *Jap. J. Appl. Phys.*, **11**, 770–1 (1972). [369]

MATTIOLI, M. (1971). Recombination processes during the expansion of a laser-produced plasma, *Plasma Phys.*, **13**, 19–28 (1971). [318]

MATTIOLI, M. & VÉRON, D. (1969). Electron–ion recombination in laser produced plasma, *Plasma Phys.*, **11**, 684–6 (1969). [429]

MATTIOLI, M. & VÉRON, D. (1971). Production of energetic highly ionized plasma by laser irradiation of Lithium Hydride pellets, *Phys. Fluids*, **14**, 717–21 (1971). [407]

MAX, C. E. (1973). Parametric instability of a relativistically strong electromagnetic wave, *Phys. Fluids*, **16**, 1480–9 (1973). [75]

MAX, C. E. & PERKINS, F. (1971). Strong electromagnetic waves in overdense plasmas, *Phys. Rev. Lett.*, **27**, 1342–5 (1971). [75]

MAXON, S. (1972). Bremsstrahlung rate and spectra from a hot gas ($Z = 1$), *Phys. Rev.*, **A5**, 1630–3 (1972). [50, 51]

MEAD, S. W. (1970). Plasma production with a multibeam laser system, *Phys. Fluids*, **13**, 1510–8 (1970). [259, 260, 261, 438]

MEAD, S. W. (1971). Reply to comments by Herman Presby, *Phys. Fluids*, **14**, 2789 (1971). [259]

MEAD, S. W. *et al.* (1972) (with KIDDER, R. E., SWAIN, J. E., RANIER, F. & PETRUZZI, J.). Preliminary measurements of X-ray and neutron emission from laser-produced plasmas, *Appl. Opt.*, **11**, 345–52 (1972). [390, 391]

MENNICKE, H. (1971a). A new ionization model for laser produced plasmas by means of high order Raman-Antistokes light, *Phys. Lett.*, **37A**, 381–2 (1971). [279]

MENNICKE, H. (1971b). Stimulated Raman scattering in the initial phase of plasma production by laser, *Phys. Lett.*, **36A**, 127–8 (1971). [426]

MENZEL, D. H. & PEKERIS, C. L. (1935). Absorption coefficients and H line intensities, *Mon. Not. R. Astron. Soc.*, **96**, 77–111 (1935). [43]

MERCIER, R. P. (1964). Thermal radiation in anisotropic media, *Proc. Phys. Soc.*, **83**, 811–17 (1964). [29]

MEYER, H. C. & SHATAS, R. A. (1972). Comments on 'Laser-induced anomalous heating of a plasma', *Phys. Fluids*, **15**, 1542–3 (1972). [62]

MEYER, J. (1972). Plasma density fluctuations in the presence of optical mixing of light beams, *Phys. Rev.*, **A6**, 2291–7 (1972). [70]

MEYER, R. L. & LECLERT, G. (1972a). Plasma scattering cross-section of an electromagnetic wave with the wave vector $\vec{k}$ nearly perpendicular to the confining magnetic field, *Nucl. Fusion*, **12**, 128–9 (1972). [98]

MEYER, R. L. & LECLERT, G. (1972b). Measurement of the magnetic-field direction in a Tokamak plasma by ruby-laser light scattering, *Nucl. Fusion*, **12**, 269–70 (1972). [98]

MEYER, R. L. *et al.* (1973) (with LECLERT, G. & FELDEN, M.). Étude numérique de la section éfficace de diffusion d'un faisceau laser par un plasma en champ magnétique. Application à la détermination expérimentale de la direction du champ magnétique de confinement. *Can. J. Phys.*, **51**, 266–76 (1973). [98]

MEYERAND, R. G. & HAUGHT, A. F. (1963). Gas breakdown at optical frequencies, *Phys. Rev. Lett.*, **11**, 401–3 (1963). [172]

MEYERAND, R. G. & HAUGHT, A. F. (1964). Gas breakdown at optical frequencies, *Proc. VI International Symposium on ionization phenomena in gases, Paris (1963)*. Vol. 2, pp. 479–90. Paris, S.E.R.M.A. (1964). [164]

MINCK, R. W. (1964). Optical frequency electrical discharges in gases, *J. Appl. Phys.*, **35**, 252–4 (1964). [167, 169, 172]

MINCK, R. W. & RADO, W. G. (1966a). Investigation of optical frequency breakdown phenomena. In *Physics of Quantum Electronics* (ed. P. L. Kelley, B. Lax and P. E. Tannenwald), pp. 527–37. New York, McGraw-Hill (1966). [172, 173]

MINCK, R. W. & RADO, W. G. (1966b). Optical frequency plasma resonance in gases, *J. Appl. Phys.*, **37**, 355–8 (1966). [237, 238]

MIRELS, H. & MULLEN, J. F. (1963). Expansion of gas clouds and hypersonic jets bounded by a vacuum, *AIAA J.*, **1**, 596–602 (1963). [318, 355]

MITCHELL, J. T. D. & HANCOX, R. (1972). A lithium cooled toroidal fusion reactor, *Proc. 7th Intersociety Energy Conversion Engineering Conference, San Diego (1972)*, pp. 1275–83, Washington D.C., Am. Chem. Soc. (1972). [144]

MITSUK, V. E. *et al.* (1966) (with SAVOSKIN, V. I. & CHERNIKOV, V. A.). Breakdown at optical frequencies in the presence of diffusion losses, *Zh. Eksp. i Teor. Fiz. Pis'ma*, **4**, 129–31 (1966). (Transl: *JETP Lett.*, **4**, 88–90 (1966).) [179]

MITSUK, V. E. *et al.* (1971) (with SAVVINA, R. M. & CHERNIKOV, V. A.). The gases mixture optical breakdown, *X International Conference on phenomena in ionized gases, Oxford (1971). Contributed papers*, p. 233. Oxford, Donald Parsons & Co. Ltd. (1971). [181]

MJOLSNESS, R. C. & RUPPEL, H. M. (1967). Contribution of inverse neutral bremsstrahlung to the absorption coefficient of heated air, *J. Quant. Spectrosc. & Radiat. Transfer*, **7**, 423–7 (1967). [53]

MJOLSNESS, R. C. & RUPPEL, H. M. (1972). Deuterium–Tritium heating to thermonuclear temperatures by means of ion–ion collisions in the presence of intense laser radiation, *Phys. Fluids*, **15**, 1620–9 (1972). [49]

MOHAN, M. (1973). Simultaneous multiphoton ionization and excitation of helium in an intense laser beam, *J. Phys. B.*, **6**, 1218–27 (1973). [153]

MOHAN, M. & THAREJA, R. K. (1972). Multiphoton ionization of atomic hydrogen in the presence of an intense electromagnetic field, *J. Phys. B.*, **5**, L134–6 (1972). [153]

MOHAN, M. & THAREJA, R. K. (1973a). Raman processes in intense fields, *Phys. Rev.*, **A7**, 34–8, (1973). [153].

MOHAN, M. & THAREJA, R. K. (1973b). Multiphoton ionization of hydrogen atom in intense laser beam, *J. Phys. B.*, **6**, 809–15 (1973). [153]

MONTGOMERY, D. (1965). On the resonant excitation of plasma oscillations with laser beams, *Physica*, **31**, 693–702 (1965). [70]

MOODY, C. D. (1973). Effect of CW power on the pulsed gas breakdown threshold in argon at $10 \cdot 6 \mu$ radiation, *Appl. Phys. Lett.*, **22**, 31–2 (1973). [192]

MOORCROFT, D. R. (1963). On the power scattered from density fluctuations in a plasma, *J. Geophys. Res.*, **68**, 4870–2 (1963). [90]

MORGAN, F. *et al.* (1971) (with EVANS, L. R. & GREY MORGAN, C.). Laser beam induced breakdown in helium and argon, *J. Phys. D.*, **4**, 225–35 (1971). [167, 174, 175, 184, 190]

MORSE, R. L. & NIELSON, C. W. (1971). Numerical simulation of the Weibel instability in one and two dimensions, *Phys. Fluids*, **14**, 830–40 (1971). [65]

MORSE, R. L. & NIELSON, C. W. (1973). Occurrence of high-energy electrons and surface expansion in laser-heated target plasmas, *Phys. Fluids*, **16**, 909–20 (1973). [330]

MORTON, V. M. (1967). Multiphoton absorption in monatomic gases, *Proc. Phys. Soc.*, **92**, 301–10 (1967). [152]

MOTT, N. F. & DAVIS, E. A. (1971). *Electronic processes in non-crystalline materials*. Oxford, Clarendon (1971). [277]

MUELLER, M. M. (1973). Enhanced laser-light absorption by optical resonance in inhomogeneous plasma, *Phys. Rev. Lett.*, **30**, 582–5 (1973). [75, 283, 285]

MUL'CHENKO, B. F. & RAIZER, YU. P. (1971). Laser breakdown of mixtures of neon with argon and the role of photoionization of excited atoms, *Zh. Eksp. i Teor. Fiz.* **60**, 643–50 (1971). (Transl: *Sov. Phys. JETP*, **33**, 349–53 (1971).) [180]

MULLER, B. W. & GREEN, T. S. (1971). X-ray temperature measurements of laser produced plasmas, *Plasma Phys.*, **13**, 73–5 (1971). [399]

MULSER, P. (1970). Hydrogen plasma production by giant pulse lasers, *Z. Naturforsch.*, **25A**, 282–95 (1970). [281, 308, 309]

MULSER, P. (1971). Electrostatic fields and ion separation in expanding laser produced plasmas, *Plasma Phys.*, **13**, 1007–12 (1971). [318]

MULSER, P. & WITKOWSKI, S. (1968). Numerical calculations of the dynamics of laser produced plasmas, *Phys. Lett.*, **28A**, 151–2 (1968). [290, 291]

MULSER, P. & WITKOWSKI, S. (1969). Numerical calculations of the dynamics of a laser irradiated solid hydrogen foil, *Phys. Lett.*, **28A**, 703–4 (1969). [308]

MULSER, P. *et al.* (1973) (with SIGEL, R. & WITKOWSKI, S.). Plasma production by laser, *Phys. Rep. (Phys. Lett. C)*, **6**, 187–239 (1973). [290, 291, 292]

NAGEL, D. J. *et al.* (1974) (with BURKHALTER, P. G., DOZIER, C. M., HOLZRICHTER, J. F., KLEIN, B. M., MCMAHON, J. M., STAMPER, J. A. & WHITLOCK, R. R.). X-ray emission from laser-produced plasmas, *Phys. Rev. Lett.*, **33**, 743–6 (1974). [394]

NAIMAN, C. S. *et al.* (1966) (with DE WOLF, M. Y., GOLDBLATT, I. & SCHWARTZ, J.). Laser-induced prebreakdown and breakdown phenomena observed in cloud chamber, *Phys. Rev.*, **146**, 133–5 (1966). [162]

NAMBA, S. *et al.* (1966) (with KIM, P. H. & MITSUYAMA, A.). Energies of ions produced by laser irradiation, *J. Appl. Phys.*, **37**, 3330–1 (1966). [341]

NAMBA, S. *et al.* (1968) (with KIM, P. H. & SCHWARZ, H.). Laser induced emission of ions from clean surfaces, *VIII International Conference on phenomena in ionized gases, Vienna (1967). Contributed papers*, p. 59. Vienna, IAEA (1968). [341]

NELSON, P. (1964). Calcul de l'ordre de grandeur des processus multiphotoniques, *C.R. Acad. Sci., Paris*, **259**, 2185–6 (1964). [152]

NELSON, P. (1965). Sur le calcul des processus multiphotoniques, *C.R. Acad. Sci., Paris*, **261**, 3089–91 (1965). [152]

NELSON, P. *et al.* (1964) (with VEYRIE, P., BERRY, M. & DURAND, Y.). Experimental and theoretical studies of air breakdown by intense pulse of light, *Phys. Lett.*, **13**, 226–8 (1964). [152]

NEMCHINOV, I. V. (1967). Steady-state motion of radiation-heated vapors of a material in the presence of lateral spreading flow, *Prikl. Mat. i Mekh.*, **31**, 300–19 (1967). (Transl: *Appl. Math. & Mech.*, **31**, 320–29 (1967).) [287, 292]

NEUMAN, F. (1964). Momentum transfer and cratering effects produced by giant laser pulses, *Appl. Phys. Lett.*, **4**, 167–9 (1964). [341]

NEUSSER, H. J. *et al.* (1971) (with PUELL, H. & KAISER, W.). Temperature measurements of plasmas produced by laser pulses in gas jets, *Appl. Phys. Lett.*, **19**, 300–2 (1971). [257].

NEW, G. H. C. & WARD, J. F. (1967). Optical third-harmonic generation in gases, *Phys. Rev. Lett.*, **19**, 556–9 (1967). [196]

NGUYEN, D. L. & PARBHAKAR, K. J. (1973). Effect of laser pulse rise time on heating of a magnetically confined plasma, *J. Appl. Phys.*, **44**, 2157–60 (1973). [331]

NICHOLSON-FLORENCE, M. B. (1971). Intensity dependence of free–free absorption, *J. Phys. A.*, **4**, 574–82 (1971). [47]

NIELSEN, P. E. *et al.* (1971) (with CANAVAN, G. H. & ROCKWOOD, S. D.). Breakdown of deuterium with a ruby laser, *Proc. IEEE*, **59**, 709–10 (1971). [169].

NIIMURA, M. *et al.* (1973) (with CHAN, P. W. & CHURCHILL, R. J.). Effects of Coulomb collisions on the dispersion relation of longitudinal plasma oscillations, *IEEE Trans. Plasma Sci.*, **1**, 46–54 (1973). [21]

NIKISHOV, A. I. & RITUS, V. I. (1966). Ionization of systems bound by short-range forces by the field of an electromagnetic wave, *Zh. Eksp i Teor. Fiz.*, **50**, 255–70 (1966). (Transl: *Sov. Phys. JETP*, **23**, 168–77 (1966).) [154]

NISHIKAWA, K. (1968*a*). Parametric excitation of coupled waves: I General Formulation, *J. Phys. Soc. Jap.*, **24**, 916–22 (1968). [55, 60]

NISHIKAWA, K. (1968*b*). Parametric excitation of coupled waves: II Parametric plasmon-photon interaction, *J. Phys. Soc. Jap.*, **24**, 1152–8 (1968). [59, 60, 62]

NODWELL, R. A. & VAN DER KAMP, G. S. J. P. (1968). Electron-density profiles from laser scattering, *Can. J. Phys.*, **46**, 833–7 (1968). [118]

NUCKOLLS, J. *et al.* (1972) (with WOOD, L., THIESSEN, A. & ZIMMERMAN, G.). Laser compression of matter to super-high densities: thermonuclear (CTR) applications, *Nature*, **239**, 139–42 (1972). [323, 324, 325, 328, 330]

OFFENBERGER, A. A. & BURNETT, N. H. (1972). CO_2 laser-induced gas breakdown in hydrogen, *J. Appl. Phys.*, **43**, 4977–80 (1972). [268]

OFFENBERGER, A. A. & BURNETT, N. H. (1973). CO_2 laser induced Compton effect in underdense hydrogen plasma, *Phys. Lett.*, **42A**, 527–8 (1973). [79]

OFFENBERGER, A. A. & KERR, R. D. (1971). Collision dominated Thomson scattering measurements using CO_2 laser radiation, *Phys. Lett.*, **37A**, 435–6 (1971). [111, 126]

OHMURA, T. & OHMURA, H. (1960). Free–free absorption coefficient of the negative hydrogen ion, *Astrophys. J.*, **131**, 8–11 (1960). [48]

OKUDA, T. *et al.* (1969) (with KISHI, K. & SAWADA, K.). Two-photon ionization process in optical breakdown of cesium vapor, *Appl. Phys. Lett.*, **15**, 181–3 (1969). [176]

OLSEN, J. N. *et al.* (1972) (with JONES, E. D. & GOBELI, G. W.). Nanosecond and picosecond laser-produced CD_2 plasmas, *J. Appl. Phys.*, **43**, 3991–2 (1972). [77, 390]

OLSEN, J. N. *et al.* (1973) (with KUSWA, G. W. & JONES, E. D.). Ion-expansion energy spectra correlated to laser plasma parameters, *J. Appl. Phys.*, **44**, 2275–83 (1973). [390]

OPOWER, H. & PRESS, W. (1966). Erzeugung energiereicher Plasmen durch Lichtimpulse, *Z. Naturforsch.*, **21A**, 344–50 (1966). [318, 366]

OPOWER, H. *et al.* (1967) (with KAISER, W., PUELL, H. & HEINICKE, W.). Energetic plasmas produced by laser light on solid targets, *Z. Naturforsch.*, **22A**, 1392–7 (1967). [366, 367, 395]

OPPENHEIMER, J. R. (1928). Quantum theory of aperiodic effects, *Phys. Rev.*, **31**, 66–81 (1928). [147]

ORAEVSKI, V. N. & SAGDEEV, R. Z. (1962). Stability of steady-state longitudinal plasma oscillations, *Zh. Tekh. Fiz.*, **32**, 1291–6 (1962). (Transl: *Sov. Phys. Tech. Phys.*, **7**, 955–8 (1963).) [60, 76]

ORLOV, R. YU. *et al.* (1971) (with SKIDAN, I. B. & TELEGIN, L. S.). Investigation of breakdown produced in dielectrics by ultrashort laser pulses, *Zh. Eksp. Teor. Fiz.*, **61**, 784–90 (1971). (Transl: *Sov. Phys. JETP*, **34**, 418–21 (1972).) [188]

OSBORN, R. K. (1972a). Nonlinear bremsstrahlung, *Phys. Rev.*, **A5**, 1660–2 (1972). [47]

OSBORN, R. K. (1972b). Atomic ionization by intense laser fields, *Nuovo Cimento*, **9B**, 414–23 (1972). [152]

OSTER, L. (1961). Emission, absorption and conductivity of a fully ionized gas at radio frequencies, *Rev. Mod. Phys.*, **33**, 525–43 (1961). [40, 41, 42]

OSTROVSKAYA, G. V. & OSTROVSKII, YU. I. (1966). Holographic investigation of a laser spark, *Zh. Eksp. i Teor. Fiz. Pis'ma*, **4**, 121–4 (1966). (Transl: *JETP Lett.*, **4**, 83–4 (1966).) [229, 248]

OTT, E. (1971). Parametric instability of a plasma driven by a high frequency, finite wavelength wave, *Phys. Lett.*, **37A**, 281–2 (1971). [60]

PACK, J. L. *et al.* (1968) (with GEORGE, T. V. & ENGELHARDT, A. G.). Retracting pedestal apparatus that presents a single solid target to a focused Q-switched laser beam, *Rev. Sci. Instrum.*, **39**, 1697–1700 (1968). [400]

PACK, J. L. *et al.* (1969) (with GEORGE, T. V. & ENGELHARDT, A. G.). Microwave diagnostics of laser produced aluminum plasmas, *Phys. Fluids*, **12**, 469–71 (1969). [404]

PALMER, A. J. (1971). Stimulated scattering and self-focusing in laser-produced plasma, *Phys. Fluids*, **14**, 2714–8 (1971). [197]

PALMER, A. J. (1972). Stimulated scattering and self-focusing processes in dense plasmas. In *Laser interaction and related plasma phenomena* (ed. H. J. Schwarz and H. Hora), Vol. 2, pp. 367–80. New York, Plenum (1972). [197]

PANARELLA, E. & SAVIC, P. (1968). Blast waves from a laser-induced spark in air, *Can. J. Phys.*, **46**, 183–92 (1968). [220]

PAPPERT, R. A. (1963). Incoherent scatter from a hot plasma, *Phys. Fluids*, **6**, 1452–7 (1963). [102]

PASHININ, P. P. & PROKHOROV, A. M. (1971). High-temperature high-density plasma from a special gas target heated by a laser, *Zh. Eksp. i Teor. Fiz.*, **60**, 1630–6 (1971). (Transl: *Sov. Phys. JETP*, **33**, 883–6 (1971).) [335]

PASHININ, P. P. & PROKHOROV, A. M. (1972). Production of a high temperature deuterium plasma by laser heating of a special gas, *Zh. Eksp. i Teor. Fiz.*, **62**, 189–94 (1972). (Transl: *Sov. Phys. JETP*, **35**, 101–3 (1972).) [335]

PASHININ, P. P. *et al.* (1965) (with MANDELSTAM, S. L., PROKHOROV, A. M. & SUHODREV, N. K.). Optical frequency electrical discharge in air, *Z. Angew. Math. & Phys.*, **16**, 125–6 (1965). [250]

PATON, B. E. & ISENOR, N. R. (1968). Energies and quantities of ions in laser-produced metal plasmas, *Can. J. Phys.*, **46**, 1237–9 (1968). [352, 353]

PATRICK, R. M. (1965). Thomson scattering measurements of magnetic annular shock tube plasmas, *Phys. Fluids*, **8**, 1985–94 (1965). [112, 113]

PAUL, J. W. M. (1970). Collisionless shock waves. In *Physics of hot plasmas* (ed. B. J. Rye and J. C. Taylor), pp. 302–45. Edinburgh, Oliver & Boyd (1970). [126, 440]

PAULING, L. (1927). The theoretical prediction of the physical properties of many-electron atoms and ions, *Proc. Roy. Soc.*, **A114**, 181–211 (1927). [85]

PEASE, R. S. (1972). Progress in nuclear fusion research, *J. Br. Nucl. Energy Soc.*, **11**, 165–82 (1972). [139]

PECHACEK, R. E. & TRIVELPIECE, A. W. (1967). Electromagnetic wave scattering from a high-temperature plasma, *Phys. Fluids*, **10**, 1688–96 (1967). [102, 103, 104]

PÉPIN, H. *et al.* (1972) (with DICK, K., MARTINEAU, J. & PARBHAKAR, K.). Plasma produced by irradiation of thin foils of aluminum by a CO_2-TEA laser, *Phys. Lett.*, **38A**, 203–4 (1972). [395]

PEREL'MAN, M. E. & ARUTYUNYAN, V. G. (1972). Multiphoton photoeffect and the theory of the delay time, *Zh. Eksp. i Teor. Fiz.*, **62**, 490–5 (1972). (Transl: *Sov. Phys. JETP*, **35**, 260–3 (1972).) [152]

PERELOMOV, A. M. *et al.* (1966) (with POPOV, V. S. & TERENTEV, M. V.). Ionization of atoms in an alternating electric field, *Zh. Eksp. i Teor. Fiz.*, **50**, 1393–1409 (1966). (Transl: *Sov. Phys. JETP*, **23**, 924–34 (1966).) [147, 148]

PERELOMOV, A. M. *et al.* (1968) (with POPOV, V. S. & KUZNETSOV, V. P.). Allowance for the Coulomb interaction in multiphoton ionization, *Zh. Eksp. i Teor. Fiz.*, **54**, 841–54 (1968). (Transl: *Sov. Phys. JETP*, **27**, 451–7 (1968).) [150]

PERESSINI, E. R. (1966). Field emission from atoms in intense optical fields. In *Physics of quantum electronics* (ed. P. L. Kelley, B. Lax and P. E. Tannenwald), pp. 499–508. New York, McGraw-Hill (1966). [150, 157]

PERKINS, F. W. & FLICK, J. (1971). Parametric instabilities in inhomogeneous plasmas, *Phys. Fluids*, **14**, 2012–8 (1971). [63, 64]

PERT, G. J. (1971). The thermodynamics of multiphoton processes, *J. Phys. B.*, **4**, L72–6 (1971). [32]

PERT, G. J. (1972). Inverse bremsstrahlung absorption in large radiation fields during binary collisions–classical theory, *J. Phys. A.*, **5**, 506–15 (1972). [47]

PERT, G. J. (1973). X-ray temperature measurements of laser produced plasmas in large radiation fields, *Phys. Lett.*, **42A**, 511–2 (1973). [47]

PESME, D. *et al.* (1973) (with LAVAL, G. & PELLAT, R.). Parametric instabilities in bounded plasmas, *Phys. Rev. Lett.*, **31**, 203–6 (1973). [71]

PETRISCHEV, V. A. & TALANOV, V. I. (1971). Transient self-focusing of light, *Kvantovaya Elektron*, **1**, (6), 35–42 (1971). (Transl: *Sov. J. Quantum Electron.*, **1**, 587–92 (1972).) [197]

PEYRAUD, J. (1969). Théorie quantique de l'interaction plasma-rayonnement: I, *J. Physique*, **30**, 307–24 (1969); II, *J. Physique*, **30**, 461–76 (1969). [73]

PHELPS, A. V. (1966). Theory of growth of ionization during laser breakdown. In *Physics of quantum electronics* (ed. P. L. Kelley, B. Lax and P. E. Tannenwald), pp. 538–47. New York, McGraw-Hill (1966). [48, 166, 167, 168, 169, 175, 176]

PHELPS, D. *et al.* (1971) (with RYNN, N. & VAN HOVEN, G.). Observation of direct nonlinear coupling of electromagnetic waves and electrostatic waves in a plasma, *Phys. Rev. Lett.*, **26**, 688–91 (1971). [76]

PHELPS, D. *et al.* (1973) (with VAN HOVEN, G. & RYNN, N.). Direct nonlinear coupling of electromagnetic waves and electrostatic waves in a plasma: experiment, *Phys. Fluids*, **16**, 1078–86 (1973). [76]

PIRRI, A. N. *et al.* (1972) (with SCHLIER, R. & NORTHAM, D.). Momentum transfer and plasma formation above a surface with a high-power CO_2 laser, *Appl. Phys. Lett.*, **21**, 79–81 (1972). [441]

PLATZMAN, P. M. *et al.* (1968) (with WOLFF, P. A. & TZOAR, N.). Light scattering from a plasma in a magnetic field, *Phys. Rev.*, **174**, 489–94 (1968). [97]

PLIS, A. I. *et al.* (1972) (with TYURIN, E. L. & SHCHEGLOV, V. A.). Heating with short laser pulses, *Zh. Tekh. Fiz.*, **42**, 2568–76 (1972). (Transl: *Sov. Phys. Tech. Phys.*, **17**, 1995–2000 (1973).) [306]

PORKOLAB, M. *et al.* (1972) (with ARUNASALAM, V. & ELLIS, R. A.). Parametric instability and anomalous heating due to electromagnetic waves in plasma, *Phys. Rev. Lett.*, **29**, 1438–41 (1972). [76]

PRESBY, H. (1971). Comments on 'Plasma production with a multibeam laser system', *Phys. Fluids*, **14**, 2788 (1971). [259]

PUELL, H. (1970). Heating of laser produced plasmas generated at plane solid targets, *Z. Natur-forsch.*, **25A**, 1807–15 (1970). [292, 293, 299]

PUELL, H. *et al.* (1970a) (with NEUSSER, H. J. & KAISER, W.). Temperature and expansion energy of laser-produced plasmas. II Experiments, *Z. Naturforsch.*, **25A**, 1815–22 (1970). [232, 383, 384]

PUELL, H. *et al.* (1970b) (ith OPOWER, H. & NEUSSER, H. J.). Experiments with two laser-produced interpenetrating plasmas, *Phys. Lett.*, **31A**, 4–5 (1970). [384, 385]

PUELL, H. *et al.* (1971) (with SPENGLER, W. & KAISER, W.). Time resolved temperature measurements of a laser produced carbon plasma, *Phys. Lett.*, **37A**, 35–6 (1971). [362, 363]

PUSTOVALOV, V. V. & SILIN, V. P. (1970). Anomalous absorption of an electromagnetic wave by a plasma, *Zh. Eksp. i Teor. Fiz.*, **59**, 2215–27 (1970). (Transl: *Sov. Phys. JETP*, **32**, 1198–1204 (1971).) [62, 65]

PYATNITSKII, L. N. *et al.* (1968) (with KHAUSTOVICH, G. P. & KOROBKIN, V. V.). Critical aperture in plasma diagnostics by the scattering method, *Zh. Eksp. i Teor. Fiz. Pis'ma*, **7**, 493–6 (1968). (Transl: *JETP Lett.*, **7**, 378–82 (1968).) [93]

PYATNITSKII, L. N. *et al.* (1971) (with KOROBKIN, V. V., MUSHINSKII, A. A. & KHAUSTOVICH, G. P.). Use of the scattering method for determining the parameters of a low-temperature plasma, *Dokl. Akad. Nauk. SSSR*, **200**, 571–4 (1971). (Transl: *Sov. Phys. Dokl.*, **16**, 745–8 (1972).) [101]

QUIGG, C. (1968). Relativistic correction to plasma bremsstrahlung, *Phys. Fluids*, **11**, 461–2 (1968). [50]

RAIZER, YU. P. (1959). Simple methods for computing the mean range of radiation in ionized gases at high temperatures, *Zh. Eksp. i Teor. Fiz.*, **37**, 1079–83 (1959). (Transl: *Sov. Phys. JETP*, **37**, 769–71 (1960).) [45]

RAIZER, YU. P. (1965a). Heating of a gas by a powerful light pulse, *Zh. Eksp. i Teor. Fiz.*, **48**, 1508–19 (1965). (Transl: *Sov. Phys. JETP*, **21**, 1009–17 (1965).) [45, 46, 201, 203, 209, 211, 214, 219, 266]

RAIZER, YU. P. (1965b). Breakdown and heating of gases under the influence of a laser beam, *Usp. Fiz. Nauk.*, **87**, 29–64 (1965). (Transl: *Sov. Phys. Usp.*, **8**, 650–73 (1966).) [151]

RAIZER, YU. P. (1970a). The feasibility of an optical plasmotron and its power requirements, *Zh. Eksp. i Teor. Fiz. Pis'ma*, **11**, 195–9 (1970). (Transl: *JETP Lett.*, **11**, 120–3 (1970).) [269]

RAIZER, YU. P. (1970b). Subsonic propagation of a light spark and threshold conditions for the maintenance of plasma by radiation, *Zh. Eksp. i Teor. Fiz.*, **58**, 2127–38 (1970). (Transl: *Sov. Phys. JETP*, **31**, 1148–54 (1970).) [269]

RAMSDEN, S. A. & DAVIES, W. E. R. (1964). Radiation scattered from the plasma produced by a focused ruby laser beam, *Phys. Rev. Lett.*, **13**, 227–9 (1964). [223, 234, 251]

RAMSDEN, S. A. & DAVIES, W. E. R. (1966). Observation of cooperative effects in the scattering of a laser beam from a plasma, *Phys. Rev. Lett.*, **16**, 303–6 (1966). [118, 120, 125]

RAMSDEN, S. A. & SAVIC, R. (1964). A radiative detonation model for the development of a laser-induced spark in air, *Nature*, **203**, 1217–19 (1964). [203, 209, 234]

RAMSDEN, S. A. *et al.* (1966) (with BENESCH, R., DAVIES, W. E. R. & JOHN, P. K.). Observation of cooperative effects and determination of the electron and ion temperatures in a plasma from the scattering of a ruby laser beam, *IEEE J. Quantum Electron.*, **2**, 267–70 (1966). [120].

RAMSDEN, S. A. (1967) (with JOHN, P. K., KRONAST, B. & BENESCH, R.). Evidence for a thermonuclear reaction in a θ-pinch plasma from the scattering of a ruby laser beam, *Phys. Rev. Lett.*, **19**, 688–9 (1967).) [121]

RAND, S. (1964). Inverse bremsstrahlung with high-intensity radiation fields, *Phys. Rev.*, **136B**, 231–7 (1964). [47]

RAYLEIGH, Baron John W. S. (1871). On the light from the sky, its polarization and colour, *Phil. Mag. Ser. 4*, **41**, 107–20 (1871). *See also* Twersky, V., *Appl. Opt.*, **3**, 1150–62 (1964). [86]

READY, J. F. (1963). Development of plume of material vaporized by giant-pulse laser, *Appl. Phys. Lett.*, **3**, 11–13 (1963). [435]

READY, J. F. (1965). Effects due to absorption of laser radiation, *J. Appl. Phys.*, **36**, 462–8 (1965). [275]

READY, J. F. (1971). *Effects of high-power laser radiation.* New York, Academic (1971). [275]

REISS, H. R. (1970a). Semiclassical electrodynamics of bound systems in intense fields, *Phys. Rev.* **A1**, 803–18 (1970). [152, 153]

REISS, H. R. (1970b). Atomic transitions in intense fields and the breakdown of perturbation calculations, *Phys. Rev. Lett.*, **25**, 1149–51 (1970). [153]

REISS, H. R. (1971). Transitions in electromagnetic fields of arbitrary intensity, *Phys. Rev.*, **D4**, 3533–42 (1971). [152]

REISS, H. R. (1972a). Approximate perturbation theory for high-order electromagnetic transitions, *Phys. Rev.*, **A6**, 817–22 (1972). [153]

REISS, H. R. (1972b). Polarization effects in high-order multiphoton ionization, *Phys. Rev. Lett.*, **29**, 1129–31 (1972). [153]

RICE, M. H. & GOOD, R. H. (1962). Stark effect in hydrogen, *J. Opt. Soc. Am.*, **52**, 239–46 (1962). [147]

RICHARDSON, M. C. & ALCOCK, A. J. (1971a). Interferometric observation of plasma filaments in a laser-produced spark, *Appl. Phys. Lett.*, **18**, 357–60 (1971). [198, 247]

RICHARDSON, M. C. & ALCOCK, A. J. (1971b). Subnanosecond interferometry of plasma filaments in a laser-produced spark, *Kvantovaya Elektron.* **1**, (5), 37–43 (1971). (Transl: *Sov. J. Quantum Electron.*, **1**, 461–5 (1972).) [198, 246, 247]

RICHARDSON, M. C. & ALCOCK, A. J. (1971c). Two wavelength interferometry of sparks produced by $10 \cdot 6 \mu$ radiation, *X International Conference on phenomena in ionized gases, Oxford (1971). Contributed papers*, p. 230. Oxford, Donald Parsons & Co. Ltd. (1971). [266]

RICHARDSON, M. C. & SALA, K. (1973). Picosecond framing photography of a laser-produced plasma, *Appl. Phys. Lett.*, **23**, 420–2 (1973). [223, 265]

RINGLER, H. & NODWELL, R. A. (1969a). Enhanced cross section for scattering of laser light, *Phys. Lett.*, **30A**, 126–7 (1969). [116]

RINGLER, H. & NODWELL, R. A. (1969b). Enhanced plasma oscillations observed with scattered laser light, *Phys. Lett.*, **29A**, 151–2 (1969). [116]

RIVIERE, A. C. & SWEETMAN, D. R. (1964). The ionization of excited hydrogen atoms by strong electric fields, *VI International Conference on ionization phenomena in gases, Paris (1963)*, pp. 105–10. Paris, S.E.R.M.A. (1964). [147]

RIZZO, J. E. & KLEWE, R. C. (1966). Optical breakdown in metal vapours, *Brit. J. Appl. Phys.*, **17**, 1137–41 (1966). [176]

ROBINSON, A. M. (1973). Laser-induced gas breakdown initiated by ultraviolet photoionization, *Appl. Phys. Lett.*, **22**, 33–5 (1973). [192, 193]

ROBOUCH, B. V. & RAGER, J. P. (1973). Filter method and silicon detectors to measure the temperature of a hot, dense thermal deuterium plasma from its quantum bremsstrahlung, *J. Appl. Phys.*, **44**, 1527–33 (1973). [232]

ROGISTER, A. & OBERMAN, C. (1968). On the kinetic theory of stable and weakly unstable plasma. Part 1, *J. Plasma Phys.*, **2**, 33–49 (1968). [102]

RÖHR, H. (1967). A 90° laser scattering experiment for measuring temperature and density of the ions and electrons in a cold dense theta pinch plasma, *Phys. Lett.*, **25A**, 167–8 (1967). [124]

RÖHR, H. (1968). A 90° laser scattering experiment for measuring temperature and density of ions and electrons in a cold dense theta pinch plasma, *Z. Phys.*, **209**, 295–310 (1968). [111, 112, 124, 125]

RÖHR, H. & DECKER, G. (1968). Measurement of the ion temperature in the Isar III theta pinch by laser forward scattering, *Z. Phys.*, **214**, 157–67 (1968). [125]

ROSE, D. J. & CLARK, M. (1961). *Plasmas and controlled fusion.* Cambridge, Mass., M.I.T. Press (1961). [133, 433]

ROSENBLUTH, M. N. (1972). Parametric instabilities in inhomogeneous media, *Phys. Rev. Lett.*, **29**, 565–7 (1972). [69, 70, 71]

ROSENBLUTH, M. N. & LIU, C. S. (1972). Excitation of plasma waves by two laser beams, *Phys. Rev. Lett.*, **29**, 701–5 (1972). [70]

ROSENBLUTH, M. N. & ROSTOKER, N. (1962). Scattering of electromagnetic waves by a non-equilibrium plasma, *Phys. Fluids*, **5**, 776–88 (1962). [87, 95]

RÖSS, D. (1969). *Lasers, light amplifiers and oscillators*. (Transl. of *Laser, Lichtverstarker und-Oszillataren*. Frankfurt-a-M., Akad. Verlag. (1966).) New York, Academic Press (1969). [108, 340]

ROSTOKER, N. (1961). Fluctuations of a plasma, *Nucl. Fusion*, **1**, 101–20 (1961). [49]

RUMSBY, P. T. & PAUL, J. W. M. (1974). Temperature and density of an expanding laser produced plasma, *Plasma Phys.*, **16**, 247–60 (1974). [318, 358, 359, 360]

RYUTOV, D. D. (1964). Theory of breakdown of noble gases at optical frequencies, *Zh. Eksp. i Teor. Fiz.*, **47**, 2194–2206 (1964). (Transl: *Sov. Phys. JETP*, **20**, 1472–9 (1965).) [164, 167]

RYTOV, S. M. (1953). *Theory of electrical fluctuations and thermal radiation*, U.S.S.R. Acad. of Sci., Moscow (1953). (Transl: *USAF Rept. AFCRC-TR-59-162* (1959).) [29]

SAKHOKIYA, D. M. & TSYTOVICH, V. N. (1968). Scattering of electromagnetic waves by the turbulent pulsations in an inhomogeneous magnetoactive plasma, *Nucl. Fusion*, **8**, 241–51 (1968). [98]

SAKURAI, A. (1953). On the propagation and structure of the blast wave I, *J. Phys. Soc. Jap.*, **8**, 662–9 (1953). [220, 240]

SALÈRES, A. *et al.* (1971*a*) (with COGNARD, D. & FLOUX, F.). Experimental studies on solid deuterium nanosecond laser produced plasma, *X International Conference on phenomena in ionized gases, Oxford (1971). Contributed papers*, p. 231. Oxford, Donald Parsons & Co. Ltd. (1971). [422]

SALÈRES, A. *et al.* (1971*b*) (with COGNARD, D. & FLOUX, F.). Étude expérimentale de l'interaction d'un faisceau laser de grande puissance avec une cible de deutérium solide, *J. Physique*, **32**, *Suppl.* **C5B**, 133–5 (1971). [422]

SALÈRES, A. *et al.* (1972) (with COGNARD, D. & FLOUX, F.). Experimental studies on solid deuterium nanosecond powerful produced plasma, *V European Conference on controlled fusion and plasma physics, Grenoble (1972)*. Vol. **1**, p. 59. Grenoble, Ass. Euratom-CEA (1972). [422, 423, 424]

SALÈRES, A. *et al.* (1973*a*) (with FLOUX, F., COGNARD, D. & BOBIN, J. L.). Experimental studies on energy transfer in laser produced plasmas, *Phys. Lett.*, **45A**, 451–3 (1973). [423]

SALÈRES, A. *et al.* (1973*b*) (with COGNARD, D., REDON, D., FLOUX, F. & BOBIN, J. L.). Étude expérimentale des mécanismes d'absorption d'un faisceau laser très intense par la matière, *J. Physique, Suppl.* **C2**, **34**, 9–13 (1973). [423]

SALPETER, E. E. (1960). Electron density fluctuations in a plasma, *Phys. Rev.*, **120**, 1528–35 (1960). [87, 88, 90, 91]

SALPETER, E. E. (1961). Plasma density fluctuations in a magnetic field, *Phys. Rev.*, **122**, 1663–74 (1961). [97, 98]

SALPETER, E. E. (1963). Density fluctuations in a nonequilibrium plasma, *J. Geophys. Res.*, **68**, 1321–33 (1963). [90]

SALZMANN, H. (1972). The applicability of Fourier's theory of heat conduction on laser produced plasmas, *Phys. Lett.*, **41A**, 363–4 (1972). [306]

SALZMANN, H. (1973). Production of hot plasmas of solid-state density by ultrashort laser pulses, *J. Appl. Phys.*, **44**, 113–24 (1973). [77, 399, 427]

SANMARTIN, J. R. (1970). Electrostatic plasma instabilities excited by a high-frequency electric field, *Phys. Fluids*, **13**, 1533–42 (1970). [60, 62]

SARACHIK, E. S. & SCHAPPERT, G. T. (1970). Classical theory of the scattering of intense laser radiation by free electrons, *Phys. Rev.*, **D1**, 2738–53 (1970). [86]

SAUNDERS, P. A. H. *et al.* (1967) (with AVIVI, P. & MILLAR, W.). Laser produced plasmas from solid hydrogen targets, *Phys. Lett.*, **24A**, 290–1 (1967). [412, 417]

SAVCHENKO, M. M. & STEPANOV, V. K. (1966). Perturbation of a magnetic field by plasma of a laser spark in air, *Zh. Eksp. i Teor. Fiz.*, **51**, 1654–9 (1966). (Transl: *Sov. Phys. JETP*, **24**, 1117–21 (1967).) [271]

SAVCHENKO, M. M. & STEPANOV, V. K. (1968). Structure of laser spark image due to reflection, *Zh. Eksp. i Teor. Fiz. Pis'ma*, **8**, 458–61 (1968). (Transl: *JETP Lett.*, **8**, 281–3 (1968).) [198, 235, 245]

SCHIRMANN, D. *et al.* (1970) (with GRELOT, P., RABEAU, M. & TONON, G.). Interaction of a laser created plasma with a strong magnetic induction, *Phys. Lett.*, **33A**, 514–5 (1970). [434]

SCHIRMANN, D. *et al.* (1974) (with BILLON, B., COGNARD, D., LAUNSPACH, J. & PATOU, C.). Compression of a cylindrical target with laser beams. Paper presented at *VIII International quantum electronics conference, San Francisco (1974)*. [408, 409, 410]

SCHWARTZ, C. (1959). Calculations in Schrodinger perturbation theory, *Ann. Phys.*, **6**, 156–69 (1959). [152]

SCHWARTZ, C. & TIEMAN, T. J. (1959). New calculation of the numerical value of the Lamb shift, *Ann. Phys.*, **6**, 178–87 (1959). [152]

SCHWARZ, H. (1971). Influence of magnetic fields on plasmas produced by laser irradiation from solid targets, *Kvantovaya Elektron.*, **1** (5), 102–5 (1971). (Transl: *Sov. J. Quantum Electron.* **1**, 512–15 (1972).) [429]

SCHWARZ, S. E. (1963). Scattering of optical pulses from a non-equilibrium plasma, *Proc. IEEE*, **51**, 1362 (1963). [113]

SCHWARZ, S. E. (1965). Plasma diagnosis by means of optical scattering, *J. Appl. Phys.*, **36**, 1836–41 (1965). [112]

SCHWIRZKE, F. & MCKEE, L. L. (1972). Investigation of self-generated magnetic fields in laser produced plasmas, *V European Conference on controlled fusion and plasma physics, Grenoble (1972)*. Vol. 1, p. 63. Grenoble, Ass. Euratom-CEA (1972). [443]

SCHWIRZKE, F. & TUCKFIELD, R. (1969). Observations of enhanced resistivity in the wave front of a laser-produced plasma interacting with a magnetic field, *Phys. Rev. Lett.*, **22**, 1284–7 (1969). [335, 431]

SCHWOB, J. L. *et al.* (1970) (with BRETON, C., SEKA, W. & MINIER, C.). Étude de l'émission dans l'ultraviolet lointain du plasma produit par focalisation d'un faisceau laser sur une cible solide, *Plasma Phys.*, **12**, 217–25 (1970). [357]

SEDOV, L. I. (1959). *Similarity and dimensional methods in mechanics.* Transl. M. Friedman from the 4th Russian ed. London, Infosearch Ltd. (1959). [220]

SEELY, J. F. & HARRIS, E. G. (1973). Heating of a plasma by multiphoton inverse bremsstrahlung, *Phys. Rev.*, **A7**, 1064–7 (1973). [47]

SEGRE, S. E. & MARTONE, M. (1971). An apparatus for the space- and time-resolved measurement of electron temperature and density in a plasma by Thomson scattering of laser light, *Nuovo Cimento*, **3B**, 45–54 (1971). [126]

SEKA, W. *et al.* (1970*a*) (with BRETON, C., SCHWOB, J. L. & MINIER, C.). Temperature measurements of a laser-produced plasma using the recombination continuum, *Plasma Phys.*, **12**, 73–7 (1970). [357, 358]

SEKA, W. *et al.* (1970*b*) (with SCHWOB, J. L. & BRETON, C.). Spectroscopic measurements on laser-produced LiH plasmas, *J. Appl. Phys.*, **41**, 3440–1 (1970). [407]

SEKA, W. *et al.* (1971) (with SCHWOB, J. L. & BRETON, C.). Multilayer targets as a diagnostic tool for laser-produced plasmas, *J. Appl. Phys.*, **42**, 315–19 (1971). [399]

SHATAS, R. A. *et al.* (1972) (with ROBERTS, T. G., MEYER, H. C. & STETTLER, J. D.). Fusion neutron and soft X-ray generation in laser assisted dense plasma focus. In *Laser interaction and related plasma phenomena* (ed. H. J. Schwarz and H. Hora), Vol. 2, pp. 527–43. New York, Plenum (1972). [222]

SHCHELKIN, K. I. & TROSHIN, YA. K. (1965). *Gas dynamics of combustion*, transl. B. W. Kuvshinoff and L. Holtschlag. Baltimore, Mono Book Corp. (1965). [207]

SHEARER, J. W. (1971). Effect of oblique incidence on optical absorption of laser light by a plasma, *Phys. Fluids*, **14**, 183–5 (1971). [282]

SHEARER, J. W. & BARNES, W. S. (1970). Mechanisms for production of neutron-emitting plasma by subnanosecond laser-pulse heating, *Phys. Rev. Lett.*, **24**, 92–4 (1970). *See also: erratum, Phys. Rev. Lett.*, **24**, 432 (1970). [302, 303, 304]

SHEARER, J. W. & BARNES, W. S. (1971). Numerical calculations of plasma heating by means of subnanosecond laser pulses. In *Laser interaction and related plasma phenomena* (ed. H. J. Schwarz and H. Hora), Vol. 1, pp. 307–37. New York, Plenum (1971). [38]

SHEARER, J. W. & DUDERSTADT, J. J. (1973). Wavelength dependence of laser-light absorption by a solid deuterium target, *Nucl. Fusion*, **13**, 401–12 (1973). [68, 282, 285]

SHEARER, J. W. *et al.* (1972) (with MEAD, S. W., PETRUZZI, J., RAINER, F., SWAIN, J. E. & VIOLET, C.E.). Experimental indications of plasma instabilities induced by laser heating, *Phys. Rev.*, **A6**, 764–9 (1972). [77, 80, 390, 391]

SHEFFIELD, J. (1972*a*). A method for measuring the direction of the magnetic field in a Tokamak plasma by laser light scattering, *Plasma Phys.*, **14**, 385–95 (1972). [98]

SHEFFIELD, J. (1972*b*). The incoherent scattering of radiation from a high temperature plasma, *Plasma Phys.*, **14**, 783–91 (1972). [104]

SHIAU, J. N. *et al.* (1974) (with GOLDMAN, E. B. & WENG, C. I.). Linear stability analysis of laser-driven spherical implosions, *Phys. Rev. Lett.*, **32**, 352–5 (1974). [328]

SHKUROPAT, P. I. & SHNEERSON, G. A. (1967). Plasma heating in a strong magnetic field by laser radiation, *Zh. Tekh. Fiz.*, **37**, 1161–5 (1967). (Transl: *Sov. Phys. Tech. Phys.*, **12**, 838–40 (1967).) [331]

SIEMON, R. E. & BENFORD, J. (1969). Ion and electron temperature and density by Thomson scattering with a low resolution monochromator, *Phys. Fluids*, **12**, 249–50 (1969). [120]

SIGEL, R. (1969). Investigation of a laser produced hydrogen plasma using holographic interferometry, *Phys. Lett.*, **30A**, 103–4 (1969). [426]

SIGEL, R. (1970). Experimental investigation of plasma production by irradiating solid hydrogen foils with an intense pulse laser, *Z. Naturforsch.*, **25A**, 488–503 (1970). [310, 426]

SIGEL, R. *et al.* (1968) (with BÜCHL, K., MULSER, P. & WITKOWSKI, S.). Laser produced plasma from solid hydrogen foils, *Phys. Lett.*, **26A**, 498–9 (1968). [426]

SIGEL, R. *et al.* (1969) (with KRAUSE, H. & WITKOWSKI, S.). Production of thin solid-hydrogen foils for use as targets in high vacuum, *J. Phys. E.*, **2**, 187–90 (1969). [412, 414]

SIGEL, R. *et al.* (1972) (with WITKOWSKI, S., BAUMHACKER, H., BÜCHL, K., EIDMANN, K., HORA, H., MENNICKE, H., MULSER, P., PFIRSCH, D. & SALZMANN, H.). Survey of studies of laser-produced plasmas at the Institut für Plasmaphysik, Garching, *Kvantovaya Elektron.*, **2** (8), 37–44 (1972). (Transl: *Sov. J. Quantum Electron.*, **2**, 117–26 (1972).) [382, 426]

SILIN, A. P. (1970). Many-photon surface photoelectric effect in metals, *Fiz. Tver. Tela.*, **12**, 3553–8 (1970). (Transl: *Sov. Phys. Solid State*, **12**, 2886–9 (1971).) [279]

SILIN, V. P. (1964). Nonlinear high-frequency plasma conductivity, *Zh. Eksp. i Teor. Fiz.*, **47**, 2254–65 (1964). (Transl: *Sov. Phys. JETP*, **20**, 1510-6 (1965).) [47]

SILIN, V. P. (1965). Parametric resonance in a plasma, *Zh. Eksp. i Teor. Fiz.*, **48**, 1679–91 (1965). (Transl: *Sov. Phys. JETP*, **21**, 1127–34 (1965).) [60]

SITENKO, A. G. (1973). Fluctuations in plasma and nonlinear susceptibilities, *Phys. Scripta*, **7**, 190–2 (1973). [54]

SJÖLUND, A. & STENFLO, L. (1967). Parametric coupling between transverse electromagnetic and longitudinal electron waves, *Physica*, **35**, 499–505 (1967). [70]

SLEEPER, A. *et al.* (1972) (with WEINSTOCK, J. & BEZZERIDES, B.). Nonlinear saturation of the ion-acoustic instability, *Phys. Rev. Lett.*, **29**, 343–5 (1972). [68]

SMIRNOV, B. M. & CHIBISOV, M. I. (1965). Ionization of atomic particles by an electric field and by electron impact, *Zh. Eksp. i Teor. Fiz.*, **49**, 841–51 (1965). (Transl: *Sov. Phys. JETP*, **22**, 585–95 (1966).) [147]

SMITH, D. C. (1970). Gas breakdown with $10·6\mu$ wavelength CO_2 laser radiation, *J. Appl. Phys.*, **41**, 4501–5 (1970). [189]

SMITH, D. C. (1971). Gas breakdown dependence on beam size and pulse duration with $10·6\mu$ wavelength radiation, *Appl. Phys. Lett.*, **19**, 405–8 (1971). [189]

SMITH, D. C. & FOWLER, M. C. (1973). Ignition and maintenance of a cw plasma in atomospheric pressure air with CO_2 laser radiation, *Appl. Phys. Lett.*, **22**, 500–2 (1973). [270]

SMITH, D. C. & HAUGHT, A. F. (1966). Energy-loss processes in optical-frequency gas breakdown, *Phys. Rev. Lett.*, **16**, 1085–8 (1966). [179, 180]

SMITH, D. C. & TOMLINSON, R. G. (1967). Effect of mode beating in laser-produced gas breakdown, *Appl. Phys. Lett.*, **11**, 73–5 (1967). [179, 185]

SOMMERFELD, A. (1951). *Atombau und Spektrallinien*, Vol. II. Braunschweig, Vieweg (1951). [41]

SOMMERFELD, A. (1964). Electrodynamics. Vol. 3 of *Lectures on theoretical physics*, transl. E. G. Ramberg. New York, Academic (1964). [83]

SOMON, J. P. (1972). The thermonuclear hot-spot, *Nucl. Fusion*, **12**, 461–74 (1972). [316]

SPALDING, I. J. (1972). Lasers and the reactor ignition problem, *Kvantovaya Elektron.*, **4** (10), 40–7 (1972). (Transl: *Sov. J. Quantum Electron.*, **2**, 329–34 (1973).) [140]

SPITZER, L. (1962). *Physics of Fully Ionized Gases.* 2nd ed. New York, Interscience (1962). [20, 263, 300, 301, 331]

STAMPER, J. A. (1972). Laser-induced sources for magnetic fields, *NRL Report 7411*. Washington, D.C., Naval Research Laboratory (1972). [336]

STAMPER, J. A. *et al.* (1971) (with PAPADOPOULOS, K., SUDAN, R. N., DEAN, S. O. & MCLEAN, E. A.). Spontaneous magnetic fields in laser-produced plasmas, *Phys. Rev. Lett.*, **26**, 1012–15 (1971). [336, 442]

STANSFIELD, B. L. *et al.* (1971) (with NODWELL, R. & MEYER, J.). Enhanced scattering of laser light by optical mixing in a plasma, *Phys. Rev. Lett.*, **26**, 1219–21 (1971). [78, 127, 128]

STEINHAUER, L. C. & AHLSTROM, H. G. (1971). One-dimensional laser heating of a stationary plasma, *Phys. Fluids*, **14**, 81–93 (1971). [222]

STENZEL, R. & WONG, A. Y. (1972). Threshold and saturation of the parametric decay instability, *Phys. Rev. Lett.*, **28**, 274–7 (1972). [76]

STERN, R. A. & TSOAR, N. (1966). Parametric coupling between electron-plasma and ion-acoustic oscillations, *Phys. Rev. Lett.*, **17**, 903–5 (1966). [76]

STERNHEIMER, R. M. (1954). Electronic polarizabilities of ions from the Hartree–Fock wave functions, *Phys. Rev.*, **96**, 951–68 (1954). [85]

STETSER, D. A. & DE MARIA, A. J. (1966). Optical spectra of ultrashort optical pulses generated by mode locked glass Nd lasers, *Appl. Phys. Lett.*, **9**, 118–20 (1966). [185]

STEVERDING, B. (1970). Pressure and impulse generation by Q-switched lasers, *J. Phys. D.*, **3**, 358–64 (1970). [275]

STEVERDING, B. (1972). Subsonic plasma motion in continuous laser light, *J. Phys. D.* **5**, 1824–8 (1972). [269]

STIX, T. H. (1962). *The theory of plasma waves.* New York, McGraw-Hill (1962). [1, 12]

STRONG, J. (1958). *Concepts of classical optics.* San Francisco, W. H. Freeman (1958). [171]

STUMPFEL, C. R. *et al.* (1972) (with ROBITAILLE, J. L. & KUNZE, H-J.). Investigation of the early phases of plasma production by laser irradiation of plane solid targets, *J. Appl. Phys.*, **43**, 902–7 (1972). [370]

SUCOV, E. W. *et al.* (1967) (with PACK, J. L., PHELPS, A. V. & ENGELHARDT, A. G.). Plasma production by a high-power Q-switched laser, *Phys. Fluids*, **10**, 2035–48 (1967). [404, 432]

SUGAYA, R. (1972). Parametric excitation of Bernstein waves in inhomogeneous magneto-plasmas, *J. Phys. Soc. Jap.*, **33**, 183–8 (1972). [76]

SUN, K. H. *et al.* (1967) (with HICKS, J. M., EPSTEIN, L. M. & SUCOV, E. W.). An attempt to measure the temperature of a laser-produced plasma using neutron-detection technique, *J. Appl. Phys.*, **38**, 3402–3 (1967). [234]

TALANOV, V. I. (1964). O samofokusirovke elektromagnitnïkh voln v nelineïnïkh sredakh, *Izv. VUZ. Radiofiz.*, **7**, 564–5 (1964). [195]

TAMOR, S. (1973). Effect of partial coherence on laser-driven plasma instabilities, *Phys. Fluids*, **16**, 1169–71 (1973). [64]

TAYLOR, G. I. (1950a). The formation of a blast wave by a very intense explosion, *Proc. Roy. Soc.*, **A201**, 159–74, 175–86 (1950). [220]

TAYLOR, G. I. (1950b). The instability of liquid surfaces when accelerated in a direction perpendicular to their planes, *Proc. Roy. Soc.*, **A201**, 192–6 (1950). [328]

TAYLOR, R. L. & CALEDONIA, G. (1969a). Experimental determination of the cross-sections for neutral bremsstrahlung, I Ne, Ar and Xe, *J. Quant. Spectrosc. & Radiat. Transfer*, **9**, 657–9 (1969). [53]

TAYLOR, R. L. & CALEDONIA, G. (1969b). Experimental determination of the cross-sections for neutral bremsstrahlung. II High temperature air species O, N and N_2, *J. Quant. Spectrosc. & Radiat. Transfer*, **9**, 681–96 (1969). [53]

TEWARI, S. P. (1972). Multiphoton ionization of hydrogen atoms in the semi-classical treatment of an intense radiation field, *Phys. Rev.*, **A6**, 1869–75 (1972). [153]

THEIMER, O. & SOLLID, J. E. (1968a). Relativistic corrections to the light scattering spectrum of a plasma, *VIII International Conference on phenomena in ionized gases, Vienna (1967). Contributed papers*, p. 521. Vienna, IAEA (1968). [102]

THEIMER, O. & SOLLID, J. E. (1968b). Relativistic corrections to the light-scattering spectrum of a plasma, *Phys. Rev.*, **176**, 198–206 (1968). [102]

THOMSON, J. J. *et al.* (1974) (with KRUER, W. L. & BODNER, S. E.). Parametric instability thresholds and their control, *Phys. Fluids*, **17**, 849–51 (1974). [387]

TIDMAN, D. A. & STAMPER, J. A. (1973). Role of magnetic fields in suprathermal particle generation by laser-produced plasmas, *Appl. Phys. Lett.* **22**, 498–9 (1973). [336]

TOLMAN, R. C. (1938). *The principals of statistical mechanics.* Oxford, O.U.P. (1938). [30]

TOMLINSON, R. G. (1965a). Multiphoton ionization and the breakdown of noble gases, *Phys. Rev. Lett.*, **14**, 489–90 (1965). [175]

TOMLINSON, R. G. (1965b). Atmospheric breakdown limitations to optical maser propagation, *Radio Science J. Res. NBS/USNC-URSI*, **69D**, 1431–3 (1965). [234]

TOMLINSON, R. G. (1969). Scattering and beam trapping in laser-produced plasmas in gases, *IEEE J. Quantum Electron.*, **5**, 591–5 (1969). [198, 245]

TOMLINSON, R. G. (1971). Plasma expansion under heating by a CO_2 laser pulse, *Appl. Phys. Lett.*, **18**, 149–52 (1971). [266]

TOMLINSON, R. G. *et al.* (1966) (with DAMON, E. K. & BUSCHER, H. T.). The breakdown of noble and atmospheric gases by ruby and neodymium laser pulses. In *Physics of quantum electronics* (ed. P. L. Kelley, B. Lax and P. E. Tannenwald), pp. 520–6. New York, McGraw-Hill (1966). [174, 181, 187]

TONDELLO, G. (1969). Grazing incidence spectra of Si XI and Si XII ions, *J. Phys. B.*, **2**, 727–9 (1969). [376]

TONON, G. (1966). Spectres de l'energie des ions émis par le béryllium, le carbone et le molybdène sous l'action du faisceau d'un laser, *C.R. Acad. Sci., Paris*, **262**, 706–9 (1966). [345, 351, 352, 353, 370]

TONON, G. & RABEAU, M. (1972). Interferometric study of a TEA–CO_2 laser produced plasma, *Phys. Lett.*, **40A**, 215–16 (1972). [355]

TONON, G. & RABEAU, M. (1973). Plasma characteristics with TEA–CO_2 laser and wavelength scaling law, *Plasma Phys.*, **15**, 871–82 (1973). [39, 355, 356]

TONON, G. *et al.* (1971a) (with PEREZ, A., FAURE, C., PARISOT, D. & AVENEAU, B.). Study of highly multiply-charged heavy ions produced in laser plasma, *X International Conference on phenomena in ionized gases, Oxford (1971). Contributed papers*, p. 232. Oxford, Donald Parsons & Co. Ltd. (1971). [370]

TONON, G. *et al.* (1971b) (with RABEAU, M. & SCHIRMANN, D.). Interaction d'un plasma créé par laser avec des inductions de plusieurs centaines de kilogauss, *J. Physique*, **32**, Suppl. **C5B**, 139–41 (1971). [434, 435]

TOZER, B. A. (1965). Theory of the ionization of gases by laser beams, *Phys. Rev.*, **137A**, 1665–7 (1965). [146]

TRUBNIKOV, B. A. & KUDRYAVTSEV, V. S. (1958). Plasma radiation in a magnetic field. *Proc. II U.N. Conference on peaceful uses of atomic energy, Geneva*, **31**, 93–8. Geneva, United Nations (1958). [28]

TSAI, H. S. *et al.* (1971) (with OKTAY, E. & AKCASU, A. Z.). Calculation of photon-absorption coefficient in laser-produced hydrogen plasmas, *J. Appl. Phys.*, **42**, 5469–77 (1971). *See also erratum,* **43**, 2490 (1972). [53]

TSYTOVICH, V. N. (1970). *Nonlinear effects in plasmas.* New York, Plenum (1970). [29, 55]

TSYTOVICH, V. N. & SHVARTSBURG, A. B. (1967). Nonlinear generation of plasma waves by a beam of transverse waves III, *Zh. Tekh. Fiz.*, **37**, 589–601 (1967). (Transl: *Sov. Phys. Tech. Phys.*, **12**, 425–32 (1967).) [60, 64]

TSYTOVICH, V. N. *et al.* (1973) (with STENFLO, L., WILHELMSSON, H., GUSTAVSSON, H. G. & ÖSTBERG, K.). One-dimensional model for nonlinear reflection of laser radiation by an inhomogeneous plasma layer, *Physica Scripta*, **7**, 241–9 (1973). [71]

TUCK, J. L. (1959). Plasma jet piercing of magnetic fields and entropy trapping into a conservative system, *Phys. Rev. Lett.*, **3**, 313–15 (1959). [433]

TUCK, J. L. (1961). Thermonuclear reation rates, *Nucl. Fusion*, **1**, 201–2 (1961). [134]

TUCKFIELD, R. G. & SCHWIRZKE, F. (1969). Dynamics of a laser created plasma expanding in a magnetic field, *Plasma Phys.*, **11**, 11–18 (1969). [430, 431]

TULIP, J. & SEGUIN, H. (1973*a*). Influence of a transverse electric field on laser-induced gas breakdown, *Appl. Phys. Lett.*, **23**, 135–6 (1973). [192]

TULIP, J. & SEGUIN, H. (1973*b*). Charge collection measurements on a plasma induced by the CO_2 laser, *Phys. Lett.*, **44A**, 469–70 (1973). [192]

TULIP, J. *et al.* (1971) (with MANES, K. & SEGUIN, H. J.). Intracavity radiation-induced air breakdown in a TEA CO_2 laser, *Appl. Phys. Lett.*, **19**, 433–5 (1971). [193, 269]

VALENZUELA, A. & ECKARDT, J. C. (1971). Preparation of extremely thin self-supporting metal foils, *Rev. Sci. Instrum.*, **42**, 127–8 (1971). [395]

VALEO, E. J. & OBERMAN, C. R. (1973). Model of parametric excitation by an imperfect pump, *Phys. Rev. Lett.*, **30**, 1035–8 (1973). [64]

VALEO, E. *et al.* (1972) (with OBERMAN, C. & PERKINS, F. W.). Saturation of the decay instability for comparable electronic and ion temperatures, *Phys. Rev. Lett.*, **28**, 340–3 (1972). [65, 76]

VANYUKOV, M. P. *et al.* (1966) (with ISAENKO, V. I., LYUBIMOV, V. V., SEREBRYAKOV, V. A. & SHOROKHOV, O. A.). Use of a laser operating in the spike mode to obtain a high-temperature plasma, *Zh. Eksp. i Teor. Fiz. Pis'ma*, **3**, 316–8 (1966). (Transl: *JETP Lett.*, **3**, 205 (1966).) [248]

VANYUKOV, M. P. *et al.* (1968) (with VENCHIKOV, V. A., ISAENKO, V. I., PASHININ, P. P. & PROKHOROV, A. M.). Production of a high temperature dense plasma by gas breakdown with the aid of a laser, *Zh. Eksp. i Teor. Fiz. Pis'ma*, **7**, 321–4 (1968). (Transl: *JETP Lett.*, **7**, 251–3 (1968).) [263]

VARDZIGULOVA, L. E. *et al.* (1967) (with KAITMAZOV, S. D. & PROKHOROV, A. M.). Laser spark in a strong magnetic field, *Zh. Eksp. i Teor. Fiz. Pis'ma*, **6**, 799–801 (1967). (Transl: *JETP Lett.*, **6**, 253–5 (1967).) [194, 195, 271]

VERBER, C. M. & ADELMAN, A. H. (1963). Laser-induced thermionic emission, *Appl. Phys. Lett.*, **2**, 220–2 (1963). [339]

VERBER, C. M. & ADELMAN, A. H. (1965). Laser-induced thermionic emission from tantalum, *J. Appl. Phys.*, **36**, 1522–5 (1965). [340]

VEYRIÉ, P. (1968). Contribution a l'étude de l'ionisation et du chauffage des gaz par le rayonnement d'un laser déclenché: résultats expérimentaux, *J. Physique*, **29**, 33–41 (1968). [163, 261]

VEYRIÉ, P. *et al.* (1967) (with FLOUX, F., GUYOT, D. & REUSS, J. D.). *Ionization and heating of gaseous deuterium by the means of a high power laser.* Paris, C.E.A. Report (1967). [263]

VILENSKAYA, G. G. & NEMCHINOV, I. V. (1969). Sudden increase in absorption of laser radiation and associated gasdynamic effects, *Dokl. Akad. Nauk. SSSR*, **186**, 1048–51 (1969). (Transl: *Sov. Phys. Dokl.*, **14**, 560–3 (1969).) [285]

VINOGRADOV, A. V. & PUSTOVALOV, V. V. (1971). Absorption of light by an inhomogeneous laser plasma, *Zh. Eksp. i Teor. Fiz. Pis'ma*, **13**, 317–20 (1971). (Transl: *JETP Lett.*, **13**, 226–8 (1971).) [75]

VINOGRADOV, A. V. & PUSTOVALOV, V. V. (1972a). Electron temperature of a plasma scattering intense light beams, *Zh. Eksp. i Teor. Fiz.*, **62**, 980–8 (1972). (Transl: *Sov. Phys. JETP*, **35**, 517–21 (1972).) [73]

VINOGRADOV, A. V. & PUSTOVALOV, V. V. (1972b). Heating of a plasma in stimulated scattering of laser radiation, *Kvantovaya Elektron.*, **2** (8), 3–22 (1972). (Transl: *Sov. J. Quantum Electron.*, **2**, 91–105 (1972).) [73]

VINOGRADOV, A. V. et al. (1973) (with ZELDOVICH, B. YA. & SOBELMAN, I. I.). Influence of saturation effects on stimulating scattering in laser heating of a plasma, *Zh. Eksp. i Teor. Fiz. Pis'ma*, **17**, 271–4 (1973). (Transl: *JETP Lett.*, **17**, 195–8 (1973).) [71]

VLASES, G. C. (1971). Heating of pinch devices with lasers, *Phys. Fluids*, **14**, 1287–9 (1971). [139]

VOLOSEVICH, P. P. & LEVANOV, E. I. (1970). Effect of thermal conductivity on the propagation of the absorption wave of laser radiation, *Dokl. Akad. Nauk. SSSR*, **194**, 49–52 (1970). (Transl: *Sov. Phys. Dokl.*, **15**, 814–16 (1971).) [286]

VORONOV, G. S. (1966). Dependence of the probability of multiphoton ionization of atoms on the photon flux intensity, *Zh. Eksp. i Teor. Fiz.*, **51**, 1496–8 (1966). (Transl: *Sov. Phys. JETP*, **24**, 1009–11 (1967).) [151]

VORONOV, G. S. & DELONE, N. B. (1965). Ionization of the xenon atom by the electric field of ruby laser emission, *Zh. Eksp. i Teor. Fiz. Pis'ma*, **1**, 42 (1965). (Transl: *JETP Lett.*, **1**, 66–8 (1965).) [155]

VORONOV, G. S. & DELONE, N. B. (1966). Many-photon ionization of the xenon atom by ruby laser radiation, *Zh. Eksp. i Teor. Fiz.*, **50**, 78–84 (1966). (Transl: *Sov. Phys. JETP*, **23**, 54–7 (1966).) [155]

VORONOV, G. S. et al. (1966a) (with DELONE, G. A. & DELONE, N. B.). Multiphoton ionization of atoms. II Ionization of krypton by ruby-laser radiation, *Zh. Eksp. i Teor. Fiz.*, **51**, 1660–4 (1966). (Transl: *Sov. Phys. JETP*, **24**, 1122–5 (1967).) [157]

VORONOV, G. S. et al. (1966b) (with DELONE, G. A. & DELONE, N. B.). Multiphoton ionization of krypton and argon by ruby laser radiation, *Zh. Eksp. i Teor. Fiz. Pis'ma*, **3**, 480–3 (1966). (Transl: *JETP Lett.*, **3**, 313–15 (1966).) [157]

VRIENS, L. (1973). Light scattering from weakly ionized nonhomogeneous plasmas, *Phys. Rev. Lett.*, **30**, 585–8 (1973). [101]

WAKI, M. et al. (1972) (with YAMANAKA, T., KANG, H. B., YOSHIDA, K. & YAMANAKA, C.). Properties of plasma produced by high power laser, *J. Appl. Phys. Jap.*, **11**, 420–1 (1972). [80, 382, 383]

WANG, C. C. & DAVIS, L. I. (1971). New observations of dielectric breakdown in air induced by a focused Nd^{3+} glass laser with various pulse widths, *Phys. Rev. Lett.*, **26**, 822–5 (1971). [188]

WANIEK, R. W. & JARMUZ, P. J. (1968). Acceleration of microscopic particles by laser-induced plasma emission, *VIII International Conference on phenomena in ionized gases, Vienna (1967). Contributed papers*, p. 57. Vienna, IAEA (1968). [400]

WARD, G. & PECHACEK, R. E. (1972). Scattering of light by relativistic electrons, *Phys. Fluids*, **15**, 2202–10 (1972). [112]

WARD, G. et al. (1971) (with PECHACEK, R. E. & TRIVELPIECE, A. W.). Scattering of a laser beam by high-temperature plasmas, *Phys. Rev.*, **A3**, 1721–3 (1971). [112]

WATSON, J. L. & BEACH, A. D. (1969). High-resolution measurements of plasma density, *J. Phys. D.*, **2**, 129–33 (1969). [115]

WATSON, R. D. & CLARK, M. K. (1965). Rayleigh scattering of 6943 Å laser radiation in a nitrogen atmosphere, *Phys. Rev. Lett.*, **14**, 1057–8 (1965). [112]

WEAVER, T. et al. (1972) (with ZIMMERMAN, G. & WOOD, L.). Prospects for exotic fuel usage in CTR systems I. $B^{11}(p, 2\alpha)He^4$: a clean high performance CTR fuel, *Lawrence Livermore Laboratory Repts. UCRL-74191 and -74352* (1972). [134, 136, 137, 144]

WEI, P. S. P. & HALL, R. B. (1973). Emission spectra of laser-supported detonation waves, *J. Appl. Phys.*, **44**, 2311–4 (1973). [441]

WEICHEL, H. & AVIZONIS, P. V. (1966). Expansion rates of the luminous front of a laser-produced plasma, *Appl. Phys. Lett.*, **9**, 334–7 (1966). [363]

WEICHEL, H. & AVIZONIS, P. V. (1968). Cooperative scattering in a laser produced plasma, *VIII International Conference on phenomena in ionized gases, Vienna (1967). Contributed papers*, p. 524. Vienna, IAEA (1968). [362]

WEICHEL, H. *et al.* (1967) (with AVIZONIS, P. V. & VONDERHAAR, D. F.). Observation of plasma ion oscillations in a laser-produced plasma, *Phys. Rev. Lett.*, **19**, 10–12 (1967). [362]

WEICHEL, H. *et al.* (1968) (with DAVID, C. D. & AVIZONIS, P. V.). Effects of radiation pressure on a laser-produced plasma, *Appl. Phys. Lett.*, **13**, 376–9 (1968). [363]

WEINBERG, F. J. (1963). *Optics of Flames*. London, Butterworths (1963). [223]

WEISBACH, F. M. & AHLSTROM, H. G. (1973). Measurements of nonresonant absorption in a fully ionized plasma, *Phys. Fluids*, **16**, 1164–6 (1973). [52]

WESTFOLD, K. C. (1950). The refractive index and classical radiative processes in an ionized gas, *Phil. Mag.*, **41**, 509–16 (1950). [40]

WHITE, R. B. *et al.* (1973) (with LIU, C. S. & ROSENBLUTH, M. N.). Parametric decay of obliquely incident radiation, *Phys. Rev. Lett.*, **31**, 520–3 (1973). [75]

WILLIAMSON, J. H. *et al.* (1966) (with NODWELL, R. A. & BARNARD, A. J.). Computed profiles for electromagnetic radiation scattered from a plasma, *J. Quant. Spectrosc. & Radiat. Transfer*, **6**, 895–8 (1966). [94, 95]

WILSON, A. H. (1966). *Thermodynamics and statistical mechanics*. Cambridge, C.U.P. (1966). [213]

WILSON, J. R. (1970). Laser-induced multiple breakdown in gases, *J. Phys. D.*, **3**, 2005–8 (1970). [247]

WINSOR, N. K. & TIDMAN, D. A. (1973). Laser target model, *Phys. Rev. Lett.*, **31**, 1044–6 (1973). [336, 337]

WINTERBERG, F. (1973*a*). Micro-fission explosions and controlled release of thermonuclear energy, *Nature*, **241**, 449–50 (1973). [143]

WINTERBERG, F. (1973*b*). The possibility of microfission explosions by laser or relativistic electron-beam high-density compression, *Lett. Nuovo Cimento*, **6**, 407–11 (1973). [143]

WINTERBERG, F. (1973*c*). Micro-fission-fusion chain reactions, *Lett. Nuovo Cimento*, **7**, 325–30 (1973). [143]

WINTERBERG, F. (1968). The possibility of producing a dense thermonuclear plasma by an intense field emission discharge, *Phys. Rev.*, **174**, 212–20 (1968). [140, 141]

WINTERLING, G. *et al.* (1969) (with HEINICKE, W. & DRANSFELD, K.). Laser-induced breakdown in liquid He⁴, *Phys. Rev.*, **185**, 285–7 (1969). [177]

WITKOWSKI, S. (1971). Free targets. In *Laser interaction and related plasma phenomena* (ed. H. J. Schwarz and H. Hora), Vol. 1, pp. 259–71. New York, Plenum (1971). [417]

WONG, A. Y. & TAYLOR, R. J. (1971). Parametric excitation in the ionosphere, *Phys. Rev. Lett.*, **27**, 644–7 (1971). [76]

WRIGHT, J. K. (1964). Theory of the electrical breakdown of gases by intense pulses of light, *Proc. Phys. Soc.*, **84**, 41–6 (1964). [164]

WRIGHT, T. P. (1971). Early-time model of laser plasma expansion, *Phys. Fluids*, **14**, 1905–10 (1971). [334]

WRIGHT, T. P. (1972). Comments on 'Demonstration of collisionless interactions between inter-streaming ions in a laser-produced-plasma experiment', *Phys. Rev. Lett.*, **28**, 268–70 (1972). [444]

WRIGHT, T. P. (1973). Reply to comments of Book and Clark, *Phys. Fluids*, **16**, 343 (1973). [334]

WUERKER, R. F. *et al.* (1959*a*) (with SHELTON, H. & LANGMUIR, R. V.). Electrodynamic containment of charged particles, *J. Appl. Phys.*, **30**, 342–9 (1959). [400]

WUERKER, R. F. *et al.* (1959*b*) (with GOLDENBERG, H. M. & LANGMUIR, R. V.). Electrodynamic containment of charged particles by three-phase voltages, *J. Appl. Phys.*, **30**, 441–2 (1959). [400]

YA'AKOBI, B. *et al.* (1972) (with GEORGE, E. V., BEKEFI, G. & HAWRYLUK, R. J.). Stark profiles of forbidden and allowed transitions in a dense, laser produced helium plasma, *J. Phys. B.*, **5**, 1017–30 (1972). [267]

YAMANAKA, C. (1972). Thermonuclear fusion by lasers. In *Laser interaction and related plasma phenomena* (ed. H. J. Schwarz and H. Hora), Vol. 2, pp. 481–502. New York, Plenum (1972). [423, 424]

YAMANAKA, C. et al. (1972a) (with YAMANAKA, T., KANG, H., SASAKI, T., YOSHIDA, K., UEDA, K., HONGYO, M. & WAKI, M.). Plasma generation and heating by lasers, *Kvantovaya Elektron.*, **2** (8), 45–52 (1972). (Transl: *Sov. J. Quantum Electron.*, **2**, 127–32 (1972).) [423]

YAMANAKA, C. et al. (1972b) (with YAMANAKA, T., KANG, H., YOSHIDA, K., WAKI, M. & SHIMAMURA, T.). Nonlinear interaction of laser radiation and plasma, *Phys. Lett.*, **38A**, 495–6 (1972). [80, 423]

YAMANAKA, C. et al. (1972c) (with YAMANAKA, T., SASAKI, T., YOSHIDA, K., WAKI, M. & KANG, H. B.). Anomalous heating of a plasma by a laser, *Phys. Rev.*, **A6**, 2335–42 (1972). [62, 63, 77, 79, 80]

YAMANAKA, C. et al. (1974) (with YOKOYAMA, M., NAKAI, S., SASAKI, T., YOSHIDA, K., MATOBA, M., YAMABE, C., YAMANAKA, T., MIZUI, J., YAMAGUCHI, N. & NISHIKAWA, K.). Thermonuclear fusion produced by lasers. Paper presented at *5th IAEA Conference on plasma physics and controlled nuclear fission research, Tokyo (1974)*. [386, 403]

YARIV, A. (1967). *Quantum electronics*. New York, Wiley (1967). [54, 195]

YASOJIMA, Y. & INUISHI, Y. (1973). Threshold minima in the superhigh-pressure gas breakdown by Q-switched lasers, *Phys. Lett.*, **44A**, 393–4 (1973). [164]

YOKOYAMA, M. et al. (1971) (with NAKATSUKA, M. & YAMANAKA, C.). Scattering diagnostics of plasma by CO_2 laser, *Phys. Lett.*, **36A**, 317–18 (1971). [111]

YOUNG, M. & HERCHER, M. (1967). Dynamics of laser-induced breakdown in gases, *J. Appl. Phys.*, **38**, 4393–4400 (1967). [167]

YOUNG, M. et al. (1966) (with HERCHER, M. & WU, C-Y.). Some characteristics of laser-induced air sparks, *J. Appl. Phys.*, **37**, 4938–40 (1966). [235, 236]

YOUNG, M. et al. (1968) (with CHIN, S. L. & ISENOR, N. R.). Laser-induced breakdown in freon-doped rare gas, *Can. J. Phys.*, **46**, 1537–8 (1968). [181]

ZAIDEL, A. N. & OSTROVSKII, YU. I. (1968). Holographic investigation of plasma, *VIII International Conference on phenomena in ionized gases, Vienna (1967). Contributed papers*, p. 508. Vienna, IAEA (1968). [229]

ZAIDEL, A. N. et al. (1966) (with OSTROVSKAYA, G. V., OSTROVSKII, YU. I. & CHELIDZE, T. YA.). Time-resolved holography of laser-produced plasmas, *Zh. Tekh. Fiz.*, **36**, 2208–10 (1966). (Transl: *Sov. Phys. Tech. Phys.*, **11**, 1650–2 (1967).) [229, 230, 248]

ZAKHAROV, S. D. et al. (1970a) (with KROKHIN, O. N., KRYUKOV, P. G. & TYURIN, E. L.). Plasma heating by ultrashort laser pulses in the process of electronic heat conduction, *Zh. Eksp. i Teor. Fiz. Pis'ma*, **12**, 47–50 (1970). (Transl: *JETP Lett.*, **12**, 36–8 (1970).) [304]

ZAKHAROV, S. D. et al. (1970b) (with KROKHIN, O. N., KRYUKOV, P. G. & TYURIN, E. L.). Electron–ion relaxation in a plasma produced by ultrashort laser pulses, *Zh. Eksp. i Teor. Fiz. Pis'ma*, **12**, 115–18 (1970). (Transl: *JETP Lett.*, **12**, 80–81 (1970).) [310, 315]

ZAKHAROV, S. D. et al. (1971a) (with KROKHIN, O. N., KRYUKOV, P. G. & TYURIN, E. L.). Enhancement of the efficiency of the laser heating of plasma by the addition of heavy impurities to the target, *Kvantovaya Elektron.*, **1** (2), 102–3 (1971). (Transl: *Sov. J. Quantum Electron.*, 1, 196–7 (1971).) [304]

ZAKHAROV, S. D. et al. (1971b) (with KROKHIN, O. N., KRYUKOV, P. G. & TYURIN, E. L.). Role of focusing in heat-conduction heating of a plasma by high-power laser radiation, *Kvantovaya Elektron.*, **1** (2), 104–7 (1971). (Transl: *Sov. J. Quantum Electron.*, **1**, 198–200 (1971).) [304, 305, 422]

ZARITSKII, A. A. et al. (1971) (with ZAKHAROV, S. D. & KRYUKOV, P. G.). Electrodynamic suspension of solid particles, *Zh. Tekh. Fiz.*, **41**, 227–34 (1971). (Transl: *Sov. Phys. Tech. Phys.*, **16**, 174–9 (1971).) [400]

ZARITSKII, A. R. *et al.* (1972a) (with ZAKHAROV, S. D., KRYUKOV, P. G. & FEDOSIMOV, A. I.). Measurements of the polarization of the radiation reflected backward from a laser-heated plasma, *Kvantovaya Elektron.*, **2** (8), 89–90 (1972). (Transl: *Sov. J. Quantum Electron.*, **2**, 162 (1972).) [388]

ZARITSKII, A. R. *et al.* (1972b) (with ZAKHAROV, S. D., KRYUKOV, P. G., MATVEETS, YU A. & FEDOSIMOV, A. I.). Changes in the spectrum of back-reflected radiation in laser heating of a plasma, *Zh. Eksp. i Teor. Fiz. Pis'ma*, **15**, 184–6 (1972). (Transl: *JETP Lett.*, **15**, 127–9 (1972).) [388]

ZEL'DOVICH, B. YA. & PILIPETSKII, N. F. (1966). Laser radiation field focused by real systems, *Izv. VUZ. Radiofiz.*, **9**, 95–101 (1966). (Transl: *Sov. Radiophys.*, **9**, 64–8 (1966).) [171]

ZEL'DOVICH, YA. B. & RAIZER, YU. P. (1964). Cascade ionization of a gas by a light pulse, *Zh. Eksp. i Teor. Fiz.*, **47**, 1150–61 (1964). (Transl: *Sov. Phys. JETP*, **20**, 772–9 (1965).) [164, 167]

ZEL'DOVICH, YA. B. & RAIZER, YU. P. (1966). *Physics of shock waves and high-temperature hydrodynamic phenomena* (2 vols.). New York, Academic (1966). [213, 214, 215, 220, 301, 319]

ZEL'DOVICH, YA. *et al.* (1972) (with LEVICH, E. V. & SYUNYAEV, R. A.). Stimulated Compton interaction between Maxwellian electrons and spectrally narrow radiation, *Zh. Eksp. i Teor. Fiz.*, **62**, 1392–1408 (1972). (Transl: *Sov. Phys. JETP*, **35**, 733–40 (1972).) [73]

ZERNIK, W. (1964). Two-photon ionization of atomic hydrogen, *Phys. Rev.*, **135A**, 51–7 (1964). [152]

ZERNIK, W. & KLOPFENSTEIN, R. W. (1965). Two-photon ionization of atomic hydrogen, II, *J. Math. Phys.*, **6**, 262–70 (1965). [152]

ZON, B. A. *et al.* (1971) (with MANAKOV, N. L. & RAPOPORT, L. P.). Multiphoton excitation of atoms, *Zh. Eksp. i Teor. Fiz.*, **60**, 1264–9 (1971). (Transl: *Sov. Phys. JETP*, **33**, 683–5 (1971).) [152, 159]

Index

503